Handbook of Research on Social Media Applications for the Tourism and Hospitality Sector

Célia M.Q. Ramos
University of Algarve, Portugal

Cláudia Ribeiro de Almeida
University of Algarve, Portugal

Paula Odete Fernandes
Polytechnic Institute of Bragança, Portugal

IGI Global
DISSEMINATOR OF KNOWLEDGE

A volume in the Advances in Hospitality, Tourism, and the Services Industry (AHTSI) Book Series

Published in the United States of America by
IGI Global
Business Science Reference (an imprint of IGI Global)
701 E. Chocolate Avenue
Hershey PA, USA 17033
Tel: 717-533-8845
Fax: 717-533-8661
E-mail: cust@igi-global.com
Web site: http://www.igi-global.com

Library of Congress Cataloging-in-Publication Data

Names: Ramos, Célia M. Q., editor. | Almeida, Cláudia Ribeiro de, 1971- editor. | Fernandes, Paula Odete, 1969- editor.
Title: Handbook of research on social media applications for the tourism and hospitality sector / Célia M.Q. Ramos, Cláudia Ribeiro de Almeida, and Paula Odete Fernandes, editors.
Description: Hershey, PA : Business Science Reference, 2020. | Includes bibliographical references and index. | Summary: "This book provides vital theoretical frameworks and the latest innovative empirical research findings of online social media in the tourism industry. It also explores user-generated content and the methods of mobile strategies"-- Provided by publisher.
Identifiers: LCCN 2019035767 (print) | LCCN 2019035768 (ebook) | ISBN 9781799819479 (hardback) | ISBN 9781799819486 (ebook)
Subjects: LCSH: Tourism--Marketing--Research. | Hospitality industry--Marketing--Research. | Social media. | User-generated content.
Classification: LCC G155.A1 H3554 2020 (print) | LCC G155.A1 (ebook) | DDC 659.19/91--dc23
LC record available at https://lccn.loc.gov/2019035767
LC ebook record available at https://lccn.loc.gov/2019035768

This book is published in the IGI Global book series Advances in Hospitality, Tourism, and the Services Industry (AHTSI) (ISSN: 2475-6547; eISSN: 2475-6555)

British Cataloguing in Publication Data
A Cataloguing in Publication record for this book is available from the British Library.

Advances in Hospitality, Tourism, and the Services Industry (AHTSI) Book Series

Maximiliano Korstanje
University of Palermo, Argentina

ISSN:2475-6547
EISSN:2475-6555

MISSION

Globally, the hospitality, travel, tourism, and services industries generate a significant percentage of revenue and represent a large portion of the business world. Even in tough economic times, these industries thrive as individuals continue to spend on leisure and recreation activities as well as services.

The Advances in Hospitality, Tourism, and the Services Industry (AHTSI) book series offers diverse publications relating to the management, promotion, and profitability of the leisure, recreation, and services industries. Highlighting current research pertaining to various topics within the realm of hospitality, travel, tourism, and services management, the titles found within the AHTSI book series are pertinent to the research and professional needs of managers, business practitioners, researchers, and upper-level students studying in the field.

COVERAGE

- Service Design
- Tourism and the Environment
- Service Management
- International Tourism
- Service Training
- Health and Wellness Tourism
- Sustainable Tourism
- Food and Beverage Management
- Destination Marketing and Management
- Cruise Marketing and Sales

IGI Global is currently accepting manuscripts for publication within this series. To submit a proposal for a volume in this series, please contact our Acquisition Editors at Acquisitions@igi-global.com or visit: http://www.igi-global.com/publish/.

The Advances in Hospitality, Tourism, and the Services Industry (AHTSI) Book Series (ISSN 2475-6547) is published by IGI Global, 701 E. Chocolate Avenue, Hershey, PA 17033-1240, USA, www.igi-global.com. This series is composed of titles available for purchase individually; each title is edited to be contextually exclusive from any other title within the series. For pricing and ordering information please visit http://www.igi-global.com/book-series/advances-hospitality-tourism-services-industry/121014. Postmaster: Send all address changes to above address.

Titles in this Series

For a list of additional titles in this series, please visit: www.igi-global.com/book-series

Global Developments in Healthcare and Medical Tourism
Sudip Paul (North-Eastern Hill University, India) and Sharad Kumar Kulshreshtha (North-Eastern Hill University, India)
Medical Information Science Reference • ©2020 • 341pp • H/C (ISBN: 9781522597872) • US $245.00

Multilevel Approach to Competitiveness in the Global Tourism Industry
Sérgio Jesus Teixeira (Higher Institute of Administration and Languages, Portugal) and João Matos Ferreira (University of Beira Interior, Portugal)
Business Science Reference • ©2020 • 393pp • H/C (ISBN: 9781799803652) • US $225.00

Ecotourism and Community Intervention Emerging Research and Opportunities
A. Vinodan (Indian Institute of Tourism and Travel Management, Nellore, India) and James Manalel (Cochin University of Science and Technology, India)
Business Science Reference • ©2020 • 253pp • H/C (ISBN: 9781799816355) • US $185.00

Strategic Business Models to Support Demand, Supply, and Destination Management in the Tourism and Hospitality Industry
Luisa Cagica Carvalho (Institute Polytechnic of Setúbal, Portugal & University of Évora, Portugal) Lurdes Calisto (University of Évora, Portugal) and Nuno Gustavo (Escola Superior de Hotelaria e Turismo do Estoril, Portugal)
Business Science Reference • ©2020 • 358pp • H/C (ISBN: 9781522599364) • US $245.00

Impact of Risk Perception Theory and Terrorism on Tourism Security Emerging Research and Opportunities
Maximiliano E. Korstanje (University of Palermo, Argentina)
Business Science Reference • ©2020 • 144pp • H/C (ISBN: 9781799800705) • US $175.00

Translation and Communication in the Promotion of Business Tourism Emerging Research and Opportunities
Elena Alcalde Penalver (University of Alcalá, Spain) and Alexandra Santamaría Urbieta (University of Alcalá, Spain)
Business Science Reference • ©2020 • 186pp • H/C (ISBN: 9781799801429) • US $165.00

Cultural Tourism in the Wake of Web Innovation Emerging Research and Opportunities
J. Scott McDonald (University of Texas at El Paso, USA) John R. Bennett Jr. (Valdosta State University, USA) Keith A. Merwin (Merwin & Associates, USA) and Gerald A. Merwin Jr. (Valdosta State University, USA)
Business Science Reference • ©2019 • 139pp • H/C (ISBN: 9781522583950) • US $165.00

IGI Global
DISSEMINATOR OF KNOWLEDGE

701 East Chocolate Avenue, Hershey, PA 17033, USA
Tel: 717-533-8845 x100 • Fax: 717-533-8661
E-Mail: cust@igi-global.com • www.igi-global.com

Editorial Advisory Board

List of Contributors

Table of Contents

Detailed Table of Contents

These network sites play a significant role in receiving customer feedback using online reviews. As a result, travel companies can utilize their customer evaluations to improve products and service offerings. The visual images and textual postings on network sites have the power to motivate others to embark on travel. This chapter is about peer influence mechanisms behind travel experience sharing on the social network sites. It uses both phenomenology and ethnographic methodological approaches to analyze some of the comments and photographs posted on SNSs by tourists in Malawi. It concludes that most of the tourists traveling to Malawi were influenced by what is being shared on social network sites, yet others do visit out of curiosity to discovering new things.

Today, tourism firms use social media to communicate and inform their clients about tourism destination offerings. Trip planners are utilizing social media to make informed decision making about tourists' destinations. With information technology and social media, travelers can get updates on their booked flights and hotels at a tourist destination. The growth of social media usage in tourism industry influences most of the firms to start adopting using social media. This chapter, therefore, focuses on the role of social media in travel planning and tourism destination decision making. It evaluates five cases whereby clients' use travel planning decisions but also changes or curtail plans as a result of social media communication. The survey study uses a simple random sampling procedure and interviewed 792 travelers using a structured questionnaire in ten tourist destinations covering developed and developing countries. The key results revealed that 92% of travelers make a booking but also change or curtail travel planning decisions for holidays through social media alerts.

With increased economic and population growth, air connectivity, and travel facilities, international travel of Chinese citizens has tremendously increased. Although usually in groups, there has been a growing trend among Chinese tourists toward travelling independently. These independent travellers or so-called 'new tourists' are highly exposed to new technologies and social media and prefer to do research about possible destinations, and new travel routes and activities that package tours usually do not include. This chapter examines how social media in general and WeChat, in particular, shapes the Chinese individual travellers' travelling decisions. As will be seen, while WeChat's popularity among the Chinese outbound tourists is largely attributed to its multi-functionality, its influences on their destination visitation choice are found in multiple levels: by providing illuminating information, allowing users to integrate their payment system easily to their bank accounts in China, and providing means to obtain discounts and promotional offers in a variety of ways.

This chapter describes how through globalisation and developments in communication technologies, the lifestyle of people has changed and developed as well. Now, events are re-designed with communication tools and are becoming an important part of social life. This study aims to determine and evaluate the factors affecting perceptions of the audience who take part in events realized via Facebook, or on another online platform. Since this study is implemented through a social media communication device, on an on-line platform, it differs from other studies and gives this area a new perspective. Data acquired from questions were prepared to test the hypothesis of the study were analysed by doing factor analysis and regression tests. The acquired results were discussed.

Millions of people with vision impairments worldwide still experience many barriers when visiting museum exhibitions, given the tenacious ocular centricity of these institutions and the lack of assistive technologies that promote multisensory exhibit design. It is urgent to give priority to issues related to the accessibility of those patrons in museums, as well as their social inclusion, considering that visual impairment is a diverse spectrum. This chapter aims to research blind and visually impaired visitors' experiences in these institutions and presents a proposal of an integrative framework, co-created based on semi-structured interviews conducted with 29 individuals from Portugal. It concluded that sensory, intellectual and physical access must be integrated into the pre-visit, on-site visit and post-visit phases of interaction with cultural institutions, in order to promote a sense of autonomy to visually impaired individuals.

This chapter describes how Tourism 4.0 is a concept that combines tourism and the fourth industrial revolution, and although the literature in this field is very scarce, this concept has been explored in some research projects, such as the government-sponsored research project in Slovenian tourism. People with various kinds of access requirements represent a combination of challenges and opportunities for the tourism industry. Tourism 4.0 set up the main goals of making tourism accessible to everyone at any time. Therefore, this new phenomenon may have an essential role in the development of accessible tourism. The adoption of technological components in accessible tourism enables the development of a new technological solution that can facilitate access to tourism products for disabled people, contributing to the development of accessible tourism. The new term Accessible@Tourism 4.0 is the answer to the role of the fourth industrial revolution in accessible tourism, emphasizing the effect of Industry 4.0 components in the tourism sector.

Foreword

Social media use has become firmly engrained in daily lives as well as tourist experiences of consumers all over the world. Even those who do not actively use social media benefit from the ratings and reviews as well as textual and visual contents as they are integrated into booking platforms, destination marketing campaigns, mobile apps, wayfinding/map functions and smart assistants such as Siri and Alexa. Recent privacy scandals, security breaches and surveillance concerns have done little to slow down social media content creation and consumption. This is especially true for the travel, tourism and hospitality contexts, where social media fulfill important informational, transactional, experiential, safety and social functions.

Importantly, social media keep changing, creating a dynamic landscape of users, uses and technologies. New applications appear, some disappear, existing platforms implement changes to their business models, interfaces and functions, and users create new uses and use cultures. Likewise, tourism and hospitality businesses constantly adapt their efforts to capitalize on the opportunities social media provide for the design, promotion and distribution of travel experiences. New intermediaries such as social media influencers have recently emerged as a result of the ability to monetize one's reputation on these platforms. Social media have also inspired governments to think about ways to harness resulting data streams for the smart management of destinations and the optimization of tourist flows and tourism resource consumption. Moreover, social media have challenged researchers to find new ways of data capture, storage, analysis, interpretation and reporting.

This book provides an important effort to capture some of these new developments. Specifically, it focuses on social media contents as data that can be transformed into valuable insights for many kinds of tourism stakeholders. The diversity of contexts the various chapters deal with illustrates the importance of social media for all aspects of tourism-related consumption, business, governance and education. The range of theories and methodologies used in the chapters makes the book a useful and interesting resource for students and researchers. In addition, the book clearly shows that existing social media research in the fields of travel, tourism and hospitality only represents the tip of the iceberg in terms of what is possible. As such, the chapters all provide important inspiration and guidance for future research in this area.

Most intriguing is the clear bridge the contents of this book lay between social media research and the emerging practice of and academic literature on smart tourism. It becomes evident from the chapters that not only is more data available but there is also ample opportunity to transform it into value propositions that can enhance the accessibility, innovativeness and sustainability of tourism destinations and lead to superior experiences for tourists. As more destinations adopt a smart tourism management mindset and more researchers seek to help move smart tourism agendas forward, the examples presented in this book will serve as useful case studies of social media data as an enabler for smart tourism development.

Ulrike Gretzel
University of Southern California, USA

Ulrike Gretzel *is a senior fellow at the Center for Public Relations at the University of Southern California. Her research focuses on human-computer interaction in the context of travel and tourism. She has published widely on the use and impacts of social media in tourism and hospitality.*

Preface

SOCIAL MEDIA APPLIED TO TOURISM AND HOSPITALITY

The society of today is characterized by creativity and innovation, whether in business or in technology, in parallel with concerns of sustainability and inclusion of all, considering accessibility concerns. This situation leads to new forms of complexity and knowledge in several areas of society and in various sectors of activity, such as: management, hospitality, marketing, tourism, among others.

One of these is social media, which have been investigated in the context of hospitality and tourism. Social media has had a huge impact in the tourism industry, in general, and in the hospitality, in specific.

Currently, tourists frequently count on social networks to provide information about a touristic product or destination as a decision support tool to contribute to make adequate decisions in the process of planning a trip. In this digital environment, tourists share their travel experiences, impressions, emotions, special moments, and opinions about the tourist services, for example hotels, restaurants, airlines, car rental services, events or auxiliary services, all of which contributes to define online reputation of a tourist destination.

There are some aspects of social media, which can be developed in scientific terms such as Business Intelligence, Business Performance, Consumer Behaviors, Customer Engagement, Data Mining, E-Business, Global Business. Influencers, Mobile Marketing, Smart Tourism, among others.

In this book, researchers have studied topics that have been grouped into three categories: Travel Decision and Tourism Experience; Social Media and Tourism Cases; and Social Media and Hospitality.

TRAVEL DECISION AND TOURISM EXPERIENCE

Nowadays, social media is widely used in the decision-making process to purchase a trip. At any stage of the trip, tourists use platforms to look for official and other tourist information to identify opportunities and challenges at their destination. Social media has now played the role of tourist destination agents, which is open 24 hours a day, 7 days a week and has the true testimony of other travelers. In this context social media can be considered for the creation of ecosystems of organizations that develop and contribute to digital branding, contribute to influence stakeholders in their choice of travel, play a relevant role in travel planning while playing the role of decision-making support for the tourist destination, help to define tourist behavior during the trip and how they use these platforms, as well as enhance research on the impacts on the performance of tourist organizations, terms of business value and performance, on the other hand, also enhance the definition of new technologies that lead to the development of new knowledge management practices about the tourism sector and all of its complementary industries that help define the associated product. to the tourist destination.

SOCIAL MEDIA AND TOURISM CASES

There is also the challenge of establishing appropriate social media utility and understanding how they can help both government decision makers and businesses, ranging from small and creative industries to large companies such as transport or tourism education.

Such challenges exist and have already been overcome and investigated in different situations and contexts, with this book, for example: in the use and analysis of traveler generated content to help understand the image associated with the destination, which tourists perceive when traveling and expressed on Twitter; using social media to offer beverage discounts through Instagram; event management with social media resources; for visitors with accessibility issues they can be used to enhance the museum experience for visually impaired tourists and how they form when associated with industry 4.0 technologies contribute to accessibility; how they can contribute to business value by linking with applications designed to be used on mobile devices; they contribute to adding value to the quality of services offered for tourism purposes, and finally, how they can be integrated into an education system associated with tourism practices.

SOCIAL MEDIA AND HOSPITALITY

Solutions to the problem of managing information associated with business and hotel activity have been resorting to social media in different ways, from the way they can be used in hotels, how TripAdvisor and the tourist feeling expressed in that environment contribute to understand the experience enjoyed by tourists, as can be combined in rural tourism, how strategies can be defined specifically for these platforms to establish local accommodation, contribute to the definition of a reputation generated online, and consequently their role in the behavior of the mainly in the establishment of relations between the tourist and the hotelier.

Social media is already part of the tools that marketers must define and implement the strategies considered to achieve the business and marketing objectives defined by the tourism company, whether it be accommodation, transportation or events. In addition, they also play a very important role in assisting tourists in planning their trip by communicating information about destinations and the testimonies of other travelers, and in providing companies with a means of publicizing their products, defining their reputation associated with the brand while allowing increased involvement with the tourist.

Social media still increases the possibility of defining metrics to support business decision makers on how to do business and defining strategies for the future.

ORGANIZATION OF THE BOOK

The book is organized into twenty-two chapters. A brief description of each of the chapters follows:

Chapter 1 proposing to analyze the utilization of Instagram as a network of smart tourism ecosystem by state tourism boards in India. In addition will give an insight into the Instagram account used by states tourism boards for the promotion of their respective tourist destinations. The findings will help to improve the utilization of Social Network Sites for online content creations, managing the social media campaigns, marketing, branding and promotions of tourism destinations.

Chapter 2 is about peer influences mechanisms behind travel experience sharing on the social network sites. It uses both phenomenology and ethnography methodological approaches to analyze some of the comments and photographs posted on SNSs by tourists in Malawi. It concludes that most of the tourists traveling to Malawi were influenced by what is being shared on social network sites, yet others do visit out of curiosity to discovering new things.

Chapter 3 focuses on the role of social media in travel planning and tourism destination decision making. It evaluates six cases whereby clients use travel planning decisions but also change or curtail plans as a result of social media communication. The key results revealed that 92% of travelers make a booking but also change or curtail travel planning decisions for holidays through social media alert.

Chapter 4 examines how social media in general and WeChat shapes the Chinese individual travelers' travelling decisions. As will be seen, while WeChat's popularity among the Chinese outbound tourists is largely attributed to its multi-functionality, its influences on their destination visitation choice are found in multiple levels: by providing illuminating information, allowing users to integrate their payment system easily to their bank accounts in China, and providing means to obtain discounts and promotional offers in a variety of way.

Chapter 5 the focus is to point out the use of social media along the travel process in order to understand the behavior of tourists before, during and after trip. Having in mind the travel decision-making process, the authors prepared a questionnaire with several questions spread through Facebook and answered by 95 persons.

Chapter 6 study the main impacts of social media on business value and performance. ICT can be a resource in resource-based perspective, becoming itself a possible source of competitive advantage (in terms of performance and value creation). Organization strategy proactively tailored can drive the people, processes and systems, harmoniously pulling all the stakeholders to unison. Social media can bring in the commitment of the workforce and ensure involved, flexible, innovative working practices, improving the quality of work. When properly used, they can build organizations to more flexible structures, sharing services, competitiveness, and interdependencies.

Chapter 7 will deal with the concepts of the innovative social technologies and branding approaches in form of different models. This is a conceptual research approach, which can be applied in the exploratory study in several countries or used for specific research purposes.

Chapter 8 proposes a new conceptual model to understand the construction of online destination images. This will lead to a more accurate and realistic portrayal of how destination images are created and allow destination managers (and other stakeholders) to better understand the images promoted, and how these comport with the actual experience of users. The model integrates a sequential and mixed methods approach, enabling a conceptualization of how User Generated Content (UGC) is utilized to formulate and construct destination images.

Chapter 9 aims to understand the effect of evident sales promotion on the number of likes on Instagram. Due to the rising use of social media to communicate with market and promotes products to guests, few academic investigations had been made in this area. Most technology research in the hospitality industry has focused on technology in service and production areas of operations or the use of web-based marketing for tourism operations and travel agents.

Chapter 10 aims to determine and evaluate the factors affecting perceptions of the audience who take part in events realized via Facebook, or on another online platform. Since this study is implemented through a social media communication device, on an on-line platform, it differs from other studies and gives this area a new perspective.

Chapter 11 aims to research blind and visually impaired visitors' experiences in these institutions and presents a proposal of an integrative framework, co-created based on semi-structured interviews conducted with 29 individuals from Portugal. It concluded that sensory, intellectual and physical access must be integrated into the pre-visit, on-site visit and post-visit phases of interaction with cultural institutions, in order to promote a sense of autonomy to visually impaired individuals.

Chapter 12 considers the Tourism 4.0 set up the main goals of making tourism accessible to everyone at any time. The adoption of technological components in accessible tourism enables the development of a new technological solution that can facilitate access to tourism products for disabled people, contributing to the development of accessible tourism. The new term Accessible Tourism 4.0 is the answer to the role of the fourth industrial revolution in accessible tourism, emphasizing the effect of Industry 4.0 components in the tourism sector.

Chapter 13 focuses on the use of mobile applications in tourism context (m-tourism). Users prefer free applications and look for the creation of itineraries. The use of the app in offline mode is the most valued feature, as well as the availability of updated and useful information.

Chapter 14 aims to investigate the long-haul low-cost carrier's business model from a customer-oriented perspective in terms of service quality and perceived value. For this purpose, the authors investigated the effect of service quality on perceived value for money for long-haul low-cost carrier business.

Chapter 15 focus in how the web 3.0 students use social networks daily, however, 1.0 higher education institutions are not taking advantage of these tools. There are some institutional constraints, pedagogical issues, and cultural resistance for its adoption. The study raises important issues and gives practical examples for effective use of social media in education. In order to address these challenges and opportunities wisely, it is necessary an effective engagement in social media by students, professors, and all the academic staff.

Chapter 16 study the impact of specific satisfaction attributes on overall satisfaction. It is used a secondary data set obtained from three of the most influential online travel platforms, being analyzed the guests' average ratings for around 130 hotel units. The three platforms present roughly the same supply of hotels, albeit presenting some differences in terms of volume of data. In terms of specific attributes, apart from "service", the three platforms present significant differences that may reflect the different user bases on these platforms.

Chapter 17 focuses on sentiment analysis of comments made on TripAdvisor regarding one resort located in the Algarve region, in Portugal. The resort has good reviews, which means that the eWOM is positive. The highest scores relate to the resort's cleanliness, location and quality of sleep, and those that were less relevant were the value for money, the rooms and the service. The most dominant emotion is joy, followed by analytical. Negative emotions, such as sadness and anger, were not very found in the online reviews.

Chapter 18 explores how rural tourism companies use social media to improve the tourist consumer experience. By assessing managers of the rural tourism businesses in the Alentejo region, Portugal, the aim is to identify entrepreneur perceptions regarding the factors that influence the tourism experience, the use of social media and their relationship with the improvement of the tourist experience and customer loyalty. It is possible to conclude that rural tourism establishments are no longer ignoring the important role of social media in promoting their tourism experiences.

Chapter 19 intends to analyze how Local Accommodation units in Aveiro use the Internet, and the applications of social media in their communication strategies, as well as the importance they give to this reality of the digital world.

Chapter 20 analyses online reputation of the hotel offer in the city of Porto (Portugal), the second most important city in the country, using information from Booking.com. Known as one of the most well-known search engines worldwide, Booking.com allows the guests to assess the hotel units they have stayed in, using seven different items, from staff to free Wi-Fi.

Chapter 21 seeks to understand the type of influence that the hotel sector in Portugal can achieve in its consumers, being the tourism sector an area in constant growth. In the investigation nine hotels were analyzed, presenting a diversified sample between the participants in terms of capacity, as well as recognition, proving that, in a general way, social networks help in the divulgation of the hotel, presenting these as a direct channel for consumers, facilitating brand exposure as well as interaction with customers..

Chapter 22 evaluate the management of social media of the hotels in the city of Santa Vitória do Palmar, Brazil, as well as to analyze the interaction between the enterprises and their clients. It was noticed that the use of social media by the managers of the research sample is carried out with the same knowledge of common users, without deepening it. In order to optimize this practice and consequently for the hotel to be more profitable, the implementation of specialized and up-to-date management is necessary.

The Editors,

Célia M.Q. Ramos
CinTurs and Universidade do Algarve, Portugal

Cláudia Ribeiro de Almeida
CinTurs and Universidade do Algarve, Portugal

Paula Odete Fernandes
UNIAG and Instituto Politécnico de Bragança, Portugal

Acknowledgment

The editors would like to acknowledge the help of all the people involved in this project. We especially would like to express our gratitude to the chapter authors' and to the colleagues who took part in the review process: Alejandro Mantecon, Alexandra Correia, Alexandra Gonçalves, Alexandra Matos, Alfred Quintano, Ana Casado-Molina, Ana Maria Ferreira, Ana Isabel Renda, Ana Rodrigues, Andreia Moura, Antonio Juan Briones-Peñalver, Beatriz Casais, Bruno Sousa, Carlos Afonso, Carlos Sousa, Celeste Eusébio, Cláudia Henriques, Cláudia Veloso, Conceição Castro, Cristiana de Muylder, Cristina Estevão, Fernanda A. Ferreira, Francieli Boaria, Georgette Andraz, Gloria Widmer, Gorete Dinis, Isabel Lopes, Isabel Sofia Brito, James Malitoni Chilembwe, Jeet Dogra, João Paulo Teixeira, João Rodrigues, Jorge Abrantes, José Agostinho Sousa Pinto, José Alvarez Garcia, José López Garcia, Juan Carlos Infante Moro, María de la Cruz del Río-Rama, Maria del Mar Alonso, Marisol Correia, Marta Amaral, Maria Mercedes Rojas de Gracia, Murat Seyfi, Nelson Matos, Noemi Araújo, Noemi Marujo, Paolo Galli, Paula Azevedo, Pedro Cardoso, Rafael Castro, Roberto Vaz, Romeu Lopes, Sandeep Basnyat, Sandra Pereira, Sandra Rebelo, Sérgio Teixeira, Vânia Costa, Victor Figueira, Vitor Roque, Zeynep A. Gedikoglu.

The editors would also like to acknowledge their research centres: CinTurs (UID/SOC/04020/2019), CEFAGE (UID/ECO/04007/2019), and UNIAG (UID/GES/04752/2019).

Section 1
Travel Decision and Tourism Experience

Chapter 1
Network Analysis of Destination Management Organization Smart Tourism Ecosystem (STE) for E-Branding and Marketing of Tourism Destinations

Jeet Dogra
https://orcid.org/0000-0002-8786-2134
Indian Institute of Tourism and Travel Management (IITTM), Gwalior, India

Sachin Shivaji Kale
Travel and Tourism Department, Government Polytechnic, Thane, India

ABSTRACT

In the current decade, there is a drastic increase in the active social network users which simultaneously raised the interest of destination management organizations (DMOs) to increase their e-visibility on social network sites (SNS), specifically to promote their tourism destinations. The main aim of DMOs is to reach the target market and create a brand image of the destination to increase the tourist footfalls. The purpose of this chapter will be to analyze the utilization of Instagram as a network of smart tourism ecosystem by state tourism boards in India. The network analysis will be done between state tourism boards by using UCINET 6 which is a software package developed for social network analysis. This chapter will give an insight into the Instagram account used by states tourism boards for the promotion of their respective tourist destinations. The findings of the study will help to improve the utilization of SNS for online content creations, managing the social media campaigns, marketing, branding and promotions of tourism destinations.

DOI: 10.4018/978-1-7998-1947-9.ch001

INTRODUCTION

Internet users' growth is rapidly increasing, whereas, in the year 2018, it had crossed the 4 billion mark with 7% growth year-on-year and more than 3 billion people around the world is now using social media platforms at their mobile devices (wearesocial.com, 2018). Internet users in India has reached 627 million in 2019 (economictimes, economictimes, 2019) and there will be more than 351 million are social media users in India in 2019 (Statista, 2019). Today, social media is the central part of smart tourism, therefore, all major State Tourism Boards (STB) and National Tourism Organizations (NTO) are focus to increase their presence on social media as a smart tourism networking. Most notably, China and Korea have been driving the smart tourism agenda forward by funding initiatives to build infrastructure and develop interactive digital technologies, all of which constitutes the foundation of a so-called STE stands for Smart Tourism Ecosystem (Gretzel et al., 2015a; Gretzel et al., 2015b; Koo et al., 2013).

Majority of the travel organizations are also making a strong presence on Social Media and used it as an effective marketing tool. Destination Management Organizations (DMO) can develop a strong network and STE by using social media to develop a destination and increase the real-time experience at tourist destination. Under STE, NTOs and DMOs are "the main vehicle to compete and attract visitors to their distinctive place or visitor space" (Pike & Page, 2014, p. 202). The use of Information technology and communication (ICT) dominates tourism industry and turns its total scenario, especially in online destination marketing aspects. Consequently, users' behavior towards using online applications has changed as a result of the new communication technology. Users are "always-on", carrying their mobile devices all the time (Weber & Rech, 2010). Technology adoption, particularly the use of mobile platforms such as tablets, smartphones and other handheld devices, is of critical importance in use of ICT to engage in dynamic dialogue between consumers and suppliers. ICT reduced uncertainty and risk of trip along with doubt about quality of trip (Fodness & Murray, 1997). ICT empowers both consumer and supplier to deal better way with each other worldwide (Buhalis, 1998). ICTs enable travelers to access reliable and accurate information as well as to undertake reservations in a fraction of time, cost and inconvenience required by conventional methods (O'Connor, 1999). Due to the popularity of Internet applications, most tourism organizations such as hotels, airlines, and travel agencies have embraced Internet technologies as part of their marketing and communication strategies.

One significant development in the evolution of the Internet is the increasing prevalence of social media platforms that enable Internet users to collaborate, communicate and publish original content such as blogs, videos, wikis, reviews, or photos (Boyd & Ellison, 2008). The traditional word-of-mouth (WOM) has moved online becoming electronic word-of-mouth (e-WOM) (Chang et al., 2015; Hennig-Thurau et al., 2004; Alalwan et al., 2017; Kapoor et al., 2018). International tourism has been growing more quickly over the past five years than the world goods trade (UNWTO, 2017) and it continues to be one of the main online commercialization sectors. Destinations are challenged to find new ways to be competitive by providing, managing and marketing positive and memorable tourist and visitor experiences (Mariani & Giorgio, 2017; Mariani, 2015; Mariani & Zan, 2011; Pike & Page, 2014; Pine & Gilmore, 1998) and building strong destination brands (Blain, Levy, & Ritchie, 2005). ICTs and the Internet have dramatically increased the number of options for consumers. Consumers can now have much more option and choices for searching and subsequently purchasing on the Internet. User-Generated Content (UGC), has proven to be more popular and effective than professional advertisements (Welbourne & Grant, 2016; Aral et al., 2013; Lipizzi et al., 2015). Destinations are challenged to find new ways to be competitive by providing, managing and marketing positive and memorable tourist and visitor experiences (Mariani & Giorgio, 2017; Mariani, 2015; Mariani & Zan, 2011; Pike & Page, 2014; Pine & Gilmore, 1998) and building strong destination brands (Blain et al., 2005). This study pays attention to the ways in which State tourism boards utilize Smart Networking Sites (SNSs) to promote tourism. More specifically, this study investigated how local State tourism boards in India utilizes Instagram for the purpose of tourism development.

INSTAGRAM - SOCIAL NETWORK SITES

Web 2.0 was introduced in 2004 by O'Reilly Media and became known as the social web' (Weber & Rech, 2010) who provides opportunities to create new online communities and connect different existing users. This allows both developers and consumers to contribute to the content of a website and offered 'peer to peer' interaction and strengthened the communication (Patel, 2013). Social media can be defined as "a group of Internet-based applications that build on the ideological and technological foundations of Web 2.0 and that allow the creation and exchange of User Generated Content" (Kaplan & Haenlein, 2010). Presence on social networks has become a crucial part of marketing strategies: the promotion of products online via business accounts and advertising on social networking platforms has proven to be efficient (De Vries & Carlson, 2014; Gensler et al., 2013).

Instagram, a social networking site was launched by Facebook in October 2010. Instagram is really an impressive platform in that respect because one can just snap a few photos [and] show what's going on (Abbott, Donaghey, Hare, & Hopkins, 2013). Christian talked about the philosophy behind the creation of Instagram which was to fulfill the desire of users to take photos using their mobile phones and to be able to connect with others "at the moment" through sharing their photos. Instagram which initially serves as a media for online photography had evolved effectively in advertising, promotion, marketing, distribution of ideas and goods, and providing information services fast, precise and accurate (Doolin et al., 2002). The contribution of Instagram's users showed a positive tendency in shaping tourism destination brand. This brand is created through eWOM (electronic word of mouth), which involve consumer comments about products and services posted on Instagram (Bronner & de Hoog, 2011). Instagram is a supporting factor in establishing the phenomenon of a tourism destination branding in public (Fantanti & Suyandnya, 2015).

In this study, the Instagram is taken and studied because it is an image and video-based social media platform, even it found good in both bridging and bonding social capital (Phua et al., 2017). As per Statista (2018), Instagram had reached 1 billion monthly active users worldwide in 2018 (Statista, 2019); India has second-highest users of Instagram 75 million after USA 120 million in the world; More than 60% Instagram social network users are between 18 to 30 age and More than 70% NTOs are using Instagram as a social media network for their promotions.

SMART TOURISM ECOSYSTEM (STE)

"Smart" is often applied as a prefix to technological terms to indicate special capabilities, intelligence and/or connectivity, as in smartphone or smart card. Smart tourism involves multiple components and layers of smart that are supported by ICTs.

Smart tourism spans three layers across these three components (Figure 1): a smart information layer that aims at collecting data; a smart exchange layer that supports interconnectivity; and, a smart processing layer that is responsible for the analysis, visualization, integration and intelligent use of data (Tu & Liu, 2014). Based on these considerations, smart tourism is defined as tourism supported by integrated efforts at a destination to collect and aggregate/harness data derived from physical infrastructure, social connections, government/organizational sources and human bodies/minds in combination with the use of advanced technologies to transform that data into on-site Experiences and business value-propositions with a clear focus on efficiency, sustainability and experience enrichment.

A smart tourism ecosystem (STE) can be defined as a tourism system that takes advantage of smart technology in creating, managing and delivering intelligent touristic experiences and is characterized by intensive information sharing and value co-creation. Collecting, processing and exchanging tourism-relevant data is a core function within the STE (Zhang, 2012). Smart tourism initiatives around the world are seeking to build viable smart tourism ecosystems (Gretzel et al., 2015) but the complexity of the sector makes it extremely difficult to go beyond very specific platform, technology or service-specific innovations. Applied to the business world, the term "ecosystem" is used to describe the relationships among economic entities (producers, distributors, consumers, government agencies, etc.) that, through competition and/or cooperation, facilitate the creation and distribution of a product or service (Investopedia, 2015). Information and communication technologies have been essential in tourism ecosystems for connecting the different players that add value to the experience. Werthner and Klein (1999) illustrated the fundamental technology supported tourism value chain and its components (Figure 2), with the Internet making it possible to completely circumvent traditional distribution channels.

STATE TOURISM BOARDS (STB) AND ICT

State tourism boards are the government organizations which are responsible for the promotion and development of tourist destination in the state and taking efforts to increase the tourist arrival which lead to generating employment in the state. In today's social media world, state tourism boards also actively making their presence on Social networking sites as one of the best economic and fastest ways to promote their destination around the world. Government involvement in tourism has many forms, such as investing resources, planning policies, and promoting tourist attractions. The previous studies

Figure 1. Components and layers of smart tourism
(Gretzel, 2015b)

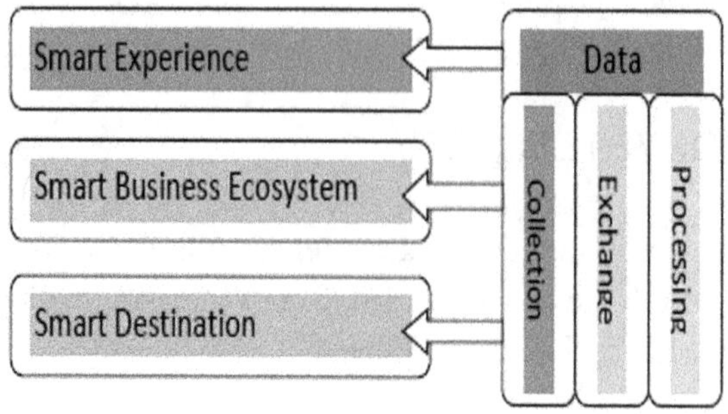

Figure 2. The tourism system as a technology-supported value chain
(Werthner & Klein, 1999)

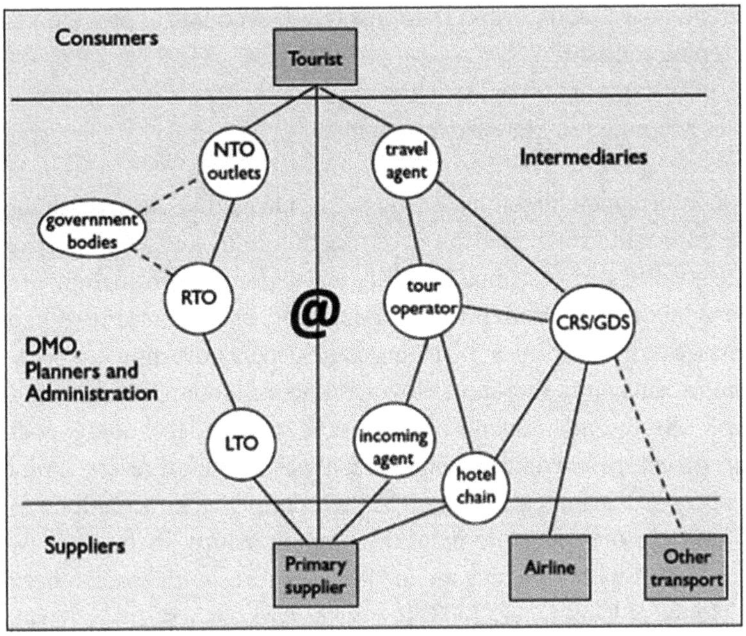

shows that economic concerns are an important interest for local governments to become involved in tourism. Some of the states are really doing well in online marketing to create awareness about their tourist destination around the world with the help of social networking sites. Residents expect that the development of tourism will result in an improvement in the local economy, better employment opportunities, infrastructure improvements, and more investment opportunities (Nunkoo & Ramkissoon, 2011). Government agencies play a role in the STE not only with respect to their typical tourism-related agendas but also in regards to ensuring data openness while at the same time regulating data privacy (Buhalis & Amaranggana, 2014). For travelers, the information provided on government tourism websites or SNSs are often the best places to obtain credible information—even better, these can be accessed 24 hours a day from anywhere in the world. Interactive design is a critical factor for tourist websites, as this allows tourist organizations to identify consumers' interests and boost their participation (Doolin, Burgess, & Cooper, 2002). The provision of virtual experiences using graphics, photos, and videos on local government websites or SNSs is also important. These features can tempt online visitors enough to prompt a real-world visit to the promoted area (Rosen & Purinton, 2004).

NTOs should become more innovative and creative in their social media strategies to differentiate digital marketing from traditional marketing methods. Hays et al. (2012) demonstrate that social media can be used as a destination marketing tool for national tourism organizations. It can offer a tool to reach a global audience with limited resources to DMO's. Tourism organizations that do not adopt social media will lack a competitive advantage (Schegg et al., 2008; Stankov et al., 2010; Wang et al., 2002). Empirical evidence on the effectiveness of social media activities at the destination level is highly desirable. This study also suggests that the state tourism boards could affect the level of Social Network Sites use for tourism, since the scale of the tourism industry and the level of ICT innovation could vary with different state tourism boards, leading to the following research questions.

Research Questions (RQ)

RQ 1: Do State tourism boards actively use their Instagram accounts to provide tourism information to users for developing tourism?

RQ 2: Do the levels of Instagram utilization in the implementation of State tourism board's smart tourism agenda vary according to State Governments?

This study focuses on information disseminated via SNSs. On an SNS, users visit the state governments' pages to obtain more information related to the government's doing and what is happening in that specific tourist location. State governments also interact with users and share information through these networks. Governments can also link content from other organizations, or from mass media outlets, to their sites. More importantly, enriched by Web 2.0 or 3.0 technologies, local governments' sites are now intertwined with other state governments' sites through following others accounts on Instagram in a smart tourism ecosystem. By using SNSs, the governments become more open and technology focused and they acts as sole providers of infrastructure and data. A network analysis is a set of research procedures that focuses on the relationships between social entities such as "who provides information to whom" and "who is linked with whom" and explores the patterns of these relationships, including how these entities "collaborate with," "subcontract with," and engage in "joint ventures with" each other (Barnett, Danowski, & Richards, 1993; Monge & Contractor, 2003; Rogers & Kincaid, 1981; Wasserman & Faust, 1994).

As per the above discusison, this study based on examine the network structure of state tourism boards' Instagram pages and the organizations "linked or followed by them as part of a smart tourism ecosystem and how this ecosystem is associated with governmental utilization of Instagram as a social media platform for activating the tourism. The following research question is also posed:

RQ3: How are State governments' tourism boards Instagram pages following other accounts as part of a smart tourism ecosystem?

METHODS

Sample

India has 28 States with 7 Union Territories (UT) and out of these 35 States / UTs, only 20 states has their Official Instagram Account and only 19 States' Instagram accounts following others Instagram Accounts as a part of Smart Tourism Ecosystem and only 16 States' Instagram Accounts are having 'active presence'. Therefore, this study examined those 19 state tourism boards' Instagram accounts which followed others only tourism-related Instagram accounts, since it is assumed that they are interconnected with third-party stakeholders with regard to the Smart Tourism Ecosystem. The research question and methodology adopted from the study done by Ji Hoon Park, et al. in 2016.

Data and Variables

For this study, the number of 'total postings,' and 'total responses to users' postings' over a three month period were collected in preparation for content analysis. The relationships among state tourism boards'

Instagram accounts and their co-liked accounts are presented as network data. However, a network based on these relationships is a 2-mode network because it has two different sets of entities viz. (1) state tourism boards Instagram accounts and (2) Following other tourism boards, tourism promotion channels, and Organization Instagram accounts only. If the rows and columns in a matrix index represent different sets of entities, the matrix is 2-mode matrix (Borgatti & Everett, 1997). In this context, a class of entities is called by the term 'mode'. The 1-mode case represents members that have social ties with other members, whereas, the 2-mode case represents members that have social ties with members of another class. Most social network analysis is concerned with the 1-mode case, in which the basic network data set is an n × n matrix S, where n equals the number of accounts followed in the network. A node might be an individual or a higher-level component such as an organization or a nation from which the system is composed. In communication research, this relationship is generally the frequency of communication among the nodes (Barnett et al., 1993; Nam, 2015; Salisbury & Barnett, 1999).The 2-mode case arises when researchers collect relations between classes of actors such as persons and events, or companies and organizations. In this case, the study collected data on which state tourism board accounts following other organization's Instagram accounts. This type of data is often referred to as affiliations. Co-memberships and co-link relationships with other organizations can be thought of as providing opportunities for social relationships among State government as actors and also as the consequences of pre-existing relationships for tourism.

Thus, this study obtained data on how State government jointly linked via certain organizations. This study used 19 × 55 matrices (each state tourism boards listed × each organization 'followed' by the state tourism boards Instagram accounts), where each cell was coded '1' for following (followed by state tourism board) and '0' for not-following. The primary dataset was pre-multiplied by its transposition to form a 19 (state tourism board) × 19 matrix of joint affiliation of state tourism boards in the Instagram network. Then, using UCINET 6 (Borgatti, Everett, & Freeman, 2002), two types of centrality were calculated to determine which state tourism boards or organizations play a significant role in the Instagram accounts: (1) degree centrality and (2) eigenvector centrality.

First, degree centrality in this study indicates the number of co-affiliated organizations between two state tourism board.To obtains a higher degree of centrality; a state tourism board should be linked with other state tourism board that is affiliated with many other organizations. Degree centrality in this study indicates that the number of co-affiliated organizations between two companies or the number of co-affiliated global companies between two organizations. The higher number of like-relationships does not simply increase the degree centrality score, rather, to obtain higher degree-centrality score. In other words, how the state tourism board build their relationships with the core co-related organization which are multi-affiliated with state tourism board are crucial for positioning themselves at the center of smart tourism ecosystem.

Second, eigenvector centrality considers the strength of ties and indirect social ties among nodes (Barnett, Danowski, Feeley, & Stalker, 2010). More central state tourism board could boost their centralities due to the inherent circularity involved in the calculation of eigenvector centrality (Nam, Barnett, & Kim, 2014). In this two-mode case, the total number of co-links with a given state tourism board was considered rather than the one-way links between an organization and a state tourism board. Therefore, eigenvector centrality determines how central (or peripheral) a state tourism board is, based on its loading on the largest eigenvector of the network socio-matrix.

RESULTS

RQ1 and RQ2: The content analysis indicate that the tourism industry dependent states has the highest average number of total postings within three months. The major difference found in the mean of per week posting by STB, the main reason for the difference is very less activeness of some STB on Instagram like Bihar, Himachal Pradesh and Karnataka's STB, as compared with rest of the STB. As per their online website, they are using the another social media platform Facebook more actively than Instagram. The most common posting posted by the STB is related with culture (festivals, handicrafts, cuisines, music, dance events, etc.), nature (mountains, valleys, wildlife's, beaches, etc), cuisine, handicraft, monuments, etc. (Table 1), to increase the active participation of tourist. The results also indicates that the Natural and cultural based strong states like Kerala, Madhya Pradesh, Rajasthan, Uttarakhand, Uttar Pradesh, etc. has the most "likes" on their tourism postings, and there is a significant difference in the number of "likes" between STB.

The use of Instagram as social media marketing varies by states as per the contribution of the tourism industry in the state overall GDP (Table 2). There was a significant difference in the number of posting and "likes" of natural and cultural posting between STB, which indicate that the experience-based tourism, is mostly focused by state which ensures the active participation of Tourist, like Rajasthan STB posted 70% posting on cultural festival and events.

RQ3: Network analyses are conducted to identify that which state tourism boards are more central in the Instagram's network. The researcher has given more emphasis on the STB Instagram accounts which following others tourism stakeholder's Instagram accounts to promote their destination worldwide as a part of smart tourism ecosystem (STE). The state has strong policies and economic dependency on tourism industry are following other International tourism related organization, international tourism TV channels, Worldwide known tourism magazines, Worldwide known celebrities, and leaders Instagram accounts to reach maximum number of users in the world. The STB especially following those Instagram accounts who have million numbers of followers. The STB also following other STB Instagram accounts and countries' important leaders and celebrities' account to promote their product within the country. The high numbers of tourism stakeholders

Table 1. Post and profile of followers and following accounts by the state tourism board's instagram accounts

Type of Post Posted by STB	Followers Profile	Profile of Accounts Followed by STB
• Festivals • Cultural events • Monuments • Nature • Wildlife • Spiritual • Cuisine • Handicraft Activities	• Individual • Travel agent • Tour operator • Hotelier • State tourism boards • Tourism Institute • Restaurant • Other Tourism Stakeholders	• Travel Channel • Travel Magazine • News Channel • Celebrity(Sport/film) • Political leader • State tourism boards • Tour Operator • Individual • Other Tourism stakeholders

Table 2. Basic information of state tourism boards' instagram accounts

State	Post Within 3 Month (Jan-Mar 2019)	Like	Total Post	Followers	Following
Uttarakhand	368	128165	1934	120000	495
Utter Pradesh	253	140730	1735	25800	31
Rajasthan	199	667316	1643	187000	31
Madhya Pradesh	163	1002200	2046	65000	46
Kerala	111	5604813	2044	191000	0
Andhra Pradesh	105	16300	544	4430	193
Delhi	69	45397	564	4414	5
Tamilnadu	58	8511	303	2511	1
Telangana	56	541	112	103	2
Odisha	53	33488	371	10600	95
Goa	46	67299	1790	67800	394
West Bengal	43	5864	885	7387	20
Sikkim	41	18346	752	10400	45
Gujrat	32	94545	1017	76000	22
Jammu & Kashmir	26	52958	2974	33000	1
Nagaland	23	3896	248	1717	20
Bihar	9	8772	208	12300	5
Himachal Pradesh	5	285	27	647	18
Karnataka	1	2350	748	3762	155

following STB Instagram account and the high number of tourism stakeholder's accounts followed by STB put them into the centrality of STE. Some STB like Kerala has highest numbers of likes and followers among all STB, but not following other tourism stockholders Instagram accounts that's why Kerala STB is not in the highest degree of centrality as a part of STE.

As per Table 3, the results shows Uttarakhand state is highly ranked in the network. Uttarakhand state has the highest degree of centrality (92) and eigenvector centrality (0.657), followed by Madhya Pradesh (60 and 0.318), Karnataka (58 and 0.344), Himachal Pradesh (57 and 0.305), Rajasthan (50 and 0.262) and Uttar Pradesh (40 and 0.203).The results of the network analysis are represented graphically in Figure 3. The graph is constructed by plotting the first two dimensions from multidimensional scaling of the network, such that closer the STB are to each other, the more similar their relationships are with other organizations.

Table 3. Utilization and centrality of state tourism boards on instagram accounts

State	No. of Posting Per Week on Tourism in 3 Month		No. of Likes on Tourism Post in 3 Month		Degree Centrality of State Government Tourism Boards		Eigenvector Centrality	
	M	SD	M	SD	DEGREE	NDEGREE	Eigenvector	NEigenvector
Uttarakhand	30.75	8	348.27	236.05	92	0.372	0.657	92.973
Utter Pradesh	21.08	5.35	210.01	350.09	40	0.162	0.203	28.645
Rajasthan	18	3	3353.34	2271.6	50	0.202	0.262	37.024
Madhya Pradesh	13.58	5.31	6141.46	10718.5	60	0.243	0.318	44.987
Kerala	9.25	5.58	50493.81	122882.5	0	0	0	0
Andhra Pradesh	8.75	5.62	155.23	95.04	17	0.069	0.094	13.357
Delhi	6.27	3.58	6657.92	2383.52	19	0.077	0.066	9.3
Tamilnadu	4.83	1.69	146.74	116.22	14	0.057	0.041	5.762
Telangana	4.66	2.77	9.66	6.18	0	0	0	0
Odisha	4.41	2.23	631.84	275.51	16	0.065	0.049	6.958
Goa	3.83	1.19	1463.02	871.55	50	0.202	0.244	34.45
West Bengal	3.58	1.5	136.37	50.08	27	0.109	0.101	14.347
Sikkim	3.41	2.84	447.46	221.22	14	0.057	0.041	5.762
Gujrat	2.66	1.3	2954.53	1747.1	40	0.162	0.231	32.697
Jammu & Kashmir	2.14	2.5	2036.84	1382.15	0	0	0	0
Nagaland	1.91	4.27	169.39	168.2	14	0.057	0.041	5.762
Bihar	0.75	1.48	974.66	563.03	14	0.057	0.041	5.762
Himachal Pradesh	0.41	0.99	57	22.66	57	0.231	0.305	43.144
Karnataka	0.07	0.277	2350	0	58	0.235	0.344	48.7

DISCUSSION

The Instagram as a social media platform is used by more than 1,000 million users in the world (Statista, 2019). That is why, most of the DMOs using Instagram as their important social media account to promote their destinations. This study finds out the utilization of Instagram by STB to promote their destinations around the world for tourism development. The results indicates that very few STB actively manage online tools, including Instagram, to communicate with the public, and promote tourism as a part of STE. In particular, the STE constitutes multiple players that interact with one another and are involved to various degrees in the system. Specifically, the STE constitutes multiple players that interact with one another and are involved to various degrees in the system. These parties include the government, travel agencies, transport and accommodation providers, tour operators, local residents and most importantly tourists (Arenas et al., 2019). Those STB are active on Instagram intentionally following that Instagram account that has good number of followers around the world, so that they can reach maximum number of Instagram users through worldwide presence on Tourism channels, Tourism Magazines, News channels, and Photography channels' Instagram accounts. In concern with marketing and promotion at domestic

Figure 3. Network visualization of state tourism boards Instagram accounts

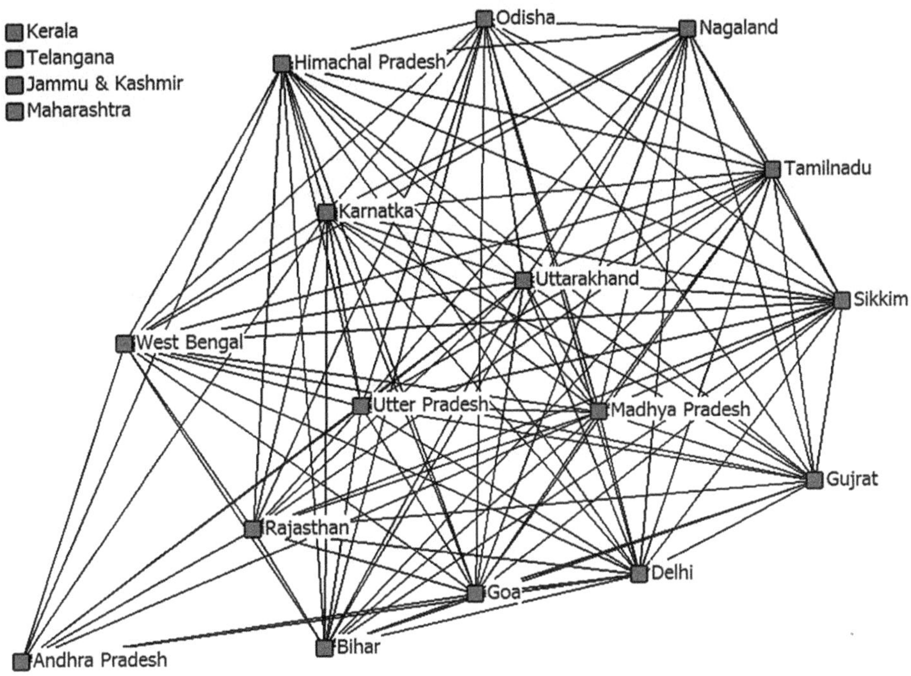

market, STB purposefully following National tourism channels, National Magazines, other STB, Celebrities from sport and film industry, important political leader's and tour operators Instagram accounts.

Figure 3 shows that the the degree of centrality of the STB indicates how much they are connected and centrally located in STE. The Eigenvector Centrality indicates that how much STB has strong bond with other STB's as a part of STE. The degree of centrality is an important part for STB to show their active online presence and reach to as much as accounts users. Not only follow the individuals accounts but strongly follow those accounts who has millions of followers and popular image or brand in the world which will take them to their target market. As a part of STE, the STB are quite focusing only on the promotion part of the tourist destination, but totally neglected to on-site services, local tourism partner networking, local tourism bodies networking and different tourism stakeholders, so tourist destination can increase the real-time experience of tourist. Based on this present study, one can foresee that social media are likely to be further integrated into the ecosystem of smart tourism in India. Network analyses have been used to show how state tourism boards Instagram accounts are co-related on Instagram as part of a STE.

The results of this study indicate that few states only are highly ranked in the network, which meant they has more co-linked accounts on Instagram and put them into the centrality of STE. This suggest that they would utilize Instagram more actively, to achieve a wider scope of government activities than the other state governments of the countries. This study suggest that the use of SNSs by STB not only benefits local residents but is also beneficial to potential tourists who live outside the state, as the STB's web-based promotion of tourism has been identified as an important factor in successful tourism development. The involvement of STB in tourism promotion has been specifically significant to achieve the ultimate goal of the country's tourism development efforts.

As compared to tourism products that rely on the natural or physical environment, cultural tourism requires more contextual information and sensitivity when promoted to tourists unfamiliar with the cultural background of a given area. In order to promote these local tourism opportunities, state governments should operate tourism websites and SNSs with the expectation that tourists seek out information about their potential destinations. It's not possible for the central / union governments to promote each product of all states which are extensively wider in their specific cultures and traditions. On the other side, as compared to other advertising media, such as television or newspapers, SNSs are relatively inexpensive and are among the most efficient promotion tools for regional tourism. In addition, through SNSs governments can showcase and store a large amount of information while also retrieving web-ecology data consisting of interconnected users' informational behaviors on SNSs. Finally, it is suggested that the ecosystem built by smart tourism will be of benefit not only to the tourism industry but also to other sectors at the national level. The significance of smart tourism achieved via social media suggests that broadband telecommunication infrastructure is a prerequisite for the ecosystem of smart tourism.

The important contribution of this study is the current status of STB in use of social networking sites but the major limitation of this study is to consider only one social networking platform which is Instagram. It might be possibility that STB may be using other social networking platforma like Facebook or Twitter as their main social media marketing.

FUTURE IMPLICATIONS AND SUGGESTIONS

The main emphasis of this study will be to increase the smart tourism network by state tourism boards and effective use of the social networking sites (SNS) for destination marketing, to create a bond between suppliers, tourism promoters and other DMOs. Future researchers may also consider how Instagram can be used to improve an organization's ability to address social concerns (Gruzd et al., 2018). The smart tourism ecosystems will provide round the clock availability and retrieval of big data at the current social network platforms, which would be utilized within the fraction of seconds by visitors, suppliers, business organizations and other national organizations. Finally, active participation of STB in SNS will significantly help to develop the tourism regionally, and then, consequently at the national level.

REFERENCES

Abbott, W., Donaghey, J., Hare, J., & Hopkins, P. (2013). An Instagram of worth a thousand words: An industry panel and audience Q&A. *Library Hi Tech News*, *30*(7), 1–6.

Alawan, A. A., Rana, N. P., Dwivedi, Y. K., & Alghatabat, R. (2017). (2017). Social Media in Marketing: A review and Analysis of the existing literature. *Telematics and Informatics*. doi:10.1016/j.tele.2017.05.008

Arenas, A., Goh, J., & Uruena, A. (2019). How does IT affect design centricity approaches: Evidence from Spain's smart tourism ecosystem. *International Journal of Information Management*, *45*, 149–162. doi:10.1016/j.ijinfomgt.2018.10.015

Barnett, G. A., Danowski, J. A., Feeley, T. H., & Stalker, J. (2010). Measuring quality incommunication doctoral education using network analysis of faculty-hiring patterns. *Journal of Communication*, *60*(2), 388–411. doi:10.1111/j.1460-2466.2010.01487.x

Barnett, G. A., Danowski, J. A., & Richards, W. D. (1993). Communication networks and network analysis: a current assessment. In W. D. Richards & G. A. Barnett (Eds.), *Progress in communication science* (Vol. 12, pp. 1–19). Norwood, NJ: Ablex.

Blain, C., Levy, S. E., & Ritchie, J. R. (2005). Destination branding: Insight and practices from destination anagement. *Journal of Travel Research*, *43*(4), 328–338. doi:10.1177/0047287505274646

Borgatti, S. P., & Everett, M. G. (1997). Network analysis of 2-mode data. *Social Networks*, *3*(3), 243–269. doi:10.1016/S0378-8733(96)00301-2

Borgatti, S. P., Everett, M. G., & Freeman, L. C. (2002). *UCINET for windows: software for social network analysis (Version 6)*. Harvard, MA: Analytic Technologies.

Boyd, D., & Ellison, N. (2008). Social Network Sites: Definition, History, and Scholarship. *Journal of Computer-Mediated Communication*, *13*(1), 210–230. doi:10.1111/j.1083-6101.2007.00393.x

Buhalis, D. (2003). eTourism: Information technology for strategic tourism management. London: Pearson. doi:10.13140/2.1.2274.0804

Buhalis, D., & Amaranggana, A. (2014). Smart tourism destinations. In Z. Xiang & I.Tussyadiah (Eds.), Information and communication technologies in tourism (pp. 553–564). Dublin:Springer.

Buhalis, D., & Law, R. (2008). Progress in information technology and tourism management: 20 years on and 10 years after the internet - the state of eTourism research. *Tourism Management*, *24*(3), 609–623. doi:10.1016/j.tourman.2008.01.005

Chang, Y. T., Lu, H., & Hu, H. (2015). Persuasive messages, popularity cohesion, and message diffusion in social media marketing. *Journal ofBusiness Research*, *68*(4), 777–782. doi:10.1016/j.jbusres.2014.11.027

Chante, K., Jessica, C., Lindsay, B., Tyler, Q., & Robert, P. D. (2014). Dermatology on Instagram. *Dermatology Online Journal*, *20*(7), 1–6. PMID:25046455

Cohen-Almagor, R. (2011). Internet history. *International Journal of Technoethics*, *2*(2), 45–64. doi:10.4018/jte.2011040104

Doolin, B. B., Burgess, L., & Cooper, J. (2002). Evaluating the use of the web for tourism marketing: A case study form New Zealand. *Tourism Management*, *23*(5), 557–561. doi:10.1016/S0261-5177(02)00014-6

Economic Times. (2019). Internet users in India to reach 627 million in 2019. Retrieved from https://economictimes.indiatimes.com:https://economictimes.indiatimes.com/tech/internet/internet-users-in-india-to-reach-627-million-in-2019-report/articleshow/68288868.cms?from=mdr

Fantanti, N. M., & Suyandnya, I. W. (2015). Beyond User Gaze: How Instagram Creates Tourism Destination Brand? *Procedia: Social and Behavioral Sciences*, *211*, 1089–1095. doi:10.1016/j.sbspro.2015.11.145

Fodness, D., & Murray, B. (1997). Tourist information search. *Annals of Tourism Research*, *24*(3), 503–523. doi:10.1016/S0160-7383(97)00009-1

Gretzel, U. (2011). Intelligent systems in tourism: A social science perspective. *Annals of Tourism Research, 38*(3), 757–779. doi:10.1016/j.annals.2011.04.014

Gretzel, U., Sigala, M., Xiang, Z., & Koo, C. (2015a). Smart tourism: Foundations and developments. *Electronic Markets, 25*(3), 179–188. doi:10.100712525-015-0196-8

Gretzel, U., Werthner, H., Koo, C., & Lamsfus, C. (2015b). Conceptual Foundations for Understanding Smart Tourism Ecosystems. *Computers in Human Behavior, 50*, 558–563. doi:10.1016/j.chb.2015.03.043

Gruzd, A., Lannigan, J., & Kevin Quigley, K. (2018). Examining government cross-platform engagement in social media: Instagram vs Twitter and the big lift project. *Government Information Quarterly, 35*(4), 579–587. doi:10.1016/j.giq.2018.09.005

Henning-Thurun, T., Gwinner, K. P., Walsh, G., & Gremler, D. D. (2004). Electronic word of mouth via consumer-opinion platform: What motivates consumers to articulate themselves n the Internet? *Journal of Interactive Marketing, 18*(1). doi:10.1002/dir.10073

Koo, C., Shin, S., Kim, K., Kim, C., & Chung, N. (2013). Smart tourism of the Korea: A case study. *Proceedings of the Pacific Asia conference on information system.* AIS Electronic Library. Retrieved from http://aisel.aisnet.org/pacis2013

Mariani, M., Di Felice, M., & Mura, M. (2016). Facebook as a destination marketing tool: Evidence from Italian regional destination management organizations. *Tourism Management, 54*, 321–343. doi:10.1016/j.tourman.2015.12.008

Mariani, M. M. (2015). Coordination in inter-network co-opetitition: Evidence from the tourism sector. *Industrial Marketing Management.* doi:10.1016/j.indmarman.2015.11.015

Mariani, M. M., Di Felice, M., & Mura, M. (2016). Facebook as a destination marketing tool: Evidence from Italian regional destination management organizations. *Tourism Management, 54*, 321–343. doi:10.1016/j.tourman.2015.12.008

Mariani, M. M., & Giorgio, L. (2017). The 'Pink Night' festival revisited: Meta-events and the role of destination partnerships in staging event tourism. *Annals of Tourism Research, 62*, 89–109. doi:10.1016/j.annals.2016.11.003

Nam, Y., Barnett, G. A., & Kim, D. (2014). Corporate hyperlink network relationships in global corporate social responsibility system. *Quality & Quantity, 48*(3), 1225–1242. doi:10.100711135-013-9831-7

Niininen, O., March, R., & Buhalis, D. (2006). Consumer centric tourism marketing. In D. Buhalis & C. Costa (Eds.), *Tourism management dynamics: Trends, management and tools* (p. xxiii). Amsterdam: Buttterworth Heinemann. doi:10.1016/B978-0-7506-6378-6.50029-9

O'Connor, P. (1999). Electronic information distribution in tourism and hospitality. Wallingford: CAB.

Park, J. H., Lee, C., Yoo, C., & Nam, Y. (2016). An analysis of the utilization of Facebook by local Korean governments for tourism development and the network of smart tourism ecosystem. *International Journal of Information Management, 36*(6), 1320–1327. doi:10.1016/j.ijinfomgt.2016.05.027

Patel, K. (2013). Incremental journey for World Wide Web: Introduced with Web 1.0 to recent Web 5.0 – A survey paper. *International Journal of Advanced Research in Computer Science and Software Engineering*, *3*(10), 410–417.

Pike, S., & Page, S. (2014). Destination marketing organizations and destination marketing: A narrative analysis of the literature. *Tourism Management*, *41*, 202–227. doi:10.1016/j.tourman.2013.09.009

Salomon, D. (2013). Moving on from Facebook: Using Instagram to connect with Undergraduates and engage in teaching and learning. *ACRL TechConnect*, *74*(8), 408–412.

Sigala, M. (2015). Applying Gamification and Assessing its Effectiveness in a Tourism Context: Behavioural and Psychological outcomes of the Tripadvisor's Gamification Users. *Asia Pacific Journal of Information System*, *25*(1), 179–210. doi:10.14329/apjis.2015.25.1.179

Smith, T. C., Coyle, J. R., Lightfoot, E., & Scott, A. (2007). Reconsidering models of influence:the relationship between consumer social networks and word-of-mouth effectiveness. *Journal of Advertising Research*, *47*(4), 387–397. doi:10.2501/S0021849907070407

Statista. (2019). Number of social network users in India from 2015 to 2023. Retrieved from https://www.statista.com/statistics/278407/number-of-social-network-users-in-india

Tu, Q., & Liu, A. (2014). Framework of Smart Tourism Research and Related Progress in China. *Proceedings of the International Conference on Management and Engineering (CME 2014)* (pp. 140-146). DES tech Publications.

Ulrike Gretzel, H. W. (2015). Conceptual foundations for understanding smart tourism ecosystems. *Computers in Human Behavior*, *50*(C), 558–563. doi:10.1016/j.chb.2015.03.043

Wang, D., Park, S., & Fesenmaier, D. (2012). The role of smartphones inmediating the tourism experience. *Journal of Travel Research*, *54*(4), 371–387. doi:10.1177/0047287511426341

Wearesocial.com. (2018). Gloval digital Report. Retrieved from https://wearesocial.com/blog/2018/01/global-digital-report-2018

Weber, S., & Rech, J. (2010). An overview and differentiation of the evolutionary steps of the web X.Y movement. Handbook of Research on Web 2.0, 3.0, and X.0, 12-39. Doi:10.4018/978-1-60566-384-5.ch002

Welbourne, D. G. (2016). Science communication on YouTube: factors that affect channel and video popularity. *public understanding of science, 25*(6), 706-718. doi:10.1177/0963662515572068

Werthner, H., & Klein, S. (1999). *Information technology and tourism: A challenging relationship*. Wien: Springer-Verlag. doi:10.1007/978-3-7091-6363-4

Zhang, L. (2012). Smart tourism: The coming age of customization and intelligent public services. *Journal of Tourism Tribune*, *27*(2), 3–5.

KEY TERMS AND DEFINITION

Degree of Centrality: It Indicates the number of co-affiliated organizations between two state tourism boards.

Destination Management Organization (DMO): An organization which is responsible of development and promotion of a tourist destination.

Eigenvector Centrality: It indicates the strength of ties and indirect social ties among organization.

Electronic Word of Mouth (e-WOM): All informal communication via the Internet addressed to consumers and related to the use or characteristics of goods or services or the sellers thereof.

Information and Communication Technology (ICT): All devices, networking components, applications and systems that combined allow people and organizations (i.e., businesses, nonprofit agencies, governments and criminal enterprises) to interact in the digital world.

Smart Tourism: A specialized type of tourism that supported by advanced technology to collect and exchange information in order to increase the real experience of the traveler.

Smart Tourism Ecosystem (STE): As a tourism system that takes advantage of smart technology in creating, managing and delivering intelligent touristic-experiences and it is characterized by intensive information sharing and value co-creation.

Social Networking Sites (STE): A different group of Internet-based applications that build on the ideological and technological foundations of Web 2.0 and that allow the creation and exchange of User Generated Content (UGC).

State Tourism Board (STB): A government organization responsible for promotion and development of tourism within its jurisdiction.

User Generated Content (UGC): It is any form of content, such as images, videos, text and audio, that have been posted by users on online platforms such as social media.

Chapter 2
Peer Influence Mechanism Behind Travel Experience Sharing on Social Network Sites

James Malitoni Chilembwe
https://orcid.org/0000-0002-3482-5073
Mzuzu University, Malawi

Victor Ronald Mweiwa
https://orcid.org/0000-0002-1382-6886
Malawi Institute of Tourism, Malawi

ABSTRACT

The way tourism consumers share travel experiences have changed tremendously in recent years. There is increased use of technology to distribute travel experiences via social networking sites. Travel companies are increasingly developing their websites in order to network with their tourism customers. These network sites play a significant role in receiving customer feedback using online reviews. As a result, travel companies can utilize their customer evaluations to improve products and service offerings. The visual images and textual postings on network sites have the power to motivate others to embark on travel. This chapter is about peer influence mechanisms behind travel experience sharing on the social network sites. It uses both phenomenology and ethnographic methodological approaches to analyze some of the comments and photographs posted on SNSs by tourists in Malawi. It concludes that most of the tourists traveling to Malawi were influenced by what is being shared on social network sites, yet others do visit out of curiosity to discovering new things.

DOI: 10.4018/978-1-7998-1947-9.ch002

INTRODUCTION

The sharing travel experiences is not a new concept (Havell & Ross, 1819). What appears to be new, is the increasing and modernised methods of distributing information to both existing and potential tourism consumers (Buhalis & Shaw, 2008; Tussyadiah & Zach, 2012; Wang et al., 2012; Xiang et al., 2014; Hjalager, 2015; Law & Wang, 2018). The digital devices and their applications have completely changed the travellers mind set because they are able to take photographs but also write instant online reviews and store share information to others through various social network sites (Tussyadiah & Fesenmier, 2009; Kim & Fesenmaier, 2017). Social network sites such as websites, Twitter, Facebook, Instagram, blogs and Trip Advisor are utilised by travellers when sharing travel experiences through messages, videos and photos (Boyd & Ellison, 2008; Volo, 2010; Munar & Jacobsen, 2014; Kim & Fesenmaier, 2017).

Sharing travel experiences through social media provides several benefits to consumers. Dedeoğlu, Taheri, Okumus and Gannon (2020) provide a summary of benefits as (a) easier access to information, (b) provision of more substantial products and services information from different viewpoints, and (c) the provision of end-user evaluation. Equally, both existing and potential travel consumers may use social media to search for information on where to go, on where to find accommodation options, leisure activities like excursions and night life (Cox et al., 2009; Amaro et al., 2016). It is acknowledged that social media plays a significant role before travelling, during information search (Gretzel & Yoo, 2008). It is also crucial for the travellers to seek for destination information when embarking on the international destination trip for the first time (Simms, 2012; Amaro et al., 2016), of which, most of the information is found and shared on the social network sites for tourism organisations and previous travellers including travel bloggers (Lin & Chen, 2012; Ho, Lin, Yuan, & Chen, 2016; Sahin, Sener & Polat, 2018).

Social media usage among top destination marketing organizations is still in its experimental stage with a high degree of variation in terms of their strategies (Hays et al., 2013). As such, there is a great need, not only for the tourism companies that market destinations but also travellers in general to learn and understand the factors that influence conversation and engagement on social network sites (Miranda & Saunders, 2003; Tussyadiah & Fesenmaier, 2009). It is important to understand both the positives and negative outcomes as a result of social media sharing as the power of information technology through social network sites cannot be underestimated. Therefore, it is critical to be more strategic and position the motivations and experiences in order to educate, inform and warn potential travellers about the status of tourist destination through social platforms (Chilembwe, Mweiwa & Mankhomwa, 2018; Chilembwe & Mweiwa, 2019). According to the blog 'Taking Africa to the World', Wodemaya (2019) advises that potential travellers should strive to check what other people comments or share online about a destination. Citing an example of China and other several cities where many people travel during holidays, hence making cities crowded. On the contrary, the cities are almost empty in some parts of the world, for example Malawi. In China, it is the opposite as there are very few public holidays resulting into many people travelling during public holidays to their families, friends and relatives. Sharing of travel experience like this one, helps to prepare a potential travel to understanding what is expected of during holidays in terms of overcrowding and delayed transport and connection systems (Wodemaya, 2019). The population of China is already high in comparison to other countries, so it also contributes to the situation in addition to other travellers to China during public holidays. Several personal and companies' websites share information about specific countries which is valuable in terms of making destination visit plans. Both visual and textual online information add value to traveller's experience at a destination.

Tourism companies also utilise social media as part of marketing their products and services. They advertise new products and services which are shared on their websites. Besides, they also receive online reviews which ultimately help to improve service and product provision. This chapter is about travel experience sharing on social network sites and the peer pressure that exists among the potential travellers, to make informed decisions, and embark on similar trips to tourism destinations and have their experiences. Drawing from Malawi case, it discusses the motivations of travel experiences on the mountain and lake or resort adventure - soft and hard activities. The examples provided are, particularly, on social network sites that travellers use in everyday when sharing travel experiences.

The next sections of the chapter are organised as follows. The first section presents the literature review of travel experience sharing on the social network sites, followed by information technology in travel and tourism industry. The second section presents the research methodology used in the chapter. The third section presents results and discussion in the context of Malawi, and the fourth section provides a conclusion.

Travel Experiences Sharing

The sharing of travel experiences on social media plays an essential role in spreading messages to potential visitors (Kim & Fesenmier, 2017). Social media shares information and can create messages about destination awareness, perceptions, expectations behaviors as well as attitudes of local residents (Kim & Tussyadiah, 2013; Narangajavana et al., 2017). Some travellers can keep the information shared on social media and it is vital to have a picture of a destination information in order to be used for future trips. Travel experiences sharing shapes the experience of new travellers because they are able to plan properly and save time for their visitation. Travellers have always shared their experiences through social media and internet have turned tourism into information-intense industry as social media enable travellers to collaborate in the production, challenge, consume but also distribute travel information through various network social sites (Yoo & Gretzel, 2009; Narangajavana et al., 2017). Moreover, the emergent of online and cloud access mobile technologies have helped creating means of searching travel information according to Xiang et al. (2015). Likewise, independent tour guides that assist in tourism destination marketing are now utilising internet and mobile devices when sharing information with their potential customers (Chilembwe & Mweiwa, 2014). Modern travellers are able to share travel experiences on social network sites even after post trip (Bae, Lee, Suh & Suh, 2017). Almost an hour cannot elapse without a posting of a travel experience on social media. Worldwide, the posting of one's selfie photographs, texts and opinions of the destination during and after trip is becoming common. Yet others can also make intimation before departure by sharing things to expect at their destinations (Wodemaya, 2019). The experiences that travelers acquire and share with others can motivate the potential travellers in various social networks. Travellers share travel stories, photos and videos that convey their overall travel experiences. People then are likely to begin preparing for their trip to get similar experiences (Önder, Gunter & Gindl, 2019). It is through peer pressure that influences people in different ways and for a variety of purposes. The majority of new generation-younger travelers are mostly influenced by peer pressure as they lead in using SNSs (Sherman et al., 2016). The SNSs drive the aspirations of travel consumers to perceive the experiences attainment differently.

There is a growing peer influence among the youths who are inspired by their friends and relatives' posts on social media networking sites resulting in imitations a curiosity of the same discoveries (Lai, Ho, & Chung, 2018). Their personal experiences, photographs, videos, and opinions through texts comments are consistently shared on social media while their trips are on-going but also after undertaking trips (Kim & Fesenmaier, 2017).

Information Technology

The widespread use of information technology (Buhalis & Shaw, 2008; Hjalager, 2015; Law & Wang, 2018) is increasingly peer influencing the mechanism behind travel experience sharing for leisure, business, education and practical experiences (Tussyadiah & Zach, 2012; Wang et al., 2012; Xiang et al., 2014). Travelers are able to share travel-related information using SNSs (Cox et al., 2009), as the 21st century has seen so many changes and innovation developments in technology that have influenced social change (Sherman et al., 2016). Technological advancement has not only changed people's characteristics and behaviour; it has also played as a key role in driving many businesses around the world (Travel Market Report, 2018). While there are many benefits accompanied by this development, technological advancement also brings changes that negative affects the social – economic performance. For instance, the positive influence of technological advancement in social context include the cost-effective ways of handling business for instance simplified production, product marketing and promotion, and general social and business communications. While the negative influence of technology in communication include presenting falsified information also known as fraud.

Internet through social media has been considered as one of the cost-effective marketing tools in this competitive business environment. It is believed to be a power tool for advertising and promotion that helps to shape consumer behaviour (Sotiriadis, 2017). Consequently, marketing concept based on referral within social network, such electronic word of mouth (e-WoM) marketing, relationship marketing and viral marketing are considered relevant strategies in this social media era (Ferguson, 2010; Helm 2010; Konznets, de Valck, Wojnicki & Wickie 2010). The approach to the marketing concept suggests that marketers can leverage the power of interpersonal network to promote their products and services by transforming the communication network into influence network (Sherman et al., 2016). In this regard, social network is not only a strategic platform to disseminate information for tourism and hospitality, but it also encourages consumers to generate and share content relevant to product and services (Medina & Hardwick, 2019) and in turn, inform and influence others to make purchase decisions (Lorenzo-Romero et al., 2014; Brandt et al., 2017).

In terms of information adoption, a study by Chilembwe, Mweiwa and Mankhomwa (2018), analysed the role of tour operators in destination marketing in Malawi and found that almost all travel firms had utilised information technology to market their tourism destination services and products.

Social Network Sites (SNSs)

Social media platforms such as Facebook and Instagram continue to dominate by users (Hutchinson, 2018, 2019; Medina & Hardwick, 2019; Travel Market Report, 2018). According to Travel Market Report Outlook (2018), about 89 percent and 44 percent of users agree that social media is very or somewhat important and very important respectively on the overall marketing plans for the travel agents. Moreover, Facebook use as a social and network platform leads by 97 percent while Instagram is trailing (Madina

& Hardwick, 2019). In terms of performance, 61.6 percent, particularly the travel marketers are of the view that Facebook is the best performing media network while Instagram follows by 15.8 percent (Eye for Travel, 2017; Önder, Gunter & Gindl, 2019). Furthermore, Hutchinson (2018) reveals that 63.7 percent like Facebook and Google in the supply chain travel trade sector because they positively influence business users. Consequently, various companies have embraced the use of information technology by developing a website for photo sharing (Lu, Wu & Sang, 2017) and incorporated user-generated technologies that can be used for information and photos tagging (Lu, Wang, Yang, Pang & Zhang, 2010) to enable others get motivated or influenced but also use the same information for their own trip plans. Social network sites do not only play a critical role in market and promotion as they assist business to forecast customer's engagement (Kiral'ova & Pavliceka, 2015), but also transforms business to a more open and collaborative (Önder et al., 2019), an approach that is considered critical in the modern business environment (Del Vecchio et al., 2018). This process enables customers to engage in a more productive process in designing and creating product shape and taste. For instance, Banghiror, Zhang, and Hashim (2019), in their study that investigated the adoption and performance of Facebook fan pages among global airlines found that airlines in developed countries adopted Facebook fan pages three years ahead and are currently performing better than developing and least developed countries. They further suggested that airline companies in developing and least developed countries to successfully use Facebook, should establish two-way communication and respond to their fan. Although, the study presented a challenge of Facebook usage utilization in developing and least developed countries, however, Facebook remains one of the major patronized social network sites in the world (Madina & Hardwick, 2019). The study can also be extended to other social network sites to compare that outcomes as proposed by Baghirov et al. (2019).

Social networking platform, in this internet era, is one of the key aspects of technological development in the world of communication (Tussyadiah, Kausar & Soesilo, 2018). It has brought a significant impact to facilitation of global social inter connections that offers unparalleled, constant connectivity that enable users to create user generate content (Kang & Schuett, 2013). In the review of social impact in hospitality and tourism industry, Kwok and Yu (2013) wrote social media messages have played a crucial role in consumer behaviour by influencing decision making about the travel and tourism product. Kand and Schuell (2013) indicated that social media is the fastest and effective marketing and communication tool for tourism and hospitality industry. In this globalised business environment, internet based online media communication has enhanced individual with common interest and goals to interact and sharing information and experiences. This practice has influenced traveller's decision about which destination to go to. One of the most important functions of social medial is that it provides a web-based application in which users can easily create and post content in form of text, photos and videos (Bae et al., 2017; Kim & Fesenmaier, 2017). So, the better understanding of what makes consumers more inclined to social media will allow tourism and hospitality businesses to make better strategic decisions in using social network channels for marketing (Amaro et al., 2016; Hawk et al., 2019). It is anticipated that identifying the determinants or factors underlying travel experience sharing behaviour would help travel and tourism organisations to gain a better understanding how to stimulate travel experience sharing on social media. Therefore, in order to embrace the benefits brought in by technology through social media, travel organisers need to define well their target audience; select and employ the proper social media marketing strategy. Again, they must encourage consumers to participate in online activities and share their travel experience on organisation websites.

RESEARCH METHODOLOGY

This chapter uses both the phenomenological and ethnographical qualitative case study. The comments based on contents and documents (both visual and texts) on the ground of the cases were thoroughly captured by researchers for guided trips that took part in the country. The feedbacks were clustered into three parts; Mulanje, Zomba and Lake Malawi (more information, see results Table 1 below) about the travel experiences sharing on the social network sites. The second part involved capturing the photographs in order to compliment to the online textual comments provided by tourists. Third, the study also considered tourism companies online reviews that provide lodging services to social network sites used in travel experience sharing. Observations of the phenomenon was made for over a period of time from travellers and visitors in tourist destinations selected for this chapter. In terms of frequency, it was thrice a year. Several photographs were captured from both domestic and international tourists after granted consent to take pictures. The a few photographs used in this study are for illustration and description purposes. According to Streubert and Carpenter (1999) indicate that there are several legitimate approaches to phenomenological inquiries. In this study, the approach was created to follow the grounded themes in the clustered parts in Malawi, namely Mulanje district, Zomba district and Mangochi district (Lake Malawi) in regard to the peer influence adventure activities. The thematic description analysis of strategies was adopted as proposed by Giorgi (1997).

Data Collection

The researchers created a team link with a tour leader where all feedback concerning travel experiences were shared through that link. Data was compiled using Microsoft Excel for processing – checking what they were commenting, similar comments and also differences. Considering that participants evaluation of the trips did not come at once, but in different stages of the tour, a tour leader used to share immediately all reviews at each and every stage. It was easy to put together the textual data using a social networking with tourism consumers. The system received over 300 texts feedbacking travel experiences for one year and three months. Photographs were also captured using smartphone and digital camera, although it was difficult to capture all the activities as tourists could also scatter within the destination depending on a place interested to be pictured and distribute on Facebook and Instagram. In order to compliment this challenge, researchers could follow-up on the updates of photos shared and tagged in the social media. Researchers also attempted to capture both negatives and positives to avoid providing unbalanced contents.

Study Area

The study took place in Malawi, particularly in Rumphi district (Nyika National Park) in the northern part of the country, lake shore resorts of Salima and Mangochi districts, Zomba district, the eastern part of the country, and Mulanje district, the southern part of the country. These districts were chosen because there are famous tourist destinations in Malawi and most of the tourists patronise these districts. There are many lodging places for tourists that have company websites where tourists can share online reviews and comments are freely readable by anyone.

Ethical Consideration

Ethical consideration was sought from tourists and also with lodging operators to use the comments applied in this study. All identities – names and images, of people who provided evaluation of their travel experiences were removed. Researchers were also allowed to capture the photographs, however, only group photographs, selected individual and other images deemed fit for analysis have been included in this study.

Data Analysis

Data was analysed based on themes and destination, both in visual form and textual in national parks (Nyika National Park and Lake Malawi Resort Park) except in mountainous areas such as Mulanje and Zomba where data was analysed textually. Due to the natural of the environment – fog, and also the time of visiting these mountains, researchers did not capture good quality photographs to use for analysis in this study.

Limitations of the Study

The major limitation of this study was that researchers were restricted using some comments from tourists. Some tourists were not comfortable to use their photographs for the study as well as direct quotes from them. They felt that some of their comments were already provided through online reviews to lodging tourism operators. Equally, some tourism operators limited the researchers in terms of reviews to use in this study. Despite that, researcher felt that using both visual and textual data could be sufficient to explain and provide a picture of a Malawi case in term of travel sharing experiences on the social media.

RESULTS AND DISCUSSION

As clarified in the limitations of the study, it was not possible to apply all contents, comments that tourists provide about their travel experiences. Several online evaluations from tourists are already in public domain, as such, they continuously share travel experiences freely about Malawi. The feedback Table 1 below has provided both positives and negatives about the three clustered tourist destinations.

Several comments are shared on the social media regarding Mulanje mountain, Zomba Plateau, and Lake Malawi national Park. As Table 1 above depicts; for example, Mulanje mountain has for so many years been associated with mountain spirits, disappearance of tourist, presence of already made food and fruit in place within the mountain for the visitors to eat. These tourists' comments on the blogs and other social networks caught the attention of many tourists, "I heard about mountain spirits which leave bananas for the climbers to eat on the summit of Sapitwa Peak – highest point of the mountain but could find one" and "I visited Mulanje mountain where travellers claimed went missing without explanation. I was afraid I would disappear too but here I' am alive and back on the foot of the mountain." Realistically, these are mere claims by the people which have been spread like wind, shared even in historical books.

Table 1. Some of the examples of comments tourists shared with others on SNSs

Mulanje Mountain	Zomba Plateau	Lake Malawi National Park
I visited Mulanje mountain where travellers claimed went missing without explanation. I was afraid I would disappear too but here I' am alive and back on the foot of the mountain.	I shared video clip of William Falls – this is one of the beautiful places on Zomba Plateau.	I shared a 15-minute video about Lake Malawi adventure. A lot of people like it and have responded positively to participate in this adventure. There is excitement during traditional boat riding on the lake.
I heard about mountain spirits which leave bananas for the climbers to eat on the summit of Sapitwa Peak – highest point of the mountain but could find one.	I visited Mulunguzi Dam, Ku Chawe Inn on Zomba plateau. I like the place, is so beautiful and perhaps close to heaven. I liked climbing the plateau on foot.	I like going for camping on the shores of Lake Malawi for adventure activities. Boat riding at the lake is so exciting activity not to miss.
I'm excited to be on the tragic expeditions where people drown in Mulanje's Likhubula Falls.	The plateau is massive with a fantastic view of Zomba city and Lake Chilwa. The weather is cool like unlike lowlands.	We spend three nights inside Cape Maclear at the Lake Malawi accommodation. It's not very far, just 2 kms offshore from the main road. We enjoyed the beach and spectacular view of lake Malawi and surrounding islands.
A guide took us to a place of mysterious old woman who live at water fall and blessed new-born infants who live long life. I thought I would see the woman but it just a historic place.	I was scared when we were climbing the plateau. I was surprise to see a very beautiful 3-star hotel there and nearby there were so many precious gold stones, tour guides displayed them freely to us to see.	I enjoyed to get experience when diving and sailing on Lake Malawi. I loved moving around the lake Malawi, camping and I also tested the famous Chambo fish, it's a nice experience. I feel lake Malawi is one of these places that people can very easily and safely travel to in Africa.
I have posted an interesting video and pictures about my travel experience. I'm motivated by what I saw here and next year, I shall come again to explore more.		

Researchers' data, 2019

Several tourists have failed to find evidence regarding these claims. Mulanje mountain is the second highest mountain in Africa with 9,849 feet above sea level. Tourists have been climbing the mountain from one side and come back using other routes without been seen by the local people. As result, the local people were conclusion that tourists disappear because of mountain spirits.

Interestingly, despite all negative comments shared on social media about Mulanje mountain, many tourists have been influenced by what they read online to go to the mountain and get first-hand information. Mulanje mountain is also shared on social networks as a place with fascinating landscape that provides unique travel experiences to the travellers. Every year, Mulanje mountain hosts 'Porters Race' that gives an opportunity to runners to participate in a competition where they climb the mountain up to Chambe Plateau. The participants come from all over the world, including host country, Malawi as others commented on social media: "since the inception of Mulanje Porters Race, I have never missed the competition, I love Mulanje mountain, it's cool weather and scenery and I get connected to fellow runners." Mulanje Race is organised as part of marketing Mulanje mountain as an attraction. The race has manged to change tourist's perception about stories of mountain spirits in the mountain. To this end, the hikers and runners shared on the social media have equally influenced many to visit Mulanje.

On Zomba plateau, most of the tourists share positive comments about the area despite that the place looks deserted. There are not many houses being the upland area and also hard to reach on the top on foot. Again, Zomba plateau surprises many tourists as they doubt that they can find a 3-star hotel on top of the mountain as depicted in these online comments in Table 1; "I was scared when we were climbing the plateau but it was surprise to me to see a very beautiful 3-star hotel there and nearby there were so many precious stones that the guides displayed them freely. They continued to say "I then took video clip of William Falls shared with you – this is one of the beautiful places on Zomba Plateau." The major attractions of Zomba plateau are William Falls, Mulunguzi Dam, Ku Chawe Inn, and Emperor View where tourists can see the beauty of the whole Zomba city and surrounding attractions. The number of tourists visiting Zomba plateau continues to grow and this could be largely due to the most positive online reviews provided by both domestic and international tourists.

On safari tourism, tourists have shared mixed comments; some positives while others negatives. For instance, some tourists shared this; "we expected to see all the big five in one national park, so we had to drive to another park to see elephants." National parks and game reserves provide a safari travel experience to spot wildlife such as; elephants, monkeys, antelopes, crocodiles, buffaloes, zebras, rhinos, warthogs, hippos and baboons. Although, Malawi as a country boosts of having five national parks and four game reserves, Nyika National Park being the largest park. However, strategically the national parks and games reserves are spread nationwide and it becomes difficult to view all the big five in one park or game reserve. Some wildlife reserves are being managed by the African Parks, a private company on concession system, for example; Majete, Liwonde, and Nkhotakota National Parks. Whereas, Vwaza Game Reserves and Nyika National Park are being managed by Wilderness Safari. The Malawi Government manages Kasungu National Park, Mwabvi Game Reserve and Lake Malawi National Park.

Comparatively, tourists share and recommend privately managed parks and game reserves than those being managed by government. The reasons provided were better park infrastructure, well abundance presence of wildlife life, and well-trained guides. Comments shared on social media said: "we were provided an excellent service at the private managed parks, with better parking and accommodation facilities, were able to see animals around us unlike the government management parks, were deserted."

In terms of tourists' numbers, the private managed park and game reserves continued to enjoy good patronage due to positive online commented shared by both domestic and international tourists. As Kwok and Yu (2013) note, the social media messages play a crucial role in consumer behaviour by influencing decision making about destination tourism product. Contents shared on social media also act as the fastest and effective marketing and communication tool for tourism operators (Kand & Schuell, 2013). Travellers who enjoy going for safari tourism share travel experiences during the trips but also after their trips. Mostly, they use their social networking sites such as Facebook and Instagram. Similarly, most of the tourism organisations have websites that link to their Facebook and Instagram pages in order to share to others some of the safari experiences their visitors received during their visits. Potential travellers can search and watch videos and see others commenting messages from all over the world using social network sites. Consequently, the behaviour of sharing travel experiences using photos and videos has motivated others to start finding out more information on how too, they can participate in safari activities. Particularly, the young ones like the youth, they easily get peer pressure from what they see from their colleagues and they request from their parents to consider putting into plans whenever the family wants to go for a vacation. Parents too are sometimes pressured by their family members and family friends to experiment other travel experiences like safari tourism. During safari tourism, travellers are involved in game drive, viewing animals, following up on animal behaviours but can also participate in animal tracking under the guidance of experienced tour guides. A good example of animal viewing is Zebra and Antelope photographs provided in Figure 1 and 2 below. Tourists comments online indicated

Figure 1. Showing zebras at Nyika National Park
Researchers' photo, 2019

Figure 2. Showing antelope at Nyika National Park
Researchers' photo, 2019

that "Zebra appears to be a quite animal but not friendly" and viewers loved its stripes colours describing as nice. While they indicated that Antelope was people friendly in the park as compared to Zebra.

Travellers or tourists to Lake Malawi National Park share their lake travel experiences on social network sites. Mostly, they love swimming, boat riding and spending holidays at the resorts. For example, one shared this message as depicted in Table 1 above (page 10); "I like going for camping on the shores of Lake Malawi for adventure activities. Boat riding at the lake is so exciting activity not to miss" A lot share their experiences through videos on Facebook and photographs on Instagram. As a lake provides plenty of activities do, some travellers do participate in the Yachting Marathon which is also shared on social network sites. Yachting Marathon, is an endurance event, requires skills and experiences, and is also more enjoyable as sailors attempt to test the resolve of other competitors throughout the race sailboats. Others share on social media about kayaking adventure. In kayaking, travellers can use paddling to move through the big waves of water and shifting winds on the Lake Malawi. Tourists who go for kayaking activities they equally post their videos and photos about experiences on their social network sites. Besides showing to the world and show off to their colleagues of what they are capable of doing, they also end up influencing others to join them and try kayaking adventure activity. Kayaking activity also offers a good experience on health apart from a being a pleasure.

Swimming, picnicking and walking on stones is another activity shared by tourists. "When we go to lake Malawi, we swim in the lake several times, we go for picnicking at the Alter-point and surrounding areas along the beach," and see also Figure 3 below. These sentiments show that Lake Malawi provides plenty of activities to do for everyone as a result, some domestic tourists get easily influenced when they see posts about their friends. Lake Malawi, apart from being is one of the largest lakes in the world, is also a UNESCO World Heritage site. The lake is unique because of the most biodiverse species of colourful fish to spot on by travellers. Therefore, the place has attracted several researchers, educationists and scientists, including students and people in general after watching videos and photos posted on social network sites by various tourism companies but also other travellers.

Figure 3. Showing domestic tourists taking pictures at Alter-point at Lake Malawi National Park
Researchers' photo, 2019

Paddling boat adventurers have as well shared their videos and photos on social network sites depicting their water sailing experiences on Lake Malawi. The travellers who enjoy sharing travelling experiences while using engine boats, sailing in deep waters, and passing through some rocky places. Sailing using engine boats is suitable to the travellers who have little or no water swimming experiences. The boats can take them to the deep waters where they cannot go on their own but they can also spend considerable time on the water while experiencing taking discovering of other places of the lake. They can travel to an island using boat engine and come back with stories and experiences of an invisible island on the Lake Malawi. Others go for scuba diving and snorkelling on Lake Malawi. These tourists or travellers spend some days enjoying themselves and they seek accommodation very close to the waters as illustrating by Figure 4 below. Besides sharing their experiences through videos and photos on Facebook and Instagram, they also share online comments which are both positive and negative about accommodation and the lake. For example, "I prefer private managed accommodations, sometimes a simple tent during summer season, with high temperature, they are good because of aircon facilities or fun. The only problem is that the rates are very high during peak season also the beach is overcrowded by the local communities." During summer seasons and also holiday festival, many domestic tourists travel to the lake, so at times, the international tourists may experience so many people at the lake. However, tourists who seek service from private operators cannot read much into crowding because there not many locals manage to take activities at the private operator's territory due high charges of fees.

The signpost shown in Figure 5 above call for a responsibility of all visitors to pay entry fees but also to keep the island clean. The signpost was planted close to the entrance of Lake Malawi National Park. Although, the signpost provides a clear message to tourists as well as excursionists, some domestic tourists shared that the signpost was misplaced. Based on their travel experience they felt that the park starts somewhere 3km before the entrance of the park where the signpost was placed. Moreover, others shared that domestic tourists pay more entrance fees as compared to international tourists while some

Figure 4. Showing Lake Malawi resorts and far away tourists are enjoying in the water
Researchers' data, 2019

Figure 5. Showing a signpost to pay entry fee and keep the park clean
Researchers' photo, 2019

felt that despite park management receiving entrance fees, little was being done to develop the park. The parking area has limited space, and the park infrastructure has not grown for some time as one tourist commented; "I visited this place two years ago, I thought I would find new infrastructure and major changes of the facilities but the park remains undeveloped". Lake Malawi National Park is managed by the government of Malawi in order to preserve water heritage. The park has over one thousand species of Mbuna fish and is also used as a place for aquatic research.

CONCLUSION

Social media has the power to influence people travel decisions (Cox et al., 2009; Munar & Jacobsen, 2014; Narangajavana et al., 2017). The digital platform has become the immediate channel of community to share travellers' experiences as more tourists are able to upload photos and video but also send online review messages after receiving a service or product from tourism provider (Bae et al., 2017; Kim & Fesenmaier, 2017). Moreover, it has become a cultural behaviour among the majority of the tourists to use social media to distribute information as one way of informing, alerting but also recommending service provision (Lin & Chen, 2012). Tourism organisations and companies have not remained behind in the use of social media (Chilembwe et al., 2018), as the social media application have also influenced customers tourism purchasing decisions (Kiral'ova & Pavliceka, 2015; Sahin et al., 2018; Baghirov et al., 2019). In case of Malawi, the use of social media by domestic tourists is highly influenced by Facebook and Instagram, Twitter as few of them own personal blogs and websites except companies. In contrast, international tourists, particularly those from overseas, have several options of sharing travel experiences. In additional to Facebook and Instagram, they also use, for example, online reviews and blogs. Malawi's tourism is dominated by natural products such as the lakes and resorts, mountains, plateaus. Therefore, tourists who participate in a kind of adventure tourism capture photographs using smartphones or video cameras, and share on social network sites. Facebook and Instagram are the highest utilized social network sites for sharing travel experiences according to Önder et al. (2019). As most of the youths use Facebook, Twitter and Instagram, they are also able to share the experiences and get each other's motivated, some through peer pressure. The travel experiences acquired by the travellers, when shared do motivate the potential tourism consumers in various social networks but can also demotivate them (Lu et al., 2010, 2017; Kim & Fesemaier, 2017). When the admiration of others' travel experiences grows, it exerts pressure to embark on travel with an expectation to get same experiences. Sherman et al. (2016) stressed that peer pressure is one of the major reasons influencing the youths to travel due to positive messages shared on social media.

ACKNOWLEDGMENT

This research received no specific grant from any funding agency in the public, commercial, or not-for-profit sectors.

REFERENCES

Amaro, S., Duarte, P., & Henriques, C. (2016). Travellers use of social media: A clustering approach. *Annals of Tourism Research, 59*, 1–15. doi:10.1016/j.annals.2016.03.007

Bae, S. J., Lee, H., Suh, E., & Suh, K. (2017). Shared experience in pre-trip and experience sharing in post-trip: A survey of Airbnb users. *Information & Management, 54*(6), 714–727. doi:10.1016/j.im.2016.12.008

Banhirov, F., Zhang, Y., & Hashim, N. H. (2019). Facebook fan page management for global airlines. *Tourism Review, 74*(3), 532–546. doi:10.1108/TR-03-2018-0045

Boyd, D. M., & Ellison, N. B. (2008). Social network sites: Definition, history, and scholarship. *Journal of Computer-Mediated Communication, 13*(1), 2010–2223.

Brandt, T., Bendler, J., & Neumann, D. (2017). Social media analytics and value creation in urban smart tourism ecosystems. *Information & Management, 54*(6), 703–713. doi:10.1016/j.im.2017.01.004

Buhalis, D., & Law, R. (2008). Progress in information technology and tourism management: 20 years on and 10 years after the internet – The state of eTourism research. *Tourism Management, 23*, 207–230. doi:10.1016/S0261-5177(01)00085-1

Chilembwe, J. M., & Mweiwa, V. R. (2014). Tour guides: Are they developers and promoters - Case study of Malawi. *Impact Journals: International Journal of Research and Business Management, 2*(9), 29–46.

Chilembwe, J. M., & Mweiwa, V. R. (2019). Responsible Travel and Tourism Adventure: Evidence from Malawi as a Tourist Destination. In A. Sharma (Ed.), *Sustainable Tourism Development: Futuristic Approaches* (pp. 31–53). New York: Apple Academic Press.

Chilembwe, J. M., Mweiwa, V. R., & Mankhomwa, E. (2018). Role of Tour Operators in Destination Tourism Marketing in Malawi. In M. A. Camilleri (Ed.), *Strategic Perspectives in Destination Marketing* (pp. 295–321). Hershey, PA: IGI Global Publishers. doi:10.4018/978-1-5225-5835-4.ch012

Cox, C., Burgess, S., Sellitto, C., & Buultjens, J. (2009). The role of user-generated content in tourists' travel planning behaviour. *Journal of Hospitality Marketing & Management, 18*(8), 743–764. doi:10.1080/19368620903235753

Dedeoğlu, B. B., Taheri, B., Okumus, F., & Gannon, M. (2020). Understanding the importance that consumers attach to social media sharing (ISMS): Scale development and validation. *Tourism Management, 76*, 103954. doi:10.1016/j.tourman.2019.103954

Del Vecchio, P., Mele, G., Ndou, V., & Secudo, G. (2018). Creating value from social big data: Implications for smart tourism destination. *Information Processing & Management, 54*(5), 847–860. doi:10.1016/j.ipm.2017.10.006

Eye for Travel. (2017) The State of Data and Analytics in Travel Report 2017. Retrieved from http://1.eyefortravel.com/LP=16751

Ferguson, R. (2010). Word of mouth and viral marketing: Taking the temperature of the hottest trends in marketing. *Journal of Consumer Marketing, 25*(3), 179–182. doi:10.1108/07363760810870671

Giorgi, A. (1997). The theory, practice, and evaluation of the phenomenological method as a qualitative research procedure. *Journal of Phenomenological Psychology, 28*(2), 235–260. doi:10.1163/156916297X00103

Gretzel, U., & Yoo, K.H. (2008) Use and impact of online travel reviews. In *Information and Communication Technologies in Tourism 2008* (pp. 35-46). Springer.

Havell, R., & Ross, J. (1819). *A voyage of discovery… made for the purpose of exploring Baffin's Bay and inquiring into possibility of North-West passage*. London: John Murray.

Hawk, A. T., van den Eijnden, R. J. J. M., van Lissa, C. J., & ter Bogt, T. F. M. (2019). Narcissistic adolescents' attention-seeking following social rejection: Links with social media disclosure, problematic social media use, and smartphone stress. *Computers in Human Behavior, 92*, 65–75. doi:10.1016/j.chb.2018.10.032

Helm, S. (2010). Viral marketing–Establishing customer relationships by word-of mouth. *Electronic Markets, 10*(3), 158–161. doi:10.1080/10196780050177053

Hjalager, A. (2015). 100 innovations that transformed tourism. *Journal of Travel Research, 54*(1), 3–21. doi:10.1177/0047287513516390

Ho, C. I., Lin, Y. C., Yuan, Y. L., & Chen, M. C. (2016). Pre-trip tourism information search by smartphones and use of alternative information channels: A conceptual model. *Cogent Social Science, 2*, 1136100.

Hutchinson, A. (2018, November 13) Social media today: Facebook releases new report into travel booking trends online [Infographic]. Socialmediatoday. Retrieved from https://www.socialmediatoday.com/news/facebook-releases-new-report-into-travel-booking-trends-online-infographic/542015/

Hutchinson, A. (2019, June 27) Social media today: The state of Instagram marketing 2019 part 4: Instagram analytics. Socialmediatoday. Retrieved from https://www.socialmediatoday.com/news/the-state-of-instagram-marketing-2019-part-4-instagram-analytics/557400/

Kang, M., & Schuett, M. A. (2013). Determinants of sharing travel experiences in social media. *Journal of Travel & Tourism Marketing*, *30*(1-2), 1–15. doi:10.1080/10548408.2013.751237

Kim, J., & Fesenmaier, D. R. (2017). Sharing Tourism Experiences: The Posttrip Experience. *Journal of Travel Research*, *56*(1), 28–40. doi:10.1177/0047287515620491

Kim, J., & Tussyadiah, I. P. (2013). Social networking and social Support in tourism experience: The moderating role of online Self-presentation strategies. *Journal of Travel & Tourism Marketing*, *30*(1-2), 78–92. doi:10.1080/10548408.2013.751220

Kiral'ova, A., & Pavliceka, A. (2015). Development of social media strategies in tourism destination. *Procedia: Social and Behavioral Sciences*, *175*, 358–366. doi:10.1016/j.sbspro.2015.01.1211

Kozinets, R. V., De Valck, K., Wojnicki, A. C., & Wilner, S. J. S. (2010). Networked narratives: Understanding word-of-mouth marketing in online communities. *Journal of Marketing*, *74*(2), 71–89. doi:10.1509/jmkg.74.2.71

Kwok, L., & Yu, B. (2013). Spreading social media messages on Facebook: An analysis of restaurant business-to-consumer communications. *Cornell Hospitality Quarterly*, *54*(1), 84–94. doi:10.1177/1938965512458360

Law, R., Chan, I. C. C., & Wang, L. (2018). A comprehensive review of mobile technology use in hospitality and tourism industry. *Journal of Hospitality Marketing & Management*, *27*(6), 626–648. doi:10.1080/19368623.2018.1423251

Lin, Y.-W., & Chen, C.-H. (2012). College students' Facebook use characteristics, critical-thinking tendencies and information verification behaviours: National Taiwan University and Technology. *Journal of Liberal Arts and Social Sciences*, *8*(1), 17–36.

Lonely Planet. (2018) Map of Malawi. Retrieved from https://www.lonelyplanet.com/maps/africa/malawi/

Lorenzo-Romero, C., Alarcón-del-Amo, M., & Constantinides, E. (2014). Determinants of use of social media tools in retailing sector. *Journal of Theoretical and Applied Electronic Commerce Research*, *9*(1), 44–55. doi:10.4067/S0718-18762014000100005

Lu, D., Wu, R., & Sang, J. (2017). Overlapped user-based comparative study on photo-sharing websites. *Information Sciences*, *376*, 54–70. doi:10.1016/j.ins.2016.10.005

Lu, X., Wang, C., Yang, J. M., Pang, Y., & Zhang, L. (2010) Photo2trip: Generating travel routes from geo-tagged photos for trip planning. *Proceedings of the 18th ACM International Conference on Multimedia*, Firenze, Italy. ACM. 10.1145/1873951.1873972

Malawi Government. (2006). *Malawi Growth and Development Strategy*. Malawi Government Press.

Medina, K., & Hadwick, A. (2019). *The state of hospitality distribution: Direct bookings – A report series between Eye for Travel and Fornova*. London: Eye for Travel Publishing.

Munar, A. M., & Jacobsen, J. K. S. (2014). Motivations for sharing tourism experiences through social media. *Tourism Management*, *43*, 46–54. doi:10.1016/j.tourman.2014.01.012

Narangajavana, Y., Fiol, L. J. C., Tena, M. A. M., Artola, R. M. R., & Garcia, J. S. (2017). The influence of social media in creating expectations. An empirical study for a tourist destination. *Annals of Tourism Research*, *65*, 60–70. doi:10.1016/j.annals.2017.05.002

Önder, I., Gunter, U., & Gindl, S. (2019). Utilizing Facebook statistics in tourism demand modeling and destination marketing. *Journal of Travel Research*. doi:10.1177/0047287519835969

Sahin, D. R., Sener, I., & Polat, A. S. (2018). Power of universe: How do you have social media effective of organisational image of airline companies? In C. Cobanoglu, L. M. Tanh, & A. Corbaci (Eds.), *Advances in business Hospitality, and Tourism Research* (Vol. 1, pp. 66–77). Florida: Anahei Publishing.

Sherman, L. E., Payton, A. A., Hernandez, L. M., Greenfield, P. M., & Dapretto, M. (2016). The power of the like in adolescence: Effects of peer influence on neural and behavioural responses to social media. *Psychological Science*, *27*(7), 1027–1035. doi:10.1177/0956797616645673 PMID:27247125

Simms, A. (2012) Online user-generated content for travel planning- Different for different kinds of trips? *e-Review of Tourism Research*, *10*(3), 76-85.

Sotiriadis, M. D. (2017). Sharing tourism experiences in social media: A literature review and a set of suggested business strategies. *International Journal of Contemporary Hospitality Management*, *29*(1), 179–225. doi:10.1108/IJCHM-05-2016-0300

Streubert, H., & Carpenter, D. (1999). *Qualitative research in nursing: advancing the humanistic perspective* (2nd ed.). Philadelphia: Lippincott Williams & Wilkins.

Travel Market Report. (2018). Travel market report's outlook on social media based on the voice of the travel advisor 2018-2019. Retrieved from https://www.travelmarketreport.com/library/TMR/TMR_SocialMediaOutlook.pdf

Tussyadiah, I., & Fesenmaier, D. R. (2009). Mediating tourist experiences: Access to places via shared videos. *Annals of Tourism Research*, *36*(1), 24–40. doi:10.1016/j.annals.2008.10.001

Tussyadiah, I., & Zach, F. (2012). The role of geo-based technology in place experiences. *Annals of Tourism Research*, *39*(2), 780–780. doi:10.1016/j.annals.2011.10.003

Tussyadiah, S. P. II, Kausar, D. R., & Soesilo, P. K. M. (2018). The Effect of Engagement in Online Social Network on Susceptibility to Influence. *Journal of Hospitality & Tourism Research (Washington, D.C.)*, *42*(2), 201–223. doi:10.1177/1096348015584441

Volo, S. (2010). Bloggers reported tourist experiences: Their utility as a tourism data source and their effect on perspective tourists. *Journal of Vacation Marketing*, *16*(4), 297–311. doi:10.1177/1356766710380884

Wang, D., Park, S., & Fesenmaier, D. R. (2012). The role of smartphones in mediating the touristic experience. *Journal of Travel Research*, *51*(4), 371–387. doi:10.1177/0047287511426341

WodeMaya. (2019, May 04) Do not come to China [YouTube video]. Retrieved from https://www.youtube.com/watch?v=Av0y_bZ_KPU

Xiang, Z., Magnini, V. P., & Fesenmaier, D. R. (2015). Information technology and consumer behaviour in travel and tourism: Insights from travel planning using the internet. *Journal of Retailing and Consumer Services*, *22*, 244–249. doi:10.1016/j.jretconser.2014.08.005

Xiang, Z., Wang, D., O'Leary, J. T., & Fesenmaier, D. R. (2014). Adopting the internet: Trends in travellers use of the web for trip planning. *Journal of Travel Research*. doi:10.177/0047287514522883

Yoo, K.H. & Gretzel, U. (2009). Comparison of deceptive and truthful travel reviews. *Information and Communication Technologies in Tourism 2009* (pp. 37-47). Springer.

ADDITIONAL READING

Barnes, S. J., Mattsson, J., Sørensen, F., & Friis Jensen, J. (2019). The mediating effect of experiential value on tourist outcomes from encounter-based experiences. *Journal of Travel Research*. doi:10.1177/0047287519837386

der Pütten, A. M. R., Hastall, M. R., Köcher, S., Meske, C., Heinrich, T., Labrenz, F., & Ocklenburg, S. (2019). "Likes" as social rewards: Their role in online social comparison and decisions to like other people's selfies. *Computers in Human Behavior*, *92*, 76–86. doi:10.1016/j.chb.2018.10.017

Hawk, A. T., van den Eijnden, R. J. J. M., van Lissa, C. J., & ter Bogt, T. F. M. (2019). Narcissistic adolescents' attention-seeking following social rejection: Links with social media disclosure, problematic social media use, and smartphone stress. *Computers in Human Behavior*, *92*, 65–75. doi:10.1016/j.chb.2018.10.032

Kim, S.-E., Lee, K. Y., Shin, S., & Yang, S.-B. (2017). Effects of tourism information quality in social media on destination image formation: The case of Sin Weibo. *Information & Management*, *54*(6), 687–702. doi:10.1016/j.im.2017.02.009

Leung, D., Law, R., van Hoof, H., & Buhalis, D. (2013). Social media in tourism and hospitality: A literature review. *Journal of Travel & Tourism Marketing*, *30*(1-2), 3–22. doi:10.1080/10548408.2013.750919

Lorenzo-Romero, C., Alarcón-del-Amo, M., & Constantinides, E. (2014). Determinants of use of social media tools in retailing sector. *Journal of Theoretical and Applied Electronic Commerce Research*, *9*(1), 44–55. doi:10.4067/S0718-18762014000100005

Phu, B., & Gow, A. J. (2019). Facebook use and its association with subjective happiness and loneliness. *Computers in Human Behavior*, *92*, 151–159. doi:10.1016/j.chb.2018.11.020

Vu, H. Q., Li, G., Law, R., & Ye, B. H. (2015). Exploring the travel behaviour of inbound tourists to Hongkong using geo-tagged photos. *Tourism Management*, *46*, 222–232. doi:10.1016/j.tourman.2014.07.003

Wang, D., Xiang, Z., & Fesenmaier, D. R. (2016). Smartphone use in everyday life and travel. *Journal of Travel Research*, *55*(1), 52–63. doi:10.1177/0047287514535847

Zhang, T., Omran, B. A., & Cobanoglu, C. (2017). Genaration Y's positive and negative eWOM: Use of social media and mobile technology. *International Journal of Contemporary Hospitality Management*, *29*(2), 732–761. doi:10.1108/IJCHM-10-2015-0611

KEY TERMS AND DEFINITIONS

Blog: A weblog of an online personal or company diary of travel experiences or encounters in one's life time documented to share with others.

Domestic Tourist: A person who travels to places of attractions or participate in tourism activity within a country of residence for at least a period of 24 hours but not more than 265 days.

Information Technology: A process of developing and maintaining computer systems, software, and networks for the processing and distributing of a company's products, services and consumers travel experiences.

International Tourist: A person who travel to another country for tourism purposes for a minimum of 24 hours and a maximum of 256 days.

Media Sharing: A system of distributing information through digital processes to keep in touch with the people in a society or community.

Online Review: An approach of evaluating the products or services provided by an organisation in a form of scale ratings or textual information.

Peer Influence: A form of consumer behaviour motivated or copied from others acting as role models.

Social Network Site: A digital platform that links people and companies to connect, communicate, market products or service, and share daily or occasional life experiences.

Tourism: The movement of people from place to place whose purpose is either pleasure or business for over short period of time.

Travel Experience: A compilation of all activities gone through from the beginning of a trip, during and after trip service offerings critical reflection.

Traveller: A person who travels for any other reason, be it business or leisure.

Chapter 3
Role of Social Media in Travel Planning and Tourism Destination Decision Making

James Malitoni Chilembwe
https://orcid.org/0000-0002-3482-5073
Mzuzu University, Malawi

Frank Wadilika Gondwe
Mzuzu University, Malawi

ABSTRACT

Today, tourism firms use social media to communicate and inform their clients about tourism destination offerings. Trip planners are utilizing social media to make informed decision making about tourists' destinations. With information technology and social media, travelers can get updates on their booked flights and hotels at a tourist destination. The growth of social media usage in tourism industry influences most of the firms to start adopting using social media. This chapter, therefore, focuses on the role of social media in travel planning and tourism destination decision making. It evaluates five cases whereby clients' use travel planning decisions but also changes or curtail plans as a result of social media communication. The survey study uses a simple random sampling procedure and interviewed 792 travelers using a structured questionnaire in ten tourist destinations covering developed and developing countries. The key results revealed that 92% of travelers make a booking but also change or curtail travel planning decisions for holidays through social media alerts.

INTRODUCTION

Social media (SM) has become a popular platform for many travelers as well as people's daily lives (Kim, Lee, Shin, & Yang, 2017). SM is used for general social networking, chatting, and information sharing (Chung & Koo, 2015), for business to grow the firm through product and services marketing (Chilembwe, Mweiwa, & Mankhomwa, 2018). Travel firms such as tour operators and travel agents

DOI: 10.4018/978-1-7998-1947-9.ch003

have embraced information technology, mainly social media platforms, to market tourism destinations (Chilembwe et al., 2018). SM plays a significant role in travel planning and destination decision making, not only by travelers but also to the travel firms as they organize trips to various places of tourists' interests. According to Lei, Nicolau, and Wang (2019), the advancement of the digital era in using social media channels has brought about new types of booking channels. It is called "SoLoMo" applications meaning; social, location, and mobile-based. Using mobile social channels, travelers can interact with the applications and have access to information they need more efficiently but also make their bookings in a simplified manner (Kim & Connolly, 2012; Thakran & Verma, 2013; Lei et al., 2019). Moreover, the SM accommodates several functions, for example; one-to-one interaction between a traveler and service provider as well as WeChat services. The chats are enabled using several options for communication such as social communities, emoji, video chat, and voice messages (Wang, Xiang, & Fesenmaier, 2016).

Besides using SM for travel planning, statistically, trends have shown that there are increased numbers of independent tourists who mainly rely on information technology to book their travel preferences. For example, Kim et al. (2017) indicate that more than half (53%) of the Chinese and United States of America (USA) travelers book their trips using digital facilities like mobile applications. More so, by January 2016, about 31% of the world population has had active social media accounts, they spend an average of over 2 hours a day using social media (Kim et al., 2017). Interestingly, the youths lead in the usage of smartphones and smartphone-embedded technologies which include social media as 89% find their access to social media thorough smartphone (Zhang, Omran, & Cobanoglu, 2017; Hawk, van den Eijnden, van Lissa, & ter Bogt, 2019). Active Facebook users alone as a social media reached 1.7 billion per month by mid-year 2016 of which 1.03 billion were recorded as average daily active users of Facebook using mobile devices alone (Kim et al., 2017). However, the number of social media active users daily, including those who use other devices like computers or combined with mobile devices is likely to be higher than 1.03 billion. Comparatively, as of January 2019, the social media statistics of consumer adoption and usage was found to be on the higher side. According to Chaffey (2019) report, the total world population was 7.676 billion (56%), unique mobile users 5.112 billion (67%), internet users 4.388 billion (57%), active media users 3.484 billion (45%), and mobile media users 3.256 billion representing 42%.

SM has played a significant impact on how travelers search information; plan for their trips, share information pre-trip, during the trip and after the trip and finally, make decisions about their destinations as well trip evaluation through memories. Travelers can share their travel memories and stories through social media such as YouTube, Facebook, Instagram, Twitter, Trip Advisor (Chilembwe et al., 2018). Besides intermediaries such as tour operators and travel agents, SM is considered more vital not only a source of information but also for communicating experiences (Xiang & Gretzel, 2010; Zeng & Gerritsen, 2014; Kim et al., 2017). So, SM continues to play a crucial role in various industries in general including the travel and tourism sector (Akehurst, 2009; Lu & Stepchenkova, 2015; Chilembwe & Mweiwa, 2019), as travelers utilize SM media in their travel planning and tourism destination decision making. Although, SM plays a role in several areas and industries, however, this chapter is limited to the role of social media in travel planning and tourism destination decision making. The objective of the chapter is to evaluate five cases whereby clients' make travel planning decision to tourists' destinations can be adopted but also curtailed as a result of social media messages. The study is guided by the framework in Figure 1 below to analyse the airlines, tour operators, hotels, car rentals, and travel agencies SM cases.

Figure 1. Study framework
Source: Authors

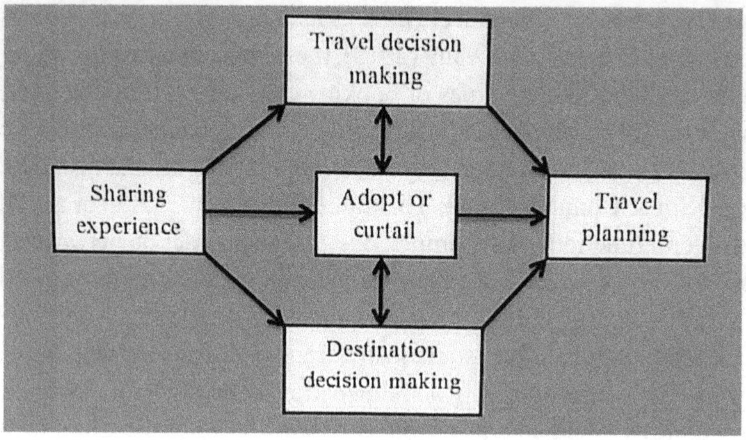

The next sections of the chapter are chronologically presented. The first section presents the literature review of travel planning and SM, followed by travel decision making and SM, and the travel planning and destination decision making and SM. The second section presents methodology for the study. The third section presents the results and discussion, and the fourth section provides further direction of study and a conclusion.

Travel Planning and Social Media

Travel planning is a series of incidents that involves information searches (Pan & Fesenmainer, 2006; Ho, Lin, & Chen, 2012; Huang, Goo, Nam, & Yoo, 2017). According to Fesenmaier, Xiang, Pan & Law (2011), users should utilize search engines properly in order to check destination preferences information. However, information search is not an easy task due to plenty of information available online (Kaplan & Haenlein, 2010). Information searches during the planning phase are not limited to tourism destination's 5A's – accessibility, accommodation, attractions, activities, and amenities. Accessibility to tourism destinations and tourists' attractions is vital for both domestic and international mobility. The knowledge and understanding of transport modes help travelers to plan ahead of their trip in terms of duration to spend on holidays and also several nights to be accommodated at one hotel (Choi, Lehto, Morrison, & Jang, 2012). They can also decide on which attractions to see and activities to participate in during their stay and when to return home.

The power of social media (SM) is evident in the searches of travel information as travelers all over the world freely connect to the internet using and check destinations of their choice. SM is a term that refers to internet-based and mobile-based online communications that assist people in connecting all over the world and searching information as part of one global village (Okazaki, Campo, Andreu, & Romero, 2015; Kaplan & Haenlein, 2010; Husain, Li, & Wang, 2018). Using SM as user-generated content, it means that users can engage in with social networking sites, web 2.0, as well as virtual worlds (Cox, Burgess, Sellitto, & Buultjens, 2009). Through SM, for example; YouTube, Facebook, WeChat, Twitter, TripAdvisor, and Instagram, people and destination marketing organizations generally share their views online by messaging, video sharing and any other online available communication (Fotis, Buhalis, &

Rossides, 2012; Xiang, 2017; Chilembwe et al., 2018). Information sharing can be done at any stage, such as pre-travel, during travel, and post travel. Moreover, many travelers share memories of their traveling with friends and relatives while on the trip (Munar & Jacobsen, 2014). It is, therefore, the behavior of sharing that brings in peer pressure and motivate others to start planning their next travel to experience the same experiences as their colleagues (Pan & Fesenmaier, 2006; Chilembwe, 2014; Munar & Jacobsen, 2014). Therefore, travelers plan it based on others experiences but also their own experiences. SM acts as one of the motivators of many leisure travelers and begin to plan for their next trips (Guiver & Stanford, 2014). Mostly, the travellers base on their tourism destination making when they plan their trips after using SM as user-generated content but also receive online messages and videos from their colleagues (Buhalis, & Law, 2008; Hjalager, 2015), and online mobile phone reminders (Chiang & Huang, 2015; Binucci, De Luca, Di Guiacomo, Liotta, & Montecchiani, 2017). The online mobile phones applications such as google maps, GPS, e-payment and photos capturing, tagging and instant messaging and photo attachments are what social media travel planners are using to prepare for their future travels (Lu, Wang, Yang, Pang & Zhang, 2010; Vu, Li, Law, & Ye, 2015). According to Raidah Asma and Daud (2012) on the perceptions of travel planning for musicians in Malaysia, they acknowledged the importance of SM in terms of attracting and promoting music. SM plays an important in spreading messages about music events in Malaysia and beyond. In a similar note, Lorenzo-Romero, Alarcón-del-Amo & Constantinides (2014) found SM tools to be useful in the retail business sector just like in tourism information services as indicated by Li, Hu, Huang & Duan (2017). This is an indication the SM is a powerful tool not only disseminating information and making promotion about the planned events in a society, but it also leads to influencing people's decision making to participate in the planned activities. Through SM promotions, the participants can shelve their plans on other programs exclusively, to take part or embark on a trip to obtain firsthand experience of the event.

Travel Decision Making and Social Media

Information communication technologies do not only bring in opportunities but also challenges. For example, Law, Chan, and Wang (2018) acknowledge that mobile travel applications have been developed, particularly for leisure travelers. However, it is also argued that only fewer mobile applications have been developed targeting business travelers who have the potential to utilize mobile technologies more than leisure travelers (Douglas, Lubbe, & van Rooyen, 2018). Due to inadequate mobile technologies for segments of travelers (Douglas et al., 2018), some destination travel decisions have suffered because the users of information have failed to verify information with other sources of information like information service providers. So, social media online messages, videos and recordings are likely to have a potential to motivate or break peoples travel plans (Li, Xu, Tang, Wang, & Li, 2018; Leung, Law, van Hoof, & Buhalis, 2013). While some information may be current, others may be outdated. It is for this reason that users-generated of online information should be careful on messages, text, and video posted on social media.

Travelers may decide to cancel or change travel plans depending on the information available on SM. Although, travelers are at liberty to cancel or change itineraries any time, however, it should also be noted that if booked through intermediaries, some cancellation fees may be applied. Therefore, decision making is vital when it comes to searching toughly on the internet but also comparing travel planning information shared or tagged on Facebook (Shang, Chen, Wei, Jesen, Wen & Kalnis, 2016; Zhou, Wang, & Li, 2017). Facebook utilization appears to be the most uncomplicated social networking for many people around the world where a potential traveler gets timely feedback through messages and videos about destination status, hence making an informed destination decision (Toh, DeKay, & Raven, 2011; Önder, Gunter, & Gindl, 2019). Travel decisions using SM can also be made throughout the trip but also after the trip to evaluate if repeat trip is necessary or not and pass memories to potential future travelers. Personal feelings, values, and host destination reception as well as general geographical evaluation in comparison with other destinations traveled to before, are critically evaluated before making recommendations to others through SM communications. Despite such recommendations, further final destination decision can be made by the potential travelers after evaluating all information at hand. Some travelers can as well discussion with colleagues about the destination in relations to issues obtained on social media that have been shared in order to make proper decision to engage on a trip to that destination or change their destination.

Travel Planning and Destination Decision Making in Social Media

Statistically, according to Kim et al. (2017), SM active Facebook users alone was 1.7 billion per month in 2016, of which, 1.03 billion were recorded as average daily active users of Facebook using mobile devices alone. This suggest that the figure may be much higher if included those who use other devices like computers. Moreover, statistics of January 2019 of SM consumers is reported that the total world population was 7.676 billion (56%), unique mobile users 5.112 billion (67%), internet users 4.388 billion (57%), active media users 3.484 billion (45%), and mobile media users 3.256 billion representing 42% (Chaffey, 2019). This suggest that, as the number of active online technologies increases, equally, the number of travelers who use online applications to plan and share experiences of their travel increases. As such, the planning for travel is critical in order to have a successful and enjoyable travel experience. Several tourism destinations have abundance of shared experiences, both positives and negatives, on the social media. As such, it is vital to carefully evaluate information first so that better destination decision is made to avoid discomfort experiences which may lead to curtail the holiday or decide to change a tourist destination abruptly. Travel planning begins with search for a destination, transport options from the generating destination to host destination. At a host destination, information is required in the planning process regarding local accessibility, accommodation, attractions, activities, and amenities, payment options, security, local transport options. According to Choi et al. (2012), stressed that the knowledge and understanding of travelers well-being at a host destination helps one to plan ahead of the trip as it becomes easy to estimate on the duration to spend on holidays, number of activities to participate, as well as, the type of accommodation to make a reservation. Ultimately, it also helps to budget effectively for the whole trip.

The sharing of experiences on certain destinations may also reduce pressure when planning to visit places but also to experiences of different social and cultural practices (Chilembwe, 2014). Travellers can make proper destination decisions through others experiences but also based on their own first hand experiences. Sometimes, SM can provide conflicting information about the destination, and reliable shared information can be disseminated by those spread latest information. It is also the most updated shared experiences that will potential and existing tourism consumers to be motivated and make destination decision about their travel plans (Pan & Fesenmaier, 2006; Munar & Jacobsen, 2014). Although SM cannot always spread the truth, many travelers depend internet and other information technologies to have gather information about tourists' destinations. To some extent, SM information can be cross-checked by others like travelers who share experiences to colleagues based on their personal relationships and knowledge. However, not everyone who travels triangulate the information before making a decision on a tourist destination as SM remains one of the greatest motivators of many leisure travelers in planning for their next trips and also making destination decisions (Guiver & Stanford, 2014). Hunag et al. (2017) note, smart tourism technologies in travel planning provides travellers with rich, diverse, and useful information as they access and make connection at various destinations. Moreover, the travelers are able to make further destination decisions on their travel plans as they use SM on user-generated content to receive online messages and allocating destination sites or attractions (Hjalager, 2015; Buhalis, & Law, 2008). Travel planning is also enhanced by the online technological applications (Vu et al., 2015). For example, Lu et al. (2010), provide SM applications used by travel planners such as google maps, GPS, e-payment and photos capturing, tagging and instant messaging and photo attachments. Travelers use technologies for their travel related transactions to obtain more accurate, rich, comprehensive and personalized information as they plan for travel as well as sharing their real-time experiences Hung et al., 2017). Therefore, in the course of using online technologies in travel planning, proper decisions about travelling to a destination have been adopted, put on hold, changed, and also curtailed completely.

Hypotheses

As provided in the literature review, several studies have already highlighted the relationships between travel planning and sharing travel experiences as well as the influences of social media in travel decisions and destinations decision making. The following are the hypotheses of the study:

1. Sharing experiences on the SM can [or cannot] lead to travel planning of a tourist destination.
2. Sharing experiences on the SM can [or cannot] lead to travel destination decision making.

STUDY APPROACH

A total of five cases were surveyed, namely; airline, travel agencies, tour operator, hotel, and car hire to understand travelers' usage of social media in the travel planning and destination decision making. The study targeted travelers: five developed countries and another five developing countries, making a total of ten tourist destinations. These were visiting tourists' destinations for leisure and holidays from 10th May 2017 to 12th January 2019. A structured questionnaire was developed and used to interview respondents in the visiting cultural and nature tourists, that is; museums centers and resorts respectively. The simple random sampling procedure was employed to interview 792 travelers.

The survey was made up of various sections. The first section covered the demographics of the respondents and seeks their gender, level of education, type of device they use, the purpose of their travel and occupation. The next section seeks responses about their frequency usage of social media in the travel planning, travel decision and destination decision making, particularly when making investigating where to make a booking and also cancel or change their booking decisions. The final section asks the perceived importance of social media in travel planning and destination decision making. In this section, respondents were requested to provide their opinions on the importance of social media usage in travel planning and destination decision making. The importance was measured on a scale from 1-3 with 1=not important and 3=very important.

In order to achieve the objective of the study, diverse data analysis techniques were applied. The descriptive methods help in describing the data in terms of gender to segment age, levels of education, the purpose of travel, and occupation. The approach enables the researchers to describe statistically but also compare the responses to the role of social media in terms of travel planning, travel decision making, and destination decision making on the larger population of tourists who use social media.

RESULTS AND DISCUSSION

Table 1 below has shown that more female (70.4%) participated in the survey than men (29.6%). Thirty-four percent of respondents fell in the 40-49-year-old category, and the majority of respondents had a graduate qualification (45%), had use a smartphone/tablet 78.9%, and their purpose of travel was a holiday (69.7%) but was also in employment (47.5%).

As the demographic information depicts on the gender participation, female dominated with 74.4% as compared to men 29.6%. This is because some data was collected during a women day when most of them had holidays and travelled for leisure activities at the resorts. As such, the majority of them were not accompanied by their spouses. Although some did, however, they confined in their hotel rooms and other leisure activities because they were not comfortable to be among women dominated function. Therefore, the dominance of women in this study may have an impact or influence on the outcome of the

Table 1. Demographic profile

Gender (n=792)	Female	70.4%
	Male	29.6%
Age (n=784)	18-39 years old	31.3%
	40-49 years old	33.8%
	50-59 years old	25.6%
	60 years and above	9.3%
Level of education (n=792)	Secondary School Qualification	13.1%
	Certificate/Diploma	27.6%
	Graduate	45.2%
	Post-Graduate	14.1%
Type of device (n=792)	Smartphone/Tablet	78.9%
	Computer	13.8
	Smartphone and computer	7.3%
Purpose of travel (n=792)	Business and leisure	22.1%
	Holiday	69.7%
	Visiting friends and relatives (VFR)	8.2%
Occupation (n=792)	Employed	47.5%
	Self-employment	34.9%
	Student	17.6%

Researchers' data, 2019

results. Despite that, the beauty of the study was that all respondents were mature in terms of age with a minimum of 18 years, but also educated with minimum qualification being secondary school level and highest post graduate level. Most of them use smartphone or tablet and were on employment and also on leisure travel. This suggest that the majority of participants in the study were knowledgeable and understand well the relationship of social media in terms of planning and travel destination decision making.

Hypothesis 1: Sharing Experiences on the SM Can [or Cannot] Lead to Travel Planning of a Tourist Destination

When questioned about their usage of social media when planning travel and also making destination decision, the majority (92%) of respondents indicated that they use smartphone and tablet to make their travel plans on airline and hotel accommodation with exception to tour operators (87%) and car rental (56%). Similarly, the majority (92%) of respondents indicated that they cancel or change their booking on airline, travel agency, hotel accommodation, and car hire. However, the exception applies to a tour operator and car rental with 52% and 44% respectively, as shown in Table 2 below.

As Table 2 depicts, travel planning was frequently applied when making bookings for airlines, tour operators, and hotels unlike travel agencies and car rentals. Likewise, travel decision making on the cancellations or changing of reservation were dominated in airlines, tour operators and hotels. The reasons could be that several airlines, tour operators and hotels have a website page where travelers easily provide evaluations or feedback regarding their products and services. Most of these companies are also large in terms of sizes, as such, there are more visible as compared to travel agencies and car rentals. Travelers and other tourism consumers also share more experiences regarding the provision of products and services in general to its customers. Moreover, the majority of travelers prefer dealing directly with the suppliers than using intermediaries which may be expensive at times, so they could utilise technological online applications to make their bookings.

Conversely, travel agencies and car hire dominated in terms of travel decision making unlike travel planning. The reasons could be that travel agencies are available almost every corner of the city and potential travellers within vicinity can approaches travel agencies to get their preliminary travel information

Table 2. The frequency of Social media usage in the travel planning and travel destination decision making

Case	Variable	Travel Planning	Travel Decision Making
Airline	Make travel booking	92%	8%
	Change/cancel travel booking	8%	92%
Tour operator	Make travel booking	87%	13%
	Change/cancel tour booking	48%	52%
Travel Agency	Make travel booking	8%	92%
	Change/cancel travel booking	36%	64%
Hotel	Accommodation booking	92%	8%
	Change/cancel hotel booking	8%	92%
Car Rental	Car hire reservation	8%	92%
	Change/cancel car hire reservation	56%	44%

Researchers' data, 2019

and compare the offers before they make their bookings. So, the distances and available of many travel agencies in a particular city add value to be considered in the travel decision making process. On car hire, most of the travelers, especially international visitors, book car hire that are recognized internationally for among reasons; privacy and also their own security. The recommendations via the social media sharing on experiences about the available car rental companies in a particular country play a significant role of ones to make a booking for their travel. This is also in line with Toh et al. (2011), Önder et al. (2019) who also said the feedback that is shared online plays a critical part in consumers travel decision making.

Hypothesis 2: Sharing Experiences on the SM Can [or Cannot] Lead to Travel Destination Decision Making

Respondents were asked to provide their perceptions regarding the importance of social media in the travel planning and destination decision making. The majority of the respondents indicated that social media were 'important' to 'very important,' during business and leisure, and holidays in travel planning, travel decision making, and destination decision making. Holiday travel had the highest respondents, with 50.6% in the destination decision making. However, visiting friends and relatives (VFR) had the majority respondents on 'not important,' that is; both in the travel planning 56.9% and destination decision making 49.4% as shown Table 3 below.

Critically, although both business and leisure as well as holiday travel, the study depicts that travel planning, travel decision-making and destination decision making were very important and important. However, the degree of planning is somehow different between business and holiday travel. For example, in business travel, it may be difficult to withdrawal from participation in a trip when a traveler weigh benefits versus risks, hence the travel decision on a destination becomes and remains important or very important. On the contrary, on holiday travel, a traveler can easily cancel the trip if seeing that the risks are greater that were shared on the social media by other travel experiences. In a such situation, the travel decision making on a destination becomes not important and destinations may as well suffer due to lack of tourists to consume products. So, the social media presents both benefits and challenges. As

Table 3. The perceived importance of social media in the travel planning, travel decision and destination decision making

Variable			Percentage		
			Very Important	Important	Not Important
The perceived importance of Social media usage in travel planning and destination decision making	Business and leisure (n=792)	Travel planning	38.8%	43.7%	17.5%
		Travel decision making	49.1%	47.4%	3.5%
		Destination decision making	39.6	48.3	12.1%
	Holiday (n=792)	Travel planning	43.1%	47.4%	9.5%
		Travel decision making	54.7%	42.5%	2.8%
		Destination decision making	48.3%	50.6%	1.1%
	VFR (n=792)	Travel planning	25.5%	17.6%	56.9%
		Travel decision making	23.8%	39.3%	36.9%
		Destination decision making	18.9%	31.7%	49.4%

Researchers' data, 2019

argued by Douglas et al. (2018) that several technological mobile applications have targeted business segments as compared to leisure segment. To this end, some destination travel decisions have also suffered because the users of information fail to verify information with other sources of information. This is also in agreement with Li et al. (2018), Leung et al. (2013) who said that social media has a potential to motivate or break peoples travel plans. Therefore, it is useful to provide shared travel experiences on the social media that is based on truth of the situation in order to have the potential travellers obtain real-time information and make better travel decisions regarding destinations.

FUTURE RESEARCH DIRECTIONS

First, the study highlights that sharing experiences on social media can lead travel planning by the travellers to tourists' destination it further confirms that the behaviour of sharing travel experiences on social media can lead to travel decision making and destination decision making of tourists' places of interest. However, the study was dominated by one gender which may act as a limitation and could also be limited by lack of statistical tests on the outcomes. As such, assumptions are that responses and the results may be biased because of one-sided gender dominance and limited at statistical tests. It is somehow difficult to ascertain the outcomes of the study if it were gender balanced. It is, therefore, recommends that a similar study needs to be conducted but taking into consideration a balanced gender or should be male dominated to confirm or disconfirm the present results. The investigation into perceptions of males regarding social media in planning and travel decision as well as destination decision making could provide more better conclusion on this study topic. Second, the study was focused more on holiday travel segment. This means that the contribution about business travel segment was not thoroughly presented. The investigation into how social media influences business travellers in planning, travel decision and destination decision making could provide insights comprehensively regarding the performance of social media on travellers. Third and final, due to the limitation of statistical tests, future study should incorporate the testing to validate the study.

CONCLUSION

The study concludes that sharing travel experiences on social media (SM) leads travel planning of tourists' destinations. SM plays a significant role in travel planning, and destination decision making because the platform is famous for many travelers (Kim et al., 2017). It is proven that SM is used for social networking, chatting, and information sharing (Chung & Koo, 2015). The various sectors such as airlines, tour operators, hotels, car rentals, and travel agencies are not exempted from SM sharing experiences as well as travel planning. The businesses at the destinations have the potential to grow as product and services keep on marketed as well as shared positively on the social media (Chilembwe et al., 2018). SM does not help travelers in travel planning and destination decision making only but also the travel firms as they organize trips to various places of tourists' interests. Lei et al. (2019) note, using social media channels have brought about new types of booking channels. People can now make their booking using mobile social channels, meaning that the booking process has been simplified, but they are also able to interact with the applications and have access to information they want (Kim & Connolly, 2012; Thakran & Verma, 2013). The chats are enabled using communication channels such as Facebook, Twitter, video chat, and voice messages (Wang et al., 2016). SM has simplified the way travelers search for information, plan for trips, share for information, and make decisions about their destinations. SM is a source of information which is communicated through experiences to help other travelers plan for their trips (Xiang & Gretzel, 2010; Zeng & Gerritsen, 2014). Travelers use SM to search for travel information online, particularly on the internet using and check destinations of their choice. The majority of them use internet-based and mobile-based online communications to search information as demographic information provided evidence that smartphones and tablets are used during a search of information on the internet (Okazaki et al., 2015; Kaplan & Haenlein, 2010; Husain et al., 2018). SM accommodates YouTube, Facebook, WeChat, Twitter, TripAdvisor, and Instagram where people and destination marketing organizations share views online through messages and videos (Fotis et al., 2012; Xiang, 2017). People share information sharing at any stage, such as pre-travel, during travel, and post travel. According to Munar and Jacobsen (2014), travelers can share memories with their friends and relatives while on the trip but also after the trip. Although the behavior of sharing memories put others on peer pressure and motivate others too to start planning for their next travel in advance, the majority of respondents found it relevant and very important to share travel experience (Pan & Fesenmaier, 2006; Munar & Jacobsen, 2014). SM motivates many other leisure travelers to begin to plan for their next trips (Guiver & Stanford, 2014), of which, most of their trip decisions and plans are made using SM as user-generated as noted by Buhalis, & Law (2008), Hjalager (2015). Furthermore, there are online mobile phone reminders (Chiang & Huang, 2015; Binucci et al., 2017) while the online mobile phones applications such photos capturing, tagging and instant messaging are highly utilized during their trips (Lu et al., 2010 Vu et al., 2015). Finally, on destination decision making, travelers may decide to cancel or change travel plans depending on the information available on SM. Decision making is vital when it comes to searching on the internet but also comparing travel planning information shared or tagged on Facebook (Shang et al., 2016; Zhou et al., 2017). Travel decisions using SM can also be made throughout the trip pass memories to potential future travelers. Personal feelings, values, and host destination reception can be compiled and communicate to potential travelers.

ACKNOWLEDGMENT

This research received no specific grant from any funding agency in the public, commercial, or not-for-profit sectors.

REFERENCES

Akehurst, G. (2009). User-generated content: The use of blogs for tourism organisations and tourism consumers. *Service Business*, *3*(1), 51–61. doi:10.100711628-008-0054-2

Binucci, C., De Luca, F., Di Guiacomo, E., Liotta, G., & Montecchiani, F. (2017). Designing the content analyser of travel recommender system. *Expert Systems with Applications*, *87*, 199–208. doi:10.1016/j.eswa.2017.06.028

Buhalis, D., & Law, R. (2008). Progress in information technology and tourism management: 20 years on and 10 years after the internet – The state of eTourism research. *Tourism Management*, *23*, 207–230. doi:10.1016/S0261-5177(01)00085-1

Chaffey, D. (2019 12 Feb). Global digital report 2019. Smart Insights. Retrieved from https://www.smartinsights.com/social-media-marketing/social-media-strategy/new-global-social-media-research/

Chiang, H.-S., & Huang, T.-C. (2015). User-adapted travel planning system for personalised schedule recommendation. *Information Fusion*, *21*, 3–17. doi:10.1016/j.inffus.2013.05.011

Chilembwe, J. M. (2014). Examination of Social Cultural Impacts of Tourism in Chembe Village in Mangochi District, Malawi. *International Journal of Business Quantitative Economics and Applied Management Research*, *1*(1), 61–92.

Chilembwe, J. M., & Mweiwa, V. R. (2019). Responsible Travel and Tourism Adventure: Evidence from Malawi as a Tourist Destination. In A. Sharma (Ed.), *Sustainable Tourism Development: Futuristic Approaches* (pp. 31–53). New York: Apple Academic Press.

Chilembwe, J. M., Mweiwa, V. R., & Mankhomwa, E. (2018). Role of Tour Operators in Destination Tourism Marketing in Malawi. In M. A. Camilleri (Ed.), *Strategic Perspectives in Destination Marketing* (pp. 295–321). Hershey, PA: IGI Global. doi:10.4018/978-1-5225-5835-4.ch012

Choi, S., Lehto, X. Y., Morrison, A. M., & Jang, S. (2012). Structure of travel planning processes and information use patterns. *Journal of Travel Research*, *51*(1), 26–40. doi:10.1177/0047287510394191

Chung, N., & Koo, C. (2015). The use of social media in travel information search. *Telematics and Informatics*, *32*(2), 215–229. doi:10.1016/j.tele.2014.08.005

Cox, C., Burgess, S., Sellitto, C., & Buultjens, J. (2009). The role of user-generated content in tourists' travel planning behaviour. *Journal of Hospitality Marketing & Management*, *18*(8), 743–764. doi:10.1080/19368620903235753

Douglas, A., Lubbe, B., & van Rooyen, A. (2018). Profiling business travellers who use mobile travel applications. *Ereview of Tourism Research*, *15*, 2–3. Retrieved from http://ertr.tamu.edu

Fesenmaier, D. R., Xiang, Z., Pan, B., & Law, R. (2011). A Framework of Search Engine Use for Travel Planning. *Journal of Travel Research*, *50*(6), 587–601. doi:10.1177/0047287510385466

Fotis, S., Buhalis, D., & Rossides, N. (2012). *Social media use and impact during the holiday travel planning process*. Berlin, Germany: Springer. doi:10.1007/978-3-7091-1142-0_2

Guiver, J., & Stanford, D. (2014). Why destination visitor travel planning falls between the cracks. *Journal of Destination Marketing & Management*, *3*(3), 140–151. doi:10.1016/j.jdmm.2013.11.005

Hawk, A. T., van den Eijnden, R. J. J. M., van Lissa, C. J., & ter Bogt, T. F. M. (2019). Narcissistic adolescents' attention-seeking following social rejection: Links with social media disclosure, problematic social media use, and smartphone stress. *Computers in Human Behavior*, *92*, 65–75. doi:10.1016/j.chb.2018.10.032

Hjalager, A. (2015). 100 innovations that transformed tourism. *Journal of Travel Research*, *54*(1), 3–21. doi:10.1177/0047287513516390

Ho, C.-I., Lin, M.-H., & Chen, H.-M. (2012). Web users' behavioural patterns of tourism information search: From online to offline. *Tourism Management*, *33*(6), 1468–1482. doi:10.1016/j.tourman.2012.01.016

Huang, C. D., Goo, J., Nam, K., & Yoo, C. W. (2017). Smart tourism technologies in travel planning: The role of exploration and exploitation. *Information & Management*, *54*(6), 757–770. doi:10.1016/j.im.2016.11.010

Husain, T., Li, B., & Wang, D. (2018). What factors influence the sustainable tour process in social media usage? Examining a rural mountain region in Pakistan. *Sustainability*, *10*(7), 2220. doi:10.3390u10072220

Kaplan, A. M., & Haenlein, M. (2010). Users of the world, unite! The challenges and opportunities of social media. *Business Horizons*, *53*(1), 59–68. doi:10.1016/j.bushor.2009.09.003

Kim, J., & Connolly, D. (2012). Hospitality in the age of empowerment: First annual customer technology study. *Hospitality Technology*. Retrieved from https://luiscodo.files.wordpress.com/2012/09/customer-engagement2012.pdf

Kim, S.-E., Lee, K. Y., Shin, S., & Yang, S.-B. (2017). Effects of tourism information quality in social media on destination image formation: The case of Sin Weibo. *Information & Management*, *54*(6), 687–702. doi:10.1016/j.im.2017.02.009

Law, R., Chan, I. C. C., & Wang, L. (2018). A comprehensive review of mobile technology use in hospitality and tourism industry. *Journal of Hospitality Marketing & Management*, *27*(6), 626–648. doi:10.1080/19368623.2018.1423251

Lei, S. S. I., Nicolau, J. L., & Wang, D. (2019). The impact of distribution channels on budget hotel performance. *International Journal of Hospitality Management*, *81*, 141–149. doi:10.1016/j.ijhm.2019.03.005

Leung, D., Law, R., van Hoof, H., & Buhalis, D. (2013). Social media in tourism and hospitality: A literature review. *Journal of Travel & Tourism Marketing*, *30*(1-2), 3–22. doi:10.1080/10548408.2013.750919

Li, J., Xu, L., Tang, L., Wang, S., & Li, L. (2018). Big data in tourism research: A literature review. *Tourism Management*, *68*, 301–323. doi:10.1016/j.tourman.2018.03.009

Li, Y., Hu, C., Huang, C., & Duan, L. (2017). The concept of smart tourism in the context of tourism information services. *Tourism Management*, *587*, 293–300. doi:10.1016/j.tourman.2016.03.014

Lorenzo-Romero, C., Alarcón-del-Amo, M., & Constantinides, E. (2014). Determinants of use of social media tools in retailing sector. *Journal of Theoretical and Applied Electronic Commerce Research*, *9*(1), 44–55. doi:10.4067/S0718-18762014000100005

Lu, W., & Stepchenkova, S. (2015). User-generated content as a research mode in tourism and hospitality applications: Topics, methods, and software. *Journal of Hospitality Marketing & Management*, *24*(2), 119–154. doi:10.1080/19368623.2014.907758

Lu, X., Wang, C., Yang, J. M., Pang, Y., & Zhang, L. (2010). Photo2trip: Generating travel routes from geotagged photos for trip planning. *Proceedings of the 18th ACM International Conference on Multimedia*, Firenze, Italy. ACM. 10.1145/1873951.1873972

Munar, A. M., & Jacobsen, J. K. S. (2014). Motivations for sharing tourism experiences through social media. *Tourism Management*, *43*, 46–54. doi:10.1016/j.tourman.2014.01.012

Okazaki, S., Campo, S., Andreu, L., & Romero, J. (2015). A latent class analysis of Spanish travellers' mobile internet usage in travel planning and execution. *Cornell Hospitality Quarterly*, *56*(2), 191–201. doi:10.1177/1938965514540206

Önder, I., Gunter, U., & Gindl, S. (2019). Utilizing Facebook statistics in tourism demand modeling and destination marketing. *Journal of Travel Research Online*. doi:10.1177/0047287519835969

Pan, B., & Fesenmainer, D. R. (2006). online information search: Vacation planning process. *Annals of Tourism Research*, *3*(3), 809–832. doi:10.1016/j.annals.2006.03.006

Raidah Asma, A. R., & Daud, D. (2012). Perceptions of travel planning for musicians: A Malaysian perspective. *Business Management Dynamics*, *2*(6), 56–63.

Shang, S., Chen, L., Wei, Z., Jesen, C. S., Wen, J., & Kalnis, P. (2016). Collective travel planning in spatial networks. *IEE Transactions on Knowledge and Data Engineering*, *28*(5), 1132–1146. doi:10.1109/TKDE.2015.2509998

Thakran, K., & Verma, R. (2013). The emergence of hybrid online distribution channels in travel, tourism and hospitality. *Cornell Hospitality Quarterly*, *54*(3), 240–247. doi:10.1177/1938965513492107

Toh, R. S., DeKay, C. F., & Raven, P. (2011). Travel Planning: Searching for and Booking Hotels on the Internet. *Cornell Hospitality Quarterly*, *52*(4), 388–398. doi:10.1177/1938965511418779

Vu, H. Q., Li, G., Law, R., & Ye, B. H. (2015). Exploring the travel behaviour of inbound tourists to Hongkong using geotagged photos. *Tourism Management*, *46*, 222–232. doi:10.1016/j.tourman.2014.07.003

Wang, D., Xiang, Z., & Fesenmaier, D. R. (2016). Smartphone use in everyday life and travel. *Journal of Travel Research*, *55*(1), 52–63. doi:10.1177/0047287514535847

Xiang, U., & Gretzel, U. (2010). Role of social media in online travel information search. *Tourism Management*, *31*(2), 179–188. doi:10.1016/j.tourman.2009.02.016

Xiang, Z. (2017). Digitalisation to the age of acceleration: On information technology and tourism. *Tourism Management Perspectives*, *25*, 147–150. doi:10.1016/j.tmp.2017.11.023

Zeng, B., & Gerritsen, R. (2014). What do we know about social media in tourism: A Review. *Tourism Management Perspectives*, *10*, 27–36. doi:10.1016/j.tmp.2014.01.001

Zhang, T., Omran, B. A., & Cobanoglu, C. (2017). Genaration Y's positive and negative eWOM: Use of social media and mobile technology. *International Journal of Contemporary Hospitality Management*, *29*(2), 732–761. doi:10.1108/IJCHM-10-2015-0611

Zhou, X., Wang, M., & Li, D. (2017). From stay to play – A travel planning tool based on crowdsourcing user-generated contents. *Applied Geography (Sevenoaks, England)*, *78*, 1–11. doi:10.1016/j.apgeog.2016.10.002

ADDITIONAL READING

Brandt, T., Bendler, J., & Neumann, D. (2017). Social media analytics and value creation in urban smart tourism ecosystems. *Information & Management*, *54*(6), 703–713. doi:10.1016/j.im.2017.01.004

Dedeoğlu, B. B., Taheri, B., Okumus, F., & Gannon, M. (2020). Understanding the importance that consumers attach to social media sharing (ISMS): Scale development and validation. *Tourism Management*, *76*, 103954. doi:10.1016/j.tourman.2019.103954

der Pütten, A. M. R., Hastall, M. R., Köcher, S., Meske, C., Heinrich, T., Labrenz, F., & Ocklenburg, S. (2019). "Likes" as social rewards: Their role in online social comparison and decisions to like other people's selfies. *Computers in Human Behavior*, *92*, 76–86. doi:10.1016/j.chb.2018.10.017

Kamis, A., & Frank, J. (2014). Advice sharing between paired users in online travel planning. *Journal of Theoretical and Applied Electronic Commerce Research*, *9*(1), 1–18. doi:10.4067/S0718-18762014000100002

Lu, D., Wu, R., & Sang, J. (2017). Overlapped user-based comparative study on photo-sharing websites. *Information Sciences*, *376*, 54–70. doi:10.1016/j.ins.2016.10.005

Phu, B., & Gow, A. J. (2019). Facebook use and its association with subjective happiness and loneliness. *Computers in Human Behavior*, *92*, 151–159. doi:10.1016/j.chb.2018.11.020

Sherman, L. E., Payton, A. A., Hernandez, L. M., Greenfield, P. M., & Dapretto, M. (2016). The power of the like in adolescence: Effects of peer influence on neural and behavioural responses to social media. *Psychological Science*, *27*(7), 1027–1035. doi:10.1177/0956797616645673 PMID:27247125

Tussyadiah, I., & Fesenmaier, D. R. (2009). Mediating tourist experiences: Access to places via shared videos. *Annals of Tourism Research*, *36*(1), 24–40. doi:10.1016/j.annals.2008.10.001

KEY TERMS AND DEFINITIONS

Booking Channel: An online system or agent that facilitate the reservation of a flight, hotel or car hire.

Information Technology: A process of developing and maintaining computer systems, software, and networks for the processing and distributing of a company's products, services and consumers travel experiences.

Social Media: A digital platform used for networking, chatting, and information sharing.

Tour Operator: A firm that arranges tourists' holidays in terms of travel, accommodation, activities, and attractions.

Tourism: The movement of people from place to place whose purpose is either pleasure or business for over short period of time.

Tourism Destination: A place or area that is dependent significantly on the attraction resources to generate its income.

Travel Agency: An intermediary firm that works on behalf of airline, accommodation, car hire, and cruise ship but can also sell its own holiday destinations

Travel Planning: The process of collecting all the relevant information and create a strategy regarding movement, lodging, sightseeing, and expenditure for the journey.

Chapter 4
The Role of Social Media in Shaping Tourists' Travel Decisions:
Multi-Functional and Multi-Level Effects of Wechat Among Individual Chinese Outbound Tourists

Sandeep Basnyat
Macau Institute for Tourism Studies, Macau, China

Yuan Jiahui
Macau Institute for Tourism Studies, Macau, China

ABSTRACT

With increased economic and population growth, air connectivity, and travel facilities, international travel of Chinese citizens has tremendously increased. Although usually in groups, there has been a growing trend among Chinese tourists toward travelling independently. These independent travellers or so-called 'new tourists' are highly exposed to new technologies and social media and prefer to do research about possible destinations, and new travel routes and activities that package tours usually do not include. This chapter examines how social media in general and WeChat, in particular, shapes the Chinese individual travellers' travelling decisions. As will be seen, while WeChat's popularity among the Chinese outbound tourists is largely attributed to its multi-functionality, its influences on their destination visitation choice are found in multiple levels: by providing illuminating information, allowing users to integrate their payment system easily to their bank accounts in China, and providing means to obtain discounts and promotional offers in a variety of ways.

DOI: 10.4018/978-1-7998-1947-9.ch004

1. INTRODUCTION

This chapter explores the role played by the social media application in shaping tourists' decision to visit a destination. The empirical data for this study was collected through in-depth semi-structured interviews with fifteen participants in 2017 and 2018 who were travelling from Shenzhen, Mainland China to Macau, the Special Administrative Region (SAR) of China, and had used WeChat, a popular social media platform, to make their travelling decisions. As will be seen, while WeChat's popularity among the Chinese outbound tourists is largely attributed to its multi-functionality, its influences on their destination visitation choices are found in multiple levels.

Traditionally, the Mainland Chinese tourists have been travelling to other destinations including Macau, using package tours. However, there has been a growing trend among Chinese tourists who enjoy travelling independently, and it is assumed that almost 70 per cent of Chinese outbound travels is being organized independently (Peltier, 2018). A 2017 report published by the Statistics and Census Service Bureau of Macau shows that the number of Chinese tourists who have travelled to Macau independently or under the Individual Visit Scheme (IVS) has increased steadily, and grown several folds since 2009 (DSEC, 2017). These independent travellers or so-called 'new tourists' are highly exposed to new technology and social media and prefer to do research about possible destinations, and new travel routes and activities that package tours usually may not include (DSEC, 2017; World Tourism Organization, 2017).

Many international travellers like to share their travel experiences on the websites, yet, others also search for and evaluate travel information online and in social media platforms such as Facebook, Sina microblog, and TripAdvisor. As a result, social media are becoming instrumental in influencing visitors in choosing attractions such as restaurants and accommodations during their travel (Tham, Croy, & Mair, 2013). As the social media platforms provide opportunities for discussions for tourists after their trips, they can potentially reinforce the image of a destination in tourists' minds, strengthen its influences, and thus, motivate them to make decisions to travel to the destination. Choi, Lehto, and Morrison (2007) have confirmed that individual traveller's perceptions about a destination play significant roles in their travel decision making, and those perceptions are highly induced and/or influenced by the social media platform they use.

Although getting online information about travel and destinations is relatively easy in many countries, accessing global major social platforms such as Facebook, Twitter and YouTube are limited in China because of the government restrictions (Chiu, Ip, & Silverman, 2012). Instead, WeChat is the most common and trusted social media application among the general Chinese population (Xu, Kang, Song, & Clarke, 2015). Because of a range of services that WeChat provides, including messaging, texting, pictures and video sharing, news and articles reading through official accounts, and payment handling among others, it is considered as 'App for Everything in China' by many (Xu, Kang, Song, & Clarke, 2015). Therefore, it is possible that the Chinese tourists who travel to a destination individually may have been influenced by the information and contents in the social media they have been using. More particularly, WeChat may have played instrumental roles in shaping the individual Chinese tourists' decisions to visit a destination, however, this issue has not gained much attention in the existing literature. Filling this gap, this chapter examines how social media in general and WeChat, in particular, shapes the travellers' travelling decisions focusing on Chinese individual travellers. The chapter will particularly explore how the distinguishing features of WeChat such as 'WeChat Pay', 'Moments' and

'Official subscription accounts' provide services and information in a way that influence the Chinese individual travellers' outbound travelling decisions. In doing so, first, the next section will explore the historical development and travel patterns of the Chinese outbound tourists which will be followed by sections that examine the use and popularity of social media and WeChat among the Chinese outbound travellers and how WeChat influences their travelling decisions. The chapter will then conclude with a short remark, implications of the findings of the chapter and direction for future research.

2. THE HISTORICAL DEVELOPMENT AND TRAVELLING PATTERNS OF THE CHINESE TOURISTS

With a GDP of US Dollar 12.24 trillion in 2017, China is not only the largest economy in Asia and the Pacific and the second largest in the world after the United States, but it is also the largest, demographically, in the world with its 1.38 billion population (The World Bank, 2018). The country has also experienced rapid growth and internationalization in its economy over time. For example, between 1991 and 2017, China's economic growth has averaged nearly 10 per cent a year (The World Bank, 2018). With increased growth economic and population growth, air connectivity, market openness and travel facilitation, the international travelling of the Chinese citizens has also tremendously increased. This section examines the historical development and patterns of Chinese outbound tourism.

Until 1978, tourism in China was considered as a part of foreign affairs, and therefore, the national tourism organization of China, SATT (The China State Administration for Travel and Tourism), was placed under the Foreign Ministry and the local tourist issues were handled by the local foreign affairs offices (Guangrui, 1989). As a result, travel services were primarily conducted as and for diplomatic activities only. However, with the socio-political change that began in 1978 and the decision to open up China to the world economy, SATT was separated from the Foreign Ministry and was set up as an independent body under the State Council in 1981 (Guangrui, 1989).

In 1983, the Chinese government allowed the Nationals to go to Hong Kong and Macau to visit their relatives. This was the beginning of international travels by the general public for visiting friends and relatives (VFR) and laid the foundation for outbound tourism. Later, in 1986, the Chinese government included tourism as an industry in the national plan that paved way for the corresponding development in the Chinese inbound as well as outbound tourism industry (Guangrui, 1989). The scope of permissible travel for the purpose of "visiting relatives" and "touring the border areas" was expanded gradually between 1983 to 1995 as the Chinese nationals were allowed to go to Thailand, Singapore, Malaysia and the Philippines (World Tourism Organization, 2017). However, visiting relatives' tours, as a special concept, came to an end since the beginning of 1995 as the Chinese National Tourism Administration (CNTA) formalized the Approved Destination Status (ADS) programme (Arita, Edmonds, Croix, & Mak, 2011). The ADS scheme aimed at establishing a control mechanism for domestic travel agencies as well as international tour operators. Under the ADS scheme, certain Chinese operators were designated to arrange, promote and sell travel to specific destinations in collaboration with approved overseas partners. These approved overseas partners provided travel and tour arrangements within the visiting destinations. New Zealand and Australia were among the first Western countries who were granted the approved destination status in 1999. With the removal of restrictions over time, the number of ADS outbound destinations has grown quickly. By 2017, a total of 150 international destinations have entered into bilateral tourism arrangements with the Chinese government under the ADS scheme (World Tourism Organization, 2017). Not only the approved destinations but also the annual total number of Chinese outbound tourists has increased rapidly from 2.9 million in 1992 to over 149 million in 2018. The Travel China Guide (2019), quoting the Chinese Ministry of Culture and Tourism, predicted that the number of annual total Chinese outbound tourists may exceed 178 million by 2022.

With the early opening and close relationship with Hong Kong and Macau, these destinations are, traditionally, the most visited outbound destinations by the Chinese nationals. Taiwan, which was opened in 2008 for an individual visit, with the implementation of the Mainland Residents to Taiwan Tourism Agreement, has become another most visited destination. Collectively, these three destinations are visited by more than 71 per cent of the total outbound Chinese tourists and form the largest outbound tourism market for China (Dai, Jiang, Yang, & Ma, 2017; World Tourism Organization, 2017). Apart from Hong Kong, Macau and Taiwan, the top ten destinations that the Chinese tourists visited in 2018 were Thailand, Japan, Vietnam, Singapore, Indonesia, Malaysia, America, Cambodia, Russia and the Philippines (Travel China Guide, 2019).

The Chinese travellers prefer to combine holiday, weekend and annual leave to make the short holiday a long one. For example, Travel China Guide (2019) reported that the Chinese travellers spent 8.5 days overseas, on average, during the 2018 Spring Festival ranging from an average of 7.8 days in closer destinations such as Japan and Thailand and 12.6 days in long haul destinations such as Australia, Europe and the United States. Although, the Chinese tourists visit outbound destinations for a variety of reasons including sightseeing, holiday making, visiting friends and relatives (VFR), medical purposes, attending conferences and other economic activities, and exploration of culture, sport and religion, a significant majority of them are attracted towards shopping. Referring to a 2011 survey by China Tourism Academy, Dai, Jiang, Yang, and Ma (2017) reported that over 32 per cent of the Chinese tourists spent the most money on shopping compared to 10 per cent who spent on entertainment and other activities. The survey also reported that in some destinations the shopping constituted a significantly larger proportion of their expenditure exceeding 70 per cent.

Although the Chinese nationals had to join tour groups organized by travel agencies appointed by the Government in the beginning, later the procedure gradually became more simplified as the Chinese travellers may now travel independently, albeit still only to ADS approved destinations (Jørgensen, King, & Law, 2018). The use of a package tour was and still a popular method of travelling for many Chinese tourists because it generally includes a planned travel itinerary and accommodation and transportation services arranged by an external agent for a charge. Therefore, Chinese tourists consider that using a package tour save their time and efforts in managing their tours (Peltier, 2018). However, Chinese non-packaged travels have also been growing significantly and in many long-haul destinations, such as the USA and Europe, now outnumber package tour travellers. For example, Peltier (2018), citing a survey report by China Luxury Advisors and Fung Business Intelligence Centre (FBIC), argued that around 70 per cent of the Chinese outbound tourists plan to either make their own travel arrangements or buy a travel package but travel on their own for their next overseas trip.

Apart from watching sports events, the motivations of independent Chinese outbound travellers vary ranging from engaging in road trips and exotic and unique tour activities and experiencing authenticity at the destinations among others (Peltier, 2018). One of the key factors that have shaped the characteristics and features of the Chinese outbound individual tourists is the development and use of information technology aided by the internet and particularly the social media platforms. How does social media influence the Chinese population in general and outbound individual travellers in particular? These issues are examined in the next section.

3. USE AND POPULARITY OF SOCIAL MEDIA AMONG THE CHINESE TOURISTS

With the increased number of outbound travels, the use and application of technologies for making travelling decisions are also changing among Chinese travellers. For example, with the emergence of high-speed Internet, online travel agencies and e-commerce platforms, the traditional distribution channels have changed and the travellers increasingly rely on travel websites, mobile applications and cashless payments (World Tourism Organization, 2017). Correspondingly, there has been a huge proliferation of websites that operate and provide travel services locally as well as internationally in China. Some internationally operating travel facilitating agencies have also established their Chinese version of websites. For example, www.skyscanner.com, the largest flight search engine, has established its Chinese local site www.tianxun.com and www.trivago.com have its own local Chinese website www.youzhan.com.

A significant number of the Chinese population likes to obtain information through numerous social media platforms where they interact. In general, social media platforms can be classified into blogs, cooperative projects, content communities, virtual social worlds and social networking sites (Men & Tsai, 2013). Among these, social networking sites that include popular platforms such as Facebook and Myspace are the most popular ones with billions of active users globally. Apart from these, domestic sites which are culturally adapted and have distinct local features are also thriving in many countries, for example, Renren and WeChat in China, Mixi in Japan, Orkut in India, and Vkontakte in Russia. Social media apps, which can be accessed through mobile phones anytime and anywhere and allow users to exchange texts, videos, video calls and voice chats provide easy access of information to its users (Men & Tsai, 2013). Because of its ability to provide unprecedented opportunities to reach the general public and build social relationships at a more personal level, travel facilitating companies also use social media to interact with their customers in China.

The Chinese travellers, particularly the new generations, make extensive use of social media platforms to share their travel experiences, especially by posting their selfies. Kang and Schuett (2013) define travel experiences sharing on social media as "behaviours or activities occurring when an individual disseminates travel-related experiences to other members through social media" (p.94). Such sharing of information may include stories, opinions, and complaints which can be shared in the forms of text, photos, audio or voice podcasts among others. Kang and Schuett's (2013) finding demonstrated that identification and internalization are among the core motivations why travellers share their travel experiences on social media. By sharing their travel experiences through social media, travellers, on the one hand, identify and feel a sense of belonging or fitting into a group of social media users, and on the other hand, internalize that their activity is congruent with their value systems and beliefs and thus find it a rewarding endeavour. Because the user-generated travel information which is shared on social media are often viewed more reliable than those provided by tourism business organizations, travellers rely on them to make their travelling decisions (Han, McCabe, Wang, & Chong, 2018; Kang & Schuett, 2013). Apart from identification and internalization, travellers also share their travelling experiences on the social media because they want to prevent others from using bad products and are willing to communicate advice on practical matters (Munar & Jacobsen, 2014). A survey of Chinese Sina Weibo users by Kim, Lee, Shin, and Yang (2017) has revealed that the degree to which "tourism information about users' travel destination (provided by social media) is relevant to ones' purpose of trips, plays an important role for users in processing the information, eventually forming both cognitive and affective destination image" (p.698). Affective image formation depends upon the "interestingness" of tourism information whereas cognitive image formation is associated with the "completeness" of the information. Consequently, as recently found by Liu, Wu, and Li (2019), shared travel experiences on social media, as a form of electronic word of mouth, play important role in "shaping tourists' awareness, expectations, perception, attitudes and behaviours" (p.355). However, instead of commenting, asking questions and creating user-generated contents, Chinese travellers are more interested in viewing pictures and reading wall posts that allow them to obtain more information about a destination and help them make their decisions (Men & Tsai, 2013). Once a destination is proposed and recommended within the peer circle in the social media, very likely others also decide to visit those destinations. This is evident as a large number of Chinese travellers, who share similar interest, purchase destinations at the last minute (World Tourism Organization, 2017).

Renren, Sina Weibo, or Tencent Weibo are some of the popular social media platforms having hundreds of millions of users in China. Studies (for example, Xu, Kang, Song, & Clarke, 2015) have shown that over half of the Chinese students use Renren which is a Chinese equivalent of Facebook to share and find information. Similarly, WeChat has been another most popular social media platform among the Chinese population. WeChat is an instant messaging App created by China's Internet Company Tencent in 2011 and provides a number of products to its users including WeChat accounts, PC desktop instant messaging service, microblogging platforms just like Twitter and mail services (Xu, Kang, Song, & Clarke, 2015). WeChat's free instant messaging, voice, text, pictures, videos and location information exchange can be enabled through a mobile phone. WeChat also allows users to transmit real-time voice intercom, video calls, group chat, post pictures in the WeChat activity timeline labelled as 'Moments', update contacts, publish a status update with illustrations, comment, retweet and forward messages just like Facebook (Xu, Kang, Song, & Clarke, 2015). WeChat's official accounts which include account names, ID, logo, introduction, verified accounts, view history, receive messages, messages and an 'unfollow button' among others (Tencent, 2019). Its official accounts are classified into two groups: service accounts and subscription accounts. Service accounts allow users to send a text message or illustrated information monthly, whereas, subscription accounts help to push a group or single information with illustrations daily (Xu, Kang, Song, & Clarke, 2015).

With over 1 billion monthly active users, WeChat has become a way of life for the Chinese people (Tencent, 2019). Most Chinese population not only interact socially through WeChat, but they also use it for staying in contact personally with family and friends and professionally with their business colleagues. Chen (2017) found that WeChat has a direct positive impact on business-consumer relationship quality particularly in terms of the trust, commitment, control mutuality and satisfaction and argued that it can be used as a great tool to manage the relationship with customers. Wu (2014) examined how WeChat improves users' wellbeing by building a strong-tie based community, selective friend list, capitalization, positive contagion, powerful visuals, and savouring. It has been found that psychological motivations such as entertainment, sociality, and information, and trust positively influence WeChat users' attitudes, and consequently influence the positive word-of-mouth (Lien & Cao, 2014). The WeChat users also perceive that the quality of services provided by the WeChat platform including interactions, overall environment and the outcome are the major sources of satisfaction for them and motivate them to continue using it (Lien, Cao, & Zhou, 2017).

One of the key ways that WeChat facilitates the Chinese outbound travellers is through its partnership alignments with different organizations across countries. For example, WeChat has partnered with tourism authorities of Australia, Dubai and Britain that produce WeChat city experiences to its users. The users are guided through the attractions in the cities such as London, Dubai and Sydney via its in-built app and its interactive Bing Maps which is primarily the Chinese version of the google map. These maps help the users to navigate around places of interests for the travellers including restaurants and shopping areas. Because of the absence of Google Play in China, these services provided by the WeChat are immensely helpful for the outbound travellers (Travel Lifestyle Network, 2018).

Similarly, WeChat helps travellers who are also a member of various loyalty programmes offered by service organizations. For example, through a WeChat account, Marriott hotel's Marriott Rewards informs members about new hotel openings, provide travel tips and information about exclusive events and offers for them. Users can also compare prices using Shangri-La Hotel's WeChat account to obtain the best deal, and they can use mobile check-in services through the China Southern Airlines's official WeChat account (Travel Lifestyle Network, 2018). It is evident that all of these services and benefits that are provided by the WeChat greatly simplify the decision making for the Chinese outbound travellers.

Considering the influence of social media and WeChat in the Chinese peoples' everyday lives, it is fairly possible that the Chinese travellers' decision making, particular those who travel individually and independently, also may have been significantly impacted by the use and application of WeChat. However, a systematic examination of this topic is lacking. The next section, in particular, examines how the Chinese tourists who travel to a destination individually are influenced by the information and contents in the WeChat. As mentioned above, the findings presented in the next section have been obtained through individual semi-structured interviews with fifteen Chinese individual travellers who were travelling from Shenzhen to Macau in 2017 and 2018. Shenzhen is one of the major cities of mainland China from where Chinese tourists can easily visit Macau through ferry roughly in an hour. This one hour-time was utilized to conduct interviews with the participants. One interview was conducted per travel between Shenzhen and Macau. Because some of the travels did not yield interviews, altogether 20 travels (ten from Shenzhen to Macau and ten from Macau to Shenzhen) were conducted. The participants were approached and briefed about the purpose of the study. Once the participants agreed to share their experiences, a consent form was signed. A general line of questions including their use of WeChat, its influence on their destination choice for travelling, its effects on their decisions making among others were asked which were further supplemented by follow-up questions. The interviews were conducted in Chinese and recorded via a voice recording device on a mobile phone. The participants' responses were later translated into English and transcribed verbatim. Thematic analysis approach was used to code and analyse interview data. Several themes emerged from the interview data that formed the basis of the findings. These findings are discussed in the next section. In the context where studies on tourism destinations are highly dominated by quantitative approaches, this qualitative research provides an in-depth understanding of the reasons behind the popularity of WeChat among the individual Chinese outbound tourists, and how they influence their destination visitation choices.

4. HOW WECHAT INFLUENCES CHINESE OUTBOUND TRAVELLER'S DECISION?

The findings of the interviews conducted with the Chinese individual travellers identified three key ways that WeChat influences their outbound travelling decisions: through the WeChat Pay, information about destinations that can be obtained from the WeChat, and promotional offers provided in the WeChat. One of the most popular uses of WeChat for the Chinese outbound travellers is making payment through WeChat using their smartphones. The Chinese travellers feel that the payment through WeChat using their smartphones is convenient and enables them to simplify their purchases as they can link both of their activities, purchase and payment, through WeChat. Paying through WeChat is convenient for them because they just have to scan the bar code using their WeChat app in their phones and the payment would be settled from their Chinese bank account. Wu (2018) reported that by 2018, WeChat payment was available in 49 different countries and regions excluding Mainland China. However, many Chinese travellers also feel that there is a limitation of WeChat payment: it has a limit for the daily transaction of around US Dollar 1,500. So, if the travellers want to spend more than the limit, then they need to use their bank card. Another limitation of WeChat pay function, as noted by the Chinese travellers, is its inflexibility of adopting the international Visa or MasterCard number to their account. Although WeChat allows an international Visa or MasterCard number to integrate into the WeChat account, it is not useable for the payment. Many Chinese outbound travellers who were interviewed for the current research reported that they like to plan their holiday in a destination where there would be easy accessibility for the WeChat payment or where their daily spending could be easily covered within the WeChat payment limits. This finding adds a dimension in the existing literature (for example, Chen, 2017; Xu, Kang, Song, & Clarke, 2015) that portrays the ways in which WeChat attracts travellers. Expanding the existing notion that WeChat's primary way of attracting travellers is through its posts (such as pictures, videos, messages, etc.), the findings of the current study revealed that the financial easiness provided by the WeChat plays an equally important role in travellers' decision-making process.

For many individual Chinese outbound travellers, WeChat functions as an inspiration by providing them with captivating information about a destination or a region and motivating them to make their travelling decisions. In general, the Chinese individual travellers obtain information about a destination when one or the other WeChat community member posts such information in the form of articles, text, messages, pictures and videos in 'Moments' and/or in official 'Subscription' accounts. Comments and reviews posted by the WeChat community also help the individual Chinese travellers to decide whether a particular destination suits their taste and preferences or not. The Chinese travellers interviewed for the current research reported that the decisions about tastes and preferences may include a variety of factors including the availability of Chinese restaurants, how Chinese people friendly the social and community environment is, the availability of comfortable accommodations and shopping centres among others. For individual Chinese travellers, obtaining such information is critical because, unlike a group which travel together and are supported/managed by the travel agents, they need to make their judgement by themselves and bear the consequences of their decisions as well. In such cases, the Chinese individual travellers find that critical information posted in the WeChat helps them to evaluate a destination and make appropriate decisions.

Some individual Chinese travellers expressed that they like to travel to exotic and adventurous destinations such as Bali or Palau to observe the nature and lifestyles of the community. Often, these destinations may not be even a popular one among the mass Chinese tourists and also not available with the local travel agents. In such cases, information obtained from the people who have already visited those destinations, particularly, pictures of the destination and areas, surrounding facilities and vicinities, daily lives of people and travelling dialog in WeChat helps the individual travellers to make their decision and plan for their visits. The Chinese individual travellers informed that many foreign travel service companies also use WeChat to create forums where they put information about these exotic and unique destinations in Chinese. Some of those companies also hire people to respond to and interact with their Chinese customers. The individual Chinese travellers find such interactions immensely helpful for them to decide about the destination they plan to visit.

It is natural for anyone to seek or follow recommendations from their family and friends when making decisions about a destination to visit. However, a number of Chinese travellers feel that user-generated contents including articles, pictures and videos in the social media from completely unknown persons are equally beneficial and reliable and help them to make decisions. A few travellers interviewed for the current research strongly reported that they like to believe in the information provided by unknown users than their family and friends because they think the users had their own personal experiences with those destinations and they do not want others, with whom they do not have any trading relationship, to be misguided. Although the majority of the individual Chinese travellers feel that information posted on WeChat are valuable for them to make their travelling decision, some feel that the information is sometimes deceiving and does not showcase the authentic pictures of the destination. Therefore, it is important that the individual travellers investigate about their travelling destination through multiple sources and validate the information before making plans for travelling. Additionally, the Chinese individual travellers suggested following the verified accounts only to avoid confusions about the authenticity of the information posted. A verified account is recognized by WeChat as official and includes high quality, creative and authentic contents. A few Chinese travellers use the feature 'BuzzFeed' in WeChat to filter the message they want to receive daily. As Wu (2014) has found, WeChat improves the wellbeing of the social community by providing selective information that is visually powerful and savouring to its users. However, it is also equally important that such information should come from authentic sources for users to consider it trustworthy and effective (Lien & Cao, 2014).

Apart from making payment through and finding a destination from the WeChat, the Chinese individual outbound travellers also frequently use WeChat searching for promotions and discount offers for a destination. Many Chinese travellers reported that they simply make their travelling decision to a particular destination after finding a promotional and or discount offer posted in the WeChat by a certain company. In general, these promotional or discount offers could be for a flight or ferry ticket to a destination or for the hotel booking and travel packages. Some individual Chinese travellers reported that they consider these offers as an opportunity not to be missed. Rather than directly posting a promotional offer, companies often include them at the end of an article that is written about a destination or a region. These articles are also usually posted in the official 'Subscription account' of companies. Therefore, a key way to attract a reader and motivate them to keep reading is by making the article interesting and useful for them. Most individual Chinese travellers felt that the articles they usually read in the subscription account are compelling enough to hold them until the end when they finally discover the promotional offer. Therefore, as revealed by Kim, Lee, Shin, and Yang (2017), a key component that binds the social media users and communities together is its ability to provide rich and complete information in interesting ways. Such interesting information provided in WeChat through various ways including "Subscription accounts" and "Moments" among others affect users emotionally as well as cognitively and thus motivate them to make their travelling decisions.

However, some companies do not require the users to go through the articles to discover the discount and promotional offer. If one is already aware of a company or companies that usually provide such offers, they can just click the bottom-menu in their official account, after which, they will receive push notifications from those companies including discounts and offers. Although not very popular and common, some Chinese individual travellers also take benefits of WeChat coupons shared by friends or sometimes companies too through their official accounts. Additionally, some companies present discounts and promotional offers in a unique, emotional and engaging ways by interacting with their users, asking them questions or seeking their opinions and ultimately presenting their promotional offers.

5. CONCLUDING REMARKS, IMPLICATIONS AND DIRECTION FOR FUTURE RESEARCH

Technology and social media have not only facilitated our everyday activities, but they have also naturally become a part of everyday lives for most people including tourists. Therefore, information and knowledge about how technologies and social media influence tourists' travelling decisions are crucial not only from the standpoint of destination management for national tourism organizations (NTOs) but also for the tourism industry stakeholders and managers who aim to promote their products and services through those social media. More particularly, social media platforms such as WeChat which has penetrated the community through all ranges and walks of lives are extremely important to examine as they reflect people's personal choices and preferences which national tourism organizations and private sector tour and travel agencies can capitalize to attract the market such as the individual Chinese travellers which have been growing tremendously across the world. As examined in this chapter, WeChat has been phenomenal in shaping individual Chinese travellers' travelling decisions by providing illuminating information through attractive means including texts, messages, pictures and videos in their interactive platforms and features including Moments and official accounts. WeChat also allows users to integrate their payment system easily to their bank accounts in China and make payment through their smartphones, and obtain discounts and promotional offers in a variety of ways. The Chinese individual travellers find this information and services fascinating and useful and take advantage of them in making a decision of their next travelling decisions.

The findings of this chapter have a number of important implications for the travel and tourism service providers and managers as well as national tourism organizations who aim to extend their marketing campaign to China focusing on the growing number of individual travellers. First, this chapter has demonstrated that the individual Chinese travellers have two distinct characteristics: one that resembles the feature of a traditional Chinese tourists who like to travel to a comfortable environment and community that are more favourable towards the Chinese people, and the other that is more sophisticated and wish to maintain its own choices and preferences and look forward to travelling in exotic and adventurous trips. Because of the duality of these characteristics of the Chinese individual travellers, the marketing organizations need to be particularly careful about promoting destinations and regions that appeal to this group in social media. Second, the individual Chinese travellers like to avoid complications in their payment system and prefer to use handy tools and platform such as WeChat Pay function to carry out their payment tasks. Additionally, they prefer to interact with the Chinese people to inquire about their planned destinations. Therefore, it is essential for the destination marketing organizations and travel tour operators to consider these aspects and include in their promotional efforts how they resolve these demands of the Chinese travellers. Third, it is evident that the Chinese individual travellers are not only interested in seeing pictures and reading articles and comments posted in 'Moments' and 'Official Subscription accounts' of WeChat, they are also equally attracted towards emotional interactions and engagement in the social media. Therefore, it would be beneficial for the tourism service providers to design their websites in a way that it not only provides information but also give ample opportunities for the potential travellers to interact with the service providers as well as other users with whom they can also clarify their doubts, ask questions, and obtain answers which are not already posted in the social media.

This chapter has attempted to unleash the importance of examining the influence of WeChat in making outbound travelling decisions. The scope of the research was limited within one social media platform: 'WeChat'. Considering the importance of social media in shaping people's choices including those of travellers in the present day, it is imperative that the future researchers explore how other social media platforms including Facebook, WhatsApp, Instagram, Twitter, Google+ etc. among others also influence traveller's decisions and make a holistic comparison which can eventually help develop an overarching theory. Such a theory will immensely helpful for the travel organizers and national tourism organizations to target this emerging group of free individual travellers and promote their destinations effectively. Furthermore, such research could also be conducted among a number of other countries to obtain cross-sectional data that can be further analysed using quantitative approaches to generalize the findings.

REFERENCES

Arita, S., Edmonds, C., Croix, S. L., & Mak, J. (2011). Impact of Approved Destination Status on Chinese travel abroad: An econometric analysis. *Tourism Economics*, *17*(5), 983–996. doi:10.5367/te.2011.0076

Chen, Y.-R. R. (2017). Perceived values of branded mobile media, consumer engagement, business-consumer relationship quality and purchase intention: A study of WeChat in China. *Public Relations Review*, *43*(5), 945–954. doi:10.1016/j.pubrev.2017.07.005

Chiu, C., Ip, C., & Silverman, A. (2012). Understanding social media in China. *The McKinsey Quarterly*, *2*, 78–81.

Choi, S., Lehto, X. Y., & Morrison, A. M. (2007). Destination image representation on the web: Content analysis of Macau travel related websites. *Tourism Management*, *28*(1), 118–129. doi:10.1016/j.tourman.2006.03.002

Dai, B., Jiang, Y., Yang, L., & Ma, Y. (2017). China's outbound tourism–Stages, policies and choices. *Tourism Management*, *58*, 253–258. doi:10.1016/j.tourman.2016.03.009

DSEC. (2017). *Tourism Statistics*. Macau: Documentation and Information Centre of Statistics and Census Service.

Guangrui, Z. (1989). Ten years of Chinese tourism: Profile and assessment. *Tourism Management*, *10*(March), 51–62. doi:10.1016/0261-5177(89)90034-4

Han, W., McCabe, S., Wang, Y., & Chong, A. Y. L. (2018). Evaluating user-generated content in social media: An effective approach to encourage greater proenvironmental behavior in tourism? *Journal of Sustainable Tourism*, *26*(4), 600–614. doi:10.1080/09669582.2017.1372442

Jørgensen, M. T., King, B. E., & Law, R. (2018). The impacts of China's policy-making and legislation on outbound tourism–perspectives from long-haul intermediaries. *Journal of Policy Research in Tourism, Leisure & Events*, 1–13. doi:10.1080/19407963.2018.1505104

Kang, M., & Schuett, M. A. (2013). Determinants of sharing travel experiences in social media. *Journal of Travel & Tourism Marketing*, *30*(1-2), 93–107. doi:10.1080/10548408.2013.751237

Kim, S.-E., Lee, K. Y., Shin, S. I., & Yang, S.-B. (2017). Effects of tourism information quality in social media on destination image formation: The case of Sina Weibo. *Information & Management*, *54*(6), 687–702. doi:10.1016/j.im.2017.02.009

Lien, C. H., & Cao, Y. (2014). Examining WeChat users' motivations, trust, attitudes, and positive word-of-mouth: Evidence from China. *Computers in Human Behavior*, *41*, 104–111. doi:10.1016/j.chb.2014.08.013

Lien, C. H., Cao, Y., & Zhou, X. (2017, March). Service quality, satisfaction, stickiness, and usage intentions: An exploratory evaluation in the context of WeChat services. *Computers in Human Behavior*, *68*, 403–410. doi:10.1016/j.chb.2016.11.061

Liu, H., Wu, L., & Li, X. R. (2019). Social Media Envy: How Experience Sharing on Social Networking Sites Drives Millennials' Aspirational Tourism Consumption. *Journal of Travel Research*, *58*(3), 355–369. doi:10.1177/0047287518761615

Men, L. R., & Tsai, W.-H. S. (2013). Beyond liking or following: Understanding public engagement on social networking sites in China. *Public Relations Review*, *39*(1), 13–22. doi:10.1016/j.pubrev.2012.09.013

Munar, A. M., & Jacobsen, J. K. S. (2014). Motivations for sharing tourism experiences through social media. *Tourism Management*, *43*, 46–54. doi:10.1016/j.tourman.2014.01.012

Peltier, D. (2018). Independent Chinese Travelers Close Gap With Tour Groups. Skift. Retrieved from https://skift.com/2018/10/29/independent-chinese-travelers-close-gap-with-tour-groups/

Tencent. (2019). WeChat is a way of life. Retrieved from https://weixin.qq.com/

Tham, A., Croy, G., & Mair, J. (2013). Social media in destination choice: Distinctive electronic word-of-mouth dimensions. *Journal of Travel & Tourism Marketing*, *30*(1-2), 144–155. doi:10.1080/10548408.2013.751272

The World Bank. (2018). China. Retrieved from https://data.worldbank.org/country/china

Travel China Guide. (2019). 2018 China Tourism Facts & Figures. Retrieved from https://www.travelchinaguide.com/tourism/2018statistics/

Travel Lifestyle Network. (2018). Using WeChat to attract more Chinese travellers. Retrieved from https://travellifestylenetwork.com/news/using-wechat-to-attract-more-chinese-travellers/

World Tourism Organization. (2017). *Penetrating the Chinese Outbound Tourism Market – Successful Practices and Solutions*. Madrid: UNWTO.

Wu, J. (2014). How WeChat, the most popular social network in China, cultivates wellbeing. *Master of Applied Positive Psychology (MAPP) Capstone Projects*. Retrieved from http://repository.upenn.edu/mapp_capstone/65

Wu, J. (2018). A surprising number of countries now accept WeChat Pay or Alipay. Techinasisa. Retrieved from https://www.techinasia.com/surprising-number-countries-accept-wechat-pay-alipay

Xu, J., Kang, Q., Song, Z., & Clarke, C. P. (2015). Applications of mobile social media: WeChat among academic libraries in China. *Journal of Academic Librarianship*, *41*(1), 21–30. doi:10.1016/j.acalib.2014.10.012

ADDITIONAL READING

Amaro, S., Duarte, P., & Henriques, C. (2016). Travelers' use of social media: A clustering approach. *Annals of Tourism Research*, *59*, 1–15. doi:10.1016/j.annals.2016.03.007

Chung, N., & Koo, C. (2015). The use of social media in travel information search. *Telematics and Informatics*, *32*(2), 215–229. doi:10.1016/j.tele.2014.08.005

Kiráľová, A., & Pavlíčeka, A. (2015). Development of social media strategies in tourism destination. *Procedia: Social and Behavioral Sciences*, *175*, 358–366. doi:10.1016/j.sbspro.2015.01.1211

Lisha, C., Goh, C. F., Yifan, S., & Rasli, A. (2017). Integrating guanxi into technology acceptance: An empirical investigation of WeChat. *Telematics and Informatics*, *34*(7), 1125–1142. doi:10.1016/j.tele.2017.05.003

Minazzi, R. (2015). *Social media marketing in tourism and hospitality*. Springer. doi:10.1007/978-3-319-05182-6

Öz, M. (2015). Social media utilization of tourists for travel-related purposes. *International Journal of Contemporary Hospitality Management*, *27*(5), 1003–1023. doi:10.1108/IJCHM-01-2014-0034

Qin, B., Strömberg, D., & Wu, Y. (2017). Why does China allow freer social media? Protests versus surveillance and propaganda. *The Journal of Economic Perspectives*, *31*(1), 117–140. doi:10.1257/jep.31.1.117

Wang, X. (2016). *Social media in industrial China*. London: UCL Press. doi:10.2307/j.ctt1g69xtj

KEY TERMS AND DEFINITIONS

Approved Destination Status (ADS): Approved Destination Status (ADS) is an arrangement between the Chinese government and another country, that lets Chinese holiday travellers visit a country in a tour group.

Authenticity: The quality of being real or true.

Distribution Channels: Network used to get a product or services from the manufacturer or creator to the end user or customer.

Individual Visit Scheme (IVS): Tourism liberalization measure adopted by the Chinese government to allow travellers from Mainland China to visit Hong Kong and Macau on an individual basis.

Instant Messaging: A real-time, text-based communication similar to chat that uses a shared software client between or among two or more people using personal computers, phones or other communicating devices.

Outbound Tourism: The activity of travelling to a foreign country by the residential citizen

Social Media: Websites and computer programmes that allow people to communicate and share information on the internet using a computer or mobile phone.

User-Generated Contents (UGC): Any form of content such as video, blogs, discussion form posts, digital images, audio files, and other forms of media created by consumers or end-users of an online system or service and is publically available to other consumers and end-users.

WeChat: A Chinese multi-purpose messaging, social media and mobile payment app developed by Tencent.

Chapter 5
Deepening the Use of Social Media and Tourism Travel Behaviour

Cláudia Ribeiro de Almeida
https://orcid.org/0000-0002-0790-3152
University of Algarve, Portugal

Célia M.Q. Ramos
https://orcid.org/0000-0002-3413-4897
University of Algarve, Portugal

Maria del Mar Alonso Almeida
Autonomous University of Madrid, Spain

Paula Odete Fernandes
https://orcid.org/0000-0001-8714-4901
UNIAG, Instituto Politécnico de Bragança, Portugal

Lucas Estrada Gamarra
University of Algarve, Portugal

ABSTRACT

The Internet has transformed people's daily lives, not only in the search of information but above all giving the possibility of buying several products and services in a quick and easy way. Social networks have emerged as the quintessential medium for sharing experiences and communicating tourist information to all players in the tourism distribution chain. Today, we can sell, search, consume, connect, and share experiences related to our travel or services on different social media platforms and some online communities that share the same interests. The focus of this chapter is to point out the use of social media along the travel process in order to understand the behaviour of tourists before, during, and after trip. Having in mind the travel decision-making process the authors prepared a questionnaire with several questions spread through Facebook and answered by 95 people. The authors present the results and main conclusions in the chapter.

DOI: 10.4018/978-1-7998-1947-9.ch005

INTRODUCTION

The Internet and all the applications that have sprung up within it have helped to change the paradigm of travel for both tourists and holidaymakers (Xiang, 2018; Xiang, Magnini & Fesenmaier, 2015), changing the challenges associated with buying and advertising tourism products.

The new paradigm contributed to the disruption in the communication and marketing of tourism products (Vargas-Sanchez, 2019), making it felt at all stages associated with making a trip, but with more impact when deciding what should be part of the trip through of your purchase.

Information and communication technologies are essential today for those who want to travel, visit a place and learn what the tourist destination offers its visitor. In addition, within ICT, social media has emerged as one of the most useful and widely used platforms for travellers to consult and learn from other tourists' views on the most visited, most recommended locations and the reasons for this recommendation with a view to support the decision making process (Femenia-Serra, Neuhofer, & Ivars-Baidal, 2019).

During the travel decision making process (Xiang & Gretzel, 2010), people use social media for different propose, as for example to collect and share information about places, services, to make comments about different realities but above all to be in contact with others. Besides the importance that social media have at this moment, there is a limited prior research investigating how the enterprises use it to influence the consumers' decision-making process.

In this chapter, the authors focus their attention in the travel decision-making process and above all in the way that a person use the social media during this all process.

Authors divide this chapter into four parts. First part focuses on the literature review that discuss some findings related to travel behaviour and travel decision-making process. Second part present the methodology used to design the proposed questionnaire and the explanation of the application through Facebook. Third part present the results of the questionnaire and the fourth part the main conclusions and some information regarding future challenges and lines of research.

1. LITERATURE REVIEW

1.1 Social Media in Tourism Sector

Social media is part of our daily life and many persons use it to search, to share and be in contact with others, and as stated by Litvin, Goldsmith and Pan (2008), transformed the nature of communication among people, particularly travellers. Munar and Jacobsen (2014) states that social media are increasingly relevant as part of tourism practices affecting destinations and businesses. Nezakati et al. (2015) refer that recently social network websites have played a significant role in marketing and tourism promotion in the travel industry worldwide.

Social media creates opportunities for tourists to participate in the production and consumption of travel experiences, the so-called "prosumers" by Sigala, Christou and Gretzel (2012, p. 7), which allow a more active relationship between industry, destination and traveller. According to Mangold and Faulds (2009), internet-based social media has made possible for one person to communicate with hundreds or even thousands of other people about products and the companies that provide them. The same authors consider social media as hybrid element of the promotion mix as it enables companies to talk directly to their customers as well as customers directly to one another.

Social media platforms allow tourists to digitise and share online knowledge (Buhalis & Law, 2008; Volo, 2010), real-time recording and sharing (Munar, 2010) emotions and experiential moments (Jacobsen & Munar, 2012), offering new ways of social interaction with more than the small group of persons belonging to the family and friends regular circle.

The term social media normally relates to some internet-based applications that provide information about products and services and help the easy interaction between travellers that share their comments about different kind of experiences during their travel process (Leung, Law, Van Hoof & Buhalis, 2013; Lund, Cohen, & Scarles, 2018). Mangold and Faulds (2009) refer that social media encompasses a wide range of online, word-of-mouth forums, including blogs, company-sponsored discussion boards and chat rooms, consumer-to-consumer e-mail, consumer product or service ratings websites and forums, blogs and social networking websites, and some others. According to Litvin, Goldsmith, and Pan (2008) these interactions can bring to discussion some personal experiences, a new way of "electronic word-of-mouth" through internet-based technology where both consumer and producers share comments and characteristics of some good or service.

The increasing popularity of social media change the way we use it in our daily life as well in the way the enterprises do business and communicate their products and services (Leung et al., 2013). Ayeh et al. (2012) refer that with the emergence of social media, consumers can now actively use different platforms like consumer review sites, blog and social networking sites, among others, to create online content and share experiences about product and services.

Bizirgianni and Dionysipoulou (2013) state that the new media introduce new possibilities providing new opportunities and challenges for those who are involved in the tourism sector. The same authors refer that for the first time, the consumer-traveller participates actively, not only as tourists who make use of tourist product but also given to him the opportunity to report his opinion or express a need or a desire.

Social media currently represent a dynamic online travel information source that can influence travel consumers (Antunes, Ramos & Sousa, 2018; Ayeh et al., 2012; Choi, Hickerson, & Kerstetter, 2018; Chung & Koo, 2015; Munar & Jacobsen, 2014; Sigala et al., 2012; Xiang & Gretzel, 2010) and above all the decision making process. Yoo and Gretzel (2012) defend that given the experiential nature of tourism, the information created by other travellers is even more important and influential in travellers' information search and decision-making processes (Hendry, 2000; Chung, & Koo, 2015).

During the last couple of years, some researchers focus their attention in social media and their impact in some areas of the hospitality and tourism sector (Leung et al., 2013; Moro & Rita, 2018; Narangajavana Kaosiri, Callarisa Fiol, Moliner Tena, Rodriguez Artola, & Sanchez Garcia, 2019; Xiang, Du, Ma, & Fan, 2017; Xiang & Gretzel, 2010). According to Ayeh et al. (2012) the literature on social media from supplier's perspective is limited as it failed to offer insights into how hospitality and tourism enterprises exploit social media in assisting customers to plane their travel and thereby generate more business for their respective companies.

In the literature, it is possible to find several researchers that present different taxonomy of social media, based in different focus or dimensions. From the several researchers the ones that present an approach more related to our study is Kaplan and Haenlein (2010) that classify social media into six types: (i) blogs; (ii) social networking (e.g. Facebook); (iii) virtual social worlds (e.g. Second life); (iv) collaborative projects (e.g. Wikipedia); (v) content communities (e.g. YouTube) and (vi) virtual game worlds (e.g. World of Warcraft). However, these authors neglect newer types of social media such as microblogging and location-based services (Fotis, Buhalis, & Rossides, 2011). The different types of social media allow the online users to be involved in various online activities classified by Shao (2009) as (i) consuming; (ii) participating and (iii) producing contents online. Yoo and Gretzel (2012) refer that only a small number of users creates online contents including text, images, audio and video, as most of the people prefer to consume the contents shared by others.

According to Statista (2019a), the number of social media users worldwide in 2018 reach 2,62 billion. The Statista estimates that in 2022 this number can be 3,06 billion users. Social networking is one of the most popular online activities with high user engagement rates and expanding mobile possibilities. Facebook is leading social network worldwide with 2.320 million active users in May 2019, YouTube is in the second position with 1.900 million active users. Whatsapp have around 1.600 users and Facebook Messenger 1.300 active users (Statista, 2019b).

1.2 Travel Decision-Making Process

Using as reference the classical five-stage consumer decision-making model proposed by Engel, Miniard and Blackwell (2006) many researchers looked in deep to travel decision making process in order to understand the different steps pre-trip, during trip and after trip, considered in several reference studies (Gyimóthy, 2013; Panwar, Anand, Ali, & Singal, 2019; Smallman, & Moore, 2010; Wolny, & Charoensuksai, 2014).

The all process starts when a traveller recognises the need to travel. Normally the search for information arise to a set of alternatives and chosen process, followed by the travel behaviour and post-trip evaluation (figure 1).

Besides the information that several web-based interfaces offer and the easy way to look for information and some comments or shared opinions, Sigala (2011) refer that enterprises must ensure that web applications provide adequate functional support to create value for their customers and meet their needs along the decision-making process. Sigala, Christou and Gretzel (2012) states that social media used during all stages of the travellers' decision-making process, and travellers realize that they have a need for social media in the tourism industry.

Fotis, Buhalis and Rossides (2011) mention that within the context of tourism, decision-making related to holiday travel purchases is a complex information intensive exercise involving not only one, but a series of purchases that are associated to the composite and experiential nature of tourism product. As a result, social media become even more important to tourism sector, as travellers share both positive and negative experiences through review sites, posting comments, photos and sometimes videos.

Cox et al. (2009) attempted to provide a comprehensive view of the role of social media in the travel planning process. They found that travellers use social media after choosing their holiday destination rather than when trying to narrow down their choices of destinations or when deciding on the destination. During and after the trip social media use was found very limited.

To analyse the questions in deep, the authors use most of the articles of the literature review to collect data and information that allow the design of a questionnaire which help to understand this process. Authors explain the methodology to prepare this questionnaire is next section of the chapter.

1.3 The Importance of Social Media in the Travel Decision-Making Process

Word of mouth is relevant when people go for is travelling, mainly because people want to know some details or information about the place and know some of their relatives and friends experiences (Alonso-Almeida & Almeida, 2016; Jalilvand, & Samiei, 2012; Litvin et al., 2008; Litvin, Goldsmith, & Pan,

Figure 1. Travel decision-making process
Source: Adapted from Engel, Miniard and Blackwell (2006) and Fotis, Buhalis and Rossides (2011)

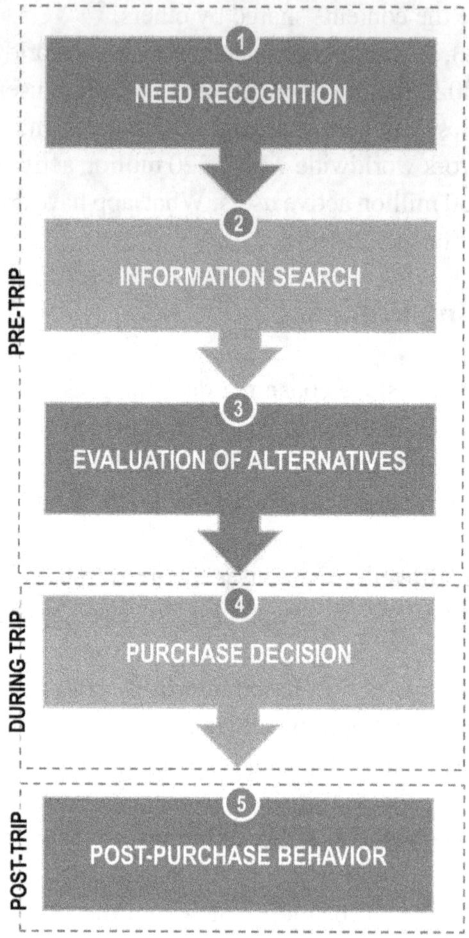

2018). Albeit that internet and social media are replacing traditional sources of travel recommendations (Alonso-Almeida & Almeida, 2016; 2018). One of main reason is that social media allow travellers to explore many more alternatives in trip planning decisions than do the traditional information sources such as brochures or travel agents (Huang et al., 2017).

Thus, it seems clear than internet and social media have a strong impact in travel behaviour. In fact, despite the important role of internet, social media is disrupting and transforming travel behaviour. Social media is transforming in an essential source for attracting potential tourists to destinations (Lee, 2017). Therefore, the influence of social networks in leisure travel increases the need for destination and tourism industry to re-appraise their competitive strategies (Phillips et al., 2015) because the recommendation to visit a destination seems to be a drive factor for go and return (Meleddu et al., 2015).

Regarding social media type, Trip Advisor was the most used social media network to look for information, followed by Instagram, Facebook and Twitter in the past years (Alonso-Almeida & Almeida, 2016; Banerjee, & Chua, 2016.). However, that trend is always changing and there are no evidences that only one social media is the leader for tourism or travel at this moment.

There are some interesting examples, like for example Instagram, that gain some market share in the last months (Alonso-Almeida & Almeida, 2018; Jovicic, 2019) or for example WhatsApp (Ascaniis, & Cantoni, 2016), especially to plan trip and during trip in order to share information or photos with travel mates. WhatsApp is an active way to contact with family and friends during the trip and to share information and media information while Facebook, Twitter and Instagram is a passive way in the sense that traveller posts the information without expecting for feedback.

Alonso-Almeida and Almeida (2016; 2018), found that social media is very popular before and after the trip, but not during the trip. During the trip, the use of WhatsApp is more usual to plan activities and get in touch with the travel group, as well as with family and friends to maintain relationships with strong ties (Alonso-Almeida & Almeida, 2016; 2018).

Alonso-Almeida and Almeida (2016; 2018), found that social media' influence in travel plans during the trip is low, creating none or just little changes in the original travel plans. Authors found that social media influence more the travel plans before the trip.

These findings emphasize that social media is a new competitor for traditional sources of travel recommendations, like a travel agency, being considered as a substitute product that is replacing the traditional mode of traveling and visiting places and destinations. French, Luo and Bose (2017) state, this happens mainly among the digital natives that use social networking technology. It is clear that social media supposes a potential threat to the future of travel agencies in their different ways. The long-term effects could be the use of alternative means for planning travel activities and accommodations by future customers.

As Kourouthanassis et al. (2017, p. 20) explain, "Travellers seeking information pertaining to a particular destination tend to initially get informed through information aggregators, which outline the highlights of the destination under scrutiny and, then, confirm or expand their information space through reviewing other travelers' experiences". These authors advocate by a hybrid profile in travel behavior composed by traditional and new way in travel decision-making process.

2. METHODOLOGY

In order to evaluate the behaviour in social media of a tourist before, during and after trip and using as reference the travel planning process proposed by Engel, Miniard, and Blackwell (2006) and Fotis, Buhalis and Rossides (2011) and the work of Alonso Almeida and Almeida (2018), the authors prepared a questionnaire with 45 questions that cover the overall process.

With several subareas, the questionnaire present questions aligned with the process of the travel planning and all the steps regarding the search for information along the trip. The authors used the Google docs to prepare the questionnaire.

The link to answer the questionnaire was active in Facebook between 3rd of May and 2nd June 2019, with 95 answers received.

The research methodology followed in this work consisted of a quantitative investigation, through the application of online questionnaire surveys, which was based on a systematic process of data collection, that can be observed and quantifiable, presupposes the observation of phenomena, the formulation of hypotheses, the control of variables, the selection of random sampling, the verification or rejection of hypotheses subject to statistical and / or mathematical analysis (Carmo & Ferreira, 1998).

In the present work the methodology followed was constituted by a set of steps: (i) considering the literature review, the study objectives were identified, (ii) the elaboration of the questionnaire, in order to investigate the proposed objectives, (iii) collection (iv) data processing and analysis with SPSS, which results will be analysed by the authors and the discussion of the results will be presented in the next section.

3. DISCUSSION OF THE RESULTS

Social media platforms offer a range of possibilities to the users, as for example sharing information about vacations, photos, videos or other type of contents (Munar & Jacobsen, 2014). To analyse in deep the use of social media before, during and after a trip, a questionnaire with 45 questions were prepared and shared in Facebook in order to collect primary data.

The questionnaire stayed active between May 3rd and June 2nd, 2019, collecting the answer of 95 respondents.

3.1 Sample Characterization

The respondents were mainly Portuguese (64,2%), female gender (69,5%), with 21 to 25 years old (44,2%), single (73,7%), having as educational background the college studies (57,9%). Around 29,5% have a social science and education occupation.

Regarding the trips, around 30,5% refer that made three to five trips in the last 12 months, mainly for international destinations (77,4%) that were reached in air transport (90%). During the trip these respondents used mainly the public transport (67,5%).

In the national trips the respondents use mainly the road transport to reach the destination and during the trip used a private car to travel within the destination.

Before the trip, 95,7% of the respondents search for information about the destination, through the internet (4,76 points out of 5), social media platforms (3,91 points) and travel blogs (3,49 points).

3.2 Information Search

The ones that look for information in social media platforms use Instagram (3,51 points out of 5), TripAdvisor (3,45 points) and Facebook (3,03 points). The information that they looked for were mainly the one related to the attractions of the destination (4,34 points), photos and videos (4,14 points), culture (4,07 points) and accommodation (3,94 points).

Only 22,8% of the respondents downloaded an APP (application) related to the destination before the trip, mainly APP's related to transports (96,4%).

During the trip the respondents used the social media from one to two hours (45,2%), through mobile devices (4,787 points out of 5). Regarding the use of WhatsApp, the respondents refer that the main motive was to contact family (4,77 points), friends (3,75 points) or travel mates (3,36 points). During the trip the respondents look for information in the social media related to attractions of the destination (3,81 points), how to reach the destination (3,55 points) and about services (3,50) or cultural details (3,46 points).

From all the respondents around 82,8% published information in social media during the trip mainly because they like to share photos and videos (3,84 points), information (3,66 points) or only because they like to help other travellers with their experiences (3,10 points). The published information was photos (4,25 points) and videos (3,19 points). The ones that mention that did not share any publications refer that it happens because they do not like to share travel information (3,16 points), opinions (3,98 points) or photos (2,66 points).

After the trip the respondents refer that they like to share travel pictures (3,87 points), travel experiences (3,54 points) or comments (2,87 points).

In relation to the classification of the use of social media before, during and after trip the respondents classify with 3,69 points the use of social media before the trip, with 3,32 points the use during the trip and with 3,48 points the use of social media after trip.

In the overall experience 67,7% of the respondents refer that due to social media they make some changes in the trip plans.

3.3 Use of Social Media During the Travel Phases

Given the age of respondents and the stages of travel, the use of social media is different for each segment considered. When preparing a trip, or the pre-trip phase, the information-seeking step is very relevant as it helps support the decision-making process about which tourist destination to visit and which tourist products to purchase and experience.

Information can be searched through different sources, depending on age, such as family, friends, travel blogs, travel agents, travel magazines, travel guides or social media, as shown in table 1.

In the phase Pre-trip, social media is the most widely used source of information for ages 18 to 30, ages 41 to 50. While for tourists between the ages of 31 and 40, the most considered source of information is Travel 'Blogs and for those over 50 prefer to consult the Travel's Guide. Young people aged 26 to 30, in addition to social media, also consult their friends for information.

With regard to the type of information sought in social media before beaming, the 18-30 age group and the 41-50 age group look more often for information on existing destination attractions; while from 31 to 40 years old and over 50 years old seek more about the culture associated with destiny. The information you are least looking for in social media before traveling for any age is about auxiliary services, as shown in table 2.

Table 1. Before the trip, did you look for information about the destination (1 it means "never" and 5 means "always")

Age	18 To 20 Years Old	21 to 25 Years Old	26 To 30 Years Old	31 To 40 Years Old	41 To 50 Years Old	Above 50 Years Old	Total
Social Media	**4,17**	**3,83**	**3,60**	3,78	**3,86**	4,25	3,91
Family	3,25	2,50	2,80	1,67	2,30	2,50	2,55
Friends	3,06	3,24	**3,60**	3,67	3,25	3,00	3,27
Travel's blogs	3,19	3,54	2,40	**4,00**	3,55	4,25	3,49
Travel's Agents	1,81	1,71	1,20	1,44	2,40	4,00	1,82
Magazines	1,63	1,91	1,40	1,67	1,82	3,33	1,83
Travel's guides	1,88	2,69	2,20	2,33	2,69	**4,50**	2,55

Source: Author's elaboration.

Table 2. What information did you search in social media before your trips (1 it means "never" and 5 means "always")?

Type of Information	18 To 20 Years Old	21 to 25 years Old	26 to 30 Years Old	31 to 40 Years Old	41 to 50 Years Old	Above 50 Years Old	Total
How to reach the destination	3,50	3,31	3,00	3,22	2,93	3,50	3,27
Services at the destination	4,17	3,58	3,00	3,89	3,54	3,00	3,68
Culture of the destination	4,24	4,13	3,40	**4,33**	3,86	**4,00**	4,08
Attractions of the destination	**4,58**	**4,37**	**4,40**	4,25	**4,14**	3,80	4,34
Accommodation	4,11	3,95	4,20	4,11	3,71	3,40	3,94
Timetables of transports	4,00	2,94	4,00	3,44	3,14	2,67	3,30
Maps of the destination	3,95	3,38	3,60	3,78	3,43	3,80	3,58
Street view to check some areas	3,89	3,61	3,40	4,00	3,31	3,75	3,66
Photos and/or video of the destination	4,41	4,34	4,20	**4,33**	3,43	3,25	4,15
Information about the language	3,47	3,19	3,40	2,44	2,31	2,20	2,99
Information about the currency and recommendations	4,12	3,56	3,80	4,11	2,77	3,60	3,62
Auxiliary Services	2,24	2,28	2,20	2,00	2,00	2,00	2,18

Source: Author's elaboration.

During the trip, when it comes to looking for information on social media, the 18-25 and 31-40 age group look more for information about the destination's attractions; while from 26 to 30 look for accommodations, from 41 to 50 for street view to check some areas, and finally the over 50 look for culture in the destination, as shown in Table 3.

It should be noted that there has been a change in interest expressed by the different types of information in the 26 to 30 age groups, which prior to travel favoured the search for information on attractions at the destination and during the trip looking for accommodation, which may mean that they buy the rest of the accommodation already during the trip, for tourists aged between 31 and 40 before the trip seek information about culture in the destination and about Photos and / or video of the destination and during the trip look for Attractions of the destination. Finally, tourists between 41 and 50 years prior to their trip seek information about Attractions of the destination and during Street view to check some areas, as presented in Table 2 and Table 3.

After the trip, the use of social media remains present in the lives of tourists, and differs from age to age. When returning to their home country, tourists between the ages of 18 and 50 like to share travel pictures on my Social Media network, while those over 50 prefer to share their social media experience. In the case of those between 26 and 30 years old, besides enjoying pictures, they also like to share the experience. Sharing opinions about the trip is of least interest to tourists, who have returned home, as shown in table 4.

In summary terms by age group, we can conclude that for any age the influence of social media has contributed to make few changes to my original trip plans because of what I have found in social media as shown in table 5.

Table 3. During the trip, how often did you look for this kind of information in social media (1 it means "never" and 5 means "always")?

Type of Information	18 To 20 Years Old	21 to 25 Years Old	26 To 30 Years Old	31 to 40 Years Old	41 to 50 Years Old	Above 50 Years Old	Total
How to reach the destination	3,69	3,66	3,40	3,89	3,00	3,40	3,56
Services at the destination	3,75	3,64	2,80	4,11	2,85	2,75	3,50
Culture of the destination	3,73	3,41	3,00	4,00	2,92	**4,00**	3,47
Attractions of the destination	**4,19**	**3,84**	3,60	**4,22**	3,08	3,80	3,81
Accommodation	3,50	3,00	**4,00**	3,56	3,00	2,80	3,20
Timetables of transports	3,53	3,17	3,60	3,44	2,33	3,00	3,16
Maps of the destination	3,40	3,50	3,20	3,33	2,62	3,40	3,31
Street view to check some areas	3,56	3,46	2,80	3,44	**3,08**	3,20	3,37
Photos and/or video of the destination	4,07	3,38	3,40	3,56	2,83	2,40	3,39
Information about the language	3,13	2,62	2,80	2,56	1,83	2,25	2,59
Information about the currency and recommendations	3,73	2,87	2,80	4,00	2,42	2,33	3,06
Auxiliary Services	2,27	2,16	2,00	1,89	2,00	1,33	2,09

Source: Author's elaboration.

Table 4. After the trip, how you use social media (1 it means "never" and 5 means "always")?

	18 to 20 Years Old	21 to 25 Years Old	26 to 30 years old	31 to 40 Years Old	41 to 50 Years Old	Above 50 Years Old	Total
I like to share my travel experiences with my Social Media network	4,11	3,50	**3,83**	3,56	2,64	**3,80**	3,54
I like to share travel comments and reviews with my Social Media Network	2,95	2,65	3,33	3,56	2,57	3,40	2,87
I like to share my travel opinions in my Social Media network	2,58	2,58	3,17	3,33	2,57	3,20	2,72
I like to share travel pictures on my Social Media network	**4,42**	**4,05**	**3,83**	**3,89**	**2,71**	3,60	3,87

Source: Author's elaboration.

Table 5. Did social media influence your trip plans?

	18 to 20 Years Old	21 to 25 Years Old	26 to 30 Years Old	31 to 40 Years Old	41 to 50 Years Old	Above 50 Years Old	Total
I did a lot of changes in my trip plans.	5,2%	2,5%	0,0%	0,0%	7,1%	0,0%	3,2%
I did make few changes to my original trip plans because of what I have found in social media.	63,1%	75,0%	66,6%	66,6%	57,1%	60,0%	67,7%
I did not make any change.	31,5%	22,5%	33,3%	33,3%	35,7%	40,0%	29,0%

Source: Author's elaboration.

4. CONCLUSION AND MAIN CHALLENGES FOR FUTURE LINES OF RESEARCH

Social media is gaining huge space in our society. People search, share, buy and interact with others and with enterprises in the same digital environment, receiving information from different sources that can be used to different proposes, being one of them the information collection regarding destinations, services and above all experiences.

Besides this importance and as stated by Xiang and Gretzel (2010), despite the phenomenal growth, there is a lack of empirical data describing the role of social media in the context of online travel information searches and above all in looking for the all process of decision-making when someone what do plan and buy a trip.

Our analysis in this article contributes to the existing literature as it analyses this decision-making process and presents the results of a questionnaire that helps the authors to collect valuable data to understand this phenomenon.

The results show that the interaction between the tourist and the social media is different along the travel-decision making process, with a different level of influence of the information collected and shared, which is different from age to age.

Considering the present study, and as a discussion of the results we can present as strategies for the tourist destinations to bet on the dissemination of the culture associated with the destination, the attractions, accommodations and bet on the dynamization of points of interest. interest through street view, such as attractions, hotels, cultural events.

For future research the authors want to collect data in different moments and above all in different countries in order to understand the differences between them, and also the evolution of travel behaviour and the new trends in terms of new social media and their importance for the respondents.

The proposed questionnaire could help to identify the behaviour in different steps of the decision-making process and add more knowledge not only in academic field but also for the enterprises that work in the tourism sector, as it explains searching and consuming behaviours before, during and after trip.

In the near future, the authors want to apply this questionnaire in different countries in order to create an international database concerning the use of social media along the decision-making process. The database could be useful to understand the main differences between countries and identify new trends that can explain some travel behaviours.

ACKNOWLEDGMENT

This chapter is financed by National Funds provided by FCT - Foundation for Science and Technology through project CINTURS (UID/SOC/04020/2019), CEFAGE (UID/ECO/04007/2019), UNIAG (UID/GES/04752/2019).

REFERENCES

Alonso-Almeida, M.M., & Almeida, C. R. (2016). Use of social networks on travel decision making. Social Networks and Travel Behaviour. *Proceedings of the 5MC & WGs meetings*, Bucharest. Academic Press.

Alonso-Almeida, M. M., & Almeida, C. R. (2018). Use of Social Media within tourism travel behaviour. *Proceedings of the Final Conference Social Networks and Travel Behaviour*, Milan. Academic Press.

Antunes, N. M., Ramos, C. M., & Sousa, C. M. (2018). Who are the tourists sharing content on social media? Behaviour and characteristics. *Journal of Spatial and Organizational Dynamics*, *6*(3), 237–257.

Ascaniis, S., & Cantoni, L. (2016). Information and Communication Technologies in Religious Tourism and Pilgrimage. *International Journal of Religious Tourism and Pilgrimage*, *4*(3), 2.

Ayeh, J., Leung, D., Au, N., & Law, R. (2012). Perceptions and Strategies of Hospitality and Tourism Practitioners on Social Media: An Exploratory Study. In M. Fuchs, F. Ricci, & L. Cantoni (eds). *Information and Communication Technologies in Tourism 2012*. Springer-Verlag. 10.1007/978-3-7091-1142-0_1

Banerjee, S., & Chua, A. Y. (2016). In search of patterns among travellers' hotel ratings in TripAdvisor. *Tourism Management*, *53*, 125–131. doi:10.1016/j.tourman.2015.09.020

Bizirgianni, I., & Dionysopoulou, P. (2013). The influence of tourist trends of Youth Tourism through Social Media (SM) & Information and Communication Technologies. *Procedia: Social and Behavioral Sciences*, *73*, 652–660. doi:10.1016/j.sbspro.2013.02.102

Buhalis, D., & Law, R. (2008). Progress in information technology and tourism management: 20 years on and 10 years after the Internet—The state of eTourism research. *Tourism Management*, *29*(4), 609–623. doi:10.1016/j.tourman.2008.01.005

Carmo, H., & Ferreira, M. M. (1998). *Metodologia de investigação. Guia para auto- aprendizagem*. Lisboa: Universidade Aberta.

Choi, Y., Hickerson, B., & Kerstetter, D. (2018). Understanding the sources of online travel information. *Journal of Travel Research*, *57*(1), 116–128. doi:10.1177/0047287516683833

Chung, N., & Koo, C. (2015). The use of social media in travel information search. *Telematics and Informatics*, *32*(2), 215–229. doi:10.1016/j.tele.2014.08.005

Cox, C., Burgess, S., Sellitto, C., & Bultjens, J. (2009). The Role of User-Generated Content in Tourists' Travel Planning Behaviour. *Journal of Hospitality Marketing & Management*, *18*(8), 743–764. doi:10.1080/19368620903235753

Engel, J., Miniard, P., & Blackwell, R. (2006). Consumer behaviour (10th ed.). Thomson/South-Western.

Femenia-Serra, F., Neuhofer, B., & Ivars-Baidal, J. A. (2019). Towards a conceptualisation of smart tourists and their role within the smart destination scenario. *Service Industries Journal*, *39*(2), 109–133. doi:10.1080/02642069.2018.1508458

Fotis, J., Buhalis, D., & Rossides, N. (2011). Social Media impact on holiday travel planning: The case of the Russian and the FSU markets. *International Journal of Online Marketing*, *1*(4), 1–19. doi:10.4018/ijom.2011100101

French, A. M., Luo, X. R., & Bose, R. (2017). Toward a holistic understanding of continued use of social networking tourism: A mixed-methods approach. *Information & Management*, *54*(6), 802–813. doi:10.1016/j.im.2016.10.006

Gyimóthy, S. (2013). Symbolic convergence and tourism social media. In *Tourism social media: Transformations in identity, community and culture* (pp. 55–71). Emerald Group Publishing Limited. doi:10.1108/S1571-5043(2013)0000018006

Hendry, J. (2000). Strategic decision making, discourse, and strategy as social practice. *Journal of Management Studies*, *37*(7), 955–978. doi:10.1111/1467-6486.00212

Huang, C. D., Goo, J., Nam, K., & Yoo, C. W. (2017). Smart tourism technologies in travel planning: The role of exploration and exploitation. *Information & Management*, *54*(6), 757–770. doi:10.1016/j.im.2016.11.010

Jacobsen, J. K. S., & Munar, A. M. (2012). Tourist information search and destination choice in a digital age. *Tourism Management Perspectives*, *1*(1), 39–47. doi:10.1016/j.tmp.2011.12.005

Jalilvand, M. R., & Samiei, N. (2012). The impact of electronic word of mouth on a tourism destination choice: Testing the theory of planned behavior (TPB). *Internet Research: Electronic Networking Applications and Policy*, *22*(5), 591–612. doi:10.1108/10662241211271563

Jovicic, D. Z. (2019). From the traditional understanding of tourism destination to the smart tourism destination. *Current Issues in Tourism*, *22*(3), 276–282. doi:10.1080/13683500.2017.1313203

Kaplan, A. M., & Haenlein, M. (2010). Users of the world, unite! The challenges and opportunities of Social Media. *Business Horizons*, *53*(1), 59–68. doi:10.1016/j.bushor.2009.09.003

Kourouthanassis, P. E., Mikalef, P., Pappas, I. O., & Kostagiolas, P. (2017). Explaining traveller online information satisfaction: A complexity theory approach on information needs, barriers, sources and personal characteristics. *Information & Management*, *54*(6), 814–824. doi:10.1016/j.im.2017.03.004

Lee, U. K. (2017). International Tourism Advertisements on Social Media: Impact of Argument Quality and Source. *Sustainability*, *9*(9), 1537–1555. doi:10.3390u9091537

Leung, D., Law, R., Van Hoof, H., & Buhalis, D. (2013). Social media in tourism and hospitality: A literature review. *Journal of Travel & Tourism Marketing*, *30*(1-2), 3–22. doi:10.1080/10548408.2013.750919

Litvin, S., Goldsmith, R., & Pan, B. (2008). Electronic word-of-mouth in hospitality and tourism management. *Tourism Management*, *29*(2), 458–468. doi:10.1016/j.tourman.2007.05.011

Litvin, S. W., Goldsmith, R. E., & Pan, B. (2018). A retrospective view of electronic word-of-mouth in hospitality and tourism management. *International Journal of Contemporary Hospitality Management*, *30*(1), 313–325. doi:10.1108/IJCHM-08-2016-0461

Lund, N. F., Cohen, S. A., & Scarles, C. (2018). The power of social media storytelling in destination branding. *Journal of Destination Marketing & Management*, *8*, 271–280. doi:10.1016/j.jdmm.2017.05.003

Mangold, W., & Faulds, D. (2009). Social media: The new hybrid element of the promotion mix. *Business Horizons*, *52*(4), 357–365. doi:10.1016/j.bushor.2009.03.002

Meleddu, M., Paci, R., & Pulina, M. (2015). Repeated behaviour and destination loyalty. *Tourism Management*, *50*, 159–171. doi:10.1016/j.tourman.2015.01.032

Moro, S., & Rita, P. (2018). Brand strategies in social media in hospitality and tourism. *International Journal of Contemporary Hospitality Management*, *30*(1), 343–364. doi:10.1108/IJCHM-07-2016-0340

Munar, A. M. (2010). Digital exhibitionism: The age of exposure. *Culture Unbound*, *2*(3), 401–422. doi:10.3384/cu.2000.1525.10223401

Munar, A. M., & Jacobsen, J. K. S. (2014). Motivations for sharing tourism experiences through social media. *Tourism Management*, *43*, 46–54. doi:10.1016/j.tourman.2014.01.012

Narangajavana Kaosiri, Y., Callarisa Fiol, L. J., Moliner Tena, M. A., Rodriguez Artola, R. M., & Sanchez Garcia, J. (2019). User-Generated Content Sources in Social Media: A New Approach to Explore Tourist Satisfaction. *Journal of Travel Research*, *58*(2), 253–265. doi:10.1177/0047287517746014

Nezakati, H., Amidi, A., Jusoh, Y., Moghadas, S., Aziz, Y., & Sohrabinezhadtalemi, R. (2015). Review of Social Media Potential on Knowledge Sharing and Collaboration in Tourism Industry. *Procedia: Social and Behavioral Sciences*, *172*, 120–125. doi:10.1016/j.sbspro.2015.01.344

Panwar, D., Anand, S., Ali, F., & Singal, K. (2019). Consumer Decision Making Process Models and their Applications to Market Strategy. *International Management Review*, *15*(1), 36–44.

Phillips, P., Zigan, K., Silva, M. M. S., & Schegg, R. (2015). The interactive effects of online reviews on the determinants of Swiss hotel performance: A neural network analysis. *Tourism Management*, *50*, 130–141. doi:10.1016/j.tourman.2015.01.028

Shao, G. (2009). Understanding the appeal of user-generated media: A uses and gratification perspective. *Internet Research*, *19*(1), 7–25. doi:10.1108/10662240910927795

Sigala, M. (2011). eCRM 2.0 applications and trends: The use and perceptions of Greek tourism firms of social networks and intelligence. *Computers in Human Behavior*, *27*(2), 655–661. doi:10.1016/j.chb.2010.03.007

Sigala, M., Christou, E., & Gretzel, U. (2012). *Social Media in Travel, Tourism and Hospitality: Theory, Practice, and Cases*. UK: MPG Books Group.

Smallman, C., & Moore, K. (2010). Process Studies of Tourists'decision-Making. *Annals of Tourism Research, 37*(2), 397–422. doi:10.1016/j.annals.2009.10.014

Statista. (2019a). *Number of global social media users 2010-2021.* Retrieved from https://www.statista.com/statistics/278414/number-of-worldwide-social-network-users/

Statista. (2019b). *Most famous social network sites 2019, by active users*. Retrieved from https://www.statista.com/statistics/272014/global-social-networks-ranked-by-number-of-users/

Vargas-Sánchez, A. (2018). The unavoidable disruption of the circular economy in tourism. *Worldwide Hospitality and Tourism Themes, 10*(6), 652–661. doi:10.1108/WHATT-08-2018-0056

Volo, S. (2010). Bloggers' reported tourist experiences: Their utility as a tourism data source and their effect on prospective tourists. *Journal of Vacation Marketing, 16*(4), 297–3111. doi:10.1177/1356766710380884

Wolny, J., & Charoensuksai, N. (2014). Mapping customer journeys in multichannel decision-making. Journal of Direct. *Data and Digital Marketing Practice, 15*(4), 317–326. doi:10.1057/dddmp.2014.24

Xiang, Z. (2018). From digitization to the age of acceleration: On information technology and tourism. *Tourism Management Perspectives, 25*, 147–150. doi:10.1016/j.tmp.2017.11.023

Xiang, Z., Du, Q., Ma, Y., & Fan, W. (2017). A comparative analysis of major online review platforms: Implications for social media analytics in hospitality and tourism. *Tourism Management, 58*, 51–65. doi:10.1016/j.tourman.2016.10.001

Xiang, Z., & Gretzel, U. (2010). Role of social media in online travel information search. *Tourism Management, 31*(2), 179–188. doi:10.1016/j.tourman.2009.02.016

Xiang, Z., Magnini, V. P., & Fesenmaier, D. R. (2015). Information technology and consumer behavior in travel and tourism: Insights from travel planning using the internet. *Journal of Retailing and Consumer Services, 22*, 244–249. doi:10.1016/j.jretconser.2014.08.005

Yoo, K., & Gretzel, U. (2012). Use and creation of social media by travellers. In M. Sigala, E. Christou & U. Gretzel (Eds.), Social media in travel, tourism and hospitality (pp. 189-205). United Kingdom: Ashgate.

KEY TERMS AND DEFINITIONS

Consumer Behaviour: This is the focus of studies of consumers and the processes they employ to choose, use (i.e., consume), and dispose of products and services.

Consumer Journey: The collection of several steps that constitute the entire experience a customer has while communicating with an organization or brand.

Social Media: These aggregates of online communications channels can be considered tools that can be used to define new business models strategically, taking into consideration analyses of community user-generated content and information shared with other members of these online communities.

Tourism Experience: This is a set of activities in which individuals engage on their personal terms, such as pleasant and memorable places, allowing each tourist to build his or her own travel experiences so that these satisfy a wide range of personal needs, from pleasure to a search for meaning.

Travel decision-making process: The process a traveller has to go through to purchase all the components of his trip, from finding destination information, selecting the most suitable products to his preferences, to the purchase decision process.

User-Generated Contents (UGC): Any form of content made by users of an online system or service that has been shared and is publicly available to other users.

Word-of-Mouth: Shared content on social media that conveys an opinion about a product, which becomes a recommendation for future users looking for testimonials about the same product to support them in the decision-making process.

Chapter 6
Impacts of Social Media on Business Value and Performance

Valentina Della Corte
https://orcid.org/0000-0001-7113-3167
University Federico II of Naples, Italy

Krishnan Umachandran
https://orcid.org/0000-0003-4934-4368
NELCAST, India

Fabiana Sepe
Federico II University of Naples, Italy

Giuliana Nevola
Federico II University of Naples, Italy

Amuthalakshmi Periasamy
Madras School of Social Work, India

ABSTRACT

The aim of this chapter is to study the main impacts of social media on business value and performance. ICT can be a resource in resource-based perspective, becoming itself a possible source of competitive advantage (in terms of performance and value creation). A proactively tailored organization strategy can drive the people, processes, and systems, harmoniously pulling all the stakeholders in unison. Social media can bring in the commitment of the workforce and ensure involved, flexible, innovative working practices, improving the quality of work. When properly used, they can build organizations to more flexible structures, sharing services, competitiveness and interdependencies. The business processes can be harmonized by standard operating procedures, automated systems of agreed quality, prime up the competency development and facilitate clear career paths. Refining suppleness to work in tune to market demands and supplementing employee satisfaction can foster the development of individual accountability leading to leadership evolution.

DOI: 10.4018/978-1-7998-1947-9.ch006

INTRODUCTION

In the last years, Internet and web services have contributed to improve business performance. Internet offers the opportunity to create social spaces where individuals can communicate and share information and knowledge through the social media (SM). Properly, the use of social exploded into its most decisive phase in 2004 with the arrival of Web 2.0. More specifically, this term was coined by Dale Dougherty of US publishing company O'Reilly Media and it was first used for the highly influential Web 2.0 conference run by the company in 2004 (Brown, 2009).

The term SM has a variety of definitions. Boyd & Ellison (2007) conceive the SM as "web-based services that allow individuals to build a public or semi-public profile in a social space where there are other users with whom they share a connection, and their list of connections and those made by others within the system". According to Kaplan & Haenlein (2010), SM refers to a group of applications based on the on-line connection of Internet networks where users can create and exchange contents. Berthon, Pitt, Plangger, and Shapiro (2012), consider SM as a series of both hardware and software technological innovations (Web 2.0) that facilitate creative online users' in expensive content creation, interaction, and interoperability.

SM, thanks to the widespread coverage of the web, facilitates and accelerates conversations and exchange of information with respect to traditional media, which offers "read-only" contents to consumers, not allowing them to be active part of the promotional process.

The use of these channels encourages the online learning, the sharing information and cooperation. SM allows users to communicate and share contents without any need to be physically present. Specifically, the use of SM positively impacts on the internal organization and on the relationship between the company and its external scenario, as well. As for the stakeholders, a pivotal role is played by consumers involved in a process of value co-creation. In particular, the increasingly widespread of digital technologies, such as SM, have changed the consumer–producer relationship dramatically. The main reason is strictly tied to the fact that these technologies have empowered consumers to create, collaborate, produce and contribute to commercialization by considerably lowering the cost of value co-creation (Rayna, 2010). In the actual context of the sharing economy, the boundaries between consumers and firms have become very blurred (Lan, Ma, Zhu, Mangalagiu, & Thornton, 2017).

Companies are increasingly using SM for organizational purposes, such as gaining competitive advantage (Grafström & Falkman, 2017), and this type of use is attracting increasing numbers of leaders (Abrahamson, 1991; see also Spicer, 2018).

Companies use SM to increase their brand image and awareness. SM influences company brand by word of mouth that attracts consumers and influences their purchase decisions (Nisar & Whitehead, 2016).

SM favours firms to obtain information and to acquire new Technology Knowledge competence. Through SM, managers and employees communicate faster and they learn deeply (Ellison, Duggan, Ellison, Lampe, Lenhart, & Madden, 2015). Developing technology competencies favours the creation of a virtual learning community that improves firm performances. Technology knowledge can be improved by increasing the connection among firm stakeholders, as suppliers, customers, distributors and logistic providers. Huge connection enables firms to acquire new knowledge, innovative opportunities and to gain a competitive advantage.

In conclusion, SM is considered as resource used by a company in combination with other resources to develop a set of capabilities, processes and strategies to create value. SM affects business performance on innovation process and production, knowledge management process, communication and promotion, customer relationship management, brand awareness, sales growth, e-commerce and social commerce (Rapp, Beitelspacher, Grewal, & Hughes, 2013)

Recent studies recognized that many potentially valuable resources are in the public domain (Luo & Buo, 2018). SM is an example of a resource that exists in the public domain and is not owned or controlled by the firm (Gao, Tate, Zhang, Chen, & Liang, 2018). We therefore ask the following question: how can SM resources be identified as strategic resources? In order to answer to this research question, we first define the theoretical framework, then, through the case study focused on the food and beverage industry, this chapter aims at studying the main impacts of SM on both business value and performance.

THEORETICAL FRAMEWORK AND CONCEPTUAL MODEL

The resource-based theory (RBT) and the capabilities-based perspective have been taken into consideration as the theoretical foundation of this book chapter. Both RBT and capabilities perspective support the assumption that performance derives from a firm's resource endowment and the firm's effectiveness at converting these resources into capabilities (Barney, 1991). Resources are a firm's assets, knowledge, and business processes used to implement a strategy. Capabilities, on the other hand, are conceived as an organization's ability to assemble, integrate, and deploy resources in combination to achieve a competitive advantage (Eisenhardt & Martin, 2000; Rapp, Trainor, & Agnihotri, 2010; Teece, Pisano, & Shuen, 1997; Krush, Agnihotri, Trainor, & Nowlin, 2013).

More specifically, while resources can be defined as single asset of an organization, capabilities can be intended as an internal process of the company, a bundle of actions to manage resources to create, produce, and offer value to a market.

Some scholars (Hooley, Greenley, Cadogan, & Fahy, 2005; Roberts, & Grover, 2012; Vorhies & Morgan, 2005) find that resources alone are not always sufficient to provide significant performance gains and, instead, have to be transformed into distinctive capabilities. Such findings suggest that improved performance occurs when distinctive capabilities are created by deploying technological resources together with other complementary organizational resources. Starting from this logic, SM technologies need to be integrated with Customer Relationship Management (CRM) processes to determine a firm-level capability that influences performance. Moreover, the extent to which SM are integrated throughout the organization will allow capability development and will improve the capability–performance relationship (Nevo & Wade, 2010; Trainor et al., 2013).

Customers have begun to adopt SM applications to connect with peers and nowadays they would like to have the same level of interactivity with their business counterparts (Berthon et al., 2012; Hanna, Rohm & Crittenden, 2011). The emergence of the so-called "social customer" is challenging practitioners and researchers to define what it means to manage customer relationships (Greenberg, 2010). In this term, the concept of CRM is wide and considers new capabilities enabled by the technological and social shifts brought by SM applications (Trainor, 2012).

According to the RBT and capabilities perspective, the attention shifts on some literature streams, focused on the explanation of how SM are considered strategic resources, able to create value and gain competitive advantage (See figure 1).

Figure 1. The impact of SM on firm value and firm performance (around here)
Sources: Our elaboration from Andriole (2010), Kietzman et al. (2011), Kane (2015), Wang and Kim (2016)

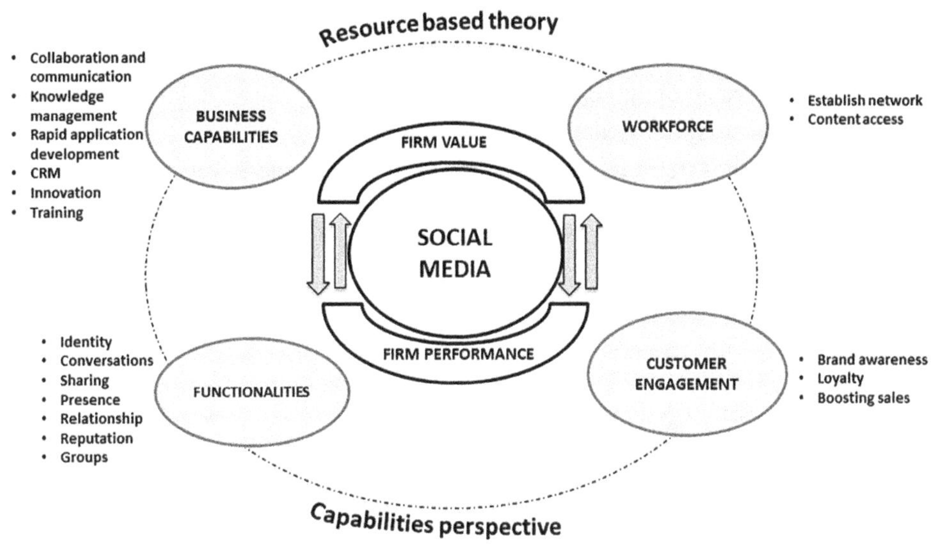

Most of papers related to the theme under investigation were collected from technical journals belonging to engineering and informatic sector in order to underline the contribution that SM can offer to different functions of a company in terms of workforce organization and training, and especially in terms of rapid application development. The literature related to the customer engagement is associated to management and marketing journals in order to offer a deeper understanding of the subject according to a managerial perspective.

SOCIAL MEDIA AND BUSINESS CAPABILITIES

Andriole (2010) identifies six business capabilities and explains how they are affected by SM:

- **Collaboration and Communication**: SM favours the communication within companies and their partners, supporting their internationalization and their market globalization process.
- SM favours the development of ability to coordinate conversations and to synchronize activities and projects. SM represents a tool to reach employees, partners and to monitor different connections within the companies and with the external environment.
- **Knowledge Management**: SM supports the knowledge management process, enabling the knowledge sharing. It represents a tool to transfer knowledge between the employees, clients and suppliers. SM allows to retrieve knowledge: user generated contents are stored, and they can be collected in any moments. And finally, it allows to improve the knowledge organization and exploitation.
- **Rapid Application Development**: Web tools represent an acceleration tool to create and develop applications, facilitating the connection among different players as customers, suppliers, company employees. SM encourages application development integrating third-party services and combining existing technologies and application into new businesses.

- **CRM**: SM facilitates firms to reach different market segments, to establish relations with them and to manage their data efficiently. Web technologies as forums, chats, wikis represent a useful tool to collect data referred to customers; they can represent an optimal CRM System to identify the customer profiles and, above all, their needs.
- **Innovation**: SMs, as open platforms, influence and accelerate innovation process speeding communication and favouring knowledge and data sharing.
- **Training**: SM influences training process, both information and knowledge are created by users, which share content, data and experiences. Training activities are not located in a specific facility and in a specific time frame, but they are provided by a series of technologies as: webinars, blogs, RSS filters, forums, wikis, podcast, videos, etc. These technologies reduce distances and costs as facilities, utilities, etc. (Mogos & Smits, 2013; Andriole, 2010).

SOCIAL MEDIA FUNCTIONALITIES

Kietzmann, Hermkens, McCarthy and Silvestre (2011), propose a theoretical approach (Honeycomb framework) based on SM functionalities. They identify 7 SM functionalities (identity, conversations, sharing, presence, relationship, reputation, and groups) and their impact on business capabilities.

- **Identity**: Users build their profile disclosing personal information as name, age, gender, profession, location and information that portray their identity. According to Kaplan and Haenlein (2010), the identity disclosure could be both conscious and unconscious. Many individuals that take part in web activities use their personal name or nickname. On the other side, the self-disclosure could be a result given from personal thoughts, pictures, feelings, “likes” and “dislikes”. An example is Facebook that can be a real timeline of the users, especially for those ones that are registered for a long time. Through SM users have developed identity strategy aimed to the self-promotion and self-branding promotion (e.g. LinkedIn).
- **Conversations**: Users, companies, customers discuss about a specific product, giving their opinions, evaluations through several SM, as Twitter and Facebook. Companies can monitor the conversation velocity in terms of rate and direction of change (McCarthy, McCarthy, Lawrence, Wixted, & Gordon, 2010).

The rate of change stands for the number of new conversations over a specific timeframe, and the direction of change means the continuity and discontinuity of a conversation.

In this context, conversations are like pieces of a continuous changing puzzle, once aggregated and combinate, produce an overall image or message.

- **Sharing**: it is the overall objective that links users to create networks and communicate. Every social platform has a common objective: e.g. through LinkedIn users can share their professional profile and brand corporate, YouTube allows users to share videos.

There are two fundamental implications for firms in terms of content sharing: on one hand, to know the common social objects of the users, on the other hand to identify new objects that can mediate shared interests.

- **Presence**: SM allows users to know if other users are available online, if they are in the real or virtual world. It is possible to verify the on-line status of a user in the virtual world, checking if it is available in chat room as Messenger, WhatsApp. In the real-life users are used to registering their local position in social platform to check where they are, allowing companies, friends and other actors to know their position in the real world. There are several functions of SM as Twitter, Tinder, Facebook that allow users to know if there are friends in close physical proximity.
- **Relationship**: by "re- late" means that users have some associations that leads them to converse, share objects of sociality, meet up, or simply just list each other as a friend. Therefore, how users of a SM platform are connected often determines the "what and how" of information exchange. In some cases, these relationships are formal, regulated, and structured.

In social platforms, the relations are often regulated: LinkedIn allows users to know how they are linked to others and the degree of relations.

Specifically, two properties are described to explain the relevance of a relationship trait.

The first one is the structural property that consists of several connections and their position in their network of relationship. A user with a grater and huger portfolio of relationship has a relevant position in the network and can be considered as an influencer.

The second one is the flow property that refers to the resources engaged in the network, how these resources are used and exchanged. Strong relations are defined long and lasting while weakness relations are "infrequent and distant". Another characteristic is the multiplicity of a relationships, when users are connected by more than one relationship (e.g. colleagues and friends).

- **Reputation**: It is the path through which each user can identify the social standing of each other. It is not easy to define what reputation is and how can be measured. It can be considered as the trustees among users and can be measured by the number of followers (e.g. in the restaurant sector a chef followed by millions of followers could be considered a good chef).

The reputation can be measured not only by the number of followers but also voting the user content with "likes".

User reputation is an important resource for firms to know trends, business performance, but it is important to identify the metrics and then choose the most appropriate evaluation tool.

There are SM services as Social Mention, a research platform, that collects user generated-contents from over 80 SM sites. This tool allows firms to measure how often they are mentioned using different metrics: strength (number of times that they are mentioned), sentiment (the ratio of positive and negative opinions), passions (how often users talk about a specific firm), and reach (the ratio between number of users talking about the firm and the total number of times in which the firm is mentioned).

- **Groups**: groups represent the way through which users can form communities and sub-communities.

Users can decide to divide the list of contacts in groups and to share with each group some specific objectives and personal information.

SOCIAL MEDIA AND WORKFORCE

One of the main successful resources a company can count on is represented by human resources. The success of any company depends on employee productivity. Digital literacy and internet connectivity are always more required mandatory skills.

According to the cognitive theory (Goldie, 2016), human resources require a constant updating of knowledge that is not only the result of a progressive process of knowledge acquisition, but it is also the result of a knowledge that is built in networking. A continuous learning process is given by the capability to be connected to digital communities, where ideas and interests are always shared.

Several studies (Pitt & Bennett, 2008; Van Zyl, 2009; Pettenati & Cigognini, 2007) have identified the benefits that SM offers in the workplace, such as improved communication channels, skills and knowledge sharing, channels for informal learning and job satisfaction improvement. (Munene & Nyaribo, 2018) built a framework that shows the effect of SM on employee productivity identifying negative effects such as; addiction, decreased presenteeism, work distraction, and absenteeism. Indeed, SM channels are used not only for job reasons but for personal reasons too, so if the utilization is not well managed, the SM can subtract time to working hours affecting the employee productivity in a negative way.

SM helps humans to satisfy specific connection needs, as the self- realization measured by the ability degree to be an "influencer", esteem needs measured by the numbers of "likes", and belonging needs measured by the number of followers.

SM era represents a shift from content owners to contributors of both ideas and contents. The knowledge is not assimilated anymore, but it's created together.

Knowledge sharing represents a way for a company to reduce costs in learning process and knowledge acquiring process.

SM allows companies to maintain personal and professional connections in an easier way. Managers, through SM, can better monitor employees and other stakeholders as customers, suppliers, shareholders.

Treem & Leonardi (2012) state that SM provides a list of hobbies and interests of workers that helps managers and recruitment departments to know more about their human resources and their behaviour in a given company. They support their theory that SM favours the visualization and monitoring of tasks, work behaviour, and the workflow of work-related activities. The access to worker profile can contribute to collaboration among workers, especially for new and part-time workers that have not a fully knowledge of their job.

According to Leonardi, Huysman, and Steinfield (2013), SM can be considered as web-based platforms serving goals in an organization. SM allows interpersonal communication between users and mass communication through sharing tools as videos, texts and pictures targeted to a mass audience and not a specific individual.

To ensure a competitive advantage, employees need tools to organize work around network communities and information aligned to them as well as ability to activate new connections. These technologies also recognize diversity of workers, contractors, consultants, partners and customers and how to share information across boundaries.

The framework proposed by Kane (2015) identifies two fundamental capabilities of SM:

- Establishing social networks among employees, managers and partners of the firm.
- Facilitating the content access within the firm.

Recently, companies are increasing the use of SM for internal collaboration, talent management and operational process.

In the real life, users are often unable to understand the structure of their social network; in this light, SM represents a useful tool to map the relations and understand the connection and to deeply identify the identity.

The access to digital content allows employees to share their contents and to store them to be searched in the future if it is needed. In this specific case SM represents a platform where employees can share their contents that are collected in the long-run and used to implement manual working.

Managers, usually act on the network boundaries, deciding who can join the platform and not. In a close platform the interaction is limited to a specific target (e.g. employees) to discuss specific internal organizational questions, while an open platform puts in connection internal employees and external business partners in order to create opportunities to work together and solve shared problems. Being characterized by transparency and network visualization, the use of SM offers an added value to employees. More specifically, transparency allows users to visualize the connections among users on a platform. It allows to create diverse networks that can be more valuable for sharing information. In addition, it can favour connection to bridge different organizational departments and remove redundant information.

With reference to network visualization, SM can provide visualization of the full network of interactions. Managers can get two benefits from that: to get a better picture of how employees interact and work together, and to promptly intervene where and if there are some difficulties.

The second social capability taken in consideration in Kalen's framework is related to the content access. Each SM influences the way people share and access to content. There are three main features through which social platforms effect user content: content support, digital trace and profile authenticity.

- Content support: every platform supports a specific type of content (tweet, videos, pictures, links, text, song, a combination of types).
- Digital trace: users leave intentionally information about their activities on platforms, such as how long a user has been member of a platform, which content has shared, commented, liked or viewed. Generally, digital traces allow firms to analyse data generated by SM. For example, the healthcare company Kaiser Permanente analysing customers' digital traces realized that the majority of their complaints was addressed to the parking area.
- Authenticity profile: it is the extension to which a user's online identity corresponds to his real life. Less anonymity entails accountability but can make user employee reluctant to share controversial ideas. A full anonymity makes users free to explain their idea, contributing to be more creative and honest but can lead to greater polarizing behaviour.

SOCIAL MEDIA AND CUSTOMER ENGAGEMENT

Nowadays, consumers are very familiar with SM, they interact with other individuals and firms, and create and share contents. Consumers use SM reviews from both organizations (provider-driven reviews) and other consumers (consumer-driven reviews) to take decisions (Ha & Lee, 2018).

Consumers are actively involved in the co-creation of their experiences with firms. This change in consumer behaviour pushes firms to implement new technologies to carry out and improve the interactions with customers.

In their model Wang & Kim (2017) explain the relationship between social CRM capabilities and customer engagement and the effect of social CRM capabilities on firm performance.

A sound knowledge of consumer behaviour can help the organization to sell ideas and concepts of social relevance.

The model defines the effect of customer engagement on firm performance and finally the moderating effects of SM usage on the relation between SM capabilities and firm performance. Buyer behaviour is influenced by motives such as rationale, emotional curiosity, attitudes, perception, social factors and personal factors.

SM capabilities engage customers in collaborative conversations. Through SM platforms, customers become "fans" and tend to be more loyal and committed to the company and to be more willing to be bombed by corporate advertising (e.g. newsletters).

Marketers when aware of social environment and internal personal interactions influencing the buyer behaviour, are very successful in today's environment.

One of the main social tools to engage customers is represented by the SM Brand Page, whose purpose is to involve customers to react and interact (e.g. linking, commenting, sharing). This tool, through the customer engagement, supports companies to improve their performance in terms of brand awareness, loyalty and boosting sales (Castronovo & Huang, 2012).

- **Brand Awareness**: SM allows firms to reach millions of people. Since people are SM users, the presence of brand's name on the social networks helps people to be reached and easily and speedily informed about the novelties a firm can introduce in its business
- **Brand Loyalty**: Brand pages represent a vehicle to achieve customer loyalty through the dynamic interaction with its fans improving the customer satisfaction and strengthening customer relations that are the main drivers to reach the customers retention (Gamboa & Gonçalves, 2014).
- **Boosting Sale**: The number of people that visit the brand page in a SM can favours the traffic for the web site and increases the online sale.

An example is given by the increasing number of food delivery apps where both restaurants and consumers interact each with other. After online orders, consumers are required to express their opinion about the received service and the food quality provided by restaurants. In this way, restaurants benefit more visibility and the increase in sales.

In this way the customer retentions increase, since they feel to be an added value for the company. Consumer socialization is an external influence affecting individual consumer's procurement and consumption of products, and their groups behaviour.

From an online brand perspective, the customer engagement is defined by as "behaviours that go beyond the transaction and may be specifically defined as a customer's behavioural manifestation that have a brand or firm focus, beyond purchase, resulting from motivational drivers" (Van Doorn et al., 2010 p. 254).

Different studies (Moro & Rita, 2018) evidence that customer engagement directly connected to CRM in terms of satisfaction, affective commitment and loyalty. Social and cultural characteristics of the consumers target and design effective marketing strategies for the specifically involved markets. Different cultures attach different meaning to things. Social contacts developed among likeminded parties gain priority over technical specifications bringing engaged and satisfied customers contribute to the brand dissemination.

An interesting example of how the interaction between firms and customers through SM applications can impact firm performance is provided using reputational channels in the food and beverage sector. Through these social channels consumers can write reviews and the number of online reviews has a significant positive impact on restaurant performance (Kim et al., 2016). Sites like Tripadvisor and Booking are the most popular and used by consumers in the hospitality and tourism sector. Twitter is considered the key tool to develop a word of mouth to advertise entertainment and tourist attractions. Instagram as a photo sharing application is playing a relevant role in the food & beverage industry where users and restaurants are used to sharing pictures of dishes and finally YouTube is used by operators and guide services (Moro & Rita, 2018). According to their previous studies Kim et al. (2016) state that an excellence certificate, a large number of reviews and a high rank provided by customers on SM platform (e.g. Trip Advisor) improve restaurant financial performance in terms of net sales, guest counts and average check. On the other side, the social platforms offer a series of services and advantages to restaurants and customers. Kapoor & Vij (2018) affirm that consumers are driven by offers and discounts, provided by online retailers. Online Food Aggregator (OFA) have alliances with affiliated restaurants offering them and advertisement space where restaurants can negotiate with OFA a discount-coupons to provide customers and allowing them to purchase on their mobile app. Consumers always expect to save money during their shop activities and the majority of them are college students that depend on pocket money of their parents. An extra discount is always appreciated and can be a useful tool to improve the customer retention too. Also, Zomato provides discount coupons on restaurants such as Dominos or Pizza hut on next subsequent purchase.

Zomato is a social network where the users share dish pictures taken in restaurants and leave opinions about their experiences. Through this channel, restaurants get different benefits such as more visibility, a targeted advertising, performance improvements in booking and presence management, in the customer service and engagement, and finally they get personal consumer data favouring the ideation of new products.

Moreover, the customer rating that includes the guest service, and the food quality offered influences restaurant performances. TripAdvisor offers a ranking service where companies, specially these belonging to tourism sector, benefit from the customer interviews. They can be evaluated in a rank from 1 star to 5 star and to get a certificate of excellence.

Restaurant managers with an excellence certificate must keep the level of customer comments high and positive to ensure that their restaurants are in the top line.

Different studies (Ha & Lee, 2018; Wang & Kim, 2017; Brodie, Hollebeek, Jurić, & Ilić, 2011) evidence that customer engagement is directly connected to CRM in terms of satisfaction, affective commitment and loyalty. Social and cultural characteristics of the consumers target and design effective marketing strategies for the specifically involved markets.

CASE DESCRIPTION

A Biryani Restaurant Chain from South of India Positions Itself Globally

In 2017, this restaurant Thalappakatti (https://thalappakatti.com/) has taken its version of biryani to the US. A familiar scent filled a new restaurant at McCarthy Ranch Plaza, in Milpitas, California. The South Indians gathered patiently inside, all hankering for a taste of home, took a deep appreciative breath. Practically every table ordered the same item off the menu: biryani chunky with tender mutton and sticky rice. Judging by SM reviews of the 400 people who ate on the first day, most were delighted about the launch of this American outpost offering typical Southern Tamil Nadu biryani, "Our favourite biryani" as one commentator put it on the restaurant's Facebook page. The growth has been phenomenal the restaurant has come a long way, from a tiny eight-seater restaurant in Dindigul to a high-powered chain of 42 restaurants serving about 0.35 million customers across the world every day. Things changed after Nagasamy took over the business. The customer is king, and customer feedback, says Nagasamy, is the ultimate driver behind business decisions (Kim, Li, & Brymer, 2016). Keeping his ear to the ground helped him learn the pulse of customers. "We are still using ancestral recipe, nothing has been changed... Well, except the method of cooking. We now make 1,500 kilos of biryani a day in the central kitchen in Chennai alone, and 2,000 kilos on weekends." Put together, all the outlets make about 25,000 biryanis a day, which, according to Nagasamy, makes them the country's largest biryani maker. He adds, "However, we still use the same spices: cinnamon, cardamom, cloves, all hand-ground on stones with a little water to maximize flavour."

To becoming a global, from a home-grown aroma in biryani to conquering food market across the world, no play of luck. "We easily adapted to the changing times," said Dhanabalan. The chain has a strong presence in SM networks such as Instagram to engage younger audience. This accidental entrepreneur Nagasamy Dhanabalan started from a humble beginning of jettisoning from being a receptionist to owning a global biriyani chain leader. It was not a planned process. I was planning to settle down in the UK, but my father was unwell, so I came to see him. While I was here, I wondered why we couldn't take this village food to Chennai. That was the idea, and slowly we journeyed from Dindigul to Chennai, says Nagasamy, a third-generation entrepreneur.

E-commerce in restaurant industry is becoming a more common feature in Indian food service industry. Online business model had become wide spread and accepted among the current trend generations of this technological era (Kapoor & Vij, 2018) Nagasamy Dhanabalan, Managing Director has relished the success with the technological upswing. He says that a unique recipe, special homemade spices and consistency helped them maintain the brand name through the years and played a huge role in its success.

Business was given an unexpected fillip by a rash of pretenders who used the chain's trademark name, and now-iconic logo of a turban, to advertise. He also claims that, the uniqueness that favoured them in the recent legal battle against other biryani outlets sporting similar-sounding names, is a milestone in the highly competitive business journey.

The company focused on increasing its online delivery share. Currently, close to 25 percent of sales come from online platforms. Players, such as Swiggy and Zomato, account for about 15 per cent and the firm's own delivery platform accounts for the rest. To increase the online share from our website, Uber eats also joined to supply online through its operators. Further to improve the online sales, the company has launched a mobile app, and its order online can be accessed through https://thalappakatti.com/order-online/.

Biryani is a popular rice dish, very nutritious as it is boiled and cooked with vegetables and meat (chicken or mutton), spiced and garnished with nuts and dried fruits. In the Indian city of Chennai, though Currently, there are over 300 brands including Anjappar, Buhari, Junior Kuppanna, Aasife Biriyani, Dindigul Thalappakatti and Behrouz as the branded or organised segment., the business is around 420 thousand of USD per annum per outlet, i.e. USD 126 million from this fully loaded non-vegetarian delicacies (Sangeetha, 2019). The market potential is very huge running to hundreds of millions in US dollars. For instance, as one of the brands Behrouz is valued at $525 million (Saritha, 2019).

The economic times states that Biryani has emerged as an unlikely champion of India's fast food scene, unseating the pizzas and burgers. Food delivery apps quote that among the most popular dish ordered and its demand had surpassed those of pizza. Swiggy found that while pizza was the most searched item around the country in 2017, biryani was the most ordered and its dominance is unchallenged. UberEats found biryani to be the dish that everyone wanted to start their New Year with. It was the most ordered dish on January 1. On Wimbledon and FIFA World Cup finals, Zomato found biryani to be the star performer (ET., 2018). (Figure 2)

Instances like these are the vanguard of a food delivery boom that is increasingly taking over established food chains by fly-by night operators and fake restaurants. After establishing themselves in the metros as the go-to apps for food ordering and delivery, Swiggy (www.swiggy.com) and Zomato have, for the past several months, begun pitched battles to win over the country's newest internet users residing in tier-II cities.

Expanding into a small town requires the food online outlets to offer laborious training to restaurants and delivery partners, later focuses on building scale in operations and increase the restaurant's reach to a larger base of consumers — including optimizing kitchens, resource planning, among others. In smaller towns, the biggest issue is that there are limited restaurants which means lesser supply. So, platforms like Swiggy partner with local brands, exclusively through Swiggy Access, which provides 10-12 kitchens a cooking space under one roof.

Zomato expanded beyond tier-I cities and convinced smaller cities to accept online food delivery, named as the "forward scout" and has launched services in close to 407 cities in June 2018. Zomato Discounts are merely a mechanism to encourage user participation and cannot explain the phenomenal demand witnessed in these cities.

Uber Eats are present in 38 cities with an increased focus on tier-II and beyond believe they still have a chance to compete with top players that are well-established in metros.

Unlike the e-commerce expansion to the hinterland, which was fraught with fears of sales cannibalization of the offline incumbents, food delivery start-ups' entry into smaller towns has been a trouble-free ride so far. As e-commerce has raised the business of prepared food delivery, at most restaurants in India, these prepared foods available for online ordering are unlimited and freshly cooked and distributed upon every order, has a great business potential still unleashed to the full extent.

Figure 2. Online order service provided by Thalappakatti restaurant

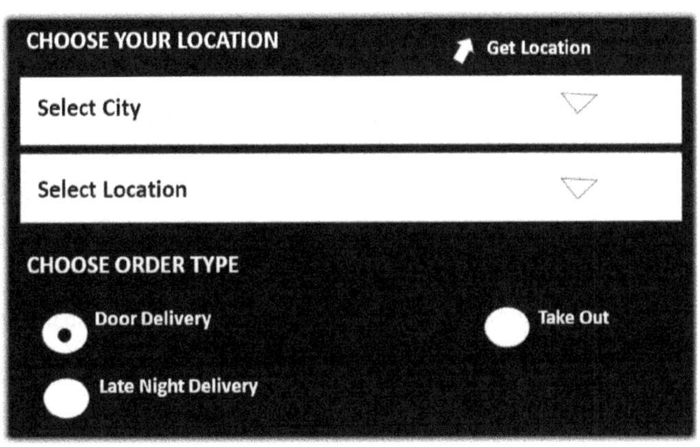

Case Analysis

Collaboration and communication: SM favour the communication within and among companies, supporting their internationalization and their market globalization through Customer relationship and friendly interdependence as social ties in the real world. Consumers seek more clarification through friends, as they would have a different perceptiveness and that such of their experience would benefit them during the business process.

"The company is focusing on increasing its online delivery share. By including online portals such as Swiggy, Zomato and Uber eats "

"Keeping his ear to the ground helped him learn the pulse of customers"

- **Knowledge Management**: SM supports the knowledge management process, favouring the sharing knowledge and creation. They represent a tool to transfer knowledge among the employees, clients, suppliers, as marketing response impacts the economics on needy consumer, and the long-term on value-orientations as strategies. The important trust factor is the consumer's behaviour being affected by informal discussion of products, services, or stores which results as an enjoyment of making the shopping process as a social fun and not as a necessary task.

"The customer is king, and customer feedback, is the ultimate driver behind business decisions"

- **Training**: SM allows the transfer of data and knowledge by several tools reducing the distance between professors and students using webinars, blogs, forums, podcast, tutorial channel. These tools are cheap and often they are free avoiding some relevant cost as location. SM and social class communication network are vital links which percolate the advantage of informal word-of-mouth communication and interpersonal relations between people on an informality associated commercial firm.

"Expanding into a small town requires the food online outlets to offer laborious training to restaurants and delivery partners"

Rapid Application Development: Web 2.0 represent an acceleration tool to create and develop applications, putting in connection different players as customers, suppliers, company employees. SM encourages application development integrating third-party services and combining existing technologies and application into new business. Involvement of consumers through SM affects and sways the consumer groups ultimately affecting products or services and its business.

"The company is focusing on increasing its online delivery share
to improve the online sales, the company has launched a mobile app"

Customer relationship management: SM as forums, chats, wikis represent an useful tool to collect data referred to customers, they can represent an optimal CRM System to identify the customer profiles, their needs, to solve customer issues, assisting customer during the use of the product, and reducing the distance among people to share information and experience on the product of service. CRM interacts with authenticity and genuinity for retaining a stable network with its customers.

"The chain has a strong presence in SM networks such as Instagram to engage younger audience"

- **Innovation**: SM, as open platform influences and accelerate innovation process speeding communication and favouring the sharing of knowledge and data, as a mass communication media leading to an impactful economic transaction. Openness in sharing of information, allows the organization to keep its ear on the feedback and course correct its process of business function.

"Unlike the e-commerce expansion to the hinterland, which was fraught with fears of sales cannibalization of the offline incumbents, food delivery start-ups' entry into smaller towns has been a trouble-free ride so far"

Web 2.0 is a read-write HTML page / XML / RSS, interactive and dynamic web which offers experience such as collaboration and sharing of information through other sites, and SM such as SM and online web communities etc., https://thalappakatti.com/ is already a web platform and connected with WhatsApp, Facebook, Twitter and Instagram. The payment is facilitated through net banking and card payments. Their order systems in addition to the restaurant services through various outlets, through web 2.0 services cater to door deliveries, takeaways and late-night deliveries as the city is economically busy even late nights. This website is compatible with all browsers and search engines.

Web 3.0 are websites created on RDF / RDFS / OWL, which are semantic with machine readability and interaction with other sites more intuitively with smart applications search functions with an intelligence that understands context rather than simply comparing keywords search for information using various search engines and offer satisfactory results. Thalappakatti has such plan in the future.

Web 4.0 is symbiotic internet of things with the entire web as operating system, in which the information flows and interacts humans with machines on an interspatial real-time. This is quite novel and may go with the course of business development at Thalappakatti and the infra structural changes that would affect the national wireless and network communication policies of the government (Choudhury, 2014).

MANAGERIAL AND THEORETICAL IMPLICATIONS

This study offers important theoretical implications to the RBT. Specifically, it extends previous studies using the RBT that have stressed the importance and value of relational resources and capabilities in supporting firm performance in international markets. Above all, through the case study, this chapter demonstrates that in SM context, resources can be an important vehicle to share informational, reputational and experiential contents.

As also demonstrated through the case study, SMs represent real strategic resources.

From a customer perspective, organizations should focus more on marketing strategies to strengthen the customer relationship built on SM, which allows more customer involvement in the firm value. Consumers seek more information from those friends, who have values or outlooks like their own. Formal social groups are more remote and serve a different function for the individuals. A person joins a formal social group to fulfil such specific goals for a specific cause. Being a member of a formal social group may influence a consumer's behaviour in several ways as they frequently jump into the opportunity on informal discussion of products, services, or stores

As for the human resource management, SM can be used to monitor employee connections, which are considered as a representation tool of identified profiles and their professional and personal skills which accesses to contribute in collaboration among co-workers, friends and their family members. Social class is connected on multiple variables and gets measured as a weighted function of one's occupation, income, wealth, education, status, prestige, etc. Thus, SM and social class as a communication networks have the advantage of the informal word-of-mouth communication and it is interpersonal in nature and takes place among people who are not directly associated with the firm.

Moreover, SM favours knowledge transfer, not only by reducing the distance among people through webinar, podcast and videos, but encourages knowledge through shared learning.

Knowledge sharing ensures that organizations do not incur extra costs reproducing existing knowledge. Opinion leaders through social networks, are activated greatly to reduce distance process for the products they have bought; may want to influence neighbours and friends. They involve themselves, to confirm their own judgment.

With specific reference to the case study, restaurant managers can obtain important advantages in terms of performance using SM tools.

Consistent with the importance of the customer engagement in terms of on-line review, restaurant managers should encourage customers to leave reviews sharing their feelings and dinner experiences implementing a customer incentive system. Such an example, managers can offer awards as membership points to customers that write reviews.

On-Iine ranking provided by SMs are very relevant tools useful to implement strategic decisions, managers should keep the tracking of their ranking in the market.

Technologies are useful to improve the operative efficiency of restaurants, e.g. waiters can use notepapers to easily find customers who have already ordered and, in this way, waiters will save time and increase the average check. Indeed, in the food and beverage sector, service efficiency is strictly connected to kitchen efficiency that is related not only to the quality of a meal, but also to the customer meal duration. Operative efficiency is positive associated with company performance, so restaurateurs must always introduce in a strategic way technology resource. According to RBT, technology resources always offer a huge contribution in terms of added value but they need to be strategically managed. In the food and beverage industry, the human resources constitute an important resource to create a friendly and warm environment to satisfy the customer experience so it is important to not replace the human staff with technologies, but they must be considered as complementary resources.

LIMITITATIONS, FUTURE RESEARCH DIRECTIONS AND CONCLUSION

Some limitations do not allow considering the conclusions of this chapter as universally valid or recognizable.

Despite an extensive literature review, the analysis is limited to the food and beverage sector, without any comparison to other sectors. Indeed, SM and web technologies can always be considered as strategic resources, but their propensity to reach a sustainable competitive advantage depends on the business areas where these resources are implemented.

Considering the web evolution from web 1.0 to web 4.0, it would be interesting to extend the research analysis to the contribution that Big Data collection can offer to the financial performance of restaurant. Nowadays, restaurants need to exploit Query System Research to profile their customers. Web-technologies will always have a huge impact on the food and beverage industry and big data offer a big contribute to restaurant marketers to strength the customer experience. Collecting data about customers' behaviour, location, order preferences and weekly routines favours a more personalized interaction with customers.

A great challenge for firms, above all those operating in the hospitality industry, as restaurants, consists in the involvement of consumers, through SM activities, in order to obtain their loyalty and promote the structure as the most welcoming possible. It is important to interact authentically and genuinely with customers, to build stable retention networks. The social networked consumers possess a wide range of information about many different types of products, retail outlets, and other dimensions of markets. They can initiate discussions with other consumers and respond to requests for market information and have accesses opinion suggestions from experienced users who have high levels of awareness and tend to influence beyond the realm of mere consumerism. Social networked consumers and prefer to consult one another 's friends and value the opinions expressed by them about new products, the quality differences among the different brands, store at which to purchase to experience shopping and so on.

REFERENCES

Abrahamson, E. (1991). Managerial fads and fashions: The diffusion and rejection of innovations. *Academy of Management Review*, *16*(3), 586–612. doi:10.5465/amr.1991.4279484

Andriole, S. J. (2010). Business impact of Web 2.0 technologies. *Communications of the ACM*, *53*(12), 67–79. doi:10.1145/1859204.1859225

Barney, J. (1991). Firm resources and sustained competitive advantage. *Journal of Management*, *17*(1), 99–120. doi:10.1177/014920639101700108

Berthon, P. R., Pitt, L. F., Plangger, K., & Shapiro, D. (2012). Marketing meets Web 2.0, social media, and creative consumers: Implications for international marketing strategy. *Business Horizons*, *55*(3), 261–271. doi:10.1016/j.bushor.2012.01.007

Boyd, D. M., & Ellison, N. B. (2007). Social network sites: Definition, history, and scholarship. *Journal of Computer Mediated Communication*, *13*(1), 210–230. doi:10.1111/j.1083-6101.2007.00393.x

Brodie, R. J., Hollebeek, L. D., Jurić, B., & Ilić, A. (2011). Customer engagement: Conceptual domain, fundamental propositions, and implications for research. *Journal of Service Research*, *14*(3), 252–271. doi:10.1177/1094670511411703

Brown, R. (2009). *Public Relations and the Social Web: How to use social media and web 2.0 in communications*. Kogan Page Publishers.

Castronovo, C., & Huang, L. (2012). Social media in an alternative marketing communication model. *Journal of marketing development and competitiveness, 6*(1), 117-134.

Choudhury, N. (2014). World wide web and its journey from web 1.0 to web 4.0. *International Journal of Computer Science and Information Technologies*, *5*(6), 8096–8100.

Duggan, M., Ellison, N. B., Lampe, C., Lenhart, A., & Madden, M. (2015). Social media update 2014. *Pew research center, 19*.

Eisenhardt, K. M., & Martin, J. A. (2000). Dynamic capabilities: What are they? *Strategic Management Journal*, *21*(10-11), 1105–1121. doi:10.1002/1097-0266(200010/11)21:10/11<1105::AID-SMJ133>3.0.CO;2-E

ET. (2018), Indians who order food online just can't get enough of this dish. Retrieved from https://economictimes.indiatimes.com/industry/cons-products/food/biryani-now-rules-indias-fast-food-landscape-and-heres-how-it-helped-emerge-a-lot-of-start-ups/articleshow/65273024.cms?from=mdr

Gamboa, A. M., & Gonçalves, H. M. (2014). Customer loyalty through social networks: Lessons from Zara on Facebook. *Business Horizons*, *57*(6), 709–717. doi:10.1016/j.bushor.2014.07.003

Gao, H., Tate, M., Zhang, H., Chen, S., & Liang, B. (2018). Social media ties strategy in international branding: An application of resource-based theory. *Journal of International Marketing*, *26*(3), 45–69. doi:10.1509/jim.17.0014

Goldie, J. G. S. (2016). Connectivism: A knowledge learning theory for the digital age? *Medical Teacher*, *38*(10), 1064–1069. doi:10.3109/0142159X.2016.1173661 PMID:27128290

Grafström, M., & Falkman, L. L. (2017). Everyday narratives: CEO rhetoric on Twitter. *Journal of Organizational Change Management*, *30*(3), 312–322. doi:10.1108/JOCM-10-2016-0197

Ha, E. Y., & Lee, H. (2018). Projecting service quality: The effects of social media reviews on service perception. *International Journal of Hospitality Management*, *69*, 132–141. doi:10.1016/j.ijhm.2017.09.006

Hanna, R., Rohm, A., & Crittenden, V. L. (2011). We're all connected: The power of the social media ecosystem. *Business Horizons*, *54*(3), 265–273. doi:10.1016/j.bushor.2011.01.007

Hooley, G. J., Greenley, G. E., Cadogan, J. W., & Fahy, J. (2005). The performance impact of marketing resources. *Journal of Business Research*, *58*(1), 18–27. doi:10.1016/S0148-2963(03)00109-7

Kane, G. C. (2015). Enterprise social media: Current capabilities and future possibilities. *MIS Quarterly Executive*, *14*(1).

Kaplan, A. M., & Haenlein, M. (2010). Users of the world, unite! The challenges and opportunities of Social Media. *Business Horizons*, *53*(1), 59–68. doi:10.1016/j.bushor.2009.09.003

Kapoor, A. P., & Vij, M. (2018). Technology at the dinner table: Ordering food online through mobile apps. *Journal of Retailing and Consumer Services*, *43*, 342–351. doi:10.1016/j.jretconser.2018.04.001

Kietzmann, J. H., Hermkens, K., McCarthy, I. P., & Silvestre, B. S. (2011). Social media? Get serious! Understanding the functional building blocks of social media. *Business Horizons*, *54*(3), 241–251. doi:10.1016/j.bushor.2011.01.005

Kim, W. G., Li, J. J., & Brymer, R. A. (2016). The impact of social media reviews on restaurant performance: The moderating role of excellence certificate. *International Journal of Hospitality Management*, *55*, 41–51. doi:10.1016/j.ijhm.2016.03.001

Krush, M. T., Agnihotri, R., Trainor, K. J., & Nowlin, E. L. (2013). Enhancing organizational sensemaking: An examination of the interactive effects of sales capabilities and marketing dashboards. *Industrial Marketing Management*, *42*(5), 824–835. doi:10.1016/j.indmarman.2013.02.017

Lan, J., Ma, Y., Zhu, D., Mangalagiu, D., & Thornton, T. (2017). Enabling value co-creation in the sharing economy: The case of mobike. *Sustainability*, *9*(9), 1504. doi:10.3390u9091504

Leonardi, P. M., Huysman, M., & Steinfield, C. (2013). Enterprise social media: Definition, history, and prospects for the study of social technologies in organizations. *Journal of Computer-Mediated Communication, 19*(1), 1–19. doi:10.1111/jcc4.12029

Luo, Y., & Bu, J. (2018). Contextualizing international strategy by emerging market firms: A composition-based approach. *Journal of World Business, 53*(3), 337–355. doi:10.1016/j.jwb.2017.01.007

McCarthy, I. P., Lawrence, T. B., Wixted, B., & Gordon, B. R. (2010). A multidimensional conceptualization of environmental velocity. *Academy of Management Review, 35*(4), 604–626.

Moro, S., & Rita, P. (2018). Brand strategies in social media in hospitality and tourism. *International Journal of Contemporary Hospitality Management, 30*(1), 343–364. doi:10.1108/IJCHM-07-2016-0340

Nevo, S., & Wade, M. R. (2010). The formation and value of IT-enabled resources: Antecedents and consequences of synergistic relationships. *Management Information Systems Quarterly, 34*(1), 163. doi:10.2307/20721419

Nisar, T. M., & Whitehead, C. (2016). Brand interactions and social media: Enhancing user loyalty through social networking sites. *Computers in Human Behavior, 62*, 743–753. doi:10.1016/j.chb.2016.04.042

Nyaribo, Y. M., & Munene, A. G. (2018). *Effect of social media predication in the workplace on employee productivity*. IJAME.

Peters, L. D., Pressey, A. D., & Greenberg, P. (2010). The impact of CRM 2.0 on customer insight. *Journal of Business and Industrial Marketing*.

Pettenati, M. C., & Cigognini, M. E. (2007). Social networking theories and tools to support connectivist learning activities. *International Journal of Web-Based Learning and Teaching Technologies, 2*(3), 42–60. doi:10.4018/jwltt.2007070103

Pitt, M., & Bennett, J. (2008). Workforce ownership of space in a space sharing environment. *Journal of facilities management, 6*(4), 290-302.

Rapp, A., Beitelspacher, L. S., Grewal, D., & Hughes, D. E. (2013). Understanding social media effects across seller, retailer, and consumer interactions. *Journal of the Academy of Marketing Science, 41*(5), 547–566. doi:10.100711747-013-0326-9

Rapp, A., Trainor, K. J., & Agnihotri, R. (2010). Performance implications of customer-linking capabilities: Examining the complementary role of customer orientation and CRM technology. *Journal of Business Research, 63*(11), 1229–1236. doi:10.1016/j.jbusres.2009.11.002

Rayna, T., & Striukova, L. (2010). Large-scale open innovation: Open source vs. patent pools. *International Journal of Technology Management, 52*(3), 477–496. doi:10.1504/IJTM.2010.035986

Roberts, N., & Grover, V. (2012). Investigating firm's customer agility and firm performance: The importance of aligning sense and respond capabilities. *Journal of Business Research, 65*(5), 579–585. doi:10.1016/j.jbusres.2011.02.009

Sangeetha. (2019, May 6). Tamil Nadu's biryani binge keeps cash registers ringing, *Life style & Food, The Hindu*. Retrieved from https://www.thehindu.com/life-and-style/food/tamil-nadus-biryani-binge-keeps-cash-registers-ringing/article25987062.ece

Saritha. (2019), Foreign investors show a growing appetite for Indian biryani. *Live mint*. retrieved from https://www.livemint.com/companies/start-ups/foreign-investors-show-a-growing-appetite-for-indian-biryani-1564637303890.html

Smits, M., & Mogos, S. (2013, June). The Impact Of Social Media On Business Performance. In ECIS 2013 Completed Research (p. 125). AIS Electronic Library.

Sophia van Zyl, A. (2009). The impact of Social Networking 2.0 on organisations. *The Electronic Library*, *27*(6), 906–918. doi:10.1108/02640470911004020

Spicer, J. (2018). 'That's their brand, their business': How police officers are interpreting County Lines. *Policing and Society*, 1–14.

Teece, D. J., Pisano, G., & Shuen, A. (1997). Dynamic capabilities and strategic management. *Strategic Management Journal*, *18*(7), 509–533. doi:10.1002/(SICI)1097-0266(199708)18:7<509::AID-SMJ882>3.0.CO;2-Z

Trainor, K. J. (2012). Relating social media technologies to performance: A capabilities-based perspective. *Journal of Personal Selling & Sales Management*, *32*(3), 317–331. doi:10.2753/PSS0885-3134320303

Treem, J. W., & Leonardi, P. M. (2013). Social media use in organizations: Exploring the affordances of visibility, editability, persistence, and association. *Annals of the International Communication Association*, *36*(1), 143–189. doi:10.1080/23808985.2013.11679130

Uber eats India. (n.d.). Retrieved from https://www.facebook.com/ubereatsindia/?brand_redir=696009167187917

Van Doorn, J., Lemon, K. N., Mittal, V., Nass, S., Pick, D., Pirner, P., & Verhoef, P. C. (2010). Customer engagement behavior: Theoretical foundations and research directions. *Journal of Service Research*, *13*(3), 253–266. doi:10.1177/1094670510375599

Vorhies, D. W., & Morgan, N. A. (2005). Benchmarking marketing capabilities for sustainable competitive advantage. *Journal of Marketing*, *69*(1), 80–94. doi:10.1509/jmkg.69.1.80.55505

Wang, Z., & Kim, H. G. (2017). Can social media marketing improve customer relationship capabilities and firm performance? Dynamic capability perspective. *Journal of Interactive Marketing*, *39*, 15–26. doi:10.1016/j.intmar.2017.02.004

KEY TERMS AND DEFINITIONS

Knowledge Sharing: Knowledge is the result of a sharing process. It is produced by users that distribute and exchange content (e.g. open platform as Wikipedia, blogs, WordPress).

Social Media: A set of web technologies aimed at favouring the connection within companies and with partners. It favours the knowledge management process and customer relation management.

User Generated Content: Any type of content, such as text, video, pictures, blog, Wikipedia content, that is created by a user and promoted on Internet through Social Media. Through SM there is a shift from content owner to content contribution.

Social Consumer Behaviour: Set of activities related no only to the transaction but related to the motivations that lead people to consume a specific brand.

Chapter 7
The Innovative Social Technologies and Their Impact on Knowledge Management Processes

Marica Mazurek
University of Zilina, Slovakia

ABSTRACT

Innovative social technologies, especially new communication technologies, allows for improving knowledge management processes. One of the managerial tools of communication in specific places are different e-technologies as well as artificial intelligence, which could be used in the consumer approach process as well as the supply side approach. The concept of ambient intelligence means a creation of the clusters of intelligence of high-tech environments—artificial intelligence, business modeling, and brand models application. This concept corresponds to the new age based on the "mobilization of ideas, knowledge and expectations" and technologies, which can be used also in branding, as for instance, experience economy concepts. The proposed chapter will deal with the concepts of the innovative social technologies and branding approaches in form of different models. This is a conceptual research approach, which can be applied in the exploratory study in several countries or used for specific research purposes.

THE OUTLINE OF A CHAPTER

The chapter´s goal is to give a reader a knowledge on the concepts of innovation applied in social technologies and how branding can be involved in this process. There exist different tools for this application, but first of all a reader has to be able to understand what the innovation is and how the theory of innovation has been created. Crucial is a symbiosis between knowledge, innovation and competitiveness, which will be explained in this chapter. The existing pressure to compete and win forces countries to apply

DOI: 10.4018/978-1-7998-1947-9.ch007

these new concepts into their development strategies and one of such area is also tourism, especially in tourism marketing, policies and branding. Branding is a concept known for couple decades, especially in product brand concepts, but also in tourism destinations as place branding. These concepts also need new communication approaches and understanding of the approach to branding from the consumer side, not only from the brand producers´ side. For this reason, we tried to underline the importance of business models, brand models (one has been selected on consumer side approach) and discussed the importance of the modern communication technologies, ambient intelligence for the improvement of knowledge management processes.

INNOVATION – A THEORETICAL FRAMEWORK

The theory of innovations is based on the Schumpeter's Theory of Economic Development created in 1911 (Schumpeter, 1934, 1939, 1943), but the first roots of innovation had started already in the industrial developments of the 18th and 19th centuries, especially in the ideas of classic economists Smith and Ricardo.

Smith criticized the mercantilist approach prioritizing production paradigms and believed that consumption is a driving force of any development and innovation in a society. This argument has been used in critiques of centrally planned economics and identified as one reason for their failure, which has been discussed by Dosi et al. (1988) and Hall and Williams (2008). A market economy and competition are widely seen as the most favorable environment for innovation. Governmental interventions in the form of protectionism (monopolies, oligopolies), unfair competition (tariffs, barriers, subsidies) constrain innovation and competition. For instance, tightly legislated entrepreneurs cannot stimulate effectiveness, efficiency, and productivity, and depend on public financing.

Nevertheless, Schumpeter, a representative of the Austrian school, was recognized as a creator of innovation theory. One of the first theorists of innovation, a French sociologist Gabriel Tarde, developed numerous important concepts and ideas on innovation. Tarde, similar to Schumpeter, agreed that entrepreneurs as individuals create change in a society, and for this reason, innovative processes are individually based (Pinchot, 1985). Thus, the role of consumers is not as important as the role of producers and individual entrepreneurs, especially inventors. Tarde and Schumpeter considered innovation to be a main source of wealth and a reason for the accumulation of capital. Sundbo (1998) added that both authors believed in "the inventions – the new ideas – solve problems and add elements that either make it possible to save on costs or constitute added value in a society" (p. 51). Despite disagreements of opinion especially concerning sources of wealth and wealth creation, and the role of individuals in a society, the first classic authors, i.e. Smith, Ricardo, Marx, Durkheim, and later Tarde, Schumpeter, made important contribution to development of innovation paradigms in a theory of changing economic cycles (Sundbo, 1998, p. 27).

Rooted in Tarde's belief that cycles are important in innovative processes, Kondratiev created a background for the innovation theory by stressing the importance of a four-stage (prosperity, recession, depression, recovery) economic cycles that begins again approximately every 45-60 years. Innovations occur most often in the period of recovery; as Schumpeter (1934) stated, as cited by Sundbo (1998, p. 40), "the innovations cluster together in some periods…boom periods".

The concept of innovation has been transferred from biology in Schumpeter's notion of economic development. Sundbo (2001, p. 22) explained that "the idea of evolution comes from biology, primarily from Darwinian theories… as the "survival of the fittest,"" means that innovation is required for continuous economic development and success and competitiveness depend on innovation and knowledge implementation. A competitive environment or changes force entrepreneurs or a society to develop and apply new ways of thinking and doing things.

Crucial is the growing impact of new strategic approaches by professional managers and the understanding of development of paradigms of competitiveness and innovation. These ideas depend on the philosophical thoughts and development in a society, economics, and especially changes. The continual development of human society and knowledge influenced views of innovation and competitiveness and confirmed symbiotic relationships; however, discrepancies also exist in both concepts, and will be examined in a tourism setting.

A SYMBIOSE OF INNOVATION, COMPETITIVENESS AND KNOWLEDGE

Innovative activities and competitiveness are influenced by knowledge and a knowledge economy. Competitiveness and innovation are interrelated in a symbiotic relationship, where in innovation is the core of destination competitiveness. The increase of competitive forces and demand to innovate is emerging in today's stressed global economy. Knowledge is an important competitive tool for innovation, in the new, post- capitalist society that is based on a knowledge economy, according to Drucker (1993). Metcalfe (2005, p. 23) stated that "modern capitalism is a particular kind of knowledge-based economic system…in which innovation, enterprise and competition are connected through systems of complementary market and non-market instituted frameworks." Codified material, for example written, or tacit knowledge (skills, general knowledge) and their transfer has incremental impact on innovation (Cooper, 2005; Malthora, 2002). Genuine local tacit knowledge could be a decisive point for tourism enterprises involved in the competitive process. In tourism, most "important seems to be the transfer of cultural knowledge" (Williams, 2007), for example cultural traditions based on history, heritage.

In competitiveness and in the deployment of existing comparative advantages, knowledge plays an important role as well, but the concept of novelty is not dominant. Innovations are similarly based on knowledge transformation, but the difference is a novelty of methods, ideas, and tools. In Latin, innovare means "to make something different" (Tidd et al., 1997). Fesenmeier and Zach (2009) support the idea of Schumpeter (1939), who argued that only completely new ideas qualify as innovations...and for this reason "the concept of innovation should incorporate incremental change" (Hage, 1999). Therefore, the critical distinctive element between innovation and competitiveness is in the character and originality of knowledge, its diffusion, and the impact of innovation.

Modern Innovation Theory is based on the industrial and technical innovation as has been mentioned by Sundbo (1998), by the authors of the "school of economical thinking" Dosi et al. (1988) and others." The domain of innovation and the source of innovation interested the authors Johannesson, Olsen, and Lumpkin (2001), of particular interest was identifying the originator of innovation and the beneficiary, for example a tourism destination, entrepreneurs, or consumers. The stress of a novelty can be a distinctive factor of the different perspectives of innovations. Schumpeter (1939) delineated a boundary between innovation and invention originating in the distinction between the use of results. Innovation is understood to be less important process of adaptation with a weaker practical use; however, inventions are more applicable, especially in a science.

A creator of concepts of invention and innovation, Tarde, similarly to Schumpeter, explained that inventions "are produced by individuals" and gave technological discoveries as an example of inventions. Tarde and Schumpeter believed in the extraordinary position of the individualistic entrepreneur, the inventor in the innovation process; however, Kanter (1983), Pinchot (1985) and especially Durkheim had doubts about the individualistic approach to invention. Sundbo (1998, p. 50) commented that "inventions are the result of combination of and internal personality factor (genius), and the external cause."

Later, Baumol (2002, p. 12) used similarly the association of entrepreneurs as the innovators; however, the perception of the innovator has been shifted from the independent person to a team of innovators. Thus, a role of the individual innovator has been dramatically changed. Society is incremental for the implication of inventions; it is a milieu in which inventions are implemented and innovative and competitive processes are socially grounded, and depend on the knowledge of humans as a part of the social environment. "Innovation is thus in principle a macrophenomenon, a matter of socioeconomic growth" (Sundbo, 1998, p. 1).

The influence and impact of innovations in a society is not identical and some innovations penetrate a society more radically: the influence of others may only be incremental. For example, in tourism a radical impact on marketing communication processes has come with the development of ICT technologies. Johanesson et al. (2001) associated revolutionary change with radical innovation, and the process of continual change and implementation of novelty considered as incremental innovation. Despite a more radical character, incremental innovations stem from market needs and are strategically easier to proceed with because a process of the implementation can be planned.

Innovative approaches of the incremental kind are used in tourism, for example, in the product mix strategies, tourism policies, branding, but some approaches are not a novelty. For instance, destination marketing organizations apply strategic partnerships cooperation, which is an effective management tool, but is not a brand-new approach. Hoelzl et al. (2005, p. 31) added that "innovation not only includes something new but also requires imitation to generate value for the destination", meaning that differences even between a type of innovations and its "newness". Abernathy and Clark (1998) recognized in their transilience model four sources of innovation, e.g., niche (which opens the opportunities for technologies), regular (incremental, revolutionary, and architectural (change of the system, entire industry). The architectural change can be explained as a change of tourism policy, especially new rules and an institutional novelty, which can significantly influence tourism.

A changing environment (social, political, economic, etc.) requires new products, processes, and technologies, which have been discussed by Tidd, Bessant and Pavitt (1977). The authors mention shortening product life cycles and replacement, innovation of products, and improvement of processes. One critical change in tourism, illustrated by how in consumer preferences in a destination choice, were depicted in Butler's lifecycle model. A change of consumers' preferences requires a change in tourism destinations development. A switch of preferences from the "first-generation destinations" with their typical high-society atmosphere and the fragmented tourism product offered in the past by individual tourism service providers to the mass "sand, sea, sun" tourism destinations (Buhalis, 2000; Poon, 1993, Claver-Cortez, Molina-Azorin & Periera-Moliner, 2007) and later to the re-adapted and more diversified destinations "tailored" to the needs of customers, exemplifies the type of changes that are fundamental to the innovative process.

Hjalager (1997) opened a discussion on the object of the innovation – was it a product or the processes, and Hjalager (1997, p. 36) defined these five innovation types: "product innovations, classical process innovations, process innovations in information handling, management innovations, and institutional innovations". The development of paradigms on innovation, changes in society and the environment have shaped the classification of the objects of innovation. For example, there has been a shift from the individualistic approaches to innovations to more mass-produced, Henry Ford thinking (Fordism) and rapid technological development, and later to increase intensification and growth of the service sector.

The typology has been taken from Schumpeter's framework, but adapted in tourism to deal with what is predominantly a service sector industry. Innovation concept has been neglected in tourism, especially in tourism services research, (Hall, 2008), (Hjalager, 2002) and Sundbo (2001) also admitted the limitation of the innovation research. For this reason, such academics as Hjalager (1997, 2002), Ioannides and Petersen (2003), Crozier at al. (1982), Naslund (1986), and Sundbo et al. (1996, 1997, 2007) started to focus on the innovations in tourism services. The development of a service sector in the post-Fordism period evoked a shift from predominantly technical and technological changes to managerial and organizational processes. The service sector became predominant in national economies (over 75% of the national product and employment is created through a service sector). The delivery of services has been influenced by their intangible character, which complicated their standardization. Problems with services and in the service sector highlight how different this sector is to be manufacturing sector. Services are offered directly to a consumer, and a consumer participation in service provision is required. The quality of services cannot be evaluated by a consumer in advance; thus, the provision of services in some cases is unpredictable, because services are not usually standardized or packed. Human factor and the skills of employees significantly influence service quality and customer satisfaction. The importance of knowledge and skills of employees and the organizational abilities used in a services sector has been confirmed by Jacob et al. (2003), Hipp et al. (2000), and Sirilli & Evangelista (1998).

Consumers' expectation and service providers' perceptions of what constitutes adequate service for the price paid are two different things. However, price might be a decisive factor in valuation of the quality of services. In tourism, quality of services has been standardized in the processes called benchmarking. Benchmarking has been used in both – private and public sectors as a tool of competitive advantage, as discussed by (Dorsch & Yasin, 1998; Lennon, Smith, Cockerrell & Trew, 2006). The method is based on continual innovation of products, services and work processes through evaluation and comparison with the best practices or leaders. In the case of benchmarking of tourism services, an innovative approach to service provision and competitiveness character are crucial elements in the improvement of service quality. Innovative management in tourism destinations means tailoring the destination tourism offer to the needs of consumers.

One problem in the process of innovation stems from the fact that a human factor is crucial, i.e. skills of employees and their ability to be up-to date, to be attentive to the changing consumer tastes and flexible enough to adapt quickly to meet those changes if they match the expectation of consumers with the latest branded tourism services offer.

Hall and Williams (2008, p. 18) called "competition, economic performance, demand- led innovation, technology, and firm-level strategy and resources" being drivers of tourism innovation. A symbiotic mutual and mostly positive relationship of competition and innovation is based on the pressure in tourism

Figure 1. Scheme 1: Keller´s CBBE model
Source: Keller (2003)

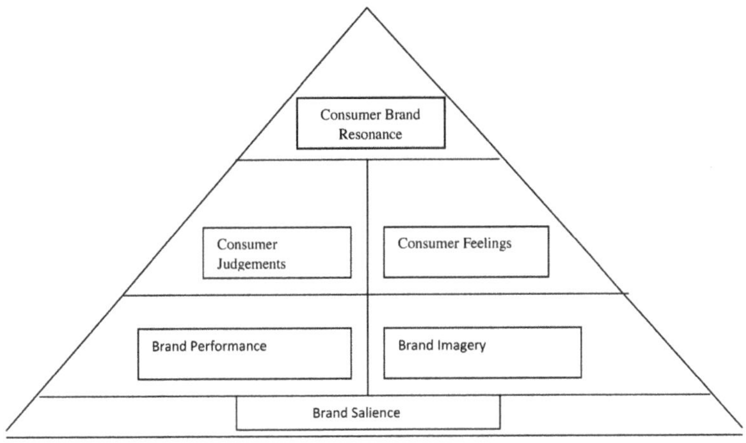

businesses to be more effective, efficient, and productive. Through the innovation comparative advantages can be utilized differently and the innovation offers a "first mover advantage" (Hall & Williams, p. 33). Some disadvantages are associated with innovative priorities in the tourism market, for example, risk, vulnerability, cost increase, new technology requirements, employees' attitudes, etc. Strategic approaches and innovative thinking influences differentiation and positioning of tourism destinations and are seen by Porter (2000a, p. 19) and Poon in the association to a success and competitiveness. The evolution of forces and paradigms shaping innovation and its impact on tourism competitiveness enables to understand the importance of innovation for competitiveness in tourism and one of the possible way is to understand clients well and use for their better understanding also some of the business models (part of the artificial intelligence approach) and the brand equity models focused on customers.

ONE OF SELECTED MODELS OF BRAND EQUITY - CONSUMER-CENTRIC MARKETING (KELLER´S CBBE MODEL)

A successful brand originates an emotion link between product and consumer, which is further developed by tools of marketing mix, especially marketing communication. Emotional attachment to tourism destinations can be fulfilled via attitudes attributes of the place, especially through the sophisticated organization of the tourism system in a place (place branding and management) and the management of brand attributes, for example, natural environment, history, culture, heritage as well as services offered in a tourism destination at the desired qualitative level. The multidimensional character of motivation of tourists seeking an experience discussed the authors Pyo, Mihalik, and Uysal (1989), especially the emotional attachment to a destination. Emotional bonds and their motivation are formed on the demand side; however, a tourism destination has to create favorable conditions to motivate and attract its visitors (supply side).

CONSUMER CENTRIC MARKETING AND THE IMPORTANCE IN THE INNOVATION PROCESS

Werther and Klein (1999) discussed new trends and shift toward user-centric models and the ambient intelligence. CRM (Consumer Relational Marketing System) is already used for the management, branding, consumer loyalty management, customer behavior monitoring, segmentation, brand awareness monitoring in business environment. The concept of ambient intelligence means a creation of the clusters of intelligence of high-tech environments (artificial intelligence, business modeling). This concept correspondents to the new age based on the "mobilization of ideas, knowledge and expectations" and technologies, which can be used also in branding, as for instance experience economy concepts discussed by Pine and Gilmore (1999). Diffusion and transformation of knowledge is crucial. Knowledge (tacit and codified) as well as cultural knowledge (history, traditions, and heritage) have an incremental impact on innovation and the quality of experiences. Experiences influence dramatically the visitors and their perception of a place, images and consequently have an impact on the reputation of place. Consumer centric marketing requires the centralization of information system, which allows to collect information on customers and to recognize their demand. The implementation of artificial intelligence and models into management practices of destinations could be desirable in a future to improve the satisfaction, reputation, competitiveness.

Moutinho, Rita, and Curry (1996) mentioned that artificial intelligence (AI) has scarcely touched today's managers; it will begin to have dramatic and widespread impacts on their activities over the next few years. Knowledge is an important competitive tool for innovation, in the new, post- capitalist society that is based on a knowledge economy. Drucker (1993, p. 38) stated that "knowledge is the only meaningful resource today"; however, this might be a little misleading, because it is not only about the possession of knowledge, but also the diffusion and use the knowledge for the company, or a region, country. Diffusion and transformation of knowledge is crucial. For this reason, it is sometimes important to focus not only on the supply side concepts, but also demand and consumers and their desire and especially implement in the new communication strategies the technologies based on the innovative principles.

CONCLUSION

Branding is a concept known for couple years, especially in product brand concepts, but also in tourism destinations as place branding. These concepts also need new communication approaches and understanding of the approach to branding from the consumer side, not only from the brand producers´ side. For this reason, we tried to underline the importance of business models, brand models (one has been selected on consumer side approach) and discussed the importance of the modern communication technologies, ambient intelligence for the improvement of knowledge management processes. Diffusion of knowledge is meaningful when it is used in accordance with a demand in society and with a sense of human needs and desires. For this reason, brand models might be helpful in some way to improve these processes and communication and the modern communication technologies are a best way to ease this process.

REFERENCES

Abernathy, W. J., & Clark, K. B. (1985). Innovation: Mapping the winds of creative destruction. *Research Policy*, *14*(1), 3–22.

Baumol, W. (2002). *The Free Market Innovation Machine: Analyzing the Growth Miracle of Capitalism*. Princeton, NJ: Princeton University Press. doi:10.1515/9781400851638

Buhalis, D. (2000). Marketing the competitive destination of the future. *Tourism Management*, *21*(1), 97–116. doi:10.1016/S0261-5177(99)00095-3

Claver-Cortes, E., Molina-Azorin, J. F., & Periera-Moliner, J. (2007). Competitiveness in mass tourism. *Annals of Tourism Research*, *34*(3), 727–745. doi:10.1016/j.annals.2007.03.010

Cleary, M. N., & Hobbs, G. D. (1983). The fifty year cycle: a look at the empirical evidence. *Long waves in the world economy*, 164-182.

Cooper, C. (2005). Knowledge management in tourism. *Annals of Tourism Research*, *33*(1), 47–64. doi:10.1016/j.annals.2005.04.005

Crozier, N. (1982). L'innovation dans les services.

Dorsch, J. J., & Yasin, M. M. (1998). A framework for benchmarking in the public sector. In J. J. Lennon, H. Smith, N. Cockerell, & J. Trew (Eds.), *Benchmarking National Tourism Organisations and Agencies – Understanding best practice*. Amsterdam: Elsevier. doi:10.1108/09513559810216410

Dosi, G. (1988). Technical Change and Economic Theory.

Drucker, P. F. (1993). *Post-Capitalist Society. Heinemann*. NY: Butterworth.

Fesenmaier, D. R., & Zach, F. (2009). Innovation in Tourism: The Case of Destination Marketing Organizations. *Ereview of Tourism Research*, *7*(1), 27–35.

Fesenmaier, D.R., & Zach, F. (2009). Innovation in Tourism: The Case of Destination Marketing Organizations.

Hage, J. T. (1999). Organizational Innovation and organizational change. *Annual Review of Sociology*, *25*(1), 597–622. doi:10.1146/annurev.soc.25.1.597

Hall, M. C., & Williams, A. M. (2008). *Tourism Innovation*. New York: Routledge. doi:10.4324/9780203938430

Hipp, C., & Grupp, H. (2000). Innovation in the service sector: The demand for service-specific innovation measurement concepts and typologies. *Research Policy*, *34*, 517–535.

Hjalager, A. M. (1996). Tourism and the Environment: The Innovation Connection. *Journal of Sustainable Tourism*, *4*(4), 201–217. doi:10.1080/09669589608667268

Hjalager, A. M. (1997). Innovation patterns in sustainable tourism: An analytical typology. *Tourism Management, 18*(1), 35–41. doi:10.1016/S0261-5177(96)00096-9

Hjalager, A. M. (2002). Repairing innovation defectiveness in tourism. *Tourism Management, 23*(5), 465–474.

Hjalager, A. M. (2002). Reparing innovation defectiveness in tourism. *Tourism Management, 23*(5), 465–474. doi:10.1016/S0261-5177(02)00013-4

Hoelzl, B., Pechlaner, H., & Laesser, C. (2005). Imitation processes of SMEs-A special form of innovation?

Ioannides, D., & Petersen, T. (2003). Tourism non-entrepreneurship in peripheral destinations: a case study of small and medium enterprises on Bornholm, Denmark. Tourism Geographies, 5(4), 408-435. In Hall, M.C. & Williams, A.M. (2008) (Eds.) Tourism Innovation. New York: Routledge.

Jacob, Tintoré, J., Aguiló, E., Bravo, A., & Mulet, J. (2003). Innovation in the tourism sector. Results from a pilot study in the Balearic Islands. *Tourism Economics, 9*(3), 279–295. doi:10.1177/135481660300900303

Johanesson, J. A., Olsen, B., & Lumpkin, G. T. (2001). Innovation as newness: What is new, how is new, and new to whom? *European Journal of Innovation Management, 4*(1), 20–31.

Johannesson, J. A., Olsen, B., & Lumpkin, G. T. (2001). Innovation as newness: What is new, how is new, and new to whom? *European Journal of Innovation Management, 4*(1), 20–31. doi:10.1108/14601060110365547

Kanter, R. M. (1983). The Change Masters.

Lennon, J. J., Smith, H., Cockerell, N., & Trew, J. (2006). *Benchmarking National Tourism Organisations and Agencies – Understanding best practice*. Amsterdam: Elsevier.

Malthora, Y. (2002). Why Knowledge Management System Fail? Enablers and Constraints of Knowledge Management in Human Entreprises. In C. Holsapple (Ed.), *Handbook of Knowledge Management (1)* (pp. 577–599). Heidelberg: Springer Verlag.

Metcalfe, J. S. (2005). Innovation, competition and enterprise foundation for economic evolution in learning economies. Manchester: University of Manchester, Centre for Innovation and Competition, Discussion Paper 71. In Hall, M.C. & Williams, A.M. (2008). *Tourism Innovation.* New York: Routledge.

Moutinho, L., Rita, P., & Cury, B. (1996). *Expert Systems in Tourism Marketing*. New York: Routledge.

Naslund, B. (1986). Financial Innovations. A comparison with R&D in Physical Products.

Pinchot, G. (1985). Intrapreneuring.

Pine, B. J., & Gilmore, J. H. (1999). *The Experience Economy: Work in Theatre and Every Business a Stage: Goods and Services are no longer enough.* Boston: Harvard Business School Press.

Poon, A. (1993). *Tourism, Technology and Competitive Strategies*. Wallingford: C.A.B. International.

Porter, M. E. (2000a). Location, competition, and economic development: Local clusters in a global economy. *Economic Development Quarterly, 14*(1), 15–34. doi:10.1177/089124240001400105

Pyo, S., Mihalik, B. J., & Uysal, M. (1989). Attraction attributes and motivations. A canonical correlation analysis. *Annals of Tourism Research, 16*(2), 277–282. doi:10.1016/0160-7383(89)90077-7

Schumpeter, J. (1939). Business Cycles.

Schumpeter, J. (1943). *Capitalism.*

Schumpeter, P. (1934). *The Theory of Economic Development.* Cambridge, MA: Harvard University Press.

Sirilli, G., & Evangelista, R. (1998). Innovation in the service sector. Results from the Italian Statistical Survey. *Research Policy, 27*, 881–899. doi:10.1016/S0048-7333(98)00084-5

Sundbo, J. (1996). Balancing Empowerment. *Technovation, 16*(8), 397–409. doi:10.1016/0166-4972(96)00029-6

Sundbo, J. (1997). Management of Innovation in Services. *Service Industries Journal, 17*(3), 432–455. doi:10.1080/02642069700000028

Sundbo, J. (1998). *The theory of innovation: enterpreneurs, technology and strategy.* Edward Elgar Publishing.

Sundbo, J. (1998). *The Theory of Innovation: Entrepreneurs, Technology and Strategy.* Cheltenham: Edward Elgar.

Sundbo, J. (2001). *The Strategic Management of Innovation.* Cheltenham: Elgar.

Sundbo, J., Orfila-Sintes, F., & Sørensen, F. (2007). The innovative behaviour of tourism firms-Comparative studies of Denmark and Spain. *Research Policy, 36*(1), 88–106. doi:10.1016/j.respol.2006.08.004

Tidd, J., Bessant, J., & Pavitt, K. (1977). *Managing Innovation.* Integrating Technological and Organizational Change.

Werthner, H., & Klein, S. (1999). *Information Technology and Tourism – A Challenging Relationship.* Springer-Verlag.

Williams, A. (2007). A lecture on "Tourism Innovations." In Lecture Series in Tourism at Waterloo University, Canada. Academic Press.

KEY TERMS AND DEFINITIONS:

Competitiveness: Divided into competitive and comparative advantage means an attempt to get the economic hegemony in the global world.

Consumer-Centric Marketing: Marketing focused on consumers and the role of consumers is changing in this period of time. Consumers and providers of services are getting closer and for this process and consumers are becoming prosumers, which means a very tight relationship among both sides – supply and demand side.

Destination Competitiveness: The rules of competitiveness applied to destinations, which could be countries, regions and cities. The existence of different sectors, users and development strategies is visible.

Emotional Attachment Created Through Branding: Important in this process are attitudes and attributes of a particular place and how this place branding product is offered to the consumer (visitor or just a user of a territory), where could be crucial as the outcome brand loyalty and brand equity of the particular place brand.

Innovation: Application of new processes, new economic and social action, which is totally new and not only a repetition of the formerly applied processes (unlike invention).

Knowledge: Important competitive tool, which belongs to comparative advantage and is one of the major factors of knowledge economy in destinations. Incremental importance has especially tacit knowledge, which cannot be easily copied. The other type of knowledge is a codified knowledge.

Social Technologies: Social technologies are able to improve a process of communication among the users of a specific place. Modern social technologies are able to use for marketing purposes not only standard devices or methods as mobiles, computers, but also tools as the artificial intelligence, business models (for managing of processes), but for marketing also tools as neuromarketing. Artificial intelligence is not so much common among managers in destinations and the reason is that some of these places are still not equipped with the appropriate devices or do not trust for instance e-marketing.

Section 2
Social Media and Tourism Cases

Chapter 8
A Conceptual Framework to Understand Online Destination Images:
A Research Model Utilizing User-Generated Content Through Twitter

Zeynep A. Gedikoglu
https://orcid.org/0000-0003-1390-2305
Clemson University, USA

Sheila J. Backman
Clemson University, USA

Joseph P. Mazer
Clemson University, USA

Kenneth F. Backman
Clemson University, USA

ABSTRACT

This chapter proposes a new conceptual model to understand the construction of online destination images. This will lead to a more accurate and realistic portrayal of how destination images are created, and allow destination managers (and other stakeholders) to better understand the images promoted, and how these comport with the actual experience of users. The model integrates a sequential and mixed methods approach, enabling a conceptualization of how user generated content (UGC) is utilized to formulate and construct destination images.

DOI: 10.4018/978-1-7998-1947-9.ch008

INTRODUCTION

The construction of image of a destination is vital in tourism marketing. In the early 2000s, Destination Marketing Organizations' (DMO) marketing of the tourism product was the preponderant research medium. Attracting visitors is a major concern for destinations, and the destination image can critically influence a traveler's destination choice (Cai, 2002). Destination marketers use the physical reality, artifacts, architecture, weather, culinary features, and the history to sell a destination to potential tourists. However, an image goes beyond the physical characteristics of a place.

The definition of a destination image (and its constructs) is contested and multi-faceted. Not surprisingly, the tourism literature does not seem to agree upon a universal definition (Martin-Santana et al., 2017). Crompton defines destination image as the sum of beliefs, ideas and impressions a person about a destination (Crompton, 1979). Given the multi-faceted definition, the concept of a destination image depends on how people get informed.

Destination branding has been the core of what marketers have communicated to travelers. However, due to Social Media Platforms, the image of a place is no longer framed by destination marketers. Alternative information sources can market a destination through a collective global outreach where brands are formed by individual, group or official sources. Existing models of the conceptualization of the destination image through studies of tourist perceptions via onsite, mail or phone surveys, focus groups, or interviews are outdated in the context of globalization and technological improvements such as WEB networks. Web technologies make it possible to assess online and global communication data without structured survey questions with pre-tested constructs. Thus, the traditional image-makers, DMOs and tourism promoters, are not the only stakeholders that can construct a destination image. Previous research has shown that destination managers may be active players through creating content and communicating by understanding images communicated by influencers (Chatzigeorgiou, 2017). This chapter presents a new approach to investigate destination images which are communicated by influencers through Twitter. An influencer is a user on social media who has access to a large audience, is mentioned by others and whose posts are re-posted so that can persuade others by virtue of their authenticity and reach. This approach will stimulate better planning for sustainable brand image in tourism practices as an exemplary model for any destination.

User Generated Content (UGC) is any type of shared communicated content that has been created and shared through online platforms by unpaid contributors. The explosion of UGC is a result of WEB and mobile technologies, which provide individuals with unprecedented power to instantaneously add digital traces when performing tasks such as reviewing and documenting travel experience (Lu & Stepchenkova, 2015). Although the term Social Media is difficult to define, a typical definition is an internet-based application that construct user generated content (Blackshaw, 2006). Social media content is produced by consumers and can challenge the content of DMOs (Xiang & Gretzel, 2010).

Perceptions about any destination might shift due to personal experiences or learned knowledge through communication. Social Media (SM) platforms, such a twitter, enables individuals to share personal experiences and/or information much quicker than word-of mouth. To understand the change in people's perceptions towards a destination, this study integrates thematic networks, cognitive and affective components of destination image, time element and sentiment analysis. This chapter offers a sequential mixed methods approach which may be operationalized in future studies. In this chapter, destination image constructs are utilized to understand the current latent variables of destination images. The model visually depicts a destination image through a deductive approach of UGC content. Qualitative methodologies reveal themes of a destination image from unstructured online communication data and further may be assessed according to the affective and cognitive constructs' properties such as time and sentiment. The objective of this chapter is to propose a new model using social media data to understand online destination images.

The study method proposes to conduct unstructured social media data and sequential mixed methods research design. This model is the first integrated multi-method framework in destination image scholarship. The first part of the chapter provides relevant background about destination image research and adaptation of online data to image studies. The second part of the chapter explains in detail the research model utilizing user generated content through twitter. Each sequential method is explained in detail in the second part. Third part is composed of issues, controversies and problems related to the methodology of the model. The fourth part provides solutions and recommendations for future studies that are going to employ the research model utilizing user generated content through twitter for destination image studies. The fifth part discusses and provides future research directions for the model.

BACKGROUND

Studies of destination images have been the subject of much attention in the academic literature. Tourism researchers have long assumed that the images people have of the world around them significantly influences their travel behavior (Walmsley & Young, 1998). The destination image literature dates to the early 1970s, and consumer marketing research was the initial territory that led scholars' research destination image as a product of tourism service industry. In the academic literature, destination image research has conducted under several alternative names: tourism image, tourist destination image, country image, holiday destination image, brand equity in tourism, destination brand image, destination perceptions, destination branding, and destination marketing.

The earliest phase of tourism studies started with the tourist as the subject and the originator of tourism. Destination image has become a fertile topic since Hunt's research on image as a factor in tourism development (Hunt, 1975). Destination image research was built on the general assumption that images as perceived by individuals have as much to do with a tourism destination's development success as the tangible tourism resources (Hunt, 1975). Image differences between prospective (non-visitors), first-time, and repeat visitors have found that each group is affected by different images such as induced or organic that are communicated to them via tradition channels or tourism promoters (Fakeye & Crompton, 1991). Gunn (1972) suggested that organic image of a destination is formed by print or visual media. An organic image evolves into an induced image with additional information gathered from tourism promoters and organizations. With the decision to travel, a prospective tourist experiences the place herself and forms a complex image in her mind. Previous studies showed that the perceived image of a place is vulnerable to myriad experiences for both visitors and citizens.

Destination image formation is a construction of a mental representation on the basis of information cues selected and delivered by the image formation agents. The image of a place is represented by the consumer's general perception of that place, based on the strong and weak points by which the respective place has made itself known (Roth & Romeo, 1992). Dichter found that image describes not individual traits or qualities, but the total impression an entity makes on others (Dichter, 1985). Telisman defined images of tourism destinations as a crucial component of a destination's tourism product (Telisman,

1989). Telisman stated that a destination image needs to be continuously assessed, as it exists in the minds of potential customers and it is a prerequisite to meaningful marketing strategy and competitive positioning (Telisman, 1989). Gartner looked at attribute measurement and his conclusions were that image does not equate to reality and image studies represent only specific time period (Gartner, 1989). Therefore, online data might reveal up-to-date image perceptions with respect to time. Embacher & Buttle has a global approach to destination image perceptions people hold. They tried to understand it as a perception construct rather than fragmenting it to socio-psychological or psychological factors (Embacher & Buttle, 1989).

Van den Berg, van der Borg, and van der Meer (1995) suggest that tourism managers should engage all available resources without losing sight of the necessity to bring about and communicate a diverse, stable, original, and attractive image to the destination. Online communication platforms are manifestations of social construction and presence for contemporary tourism destinations. These platforms provide relevant data as a representation for researchers and for managers to mold their marketing strategies.

The main concern of researchers is the association of tangible physical components of place perception commonly associated with the destination image (Pearce 1977, 1982; Walmsley & Jenkins, 1992). Echtner & Richie examined the literature in order to understand the unique characteristics of the destination image concept, and developed a measure and framework (Echter & Richie, 1991; Echter & Richie, 1993). Their framework consists of three characteristics; attribute-holistic, functional-psychological, and common-unique, which is then used to measure the destination image (Echter & Richie, 1991). Fakeye & Crompton developed a model in order to understand the components of destination image. Their research described the relationship between induced, organic and complex images that amalgamated experiences to that destination (Fakeye & Crompton, 1991). Walmsley & Young examined the applicability of the local scales of destination image to the international scale. Their results support a common pattern to be used to evaluate destination images at the local and international levels (Walmsley & Young, 1998). The existence of a basic schema to evaluate intangible characteristics of a destination image has been important for future tourism scholarship about destination image.

Chon found that destination image is the most important influence in tourist buyer behavior and satisfaction (Chon, 1990). Building on Chon's analysis, Echtner and Ritchie examined the conceptualization and operationalization of 15 previous destination image studies to find a pattern for future studies (Echtner & Ritchie, 1991). Their 1991 review concluded that researchers still had not successfully operationalized the destination image construct. Echtner and Ritchie also found that researchers held a strong preference for quantitative research techniques (Pike, 2002). Several authors have looked at what visitors know about a destination, the cognitive image, and how visitors emotionally feel about that destination, the affective image (Baloglu & McCleary, 1999b; San Martin & Del Bosque, 2008; Beerli & Martin, 2004). Cognitive components of destination image are an individual's unique knowledge, perceptions and beliefs about the attributes of a destination which are observable, descriptive and measurable. On the other hand, affective components of destination image are an individual's unique emotional interpretation of a destination which is composed of feelings and emotions. In the context of tourism, research empirically indicates that cognitive and affective evaluations directly influence the overall image (Baloglu & McCleary, 1999a, 1999b; Stern & Krakover, 1993). Previous research recognized multidimensionality and complexity (Ryan & Cave, 2005) of the destination image concept and applied different methods for measuring different components (Pezenka, 2016).

The extent to which consumers use an overall or composite image is not yet understood (Dichter, 1985; Stern & Krakover, 1993; Baloglu & McCleary, 1999a). The sources, through which individuals receive information such as promotions and recommendations, have been found to influence destination image (Baloglu and McCleary 1999b). Pike in his analysis of 142 papers from 1973 to 2000 found that in destination image analysis cognitive image attributes are generally generated from previous literature (Pike, 2002). Sometimes, a content analysis of destination guidebooks (or brochures) is also used to generate cognitive image attributes (Di Marino, 2008; Tasci, Gartner, & Cavusgil, 2007; Y. Wang & Fesenmaier, 2007). Affective evaluations, on the other hand, are generally measured using the bipolar scales developed by. Russell and his colleagues (Russell & Pratt, 1980; Stepchenkova & Mills, 2010). Previous research shows that some image components may be universal to all destinations, while others are destination specific (Echtner & Ritchie, 1993). Therefore, a new model may extend understanding of the components and attributes of destination image.

According to Woodside and Lysonski's (1989) destination choice model, marketing variables or information sources influence the formation of perceptions or cognitive evaluations but not the affective component of image (Woodside & Lysonski, 1989). The image and brand of a destination are two different concepts but not mutually exclusive (Munar, 2012), meaning that the brand's existence depends on the image formation (Cai, 2002). Image represents the sum of beliefs, attitudes and impressions that a group has of an object (Nadeau et al., 2008). Previous research has shown that, the destination image evolves organically through word-of-mouth and other communication received about the destination (Gartner, 1986). While numerous studies contributed significantly to the understanding of destination image formation, more recent communication technologies, especially the influence of Social Media, haven't been considered in these frameworks (Kislali, Kavaratzis & Saren, 2016). Social Media is an important means to access information from trusted sources and a vital new platform to share personal experiences. Globalization processes of how potential visitors choose, decide, and behave have enriched options through information communication technologies (ICTs).

Every social networking site has millions of users, and each provides worldwide connectivity. The connections provide, among other things, free advice and free advertising. Social networking sites have revolutionized real time sharing of information. The sharing of personal stories through social networks alter perceptions and influence tourism behaviors. Thus, people interactively co-construct and re-create destination images by digital storytelling verbally and/or visually; sharing their own narratives about tourism destinations through their experiences (Amersdorffer, Bauhuber, & Oellrich, 2012) as well as perceptions and expectations.

Recently, tourism has become the largest industry for online transactions (Mack et al., 2008). The rapid increase in internet usage and Social Media adaptation brings new challenges and opportunities both for tourism destinations and travelers (Schmallegger & Carson, 2008). Social Media has changed the information search and purchase behaviors of all tourism stakeholders (Lo et al., 2011) and ultimately has affected how destination images are formed (Kislali, Kavaratzis, & Saren, 2016). Research has not investigated if destination images may be sustained via incorporating contemporary communication platforms such as Social Media, which are global and instant.

Social Media (SM) is affecting tourists' consumption behavior by facilitating conversations. Social Media websites, representing various forms of UGC such as blogs, virtual communities, wikis, social networks, collaborative tagging, and media files shared on sites like YouTube and Flickr, have gained substantial popularity with online travelers (Gretzel, 2006; Pan, MacLaurin, & Crotts, 2007). UGC is a mixture of fact and opinion which are sentimental impressions on various social media platforms. This is shifting the broadcast medium from one-to-many to a many-to-many model, rooted in conversations between authors, people, and peers (Solis, 2010). Social Media platforms such as YouTube, Twitter and Flickr, play a central role in empowering enhanced levels of interactions among multiple parties (Neuhofer, Buhalis, & Ladkin, 2014).

Social Media marketing leverages specific Social Media platforms (places where people connect and communicate) to promote a product or a service to increase sales (Sotiriadis & Van Zyl, 2013). Tourism companies use Social Media as a platform to apply their marketing strategies and spread their destination images through images and texts. Whatever these companies are communicating is just a fraction of communication data that is produced about a destination on Social Media. Fotis, confirmed the importance of Social Media as a platform for tourism providers and tourist consumers to engage, interact and share experiences (Fotis et al., 2011).

Online/virtual destination image concepts has been researched by few scholars until now such as Govers & Go, 2004; Dwivedi, 2009 and Stepchenkova & Morrison, 2006. With online technologies tourism destination images are becoming increasingly fragmented and ephemeral in nature (Govers et al., 2007). One of the first papers published in Information Technology and Tourism Journal, tried to deconstruct the destination image concept into attribute-based components using structured methodologies (Govers & Go, 2003). The article demonstrated for the first time that the multi-attribute destination image research is inadequate in predicting destination choice behavior. Govers, Go, and Kumar (2007) utilized a qualitative approach to study seven case studies with an online survey; the respondents of the online survey were not current or potential tourists, but anyone who provided an input about a destination.

Dwivedi (2009) researched the use of WEB 2.0 in destination image formation, specifically by exploring India's online image through consumer queries posted on travel message boards. The results of the study showed that consumers are not only perceivers of destination image information, but they actively construct and share their own images via the WEB 2.0 platforms (Dwivedi, 2009). Traditional internet-based travel studies have focused on supply side sources such as tour Websites and government travel websites, while this paper provides a much-needed and different perspective.

Hunter showed that an image is an aggregate online product and it is not entirely in the hands of tourism marketers (Hunter 2013). The first study which compared national tourism organizations' and tourists' perspectives of an online destination image was done by Mak (2017. The study examined perceived and projected online destination images that are manifested in tourist and national tourism organization generated content (Mak, 2017). Findings revealed that tourist generated textual content tended to outperform that of national tourism organizations in reflecting affective destination image attitudes (Farías et al., 2016).

Globalism has affected the tourism industry and how people perceive tourism products. The Internet has fundamentally reshaped how tourism-related information is distributed, and how people plan for and consume travel (Buhalis & Law, 2008). Before the Internet, DMOs controlled the media messages, and they tried to create desired destination images through their marketing campaigns. Generalizations of destination images were prominent within the marketing products of tourism promoters when mass tourism peaked in 1980s. However, today tourism is a very fragmented industry and an information-rich business, which makes it especially receptive for the benefits offered by the WEB 2.0 (Schwartz, 1998). Therefore, destinations compete through images held in the minds of potential tourists (Baloglu & McClearly, 1999a).

Waters's case study demonstrated thematic analysis to be a rigorous and valid means of analyzing tourism texts, particularly where it is recognized that the meaning or significance of the whole may be more than the sum of its parts (Walters, 2016). The most frequent use of thematic analysis in tourism research has been its application to the interpretation of written documents (Walters, 2016). Although not commonly utilized in tourism research, thematic analysis has been applied to big data in various social sciences. In tourism scholarship, several recent examples which utilize thematic analysis in destination image research include analysis of information sources construct and its relevance to destination image formation (Llodra-Riers et al., 2015) and the social construction of tourism online destination images (Hunter, 2016).

Use of social media on the Internet has changed DMOs use social media, especially blogs, as part of their business strategy for each of the five key marketing functions—promotion, product distribution, communication, management, and research above-listed functions (Schmallegger & Carson, 2008). However, authors have not yet sufficiently addressed the stability of a destination image, in predicting destination choice across different tourists and destinations, how to sustain an image through communication actors (Lin, Morais, Kerstetter, & Hou, 2007). A review of the literature shows that there is a gap in research which integrates new communication technologies to destination image research. The destination image critically influences the traveler's destination choice process (Cai, 2002). Therefore, in this chapter a new model is proposed which utilizes data from technologies that have and will have a potent influence on destination image formation. This challenges long-standing assumptions about information-technology image measurement paradigms. The main focus of the proposed model is to deconstruct the image into affective and cognitive components looking at the influencers' communication.

A RESEARCH MODEL UTILIZING USER GENERATED CONTENT THROUGH TWITTER

The objective of this chapter is to explain a research model which relies on unstructured online communication data to understand online destination images. Proof of concept is explored in the case study by looking at how the destination image of the capital of France, Paris, is affected by the terror crisis after the November 2015 terror attacks in Paris (Gedikoglu, 2018). Influencers diffuse information that shapes perceptions in online communication platforms such as twitter. Twitter is possibly the best-known microblogging site and was launched in 2006 (Sotiriadis & Van Zyl, 2015). In a 24-hour period, 140 million active Twitter users send out more than 340 million tweets (Gašpar & Mabić, 2017). Therefore, through Twitter data, it is possible to evade the problem of ensuring a representative sample using the population of all users of Twitter with respect to search query. Therefore, Twitter is chosen as the data collection platform for this study,

The purpose is to ascertain basic and organizing themes about a destination through thematic analysis, and then model a further quantitative analysis framework. This will fill the gap in literature for exploring the Social Media sites as agents and online identities as influencers of cognitive and affective components of destination image. At the same time, it uniquely supports a holistic approach which embraces cognitive and affective factors with a qualitative analysis methodology: thematic network analysis.

Additionally, the sequential approach of the model utilizes the themes generated in thematic analysis to further operationalize the model by a quantitative analysis to understand the relationship between time, sentiments, and cognitive and affective constructs. There is copious research that focuses on analyzing user sentiment towards different products, services or even intangible aspects on Twitter (Mochón & Martínez, 2014). It is important to note that online communication networks such as Twitter are only a tool in analysis, not the analysis itself. However, there is a lack of related work on the analysis of users' feelings towards tourist applications (Fujita & Herrera-Viedma, 2018). Such knowledge could be used as part of a recommendation system for tourism.

This model is the first integrated multi-method framework in destination image scholarship. The first part of the model explores the thematic network of online destination image themes. This fills the gap in literature for exploring Twitter as the platform for influencers as agents of cognitive and affective components of destination image. At the same time, it will present a novel analysis approach to destination image literature by utilizing a qualitative analysis methodology: thematic network analysis. The second part of the model investigates the patterns of change in destination image themes with respect to time and sentiments incurring. Specifically, it will investigate the relationship between online destination image themes, time and sentiments.

The nature of online data can potentially incorporate mixed methods. This model incorporates a mixed methods approach. Classification of the mixed methods design is qualitative methods to develop quantitative variables (Figure 1). The first step in analyzing a destination image through UGC content is to access the social media communication cloud with respect to the destination and particular time period in which the destination image is analyzed. Several studies have accessed online data through Social Studio Software (Boatwright & Mazer, 2019; Gedikoglu, 2018; Jenkins & Mazer, 2018). Social Studio Software is a social media listening software which mines the data from the internet. This allows access to online textual communication data through unobtrusive observation. The data is composed of people's interactive online communication, which are interpersonal, and/or intrapersonal. Data sources, which are Tweets, are public posts (Gašpar & Mabić, 2017). Tweets are driven from people's intended actions and perceptions rather than a prepared answer for a survey question. In other words, this model uses unstructured secondary data, which creates a big data set composed of Tweets.

The detection of UGC of influencers is the second step in the model. Today's new type of opinion leaders, influencers, persuade their followers (Martensen et al., 2018). Developing a model of destination image via UGC with important technological and personal characteristics poses a challenge due to the large number of potential constructs. To overcome this challenge, this model utilizes previous findings in literature using three influence measures to ascertain the online influencers: Influence of re-posts (Oh & Nguyen, 2010); the number of followers (indegree) (Creamer, 2011), and the number of mention names (Sameh, 2013). Indegree represents the popularity of a user, how many followers that account

Figure 1. Sequential mixed methods research design
(Baran, 2016)

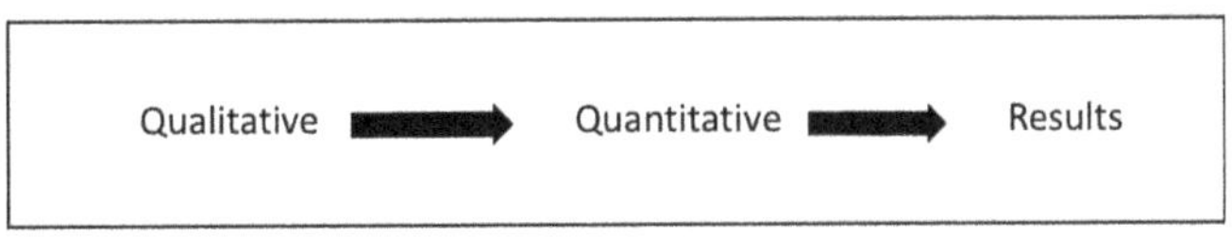

has; re-posts represent the content value of one's tweets; and mention names represent the trustability of a user. First, an UGC influencer must have at least 1 follower; second, the posts should have been re-posted at least once in the relative time period; and third, the online name of the influencer should have been mentioned at least once in the big data during the particular time period in which destination image is analyzed.

The third step in the model details a technique for conducting thematic analysis of textual data, presenting a step-by-step guide of the analytic process (Guest et al., 2011). The thematic analysis method employs established, well-known techniques, and presented as thematic networks. Thematic networks are web-like illustrations that summarize the main themes constituting a piece of text (Attride-Stirling, 2001). The thematic networks technique is a robust and highly sensitive tool for the systematization and presentation of qualitative analyses (Attride-Stirling, 2001). Thematic network modelling is the qualitative method this research utilizes to ascertain relevant destination image themes. Thematic analysis is done by inductively coding themes of a destination image for each post. Open coding is utilized to identify the organizing themes. Word counts and word trees retrieve relevant themes and networks. The purpose of the qualitative methodology is to understand the meanings in conversation beyond word counts, and to reveal a network of global, organized and basic themes.

This technique enables a methodical systematization of textual data, facilitates the disclosure of each step in the analytic process; aids the organization of an analysis and its presentation; and allows a sensitive, insightful and rich exploration of a text's overt structures and underlying patterns. It is important that the researcher has a comprehensive understanding of the content of the interaction and has familiarized him/herself with all aspects of the data.

Thematic networks systematize the extraction of: (i) lowest-order premises evident in the text (Basic Themes); (ii) categories of basic themes grouped together to summarize more abstract principles (Organizing Themes); and (iii) super-ordinate themes encapsulating the principal metaphors in the text as a whole (Global Themes). Figure 3 shows the web-like maps depicting the salient themes at each of the three levels, illustrating their relationships (Figure 2). The procedure of thematic networks. Web-like representations do not aim or pretend to discover the beginning of arguments or the end of rationalizations; but simply provide a technique for breaking up text and finding its explicit rationalizations and implicit signification.

Figure 2. A social media model for destination image formation
(Gedikoglu, 2018)

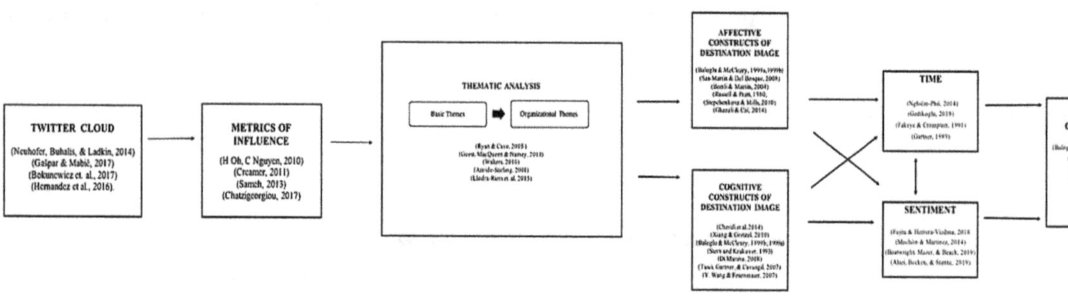

Figure 3. Structure of a thematic network
(Attride-Stirling, 2001)

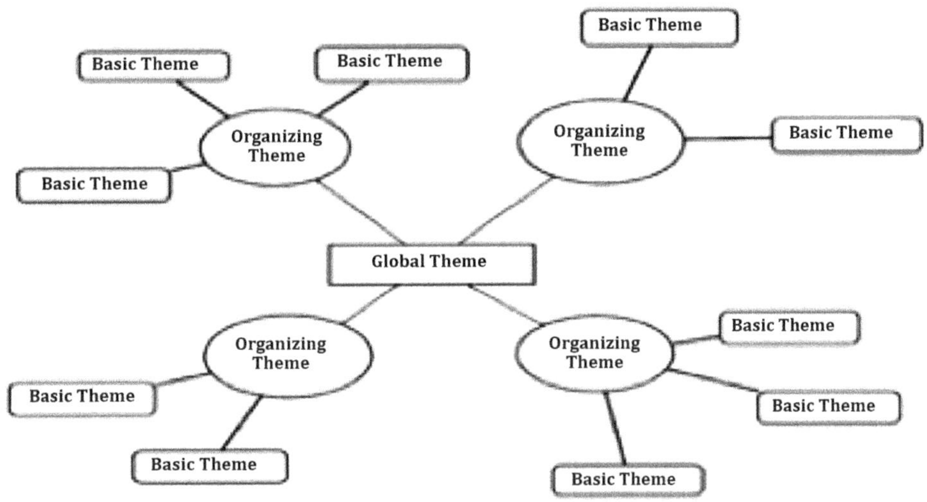

Basic themes are the lowest-order themes that are derived from the textual data. Once familiar with the data, the researcher must then start identifying preliminary codes, which are the data features that appear interesting and meaningful. These codes are more numerous and specific than themes, while indicating the context of the conversation. Basic Themes are simple premises characteristic of the data, and on their own indicate very little about the text or group of texts as a whole. In order for a Basic Theme to make sense beyond its immediate meaning it needs to be read within the context of other Basic Themes.

The next step requires a deeper review of identified themes where the researcher needs to question whether to combine, refine, separate, or discard initial themes. Data within themes should coalesce meaningfully, while there should be clear and identifiable distinctions between themes. This is done over two phases, where the themes need to be checked in relation to the coded extracts (phase 1), and then for the overall data set (phase 2). A thematic 'map' can be generated from this step.

Organizing themes are the clusters of the Basic Themes in similar issues. Relevant data extracts are sorted, combined or split according to overarching themes. The researcher's thought process should allude to the relationship between codes, subthemes, and themes. Clusters summarize the principal assumptions of a group of Basic Themes, so they are more abstract and more revealing than the texts. They are the principles on which a super-ordinate claim is based (Attride-Stirling, 2001). Thus, Organizing Themes simultaneously group the main ideas proposed by several Basic Themes. A group of Organizing Themes then constitutes a Global Theme, which is like a claim in that it is a concluding or final tenet. As such, a Global Theme groups sets of Organizing Themes that together present a position about a reality. A pool of attributes related to each global theme is developed, based on a review of the previously used items in measuring destination image (Beerli & Martin, 2004; Chen & Tsai, 2007; Chi & Qu, 2008; Wang & Hsu, 2010). It must go beyond a mere description of the themes and portray an analysis supported with empirical evidence that addresses the destination image. For example, in the research of Gedikoglu (2018) the thematic analysis revealed four organizing themes around the global theme which is the destination itself (Gedikoglu, 2018).

Past research has delineated a number of destination image dimensions and concluded that image is a complex and multifaceted concept (e.g., Beerli & Martin, 2004; Bonn et al., 2005; Lin et al., 2007). The fourth step of the model is to assign affective or cognitive themes. Baloglu and McClearly (1999a) have provided a formation model in which information sources are factors of perceptual/cognitive and affective global destination images. The coding of posts according to cognitive and affective constructs of destination image in line with previous studies is preferred.

The fifth step of the model uses the themes generated in thematic analysis. The purpose of this step is to take the model a step further by quantitative analysis to understand the relationship between time, sentiments towards the cognitive and/or affective themes. Quantitative analysis unfolds the factors of the holistic destination image construct. Destination image is formed over time from a variety of influential information sources (Nghiêm-Phú, 2014). To operationalize this analysis to empirical data, depending on the nature of the data several quantitative methods may be applied to understand the hierarchical relationship and the clusters of factors of online destination images. The main purpose is to understand the relationship between sentiments (Alaei et al., 2019) and time. In this step, hypothesis may be analyzed with SPSS analytical software. The results of F-tests will show the significance of predictors for each hypothesis for any regression model the researcher wants to analyze. For example, results of significance for Null Hypothesis will show the probability of getting negative, neutral and positive sentiments differ for each theme with respect to time.

Global evaluations of image are a function of cognitive and affective constructs of image (Baloglu & McCleary, 1999a). The objective of this research is to evaluate the formation of destination image through a global approach and thus evaluations are deemed necessary in which non-visitors and visitors are included (Cherifi et al., 2014). UGC possesses a global approach to understand destination image. The measurement of destination image in this model involves cognitive and affective evaluations with respect to time and sentiment variables.

ISSUES, CONTROVERSIES, PROBLEMS

Collecting, sampling and analyzing Social Media data for qualitative and/or quantitative research presents several challenges. First, a central issue with methods of data collection have their own limitations. For one thing, the researcher must have the computational skills in order to collect relevant data from Social Media platforms. It is important to note that the specific software such as Social Studio Software is an interphase between Twitter and the researcher, therefore big data gathered from Twitter may influence research questions and bias the selection of relevant search queries.

Social Studio Software was not able to yield an influencer analysis because the analytical tools cannot analyze past data to ascertain influential digital identities. Therefore, for this model previous studies and research were incorporated from the communication sciences literature. Another challenge is that Social Media users are not a homogeneous group, and thus users are hardly a representative group of the entire population. However, taking mixed methods research online produces a more representative sample that improves researchers' ability to generalize overall results. The data collected through Twitter may have an in-built bias toward users who are the most active content contributors, whereas the data say very little about the users who may read their Twitter streams with great interest on a daily basis, but who barely post anything to the stream (Giglietto, Rossi, & Bennato, 2012; Gonzales-Bailon et al., 2013).

Finally, in addition to the methodological limitations and challenges of sampling, generalization and validity, there are ethical implications of researching online data. On Twitter, by the time content is posted, it is by default public. As argued by Markham and Buchanan (2012), ethical issues vary by scope of the research questions. Ethical judgments must be made according to the research questions at hand and entirety of the data. This kind of research acknowledges that public Social Media data may be personal and any empirical research performed by utilizing this model should ensure that it is handled according to the privacy laws.

SOLUTIONS AND RECOMMENDATIONS

Scholars have increasingly recognized the advantages of applying quantitative and qualitative data analysis methods in a single study. A number of destination image studies employed sorting and categorization techniques to identify the frequencies of certain words, concepts, objects, or people, and treated the most frequent ones as variables (or dimensions) of the destination image construct (Echtner & Ritchie, 1993; Dann, 1996; MacKay & Fesenmaier, 1997; Tapachai & Waryszak, 2000; Andsager & Drzewiecka, 2002; Echtner, 2002). Proposed destination image analysis model suggests that portrayals through social media platforms are global depictions of destination image. In other words, online users share their own narratives about tourism through their perceptions and expectations.

Operationalizing the proposed online destination image model will show the change with respect to time and sentiments of organizational themes. A distinct advantage of this model in the tourism literature is the application of thematic network analysis to destination image research using social media data. The approach of utilizing unstructured data is a novel approach. Social media is a continuous, global and instantaneous meeting space for friends and strangers where people share content to social networks by exporting their information, knowledge and experience. A positive aspect of data collected from social media is that people post about issues by their own will and motivation. Therefore, the researcher bias disappears from the data collection stage. Additionally, findings indicate that online WOM communications have a significant impact on attitudes toward visitation, subjective norms, perceived behavioral control, and intention to travel (Jalilvand & Samiei, 2012). To overcome challenges of representation and sampling, this model should use the purposeful sampling technique rather than random sampling to select the Twitter accounts.

This study unites several approaches from the literature to identify online influencers. Informational influence may occur in online travel communities when facts, evidence, and other travel related information is discussed via postings (Arsal, Woosnam, Baldwin, & Backman, 2010). This study defines a new metric to identify online influencers with three influence measures, which provide a better understanding of different roles users play in Social Media platforms.

The contributions to tourism research of this study include: the utilization of social media data in destination image studies; a mixed methods approach to big data analysis, the singular approach to first-time or repeat visitors' perceptions; and expansion of influencers studies in tourism literature. Additionally, there are some unique approaches in this model that are worth to mention such as the use of time and sentiment variables.

In conclusion, the implications of this research go beyond academia. The empirical findings utilizing this model may support professional tourism planning. To develop effective marketing campaigns and promotions this model might be utilized to understand the dynamics in the market and respond to change promptly. This study suggests that future tourism players might be proactive and reactive by understanding online the formation of destination images. Therefore, this model contributes not only to the academic literature, but offers professional tourism players a model that might be applied to future marketing strategic planning.

DISCUSSION AND FUTURE RESEARCH DIRECTIONS

This study will initiate future research, which may extend o understanding of destination image, using social media platforms as mass medium which can influence formation of community opinion. Acceptance of the social media platforms as the new global communication landscape brings new research areas. Traditional media is subjectively selective in setting the agenda about other places. This study suggests that portrayals through a social media platform also create an aggregate image of a place.

This study brings together many unique methods and approaches to destination image research. Future studies may operationalize approaches explained to look at case studies. Research might look at alternative social media platforms other than Twitter. Recently, Social Media has moved towards more visual than social content, e.g., such as Instagram and Snapshot. Even though the methodological approaches to visual analysis are limited, there is a dearth of research using online depictions of destination images. This model might integrate visual data in future studies.

Another orientation may be a comparative analysis of destination images as projected by user generated content (UGC) creators and various travel intermediaries, government sources, and general media. Such a comparison analysis has consistently captured differences in destination images transmitted via various information channels (e.g., Choi, Lehto, & Morrison, 2007; Dwivedi, Yadav, & Patel, 2009; Law & Cheung, 2010); however, it has never utilized UGS content. Comparison of online sources to traditional sources of destination information may provide beneficial implications to tourism literature. Additionally, the literature lacks comparative analysis of online sources with each other.

Previous destination image studies have looked at perceptions of tourists through structured questionnaires. It is important to note that online communication networks such as Twitter, are only a tool, not the analysis itself. Once a thematic network has been constructed, it will then serve as an organizing principle and an illustrative tool in the interpretation of the text, facilitating disclosure for the researcher and understanding for the reader. Therefore, future studies may apply the qualitative methodology of this research in a different context.

CONCLUSION

The main purpose of this chapter is to propose a model to ascertain the destination image themes communicated by influencers on Twitter. Additionally, the model may be used to further investigate relationships between sentiments of themes and time constructs such as seasons or days. Thematic analysis is a useful tool in measuring the multi-dimensional nature of destination image through online textual data. Applying quantitative methods to the results of thematic analysis may further provide insights on online destination image. In sum, this model suggests that social media is a communication data source that reveals standard schemas on the dynamics of destination image for future studies. In conclusion, this model particularly fills the gap in the literature by applying a variety of novel analysis methods and offers a new approach to understanding destination image.

ACKNOWLEDGMENT

This research received no specific grant from any funding agency in the public, commercial, or not-for-profit sectors.

REFERENCES

Alaei, A. R., Becken, S., & Stantic, B. (2019). Sentiment Analysis in Tourism: Capitalizing on Big Data. *Journal of Travel Research*, *58*(2), 175–191. doi:10.1177/0047287517747753

Amersdorffer, D., Bauhuber, F., & Oellrich, J. (2012). The Economic and Cultural Aspects of The Social Web: Implications for The Tourism Industry. *Journal of Vacation Marketing*, *18*(3), 175–184. doi:10.1177/1356766712449364

Andsager, J. L., & Drzewiecka, J. A. (2002). Desirability of Differences in Destinations. *Annals of Tourism Research*, *29*(2), 401–421. doi:10.1016/S0160-7383(01)00064-0

Arsal, I., Woosnam, K. M., Baldwin, E. D., & Backman, S. J. (2010). Residents as Travel Destination Information Providers: An Online Community Perspective. *Journal of Travel Research*, *49*(4), 400–413. doi:10.1177/0047287509346856

Attride-Stirling, J. (2001). Thematic Networks: An Analytic Tool for Qualitative Research. *Qualitative Research*, *1*(3), 385–405. doi:10.1177/146879410100100307

Baloglu, S., & McCleary, K. W. (1999a). A Model of Destination Image Formation. *Annals of Tourism Research*, *26*(4), 868–897. doi:10.1016/S0160-7383(99)00030-4

Baloglu, S., & McCleary, K. W. (1999b). US International Pleasure Travelers' Images of Four Mediterranean Destinations: A Comparison of Visitors and Nonvisitors. *Journal of Travel Research*, *38*(2), 144–152. doi:10.1177/004728759903800207

Baran, M. L. (2016). It Is All in The Design: Creating the Foundations of a Mixed Methods Research Study. In Mixed methods research for improved scientific study (pp. 66-78). Hershey, PA: IGI Global.

Beerli, A., & Martin, J. D. (2004). Factors Influencing Destination Image. *Annals of Tourism Research*, *31*(3), 657–681. doi:10.1016/j.annals.2004.01.010

Berg, L. V. D., Borg, J., & Meer, J. V. (1995). *Urban Tourism: Performance and Strategies in Eight European Cities*. Avebury.

Blackshaw, P., & Nazzaro, M. (2006). Word of Mouth in The Age of The Web-Fortified Consumer. *Consumer-generated media (CGM), 101*.

Boatwright, B., Mazer, J. P., & Beach, S. (2019). The 2016 US Presidential Election and Transition Events: A Social Media Volume and Sentiment Analysis. *The Southern Communication Journal*, 1–14.

Bonn, M. A., Joseph, S. M., & Dai, M. (2005). International versus domestic visitors: An examination of destination image perceptions. *Journal of Travel Research*, *43*(3), 294–301. doi:10.1177/0047287504272033

Buhalis, D. (1998). Strategic Use of Information Technologies in The Tourism Industry. *Tourism Management*, *19*(5), 409–421. doi:10.1016/S0261-5177(98)00038-7

Buhalis, D., & Law, R. (2008). Progress in Information Technology and Tourism Management: 20 Years on and 10 Years After the Internet—The State of Etourism Research. *Tourism Management*, *29*(4), 609–623. doi:10.1016/j.tourman.2008.01.005

Butowski, B. (2011). Tourism-An Academic Discipline. *Tourism (Zagreb)*, *21*(1-2), 17–24.

Cai, L. A. (2002). Cooperative Branding for Rural Destinations. *Annals of Tourism Research*, *29*(3), 720–742. doi:10.1016/S0160-7383(01)00080-9

Chatzigeorgiou, C. (2017). Modelling the Impact of Social Media Influencers on Behavioral Intentions of Millennials: The Case of Tourism in Rural Areas in Greece. *Journal of Tourism. Heritage & Services Marketing*, *3*(2), 25–29.

Chen, C. F., & Tsai, D. (2007). How Destination Image and Evaluative Factors Affect Behavioral Intentions? *Tourism Management*, *28*(4), 1115–1122. doi:10.1016/j.tourman.2006.07.007

Cherifi, B., Smith, A., Maitland, R., & Stevenson, N. (2014). Destination Images of Non-Visitors. *Annals of Tourism Research*, *49*, 190–202. doi:10.1016/j.annals.2014.09.008

Chi, C. G. Q., & Qu, H. (2008). Examining the Structural Relationships of Destination Image, Tourist Satisfaction and Destination Loyalty: An Integrated Approach. *Tourism Management*, *29*(4), 624–636. doi:10.1016/j.tourman.2007.06.007

Choi, S., Lehto, X. Y., & Morrison, A. M. (2007). Destination Image Representation on The Web: Content Analysis of Macau Travel Related Websites. *Tourism Management*, *28*(1), 118–129. doi:10.1016/j.tourman.2006.03.002

Chon, K. S. (1990). The Role of Destination Image in Tourism: A Review and Discussion. *Tourism Review*, *45*(2), 2–9.

Creamer, M. (2011). Your Followers Are No Measure of Your Influence. *Advertising Age*, *82*(1), 1–22.

Crompton, J. L. (1979). An Assessment of The Image of Mexico As A Vacation Destination and The Influence of Geographical Location Upon That Image. *Journal of Travel Research*, *17*(4), 18–23. doi:10.1177/004728757901700404

Dann, G. M. (1996). Tourists' Images of a Destination-An Alternative Analysis. *Journal of Travel & Tourism Marketing*, *5*(1-2), 41–55. doi:10.1300/J073v05n01_04

Di Marino, E. (2008, May). The Strategic Dimension of Destination Image: An Analysis of The French Riviera Image from The Italian Tourists' Perceptions. *Proceedings of the 17th International Tourism and Leisure Symposium* (pp. 17-18). Academic Press.

Dichter, E. (1985). What's in An Image. *Journal of Consumer Marketing*, *2*(1), 75–81. doi:10.1108/eb038824

Dwivedi, M. (2009). Online Destination Image of India: A Consumer Based Perspective. *International Journal of Contemporary Hospitality Management*, *21*(2), 226–232. doi:10.1108/09596110910935714

Dwivedi, M., Yadav, A., & Raghavjibhai Patel, V. (2009). The Online Destination Image of Goa. *Worldwide Hospitality and Tourism Themes*, *1*(1), 25–39. doi:10.1108/17554210910949869

Echtner, C. M. (2002). The Content of Third World Tourism Marketing: A 4A Approach. *International Journal of Tourism Research*, *4*(6), 413–434. doi:10.1002/jtr.401

Echtner, C. M., & Ritchie, J. B. (1991). The Meaning and Measurement of Destination Image. *Journal of Tourism Studies*, *2*(2), 2–12.

Echtner, C. M., & Ritchie, J. B. (1993). The Measurement of Destination Image: An Empirical Assessment. *Journal of Travel Research*, *31*(4), 3–13. doi:10.1177/004728759303100402

Embacher, J., & Buttle, F. (1989). A Repertory Grid Analysis of Austria's Image as A Summer Vacation Destination. *Journal of Travel Research*, *27*(3), 3–7. doi:10.1177/004728758902700302

Fakeye, P. C., & Crompton, J. L. (1991). Image Differences Between Prospective, First-Time, And Repeat Visitors to The Lower Rio Grande Valley. *Journal of Travel Research*, *30*(2), 10–16. doi:10.1177/004728759103000202

Farías, D. I. H., Patti, V., & Rosso, P. (2016). Irony Detection in Twitter: The Role of Affective Content. *ACM Transactions on Internet Technology*, *16*(3), 19. doi:10.1145/2930663

Fotis, J., Buhalis, D., & Rossides, N. (2011). Social Media Impact on Holiday Travel Planning: The Case of The Russian And the FSU Markets. *International Journal of Online Marketing*, *1*(4), 1–19. doi:10.4018/ijom.2011100101

Fujita, H., & Herrera-Viedma, E. (2018). A Proposal for Sentiment Analysis on Twitter for Tourism-Based Applications. *A Proposal for Sentiment Analysis on Twitter for Tourism-Based Applications*. Academic Press.

Gartner, W. C. (1986). Temporal Influences on Image Change. *Annals of Tourism Research*, *13*(4), 635–644. doi:10.1016/0160-7383(86)90006-X

Gartner, W. C. (1989). Tourism Image: Attribute Measurement of State Tourism Products Using Multidimensional Scaling Techniques. *Journal of Travel Research*, *28*(2), 16–20. doi:10.1177/004728758902800205

Gašpar, D., & Mabić, M. (2017). Strengths and Limitations of Social Media Analytics Tools. In Applying Predictive Analytics Within the Service Sector (pp. 198–219). Hershey, PA: IGI Global. doi:10.4018/978-1-5225-2148-8.ch012

Gedikoglu, Z.A. (2018). Exploring Destination Image Themes on Twitter, Before, During, and After Terror Attacks in Paris: An Application of Agenda Setting Theory.

Giglietto, F., Rossi, L., & Bennato, D. (2012). The Open Laboratory: Limits and Possibilities of Using Facebook, Twitter, And YouTube As A Research Data Source. *Journal of Technology in Human Services*, *30*(3-4), 145–159. doi:10.1080/15228835.2012.743797

González-Bailón, S. (2013). Social Science in The Era of Big Data. *Policy and Internet*, *5*(2), 147–160. doi:10.1002/1944-2866.POI328

Govers, R., & Go, F. M. (2003). Deconstructing Destination Image in The Information Age. *Information Technology & Tourism*, *6*(1), 13–29. doi:10.3727/109830503108751199

Govers, R., & Go, F. M. (2004). Projected Destination Image Online: Website Content Analysis of Pictures and Text. *Information Technology & Tourism*, *7*(2), 73–89. doi:10.3727/1098305054517327

Govers, R., Go, F. M., & Kumar, K. (2007). Promoting Tourism Destination Image. *Journal of Travel Research*, *46*(1), 15–23. doi:10.1177/0047287507302374

Govers, R., Go, F. M., & Kumar, K. (2007). Virtual Destination Image A New Measurement Approach. *Annals of Tourism Research*, *34*(4), 977–997. doi:10.1016/j.annals.2007.06.001

Gretzel, U. (2006). Consumer Generated Content–Trends and Implications for Branding. *Ereview of Tourism Research*, *4*(3), 9–11.

Guest, G., MacQueen, K. M., & Namey, E. E. (2011). *Applied Thematic Analysis*. Sage Publications.

Gunn, C. A. (1972). *Vacationscape: Designing Tourist Regions (Vol. 1.)*. Austin: Bureau of Business Research, University of Texas at Austin.

Hunt, J. D. (1975). Image as A Factor in Tourism Development. *Journal of Travel Research*, *13*(3), 1–7. doi:10.1177/004728757501300301

Hunter, W. C. (2013). China's Chairman Mao: A Visual Analysis of Hunan Province Online Destination Image. *Tourism Management*, *34*, 101–111. doi:10.1016/j.tourman.2012.03.017

Hunter, W. C. (2016). The Social Construction of Tourism Online Destination Image: A Comparative Semiotic Analysis of The Visual Representation of Seoul. *Tourism Management*, *54*, 221–229. doi:10.1016/j.tourman.2015.11.012

Jalilvand, M. R., & Samiei, N. (2012). The Impact of Electronic Word of Mouth on A Tourism Destination Choice: Testing the Theory of Planned Behavior (TPB). *Internet Research: Electronic Networking Applications and Policy*, *22*(5), 591–612. doi:10.1108/10662241211271563

Jenkins, A. S., & Mazer, J. P. (2018). #Notokay: Stories of Sexual Assault in The Midst of the 2016 US Presidential Election. *Qualitative Research Reports in Communication*, *19*(1), 9–17. doi:10.1080/17459435.2017.1404487

Kislali, H., Kavaratzis, M., & Saren, M. (2016). Rethinking Destination Image Formation. *International Journal of Culture, Tourism and Hospitality Research*, *10*(1), 70–80. doi:10.1108/IJCTHR-05-2015-0050

Law, R., & Cheung, S. (2010). The Perceived Destination Image of Hong Kong As Revealed in The Travel Blogs of Mainland Chinese Tourists. *International Journal of Hospitality & Tourism Administration*, *11*(4), 303–327. doi:10.1080/15256480.2010.518521

Leung, D., Law, R., Van Hoof, H., & Buhalis, D. (2013). Social Media in Tourism and Hospitality: A Literature Review. *Journal of Travel & Tourism Marketing*, *30*(1-2), 3–22. doi:10.1080/10548408.2013.750919

Lin, C. H., Morais, D. B., Kerstetter, D. L., & Hou, J. S. (2007). Examining the Role of Cognitive and Affective Image in Predicting Choice Across Natural, Developed, And Theme-Park Destinations. *Journal of Travel Research*, *46*(2), 183–194. doi:10.1177/0047287507304049

Lin, C. H., Morais, D. B., Kerstetter, D. L., & Hou, J. S. (2007). Examining the Role of Cognitive and Affective Image in Predicting Choice Across Natural, Developed, And Theme-Park Destinations. *Journal of Travel Research*, *46*(2), 183–194. doi:10.1177/0047287507304049

Llodrà-Riera, I., Martínez-Ruiz, M. P., Jiménez-Zarco, A. I., & Izquierdo-Yusta, A. (2015). A Multidimensional Analysis of The Information Sources Construct and Its Relevance for Destination Image Formation. *Tourism Management*, *48*, 319–328. doi:10.1016/j.tourman.2014.11.012

Lo, I. S., McKercher, B., Lo, A., Cheung, C., & Law, R. (2011). Tourism and Online Photography. *Tourism Management*, *32*(4), 725–731. doi:10.1016/j.tourman.2010.06.001

Lu, W., & Stepchenkova, S. (2015). User-Generated Content as A Research Mode in Tourism and Hospitality Applications: Topics, Methods, And Software. *Journal of Hospitality Marketing & Management*, *24*(2), 119–154. doi:10.1080/19368623.2014.907758

MacKay, K. J., & Fesenmaier, D. R. (1997). Pictorial Element of Destination in Image Formation. *Annals of Tourism Research*, *24*(3), 537–565. doi:10.1016/S0160-7383(97)00011-X

Mak, A. H. (2017). Online Destination Image: Comparing National Tourism Organization's and Tourists' Perspectives. *Tourism Management*, *60*, 280–297. doi:10.1016/j.tourman.2016.12.012

Markham, A., & Buchanan, E. (2012). Ethical Decision-Making and Internet Research: Version 2.0. The Aoir Ethics Working Committee. Retrieved from aoir.org/reports/ethics2.pdf

Martensen, A., Brockenhuus-Schack, S., & Zahid, A. L. (2018). How Citizen Influencers Persuade Their Followers. *Journal of Fashion Marketing and Management: An International Journal*, *22*(3), 335–353. doi:10.1108/JFMM-09-2017-0095

Martín-Santana, J. D., Beerli-Palacio, A., & Nazzareno, P. A. (2017). Antecedents and Consequences of Destination Image Gap. *Annals of Tourism Research*, *62*, 13–25. doi:10.1016/j.annals.2016.11.001

Mochón, F., & Martínez, O. S. (2014). A first approach to the implicit measurement of happiness in Latin America through the use of social networks. *IJIMAI*, *2*(5), 16–22. doi:10.9781/ijimai.2014.252

Munar, A. M. (2012). Social media strategies and destination management. *Scandinavian Journal of Hospitality and Tourism*, *12*(2), 101–120. doi:10.1080/15022250.2012.679047

Nadeau, J., Heslop, L., O'Reilly, N., & Luk, P. (2008). Destination in A Country Image Context. *Annals of Tourism Research*, *35*(1), 84–106. doi:10.1016/j.annals.2007.06.012

Neuhofer, B., Buhalis, D., & Ladkin, A. (2014). A Typology of Technology-Enhanced Tourism Experiences. *International Journal of Tourism Research*, *16*(4), 340–350. doi:10.1002/jtr.1958

Nghiêm-Phú, B. (2014). A Review of Destination Image Studies From 2008 to 2012. *European Journal of Tourism Research*, *8*, 35.

Oh, H., & Nguyen, C. (2010). Influence of Retweets.

Pan, B., MacLaurin, T., & Crotts, J. C. (2007). Travel Blogs and The Implications for Destination Marketing. *Journal of Travel Research*, *46*(1), 35–45. doi:10.1177/0047287507302378

Pearce, P. L. (1977). Mental souvenirs: A study of tourists and their city maps. *Australian Journal of Psychology*, *29*(3), 203–210. doi:10.1080/00049537708255282

Pearce, P. L. (1982). Perceived changes in holiday destinations. *Annals of Tourism Research*, *9*(2), 145–164. doi:10.1016/0160-7383(82)90044-5

Pezenka, I. (2016). Paired Comparisons or Sorting? Comparing Web-Based Methods for Collecting Similarity Data for Large Stimulus Sets for Destination Image Positioning. *Journal of Travel Research*.

Pike, S. (2002). Destination Image Analysis—A Review of 142 Papers from 1973 to 2000. *Tourism Management*, *23*(5), 541–549. doi:10.1016/S0261-5177(02)00005-5

Roth, M. S., & Romeo, J. B. (1992). Matching Product Category and Country Image Perceptions: A Framework for Managing Country-Of-Origin Effects. *Journal of International Business Studies*, *23*(3), 477–497. doi:10.1057/palgrave.jibs.8490276

Russell, J. A., & Pratt, G. (1980). A Description of The Affective Quality Attributed to Environments. *Journal of Personality and Social Psychology*, *38*(2), 311–322. doi:10.1037/0022-3514.38.2.311

Ryan, C., & Cave, J. (2005). Structuring Destination Image: A Qualitative Approach. *Journal of Travel Research*, *44*(2), 143–150. doi:10.1177/0047287505278991

Sameh, A. (2013). A Twitter Analytic Tool to Measure Opinion, Influence and Trust. *Journal of Industrial and Intelligent Information*, *1*(1), 37–45. doi:10.12720/jiii.1.1.37-45

San Martín, H., & Del Bosque, I. A. R. (2008). Exploring the Cognitive–Affective Nature of Destination Image and The Role of Psychological Factors in Its Formation. *Tourism Management*, *29*(2), 263–277. doi:10.1016/j.tourman.2007.03.012

Schmallegger, D., & Carson, D. (2008). Blogs in Tourism: Changing Approaches to Information Exchange. *Journal of Vacation Marketing*, *14*(2), 99–110. doi:10.1177/1356766707087519

Sotiriadis, M. D., & Van Zyl, C. (2013). Electronic Word-Of-Mouth and Online Reviews in Tourism Services: The Use of Twitter by Tourists. *Electronic Commerce Research*, *13*(1), 103–124. doi:10.1007/10660-013-9108-1

Stepchenkova, S., & Mills, J. E. (2010). Destination Image: A Meta-Analysis Of 2000–2007 Research. *Journal of Hospitality Marketing & Management*, *19*(6), 575–609. doi:10.1080/19368623.2010.493071

Stepchenkova, S., & Morrison, A. M. (2006). The Destination Image of Russia: From the Online Induced Perspective. *Tourism Management*, *27*(5), 943–956. doi:10.1016/j.tourman.2005.10.021

Stern, E., & Krakover, S. (1993). The Formation of a Composite Urban Image. *Geographical Analysis*, *25*(2), 130–146. doi:10.1111/j.1538-4632.1993.tb00285.x

Tapachai, N., & Waryszak, R. (2000). An Examination of The Role of Beneficial Image in Tourist Destination Selection. *Journal of Travel Research*, *39*(1), 37–44. doi:10.1177/004728750003900105

Tasci, A. D., Gartner, W. C., & Tamer Cavusgil, S. (2007). Conceptualization and Operationalization of Destination Image. *Journal of Hospitality & Tourism Research (Washington, D.C.)*, *31*(2), 194–223. doi:10.1177/1096348006297290

Telisman-Kosuta, N. (1989). Tourist Destination Image. In *Tourism Marketing and Management Handbook* (pp. 557-561). Academic Press.

Walmsley, D. J., & Jenkins, J. M. (1992). Tourism Cognitive Mapping of Unfamiliar Environments. *Annals of Tourism Research*, *19*(2), 268–286. doi:10.1016/0160-7383(92)90081-Y

Walmsley, D. J., & Young, M. (1998). Evaluative Images and Tourism: The Use of Personal Constructs to Describe the Structure of Destination Images. *Journal of Travel Research*, *36*(3), 65–69. doi:10.1177/004728759803600307

Walters, T. (2016). Using Thematic Analysis in Tourism Research. *Tourism Analysis*, *21*(1), 107–116. doi:10.3727/108354216X14537459509017

Wang, C. Y., & Hsu, M. K. (2010). The Relationships of Destination Image, Satisfaction, And Behavioral Intentions: An Integrated Model. *Journal of Travel & Tourism Marketing*, *27*(8), 829–843. doi:10.1080/10548408.2010.527249

Wang, Y., & Fesenmaier, D. R. (2007). Collaborative Destination Marketing: A Case Study of Elkhart County, Indiana. *Tourism Management*, *28*(3), 863–875. doi:10.1016/j.tourman.2006.02.007

Woodside, A. G., & Lysonski, S. (1989). A General Model of Traveler Destination Choice. *Journal of Travel Research*, *27*(4), 8–14. doi:10.1177/004728758902700402

Xiang, Z., & Gretzel, U. (2010). Role of Social Media in Online Travel Information Search. *Tourism Management, 31*(2), 179–188. doi:10.1016/j.tourman.2009.02.016

ADDITIONAL READING

Garay Tamajón, L., & Cànoves Valiente, G. (2017). Barcelona Seen Through the Eyes of TripAdvisor: Actors, Typologies and Components of Destination Image in Social Media Platforms. *Current Issues in Tourism, 20*(1), 33–37. doi:10.1080/13683500.2015.1073229

Kirilenko, A. P., Stepchenkova, S. O., Kim, H., & Li, X. (2018). Automated Sentiment Analysis in Tourism: Comparison of Approaches. *Journal of Travel Research, 57*(8), 1012–1025. doi:10.1177/0047287517729757

Kladou, S., & Mavragani, E. (2015). Assessing Destination Image: An Online Marketing Approach and The Case of TripAdvisor. *Journal of Destination Marketing & Management, 4*(3), 187–193. doi:10.1016/j.jdmm.2015.04.003

Marine-Roig, E., & Anton Clave, S. (2016). Affective Component of The Destination Image: A Computerized Analysis. In *Destination Marketing: An International Perspective* (pp. 49-58). Academic Press.

Marine-Roig, E., & Clavé, S. A. (2016). A Detailed Method for Destination Image Analysis Using User-Generated Content. *Information Technology & Tourism, 15*(4), 341–364. doi:10.100740558-015-0040-1

Park, S. B., Ok, C. M., & Chae, B. K. (2016). Using Twitter Data for Cruise Tourism Marketing and Research. *Journal of Travel & Tourism Marketing, 33*(6), 885–898. doi:10.1080/10548408.2015.1071688

Uysal, M., Harrill, R., & Woo, E. (2011). Destination Marketing Research: Issues and Challenges. In *Destination Marketing and Management: Theories and Applications* (pp. 99-112). Academic Press.

Yoon, S., Petrick, J. F., & Backman, S. J. (2017). Twitter Power and Sport-Fan Loyalty: The Moderating Effects of Twitter. *International Journal of Sport Communication, 10*(2), 153–177. doi:10.1123/IJSC.2017-0010

KEY TERMS AND DEFINITIONS

Affective Components of Destination Image: An individual's unique emotional interpretation of a destination which is composed of feelings and emotions.

Cognitive Components of Destination Image: An individual's unique knowledge, perceptions and beliefs about the individual attributes of a destination which are observable, descriptive and measurable.

Destination Image: The sum of attitudes, experiences, beliefs, ideas and impressions a non-visitor, visitor, or repeat-visitor has of a tourism destination.

Destination Image Formation: A construction of a mental representation on the basis of information cues selected and delivered by the image formation agents.

Influencer: An influencer is a user on social media who has access to a large audience, is mentioned by others and whose 's posts are re-posted so that can persuade others by virtue of their authenticity and reach.

Online/Virtual Destination Image: The sum of attitudes, experiences, beliefs, ideas and impressions a non-visitor, visitor, or repeat-visitor has and has communicated through an online communication platform of a tourism destination.

Sentiment Analysis: The Process of analyzing various sources of unstructured online communication data to mine conversations online and to determine deeper context as they apply to a topic, brand, or theme.

User Generated Content: Any type of shared communicated content that has been created and shared through online platforms by unpaid contributors.

Chapter 9
The Discount Effect in Food and Beverage Advertising and Instagram's Likes Power:
An Experimental Study

Nilton Gomes Furtado
https://orcid.org/0000-0002-6372-4693
University of Aveiro, Portugal

Pedro H. Drudi
Anhembi Morumbi University, Brazil

Julia Vasconcelos Furtado
University of Aveiro, Portugal

Rafael de Vasconcelos Silva
University of Aveiro, Portugal

Lauro César Vieira Filho
University of Aveiro, Portugal

ABSTRACT

This study aims to understand the effect of evident sales promotion on the number of likes on Instagram. Due to the rising use of social media to communicate with the market and the promotion of products to guests, few academic investigations have been made in this area. Most technology research in the hospitality industry has focused on technology in service and production areas of operations or the use of web-based marketing for tourism operations and travel agents. The two most effective sales offer patterns were used in a 2×2 experiment, with the intention of finding out if a large amount of likes on Instagram influences the purchase intention of the product offered. Four randomized questionnaires were distributed among 4 university groups, obtaining 923 respondents. To validate the hypotheses the study proceeded with the descriptive technique and analysis of variances (ANOVA).

DOI: 10.4018/978-1-7998-1947-9.ch009

INTRODUCTION

The food and beverage industry is a critical component of the Brazilian economy, even when facing a period of economic recession in 2016, with a decrease of -3.75% on revenues in 2015 (Brazilian Association of Bars and Restaurants - ABRASEL), this sector contributed with approximately 10.1% of the Brazilian Gross Domestic Product (ABIA, 2017). On the other hand, bar's and restaurants' segment found that at least 32.9% of Brazilians have some kind of meal outside - not home cooking - on a daily basis. IFB's (Food Service Brazil Institute) recent studies indicates that 64% of sales are from small and medium-sized establishments.

The industry's competitiveness makes the use of discount advertisement something common among companies, as means to attract consumer's attention (Aureliano-Silva, Strehlau, & Strehlau, 2017). Discounts are effective in selling products directly to consumer. It is widely used and can be presented in several formats, such as in percentage or monetary value (Aureliano-Silva, Brandão, Strehlau, Pimenta, & Pimenta, 2015); offering a second product as a gift at the purchase of the product on sale, is reported as well in some cases. Such offers have already been used in retail for years, leading to academic research on the matter (Santini, Lübeck, & Hoffmann, 2015). However, over the past two decades there had been an increase in the use of digital media and social networks for marketing purposes, such as the sale/ discount advertising (Needles & Thompson, 2013).

Due to food and beverage industry size and relevance, it is essential to understand how to use components of foodservice operations and technology to increase effectiveness in several business aspects such as marketing, recruitment, training, guest service, and global operations. With increasing availability and use of technology and social media, social networking platforms had become a new media to be used by restaurants and other service organizations (DiPietro, Crews, Gustafson, & Strick, 2012; Kaplan & Haenlein, 2009). Social media can be used in many ways, including content, social networking websites or blogs (Facebook and Twitter), and collaborative projects. Social media can be understood as "[i]nternet applications that help consumers share insights, experiences and perspectives" (Kaplan & Haenlein, 2009).

The social media is defined as a web-based service that allows individuals to build a public network or public profile within a delimited system; create a list of other users with whom they share a connection and view/cross its own list of connections, as well as those made by others within the system (Boyd & Ellison, 2008). These platforms allow people to find others with similar interests and to share those interests through the Internet. The survey by the Global Web Index found that the Brazilian public spent about 40 minutes online every day, in 2015, making Brazil the third country in the world (only behind Thailand and Saudi Arabia) in the use of smartphone on the internet (Bayindir & Kavanagh, 2018).

The smartphones' use as a means of access to social media continues to increase, and small and medium-sized companies, especially in bars and restaurants industry, have little advertising and marketing strategy, once only the big corporations actually invest in advertising and branding. Social media like Facebook, MySpace, Twitter, LinkedIn and others can be used by the industry to recruit staff, advertise and solicit guest feedback (Furtado & Oliveira, 2016).

The current study aims to understand the discount effect on consumer's opinion/decision, when it is advertised with the number of likes on Instagram. Due to the recent advent of social media use to communicate with market and clients, few academic investigations in this field have been held. Most studies on technology in hospitality industry has focused on technology in service and production operations areas, or in the web-based marketing use for tourism operations and travel agents (DiPietro et al., 2012;

Furtado, Canto, Oliveira, & Moretti, 2017; Gregory, Wang, & DiPietro, 2010; Lovelock & Gummesson, 2004). The two most effective promotion patterns were used in a 2×2 experiment, aiming to find out if a large number of likes on Instagram influences on intended purchase of the product being promoted.

The present chapter is divided in:

1. Theoretical reference on sales promotions.
2. Theoretical reference on advertising techniques.
3. Method and control procedures.
4. Experimental data analysis.
5. Conclusion.

SALES PROMOTIONS

Several studies have shown the use of discounts as a marketing tool to consumer's attraction and influencing purchase intention (Aureliano-Silva et al., 2017; Santini et al., 2015). The relationship between price and purchase intention is directly linked to two factors: product quality and price aversion (Oliveira, 2007). In Price Search strategy, the buyer chooses the most expensive product to avoid the risk of buying a low-quality product. In Price Aversion strategy, on the other hand, the buyer opts for a cheaper product in order to avoid the risk of making great purchase sacrifice. By relating the assertion to pricing in promotions, it can be understood that the price presentation in promotions can be related to consumer's behavior and purchase intention (Aureliano-Silva et al., 2017; Kukanja, Gomezelj Omerzel, & Kodrič, 2017; Namin, 2017): The use of dollar signs or any other element referring to the payment will be less effective to the purchase intention. However, according Aureliano-Silva et al. (2015), when using the discount in currency, the relation with purchase intention is greater.

Another attribute used in promotional actions is the use of discount by percentage. Based on the research carried out by (Medeiros, Gosling, Alves, & Vera, 2015) discounts up to 25% are effective in influencing purchase intention. Above this value, the product's perception of quality is affected, leading the consumer to characterize the product as 'low quality'. The work of Aureliano-Silva et al. (2015) shows a lower purchase intention when a percentage discount is used, when compared to monetary discount. Furthermore, when the attributes of the product are omitted such as the discount value, it can lead to a negative interpretation by the consumer over the product's discount, particularly due to the connection to the need of payment (Aureliano-Silva et al., 2015; Santini et al., 2015). However, such research had been carried out on products from different industries, demanding a check on the results in food and beverage industry, and the study indicates only the consumer's purchase intention, as well as the quality and value perceptions of the product, not measuring the attractiveness of the discount in percentage when compared to the currency discount. Thus, hypotheses 1, 2, and 3 were created:

H1: The currency discount will lead to greater purchase intention, when compared to percentage discount.

H2: The influence of the higher number of likes on the sales with a monetary discount of 7 Reais (BRL), will have greater attractiveness than the same offer with lower number of likes.

H3: The influence of the highest number of likes on the 20% discount offer will be more attractive than the same offer with lower number of likes.

It is also common to offer free items in addition to the product offered. Such offers may be in the form of a bonus-pack, popularly used as an example, in the lightweight format buy 2 pay 1 (Twenge, Campbell, Hoffman, & Lance, 2010) or freebie items, which are products other than the one sold but offered as a prize for the purchase of the product. Bokunewicz and Shulman (2017) present the effectiveness of freebie items in fast-food industry on purchase intention and in offer attractiveness', considering that the product offered as a gift should not be much higher value nor much lower than the currency discount, which would cause consumer mistrust (Ha, 2015).

ADVERTISING TECHNIQUES

Instagram is a photo and video sharing site with followers and friends, and even links to other social networks such as Facebook and Twitter. It is no longer a secondary social network, and it should be part of the marketing planning of any Business to Consumer (B2C) company (Nobre & Silva, 2014). It is even estimated that about 67% of the world's top brands are using Instagram, reaching over 100 million active users on the network (Nuno, 2018).

Social media marketing strategy is useful in market and academics resources (Nobre & Faria, 2017; Nobre & Silva, 2014). Social media with Facebook and Instagram should be used by companies, small and middle business. (Nobre & Silva, 2014; Soares, Pinho, & Nobre, 2012) The influence of the social media show that it can be used as a new method to increase the number of clients, and to bring consumers closer to the company.

There is an increased use of social media for advertising and feedback (Singh, Sarfaraz, Sarfaraz, & Jenab, 2015). Beyond that, it influences increased or decreased revenue management. Furthermore, this phenomenon has lowered the product lifecycle at restaurants, that in order to make profit, must opt for technology that responds to changing customer requirements - both implicit and explicit, efficiently and profitably (Furtado, Pereira, Pereira, & Moreira, 2019; Furtado & Cunha, 2017; Furtado, Furtado, Vieira, & Cunha, 2019).

The social media use brings satisfaction to the customer, and in some cases, may be able to generate loyalty to a restaurant (Furtado & Cunha, 2017; Furtado et al., 2019; Helena Nobre & Silva, 2014; Soares et al., 2012). However, this factor is not fundamental to generate loyalty, while dining experience, environment and other factors directly influence customers loyalty in catering.

We have to pay special attention to gender, since its influence on social media participation stands out, showing that the attitude of technology usage may have a relation to this demographic variable (Venkatesh, Morris et al., 2003).

Based on these studies, Hypothesis 4 and 5 were elaborated:

H4: The gender influence in purchase intention of product with greater number of likes.
H5: The gender influence in purchase intention of products that presents monetary discount value on percentage.

METHOD AND CONTROL PROCEDURES

The survey used questionnaires as data collection procedure, with the purpose of collecting data within the subjects to generate the information necessary to verify and attain the study objectives.

The sample was composed of undergraduate and graduate students from four universities: two public and two private universities, located in the city of São Paulo (BRAZIL). Four randomized questionnaires were distributed to four groups from each university, that obtained 923 respondents.

The survey presented one advertisement with seven questions, being two demographics and five questions about purchase intention related to the picture presented and the perception of the number of likes. Following are figures of each survey. The figure 1 shows the variations of the picture used at the surveys, there were 4 variations - each one had a promotion type (percentage or monetary discount) and number of likes - 20% discount with lower likes, 7 reais discount with lower likes, 20% discount with higher likes and discount of 7 reais with higher number of likes.

As a research technique, a convenience sample had been used, a common technique in which one chooses a population sample that is accessible. Therefore, subjects involved in the research were chosen because of their availability, not through statistical criteria. In general, convenience sample allows greater operational accessibility and lower sample costs, however, those results may be limited in providing general statements with statistical accuracy regarding population sample.

EXPERIMENTAL DATA ANALYSIS

In the first questionnaire, the lower number of likes was analyzed when compared to the monetary discount of 7 Reais (BRL) presented in the advertisement. There were 137 respondents, in which 62 (45.3%) were male and 75 (54.7%) female, with an average age of 25 years. The second questionnaire evaluated the influence of the lower number of likes compared to the percentage discount (20%) in the advertisement. There were 139 respondents participating, being 42 (30.2%) males, 92 (66.2%) females and 5 (3.6) chose not to declare their gender, with a mean age of 23 years. Moreover, a third question-

Figure 1. Survey promotions images

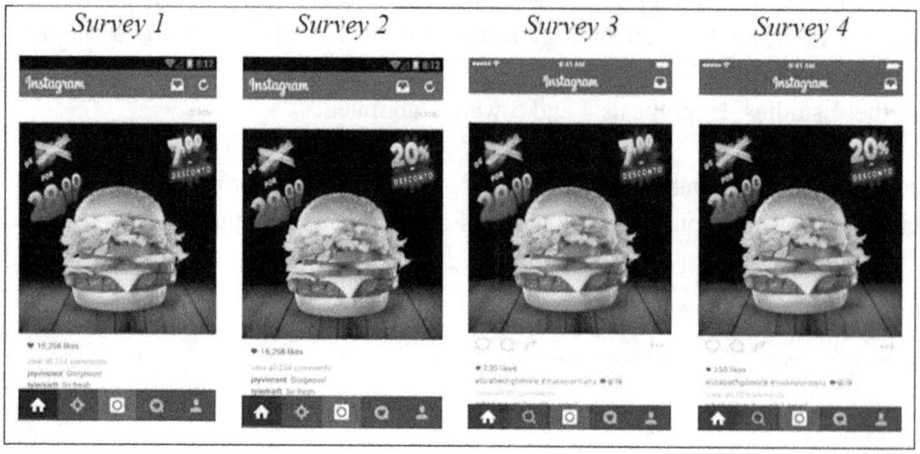

naire was elaborated, to evaluate the influence of the higher number of likes compared to the monetary discount of 7 Reais (BRL) presented in the advertisement. There were 295 respondents, being 131 (44.6%) male, 161 (54.8%) female and 2 (0.7) preferred did not declare their gender, with a mean age of 28 years. Finally, with the fourth questionnaire, we established the criterion of 20% discount with a higher number of likes in the ad. There were 342 respondents, which 142 (41.6%) were male, 194 (56.9%) female and 5 (1,5) preferred did not declare their gender, with mean age of 23 years.

In order to validate the hypotheses, we proceed with descriptive technique and analysis of variances (ANOVA). P <0.05 was used as a parameter of significance (Hair, 2010; Hair, Black, Babin, & Anderson, 2014). The statistical software SPSS version 24 was used for statistical analysis.

The Variance Homogeneity Test (table 1) showed that the purchase intention with monetary discount of 7 Reais (BRL) presented an average significance of p-value (sig 0.575), while the purchase intention discount with 20% percentage discount, obtained a significance mean p-value (sig 0.206), showing that the monetary discount value is more representative than the percentage value.

The ANOVA results presented on table 2 above, determined that there is a statistically significant difference (z = 0.680, p = 0.411) between the monetary discount and the percentage discount variables (z = 0.019, p = 0890). The following are the results of the TukeySchefféPost-Hoc test (table 3), which indicates the presence or absence of a statistically significant difference in the number of Likes at an individual level and gender difference.

Although the ANOVA test had indicated the presence of a statistically significant difference (p = 463) in number of likes between the two categories (monetary and percentage), the TukeySchefféPost-Hoc test did not statistically identify any significant differences at individual level.

The results of TukeySchefféPost-Hoc test on table 3 also indicate that no homogeneous subset is present at a statistically significant combined level (p = .164) between the two variables, analyzing that H1 - The currency promotion will have greater purchase intention, compared to the promotion in percentage, is supported.

Through the multiple comparisons' observation, it shows the influence of gender in the purchase of the product with greater monetary discount to the percentage with the male gender significance (p = 0.224), while the female gender has (p = 983). There was no significance related to those who do not

Table 1. Test of Homogeneity of Variance

		Levene' Statistics	df1	df2	*Sig.*
Purchase intention with monetary discount of 7 Reais (BRL)	Based on mean	,555	2	136	*,575*
	Based on median	,515	2	136	*,599*
	Based on median and with df adjusted	,515	2	135,944	*,599*
	Based on trimmed mean	,472	2	136	*,625*
Purchase Intention with 20% percentage discount	Based on mean	1,599	2	136	*,206*
	Based on median	1,136	2	136	*,324*
	Based on median and with df adjusted	1,136	2	135,640	*,324*
	Based on trimmed mean	1,537	2	136	*,219*

Table 2. Test ANOVA

		Between Groups				In Groups	Total
		Combined	Linear Term				
			Inconsiderate	Considerate	Desviation		
Purchase intention with monetary discount of 7 Reais (BRL)	Sum of Squares	11,726	1,732	2,175	9,552	434,921	446,647
	Df	2	1	1	1	136	138
	Mean Square	5,863	1,732	2,175	9,552	3,198	
	Z	1,833	,542	,680	2,987		
	Sig.	,164	,463	,411	,086		
Purchase Intention with 20% percentage discount	Sum of Squares	,701	,559	,048	,653	343,745	344,446
	Df	2	1	1	1	136	138
	Mean Square	,350	,559	,048	,653	2,528	
	Z	,139	,221	,019	,258		
	Sig.	*,871*	*,639*	*,890*	*,612*		

Table 3. Multiple comparisons

Gender			Mean Difference (I-J)	Std. Error	Sig.	95% Confidence Interval	
						Lower Bound	Upper Bound
Purchase intention with monetary discount of 7 Reais (BRL)	Female	Male	,553	,333	,224	-,24	1,34
		Not to declare	-,604	,821	,743	-2,55	1,34
	Male	Female	-,553	,333	,224	-1,34	,24
		Not to declare	-1,157	,846	,361	-3,16	,85
	Not to declare	Female	,604	,821	,743	-1,34	2,55
		Male	1,157	,846	,361	-,85	3,16
Purchase Intention with 20% percentage discount	Female	Male	-,052	,296	,983	-,75	,65
		Not to declare	,343	,730	,885	-1,39	2,07
	Male	Female	,052	,296	,983	-,65	,75
		Not to declare	,395	,752	,859	-1,39	2,18
	Not to declare	Female	-,343	,730	,885	-2,07	1,39
		Male	*-,395*	*,752*	*,859*	*-2,18*	*1,39*

declare the gender x female (p = 885), do not declare gender x male (0.361). Those comparisons sustain the H5 - Gender influence in the intention to purchase products that presents monetary discount value.

The descriptive statistics on table 4 also indicate an average value of 2.85 (95% confidence interval: lower bound of 2.40 upper bound of 3.31) in purchase intention with monetary discount of 7 Reais (BRL) (greater number of likes) males. Also, mean value of 3.17 (95% confidence interval: lower bound of 2.76 upper bound of 3.59) in purchase intention with a monetary discount of 7 Reais (BRL) (greater number of likes) female, with a standard deviation of 1,782 males and the deviation standard 1,804. Therefore, the statistics sustains hypothesis H3 - The influence of gender in the intention to buy the product with a greater number of likes.

Then, a One-Way ANOVA test was performed to test the hypothesis and to identify the presence or absence of a statistically significant difference in the content level of likes on Instagram among the four categories: purchase intention with monetary discount of 7 Reais (BRL) (higher number of likes), purchase intention with 20% discount (higher number of likes), purchase intention with monetary discount of 7 Reais (BRL) (lower number of likes) and purchase intention with 20% discount (lower number of likes).

Following the contextual variables: number of likes, purchase intention, money discount and percentage discount, hypothesis H2 - The influence of the higher number of likes on the promotional discounted money - had been tested using the One-Way ANOVA test to identify the presence of differences statistically between categories.

The ANOVA results presented at table 5 above, can determine that there is a statistically significant difference (z = 1,070, p = 0.303) between the monetary discount variables with the higher number of likes and discount with percentage with the higher number of likes (z = 0.183, p = 0.670). Among the variables with the lower number of likes, purchase intention with monetary discount of 7 Reais (lower number of likes), the significance was (z = 0.028, p = 0.867), and also purchase intention with 20% percentage discount (lower number of likes) its significance was (z = 0.629, p = 0.429).

Table 4. Descriptive analysis

		N	*Mean*	*Std. Error*	*Error*	*95% Confidence Interval*	
						Lower Bound	**Upper Bound**
Purchase intention with monetary discount of 7 Reais (BRL) (higher number of likes)	Female	62	2,85	1,782	,226	2,40	3,31
	Male	75	3,17	1,804	,208	2,76	3,59
	Total	137	3,03	1,794	,153	2,73	3,33
Purchase Intention with 20% percentage discount (higher number of likes)	Female	62	2,60	1,541	,196	2,21	2,99
	Male	75	2,48	1,630	,188	2,10	2,86
	Total	137	2,53	1,586	,136	2,26	2,80
Purchase intention with monetary discount of 7 Reais (BRL) (lower number of likes)	Female	62	1,53	,503	,064	1,40	1,66
	Male	75	1,55	,501	,058	1,43	1,66
	Total	137	1,54	,500	,043	1,46	1,62
Purchase Intention with 20% percentage discount (lower number of likes)	Female	62	1,53	,503	,064	1,40	1,66
	Male	75	1,60	,493	,057	1,49	1,71
	Total	*137*	*1,57*	*,497*	*,042*	*1,49*	*1,65*

Table 5. ANOVA

		Sum of Squares	df	Mean Square	Z	Sig.
Purchase intention with monetary discount of 7 Reais (BRL) (greater number of likes)	Between Groups	3,443	1	3,443	1,070	,303
	Within groups	434,440	135	3,218		
	Total	437,883	136			
Purchase Intention with 20% percentage discount (greater number of likes)	Between Groups	,463	1	,463	,183	,670
	Within groups	341,639	135	2,531		
	Total	342,102	136			
Purchase intention with monetary discount of 7 Reais (BRL) (a smaller number of likes)	Between Groups	,007	1	,007	,028	,867
	Within groups	34,022	135	,252		
	Total	34,029	136			
Purchase Intention with 20% percentage discount (a smaller number of likes)	Between Groups	,156	1	,156	,629	,429
	Within groups	33,435	135	,248		
	Total	*33,591*	*136*			

The presence of statistically significant results in the One-Way ANOVA (p = 0.867) in the purchase intention with monetary discount with greater number of likes, as presented above, indicates that H2 is supported when related to the level of likes contained in Instagram's posts. That is the case, the results are statistically significant, since the results are more sporadic due to the homogeneity of the group of students.

To support H2, the results should indicate a higher level of likes in the monetary promotion when compared to the lower number of likes in the monetary discount, so these variables could be statistically significant.

CONCLUSION

The role of emotions in a marketing and consumer decision-making context has been highlighted by several authors, in many contexts (Israeli, Lee, & Karpinski, 2017; Campbell, 2013; Leung, Bai, & Stahura, 2015). The reference to emotion was also presented when related to purchases of products (goods) and services (Berry, Zeithaml, & Parasuraman, 1990; Campbell, 2013; Soares et al., 2012). In a more specific sense of service, (Becker, Nobre, & Kanabar, 2013; Nogueira Tomas, Pierre Meschgrahw, & Lúcia Chicarelli Alcantara, 2012; Soares et al., 2012) leverage the intimate and intangible nature of services as a means of highlighting why emotion may be a particular influence on the services' purchase.

It had been shown that the presence of the attributes of higher likes influences the consumer and the presence of discount in percentage is significant. The correspondence between the 4 questionnaires with the ads was considered in the heterogeneous groups of students (Ha, 2015; Souza et al., 2015), supporting H2 hypotheses - The influence of the greater number of likes on the offer with monetary discount of 7 Reais, will have greater attractiveness than the same offer with less number of likes; H3 - The influence of the greater number of likes in an offer with a 20% discount will have greater attractiveness than the same offer with less number of likes; and H4 - The influence of gender in the intention to buy the product with greater number of likes.

The promotional images production only performs in a sense of creating forms of action that brands and platforms modulate and calibrate openly (Brandau, 2010; Furtado & Oliveira, 2016). Brands' efforts to experiment with mobile media imaging machines, our bodies, and cultural spaces draw attention to the need to critically examine how modes of participatory, discursive, and algorithmic control are interrelated (Frazer & Merrilees, 2012; Joachimsthaler & Aaker, 1997; Singh & Pattanayak, 2016). The hypothesis H1 - The promotion in currency will present greater intention to buy, compared to promotion in percentage, was supported, and it is important to understand how companies can improve their communication and marketing for the consumer, and if the social environment plays a role in the success. The effectiveness of social media marketing is difficult to derive, because the amount of advertising that an ad or business can receive on the Internet can depend heavily on word of mouth (Campbell, 2013; Soares et al., 2012).

As mobile social media like Instagram become more algorithmic, an important issue concerns the images produced under the trademark hashtags or promotions can influence how gender norms are reproduced in mobile social media (Aureliano-Silva et al., 2015, 2017; Berta, 2009). Supporting, therefore, hypothesis H5 - The influence of gender in the intention to buy products that presents value of monetary discount to the percentage.

In future works, it would be interesting to analyze how the algorithms recognize the performance of gender norms as associated patterns. The analysis of how brands reinforce, and exploit gender norms needs to take into account the interdependence between the discursive and algorithmic aspects of mobile social media platforms.

REFERENCES

ABIA. (2017). Números do setor de alimentos e bebidas, Brasil, 2017. Retrieved from https://www.abia.org.br/vsn/anexos/faturamento2017.pdf

Aureliano-Silva, L., Brandão, M. M., Strehlau, S., Pimenta, R. F., & Pimenta, R. F. (2015). O efeito do formato de apresentação de desconto em ambiente crowding: Um estudo experimental. *RACE - Revista de Administração Contabilidade e Economia*, *14*(2), 653. doi:10.18593/race.v14i2.6735

Aureliano-Silva, L., Strehlau, S., & Strehlau, V. (2017). Please! Wrap this gift for me: The brands effect with different promotional frames. *ICTACT Journal on Management Studies*, *3*(1), 452–456. doi:10.21917/ijms.2017.0061

Bayindir, N., & Kavanagh, D. (2018). *Global Web Index - Social*. Retrieved from www.globalwebindex.com

Becker, K., Nobre, H., & Kanabar, V. (2013). Monitoring and protecting company and brand reputation on social networks: When sites are not enough. *Global Business and Economics Review*, *15*(2/3), 293. doi:10.1504/GBER.2013.053075

Berry, L. L., Zeithaml, V. A., & Parasuraman, A. (1990). Five Imperatives for Improving Service Quality. *Sloan Management Review*, *31*(4), 29–38. Retrieved from https://search.proquest.com/openview/8b3a8e9fbf82128a35d092cf3a81003f/1?pq-origsite=gscholar&cbl=26142

Berta, D. (2009). Brands say employees atwitter for social sites. *Nation's Restaurant News*, *43*(15), 1+14.

Bokunewicz, J. F., & Shulman, J. (2017). Influencer identification in Twitter networks of destination marketing organizations. *Journal of Hospitality and Tourism Technology*, *8*(2), 205–219. doi:10.1108/JHTT-09-2016-0057

Boyd, M. D., & Ellison, B. N. (2008). Social Network Sites: Definition, History, and Scholarship. *Journal of Computer-Mediated Communication*, *13*(3), 210–230. doi:10.1111/j.1083-6101.2007.00393.x

Brandau, M. (2010). *Chipotle, McD sign on with Facebook Deals*. Nation's Restaurant News.

Campbell, A. (2013). Word-of-Mouth Communication and Percolation in Social Networks. *The American Economic Review*, *103*(6), 2466–2498. doi:10.1257/aer.103.6.2466

de Oliveira, S. L. I. (2007). *Desmistificando o marketing*. São Paulo: Novatec Editora.

DiPietro, R. B., Crews, T. B., Gustafson, C., & Strick, S. (2012). The Use of Social Networking Sites in the Restaurant Industry: Best Practices. *Journal of Foodservice Business Research*, *15*(3), 265–284. doi:10.1080/15378020.2012.706193

Frazer, L., & Merrilees, B. (2012). Pioneering Asian Franchise Brands: Pho24 in Vietnam. *Journal of Marketing Channels*, *19*(4), 295–309. doi:10.1080/1046669X.2012.700264

Furtado, J. V., Pereira, A. R., Pereira, I., & Moreira, A. C. (2019). Does Theory Really Fit Real Life Situations? In Handbook of Research on Entrepreneurship, Innovation, and Internationalization (pp. 419–438). Hershey, PA: IGI Global. doi:10.4018/978-1-5225-8479-7.ch016

Furtado, N. G., Canto, R. S., de Oliveira, S. L. I., & Moretti, S. L. A. (2017). Perceptions in the use of technology for payments: A study of customer behavior in food and beverage sector. *Ágora - Revista de Divulgação Científica*, *22*(2), 4–23.

Furtado, N. G., & Cunha, R. da S. (2017). *A influência da adoção de technologia de meios de pagamentosna satisfação do consumidor: Um estudo com consumidores de restaurantes*. Universidade Anhembi Morumbi.

Furtado, N. G., & de Oliveira, S. L. I. (2016). *Expectativas e percepções no uso da tecnologia de pagamentos voltada ao cliente*. São Paulo: III - Encontro Dos Mestrados Profissionais Em Administração.

Furtado, N. G., Furtado, J. V., Vieira, L. C., & Cunha, R. da S. (2019). The influence of technologic payment adoption in satisfaction: a study with restaurant costumers. *International Journal of Business Excellence*. doi:10.1504/IJBEX.2019.10020420

Gregory, A., Wang, Y., & DiPietro, R. B. (2010). Towards a functional model of website evaluation: A case study of casual dining restaurants. *Worldwide Hospitality and Tourism Themes*, *2*(1), 68–85. doi:10.1108/17554211011012603

Ha, A. (2015). An Experiment: Instagram Marketing Techniques and Their Effectiveness. *Communication Studies*. Retrieved from http://digitalcommons.calpoly.edu/comssp/185

Hair, J. F. (2010). *Multivariate data analysis*. Prentice Hall.

Hair, J. F., Black, W. C., Babin, B. J., & Anderson, R. E. (2014). *Multivariate Data Analysis* (7th ed.)., Retrieved from www.pearsoned.co.uk

Israeli, A., Lee, S. A., & Karpinski, A. C. (2017). Investigating the Dynamics and the Content of Customers' Social Media Reporting after a Restaurant Service Failure. *Journal of Hospitality Marketing & Management*, *26*(6), 606–626. doi:10.1080/19368623.2017.1281193

Joachimsthaler, E., & Aaker, D. A. (1997). Building brands without mass media. *Harvard Business Review*, *75*(1). PMID:10174453

Kaplan, A. M., & Haenlein, M. (2009). The fairyland of Second Life: Virtual social worlds and how to use them. *Business Horizons*, *52*(6), 563–572. doi:10.1016/j.bushor.2009.07.002

Kukanja, M., Gomezelj Omerzel, D., & Kodrič, B. (2017). Ensuring restaurant quality and guests' loyalty: An integrative model based on marketing (7P) approach. *Total Quality Management & Business Excellence*, *28*(13–14), 1509–1525. doi:10.1080/14783363.2016.1150172

Leung, X. Y., Bai, B., & Stahura, K. A. (2015). The Marketing Effectiveness of Social Media in the Hotel Industry: A Comparison of Facebook and Twitter. *Journal of Hospitality & Tourism Research (Washington, D.C.)*, *39*(2), 147–169. doi:10.1177/1096348012471381

Lovelock, C., & Gummesson, E. (2004). Whither services marketing? *Journal of Service Research*, *7*(1), 20–41. doi:10.1177/1094670504266131

Medeiros, S. A. De, Gosling, M., Alves, L., & Vera, R. (2015). *Emoções em Experiências Negativas de Turismo: um estudo sobre a influência na insatisfação, 26*, 188–215.

Namin, A. (2017). Revisiting customers' perception of service quality in fast food restaurants. *Journal of Retailing and Consumer Services*, *34*, 70–81. doi:10.1016/j.jretconser.2016.09.008

Needles, A., & Thompson, G. (2013). Social Media Use in the Restaurant Industry: A Work in Progress. *Center for Hospitality Research Publications, 13*(7), 6–16. Retrieved from http://scholarship.sha.cornell.edu/chrpubs/101

Nobre, H., & Faria, J. (2017). Exploring marketing strategies in architectural services: The case of the architecture firms in Portugal. *International Journal of Business Excellence*, *12*(3), 273–293. doi:10.1504/IJBEX.2017.084438

Nobre, H., & Silva, D. (2014). Social Network Marketing Strategy and SME Strategy Benefits. *Journal of Transnational Management*, *19*(1), 138–151. doi:10.1080/15475778.2014.904658

Nogueira Tomas, R., Pierre Meschgrahw, R., & Lúcia Chicarelli Alcantara, R. (2012). *As redes sociais e o comportamento de compra do consumidor: O reinado do & BOCA-A-BOCA & ESTÁ DE VOLTA?* doi:10.5585/remark.v11i2.2325

Nuno, M. (2018). As 6 redes sociais mais usadas do mundo. E-Konomista. Retrieved from https://www.e-konomista.pt/artigo/redes-sociais-mais-usadas/

Santini, F. O., Lübeck, R. M., & Hoffmann, C. S. (2015). Promoção de vendas: Uma análise dos fatores influenciadores da intenção de compra de um produto em desconto. *Revista Pensamento Contemporâneo Em Administração*, *9*(4). doi:10.12712/rpca.v9i4.490

Singh, M., Sarfaraz, A., Sarfaraz, M., & Jenab, K. (2015). Analytical QFD model for strategic justification of advanced manufacturing technology. *International Journal of Business Excellence*, *8*(1), 20–37. doi:10.1504/IJBEX.2015.065979

Singh, P. K., & Pattanayak, J. K. (2016). Study of the Relationship among the Factors of Brand Equity: A Study on Fast-food Brands. *Global Business Review*, *17*(5), 1227–1239. doi:10.1177/0972150916656694

Soares, A. M., Pinho, J. C., & Nobre, H. (2012). From Social to Marketing Interactions: The Role of Social Networks. *Journal of Transnational Management*, *17*(1), 45–62. doi:10.1080/15475778.2012.650085

Souza, F., De Las Casas, D., Flores, V., Youn, S., Cha, M., Quercia, D., & Almeida, V. (2015). Dawn of the selfie era: The whos, wheres, and hows of selfies on Instagram. In *COSN 2015 - Proceedings of the 2015 ACM Conference on Online Social Networks* (pp. 221–231). ACM. doi:10.1145/2817946.2817948

Twenge, J. M., Campbell, S. M., Hoffman, B. J., & Lance, C. E. (2010). Generational differences in work values: Leisure and extrinsic values increasing, social and intrinsic values decreasing. *Journal of Management*, *36*(5), 1117–1142. doi:10.1177/0149206309352246

Venkatesh, V., Morris, M. G., Davis, G. B., & Davis, F. D. (2003). User acceptance of information technology: Toward a unified view. *Source: MIS Quarterly*, *27*(3), 425–478. doi:10.2307/30036540

Chapter 10
Event Management in Social Media

Murat Seyfi
Gaziosmanpaşa University, Turkey

ABSTRACT

This chapter describes how through globalisation and developments in communication technologies, the lifestyle of people has changed and developed as well. Now, events are re-designed with communication tools and are becoming an important part of social life. This study aims to determine and evaluate the factors affecting perceptions of the audience who take part in events realized via Facebook, or on another online platform. Since this study is implemented through a social media communication device, on an on-line platform, it differs from other studies and gives this area a new perspective. Data acquired from questions were prepared to test the hypothesis of the study were analysed by doing factor analysis and regression tests. The acquired results were discussed.

INTRODUCTION

Event management is an applied science in which inter-disciplinary studies are used. It is a format that is continuously renewing and developing its practices through technology that is constantly changing and developing. In particular, event practices now cease to be ordinary activities and have become a structure that has horizontal and vertical components through recent developments in communication technologies and multi-functional globalization effects. The base for this is that events are creating their own stories with creative, interesting and transmediatic practices. In relation to this, event management and event marketing caused the rise of a new dimension by using technological instruments in all the sub-elements of the process of "strategical communication management".

Events have appeared in different formats and styles throughout the history of humankind. In the ancient era; wars, occupations, religious celebrations and agricultural harvest seasons were bases for events. Today, formats of events have changed through the developments in communication technolo-

DOI: 10.4018/978-1-7998-1947-9.ch010

gies and effects of globalization. Now, events have gained a new memory, place, and space which are different from traditional events. Therefore, factors affecting perceptions of the audience have changed and developed.

Every day, thousands of events are held on social media websites that have billions of users today. Social media and the events held on such websites acquired such a vital place in social life that they have become an important research area. There are almost any research in the literature about the evaluation of the events held on social media. Therefore, focusing on pilot studies rather than studies producing hypothesis will be the right thing to do since various dimensions are in question when the audience is evaluating Facebook events and traditional events. This study conducts a pilot study to determine sub-dimensions of audience perceptions about the event on Facebook. This study analyzes yoga and recovery events held by Seçil Tezgel in Turkey on a Facebook page which is created by herself. The participants of the events are people who want to take online yoga courses and to recover through spiritual methods. This study differs from other studies in that the event was held on a social media website. McLuhan (1964) focuses on the fact that mediums change people's frame of mind rather than the message itself by saying "medium is message". Therefore, people's lifestyle and way of perceiving life are directly influenced by the mediums. In other words, social media devices which came into our lives through communication technologies are not only the inventions people use but also the mediums that re-invent people and their lifestyle. Hence, participant's adaptation process to the digital platform is also analyzed apart from their loyalty and satisfaction.

There are scarcely any studies in the literature about the topic. Therefore, focusing on pilot studies rather than studies producing hypothesis will be the right thing to do since various dimensions are in question when the audience is evaluating Facebook events and traditional events. This study conducts a pilot study to determine sub-dimensions of audience perceptions about the event on Facebook. Within this context, this study differs from other studies and is thought to contribute to the literature.

BACKGROUND

Event is described as a remarkable activity taking place in a specific place and a period of time (Getz, 2007, p.19). Events, from the perspective of the audience, are seen as a medium for recreation and entertainment. Individuals expect to experience happiness and joy-oriented experiences by participating in the events (McLean, 2006, pp.40-41). According to Pira (2004), each event has its own purposes. These purposes might be sustaining social communication or reaching related purposes by creating an environment for the benefit of the institution, affecting the people and carrying the mission and the needs of the event into a general area. However, an event which cannot achieve its aims, even if it is profit-oriented or a charity, becomes insignificant and a loss of resources (Pira, 2004, p.31). Therefore, researching perceptions of the audience in the events is vital in terms of determining sub-factors of participant satisfaction and loyalty.

Event management is the sector that has grown biggest in the leisure industry in the recent years. Therefore, it has been researched exhaustively by both the academicians and experts of the profession (Nicholson & Pearce, 2001). Within this scope, there have been many studies conducted which particularly consider motivation, satisfaction, loyalty, social and cultural effects of participants (Visitors loyalty Gandhi & Shaw 2002; social impacts of event Kim et al., 2010; event motivation Li & Petrick, 2006;

building brand image Gwinner et al., 1999; loyalty Gedenk & Neslin, 1999; effects of events Xing & Chalip, 2006). The basis for these studies is to develop communication strategies which will create a behavioral change on event participants, reinforce consumer behaviors, and which are persuasive and encouraging. Potential consumers are motivated to purchase this free-time experience thanks to these communication studies. However, qualified an event product is designed, the event in question will be sealed behind closed doors unless there is a successful advertisement of the event with promotional efforts. The success of the most well-known events (Super Bowl, Olympic Games) depends upon a well-developed strategic communication plan (Goldblatt, 2010). Through globalization and the developments in communication technologies, communication studies have changed and developed. Now, people's participation process to the events begins with their cellphones and on their accounts of social media without leaving their homes thanks to social media devices. Therefore, the relationship between social media and event management deepens each passing day.

DIGITALIZATION OF EVENT MANAGEMENT

Through developments in communication technologies, the lifestyle of people has changed (McLuhan, 1964). The flow of daily life has also undergone a transformation within the post-modern world. Virtual or non-virtual communities which appeared as a result of this transformation developed many opportunities in individual mobility via the internet. Particularly, virtual communities can benefit from high-level education and technology opportunities when compared to conventional communities. Therefore, it is now an obligation to organize events which are unique to individuals' lives. All those free times make free time events acquire a new dimension. This feature formed the basis for large individualization of events, which makes free time experiences a catalyst for individual cultural values and cultural capital. Now, the cultural capital of people and free time events have started to nourish each other. That is to say, events caused societies and individuals to find an important place within their socio-cultural and economic context. For this reason, globalization and technological developments have become driving forces for modern-day events (Argan & Yüncü, 2015, pp. 4-5). In particular, various vital habits of people, the habit of recreation through applications on the internet is in the first place, have entered into an evolution process. In relation to this, social media devices have started to bring people together for different purposes on different platforms (Kaplan & Haenlein, 2010). Developments in communication technologies have particularly brought about practices like integration with social media in terms of event management. Social media has started to be seen as the most critical platform in terms of the function it has; that is, it gives the opportunity to reach the individual directly and enables feedback from the individual. In addition, social webs which have a significant number of followers in the world have acquired the position of webs which are included in plans of event management. Social media integration in events varies each passing day through the developments in communication technologies. The purpose of web-based social media applications that are peculiarly designed for events is including more participants in the application with ease, understandability and quickly, and thus to reach the aimed number of participants. The most frequently used social media based digital platforms are applications like QR code, RFID Radio Frequency Identification, TweetWall, 360 Selfie, Prestagram, SocialBreak, and SocialGoal. In addition, Hashtags, Likes, and Tweets are widely used for events to reach large masses. The more social media develops the more these kinds of applications used in the events will develop, and their numbers will increase (Akay, 2014, pp. 64-68). In the recent years, Facebook, YouTube and especially Twitter have become highly efficient devices to observe event processes. In addition, they have many functions in terms of lead generation. In particular, they make potential customers involve in the process for each stakeholder of the event beyond just simple sales regarding the event. This makes event management acquire a dynamic structure (Rothschild, 2011). Social media has become an important platform as a result of the constant development of the technology and increasing number of the participants. Consequently, each social web appearing on the internet has developed its own format of the event.

There are many studies in the literature about the classification of events. Getz (1997) classified events by considering their contents and extent, Roche (2000) made a classification from the perspective of modernity and mega events, Armstrong (2001) conducted studies to classify events within themselves, Shone and Parry (2004) classified events as free time, cultural, private and personal events, Tassiopoulus (2005) classified them whether they are planned or not, and whether they have a specialty or not. Especially with technological developments in the 2000s, management and practice of events have changed and developed. Therefore, new classification studies and developing new theories within this context have become a necessity. As a result of the literature review, sectoral analyses and in-depth interviews with people who work in this sector by Seyfi (2017), digital events are classified as technology-based events, virtual-on-line events, and transmediatic events. While technology-based events describe people according to their condition of activity or passivity, in virtual on-line events the focus is on on-line or off-line processes of the participants. Within this framework, Facebook events fall into the group of virtual-on-line events.

Facebook and Event Management

Today, Facebook is the most popular social media device having the highest number of users. Facebook was established by Mark Zuckerberg and his friends in 2004 with the purpose of gathering people in the campus together on an on-line platform. Facebook is a profit-oriented, the USA originated institution providing on-line social media and social web service (Carlson, 2010). In 2006, when Facebook was brought into service of everybody, it grew so quickly. Facebook created new applications to develop itself and focused on satisfying the needs of both the business world and the individuals with the increase in the number of its users (Richmond, 2007). By the year 2011, users had the feature of creating their "memorable life events" that bring their characteristics to the forefront. Facebook has become an important event management area with its development each passing day. In particular, Facebook became an important source of information in times of crisis beyond being a communication device (Seyfi & Güven, 2016). Facebook was frequently used when people were gathering together and events were being organized during social events like the Arab Spring (Huang, 2011). As a consequence of the frequent use of Facebook in social, political and cultural events, Facebook event application has started to be often used both in individual and commercial practices. For this reason, Facebook prepared briefing pages for its users on how to organize an event.

According to a 2014 dated report by Pew Research Center, Internet usage rate is 87% in America, and 97% of the users are between the age of 18 and 29. Therefore, re-designing events and in fact, the social life as a whole according to this circumstance have become a necessity. Studies conducted on the topic suggest that participation rate to events has increased more than ever thanks to announcements through social media channels. It is determined that use of Facebook before and about the event is 77%, use of Twitter during the event is 73% and use of Facebook after the event is 55%. Besides, technology orientation directly affects not only the process of the event but also the planning process of the event. Due to all these reasons, the sector is growing each passing day and the evolutionary process continues (Pew Research Center, 2014). Furthermore, the number of the events realized on virtual platforms are also increasing apart from the use of social media. The report of Market Research Media demonstrates that this sector will worth 18 billion dollars by experiencing a 5% growth by the year 2023 (Market Research Media, 2017). Thus, the importance of new communication devices and technological applications in the event management process are increasing day by day.

Through technological developments and social media practices, event management processes have also changed and developed constantly. While conventional events have a specific time and place, events realized on digital platforms have started to create their own memories and places. According to Halbwachs (1992), memory creates the process of an individual's socialization and it is determined by social elements. Since memory cannot stay the same, it has a "reconstruction" function in a community. This explains the organic link between the memory and communities (Huyssen, 2003). Events held on social media platforms also contribute to the memory of the virtual community, which is created by those events. Thus, these kinds of events become a part of social life. On the other hand, constant update of knowledge with events by means of social media puts forward a new concept of place because sharing made on different social media devices adds a new dimension to the events.

Harmonization of Technology and Humans

The event management process has encountered a new function through events organized on digital platforms like Facebook. Different from conventional events, human-technology adaptation seems to be an important function within this new platform. Although the intense use of technology in events supports factors like creativity, speed, and entertainment, participants might feel like bullied unless they can adapt these technological devices. According to a report prepared by AARP Foundation and Microsoft in 2009, participants desire security and accessibility in the environment during the process of technology usage. In addition, participants seek an adaptation process in which they can take responsibilities rather than technology's managing all of the communication process (Rogers, 2009). In the events realized on on-line platforms, it is vital to know from which aspects technological innovations adapt to the society and culture, how the innovations spread in the society, how the individuals perceive and embrace these events and the factors increasing or decreasing the use of technology. Therefore, there have been many studies conducted to explain the relationship between technology and people. Rogers (2003) focuses on the concept of innovation when he develops the Theory of Diffusion of Innovation. According to Rogers, innovation is not only newly discovered technological devices but also a process. For this reason, if individuals come across an object or technology for the first time, which was previously discovered, they perceive it as innovation. Thus, from that moment on, individuals exert effort to embrace and adapt this innovation they come across. When event areas are handled at this point since the event area constantly renews itself, it is a new area for participants, and they must use their communication channels actively in order to embrace and adapt to these innovations.

Fishbein and Azjen (1975) stated that available attitudes in social science studies are prerequisites for determining specific behaviors of individuals. In other words, they drew attention to the relationship between attitude and behavior. Attitude is the tendency to react positively or negatively and is an important element influencing the intention to use or not to use the computer. In addition, attitudes might change on digital platforms and may be re-shaped according to the features of the digital platform (Ma et al., 2005). The basis for this is perceived ease of use and perceived usefulness. Perceived ease of use and perceived usefulness are important elements influencing individuals' intention about using a computer (Davis, 1989). While perceived usefulness was defined by Davis (1989) to point out individuals' tendencies and opinions about increasing their performance in a work that they are doing with the use of technology, perceived ease of use points out easy use of a specific technology and ease of learning how to use this technology. Therefore, perceived ease of use and perceived usefulness of participants in the events realized in digital platforms are vital. It can be anticipated that people may develop many positive attitudes in terms of ease of use against the events to be held on a digital platform, which is widely used by people like Facebook. On the other hand, when the topic is handled from an organizational perspective, a technological change that causes re-shaping of production system will bring about an essential organizational change (Woodward, 1965, p.72). When we consider that events are generally organized by companies, the relationship between organizational change and technological adaptation should be questioned by event managers. Organizations must have the ability to adapt to this change because technology makes the organizational change necessary by changing the structure of organization and management philosophy (Liker et al., 1999, p. 582). This will directly influence event management process.

Hypothesis of the Study

When the evaluation studies about the events are examined, it can be seen that studies are most often conducted to determine people's satisfaction from the event and their loyalty, and events' cultural influences and social benefits (Kim & Petrick, 2005; Parasuraman et al., 1988; Pons, Mourali & Nyeck 2006; Kim et al., 2015; Delamere et al., 2001; Lee & Cho, 2012; Pierce, Stacey & Barkatsas, 2007). Now it is an obligation to conduct different evaluation studies since the technological devices are frequently used in event platforms and the number of the event practices on virtual platforms is increased. The reason behind this is that the events that take place on virtual platforms and conventional events differ greatly from each other in terms of both usage requirements and environmental influences. Therefore, developing hypotheses that analyses the effects of technological devices on both the event and the people' behavior is vital for understanding the topic. Within this framework:

H1: On Facebook events, participant satisfaction influences participant behaviors. Besides, there is a substantial relationship between participant satisfaction and loyalty.

H2: Loyalty is defined as "constant purchasing, frequent purchasing, and word-of-mouth recommendations" (Lee, Kim, and Kim, 2006). Every event organizer wants participants to re-participate in the event. In online events held on Facebook, participant loyalty influences participant's behaviors.

H3: In particular, the place that the event was held directly affected the perception of participants in the event evaluations. Therefore, in events held on Facebook, the behaviors and perception of participants will be directly influenced by Facebook's factors (being online and sharing online etc.).

H4: Individual and social events have always had an important role in the event evaluations. The behaviors and perception of participants on Facebook events are directly influenced by individual and social effects of the event.

Methodology of the Study

The purpose of this study is to conduct a pilot study which aims to determine sub-factors for evaluation of perception of participants in events held on Facebook. As a result of the field research made within this scope, questions developed in studies which aim to evaluate satisfaction and loyalty of participants (Kim et al., 2015; Yürük et al., 2017; Petrick et al., 2013; Parasuraman et al., 1988) were based in this study. Moreover, questions regarding the adaptation of human to computers (Thüring & Mahlke, 2007; Rosen et al., 2013) and aiming to put forward effects of the Facebook environment were included in the study due to the fact that it was conducted on an online platform. These questions were checked by 5 people who conduct studies in this field, and structural changes were made. As a result of the evaluations, a survey questionnaire is created consisting of 18 questions which are of 5 Likert type. Before factor analysis, whether the data is suitable for factor analysis is examined with KMO (Kaiser-Meyer-Olkin) and Barlett's test. After acquiring meaningful results, it is determined that there are four factors among the 18 questions analyzed whose eigenvalue is higher than 1.

Participants of an event page titled "Şifanın Yolu" which was realized by Seçil Tezgel in Turkey make up the sample of the study. Seçil Tezgel organizes yoga, meditation and recovery events on Facebook being sometimes free of charge and sometimes not. The average number of participants of these events is 150-250. 210 people took part in the event organized in December, before the New Year, and the study was conducted on these people. 106 of the 210 people accepted to take part in the research and answered the questionnaire on Google form. Acquired data were analyzed with the program SPSS. Within the scope of the analyses, factor and regression analyses were conducted

Findings

Descriptive analyses were conducted in order to describe demographical features of the participants. Within this framework, it is determined that 90.6% of the participants are women. Since there is a huge difference between the number of women and men participants there was not an analysis conducted on gender difference. When participation frequency of participants to events on Facebook is observed, it can be seen that 69.8% of the participants said that they participated in more than 7 Facebook events.

Factor Analyses

Factor analysis can be described as a multivariate statistic that aims to find and explore few conceptually meaningful new variables (factors, dimensions) by gathering together various variables that are related to one another (Dunteman, 1989; Renie, 1997; Stevens, 2012). In this study, since the researcher does not have any opinion or prediction about the relationship between the variables regarding the topic that he researches, Exploratory Factor Analysis was conducted in order to reveal the probable relationship between the variables.

In order to test whether the factor analysis was to be made or not, firstly KMO and Barlett tests were made. Since the KMO value was higher than 0.60, that is 0,741, factor analysis could be made. In addition, according to Bartlett test results, X2=1391,584, df=153 and p=0.000 were found significantly at a confidence level. They show that there was a relationship between the variables and so determined that the data have been in conformity with the factor analysis.

Table 1. Demographical features of the participants

	Frequency	Percentage
Where do you live for the most part of your life?		
Big Cities	74	69.8
Small and Medium Scaled Cities	32	31.2
How often do you participate in the events on Facebook?		
I have participated only for 1 time up to now.	6	5.6
1-3 times	18	16.9
4-7 times	8	7.5
More than 7	74	69.8
Your Educational Status		
Primary School	6	5.6
Secondary School	26	24.5
Associate Degree	18	16.9
Bachelor's Degree	40	37.7
Master's Degree	16	15
Your Age?		
21-30	4	3.7
31-40	36	33.9
41-50	44	41.5
51-60	16	15.1
+60	6	5.6

Table 2. KMO and Bartlett's test

Kaiser-Meyer-Olkin Measure of Sampling Adequacy.		,741
Bartlett's Test of Sphericity	Approx. Chi-Square	1391.584
	df	153
	Sig.	,000

Table 3. Total variance explained

Component	Initial Eigenvalues			Extraction Sums of Squared Loadings	
	Total	% of Variance	Cumulative %	Total	% of Variance
1	6,980	38,779	38,779	6,980	38,779
2	2,421	13,447	52,227	2,421	13,447
3	1,712	9,511	61,738	1,712	9,511
4	1,252	6,956	68,694	1,252	6,956

Table 4. Dimensions of the questions

	Factor Loading	Cronbach Alfa
Dimension1: Social-Individual Effect		,735
I think I found health, luck, and recovery through participation in this event.	,650	
I can implement the things that I learned from this event.	,653	
My self-confidence increases in Facebook events.	,681	
I really feel good during the event.	,883	
I think this event is beneficial for the society.	,789	
Dimension2: Satisfaction		,702
I am pleased to have participated in recovery events of Seçil Tezgel on Facebook.	,701	
I think this event's being online, on Facebook is economically more suitable for me.	,689	
I think this event's being online, on Facebook saves time for me.	,653	
It is important for me to participate in this event.	,892	
I enjoy Facebook events as much as real-life events.	,782	
Dimension3: Effects of Social Media Environment		,769
I can easily adapt the event on Facebook.	,703	
I like participating in questionnaires and discussions on the event page.	,746	
I like sharing things with other participants. (I like their comments and I answer them)	,701	
It makes me feel happy when I see participants on Facebook.	,708	
It makes me feel confident when I see participants on Facebook.	,682	
Dimension4: Loyalty		,805
I recommend people around me to participate in this kind of events.	,643	
I will participate in this kind of events again.	,830	
I will mention about these events to my friends.	,796	

Table 5. Regression analysis

Independent Variables	Unstandardized Coefficients		Standardized Coefficients	t	Sig.
	B	Std. Error	Beta		
Social and Personal Effect	,278	,000	,316	7.310	,000
Satisfaction	,278	,000	,289	5.595	,000
Effects of Social Media	,278	,000	,363	8.074	,000
Loyalty	,167	,000	,264	9.309	,000

After it was seen that the expressions in the research were suitable, factor analysis was made one more time and the dimensions of the factor were determined. As it can be seen from the table, research questions consist of four dimensions.

Through Total Variance Analysis explained, four factors appeared whose eigenvalue was higher than 1. This shows that the scale developed has four dimensions.

With acquired results through Total Variance Analysis, factor loads were determined in order to determine questions forming four sub-dimensions. From 18 questions that were prepared within the framework of the study, questions whose factor load was lower than 0.5 were considered invalid. Thus, factor loads that make up the sub-dimension of the scale developed were determined in the table.

Since there are not any studies aiming to develop a scale regarding the events held on Facebook, it is not possible to foresee implications. Therefore, Exploratory Factor Analysis was conducted within the context of the study. Four dimensions were determined in the study of determining the dimensions. Since there are not any studies aiming to determine dimensions in the literature, these four dimensions were classified according to questions' conceptuality levels. In addition, the cumulative percentage of these four dimensions is 68,455%, which demonstrates that these dimensions widely involves the research.

Linear regression analysis was made in order to determine how sub-dimensions of the event held on Facebook affect the perception of participants. Results of this regression analysis that was conducted to test the hypothesis of the study are as follows:

According to the results of multiple regression analyses that were aimed to explain the effects of sub-factors which are social and individual effect, satisfaction, social media effects and loyalty on total satisfaction and loyalty ($p< 0,01$), it is determined that sub-factors are substantially efficient. When Beta values that are standardized according to regression analysis, it is determined that Social and Individual effect ($\beta=0,316$; $p< 0,01$), Satisfaction ($\beta=0,289$; $p< 0,01$), Effects of Social Media ($\beta=0,363$; $p< 0,01$), Loyalty ($\beta=0,264$; $p< 0,01$) have effects on perception of participants.

CONCLUSION

Events that have been at the center of the social life for centuries is changing and developing each passing day. Especially through the developments in communication technologies, types of events and event management have acquired a new dimension; therefore, studying the dimensions of this new social media-based digital event world has become an obligation. Facebook stands out as a Virtual-On-line Event platform in terms of the abundance of participants and its being prevalent in social media-based digital events. The number of events held on Facebook is increasing each passing day. In relation to this, it is an important field of practice for both individual marketing and entertainment and commercial marketers.

There have been many studies conducted to analyses satisfaction and loyalties of event participants. The main reason underlying these studies' being conducted is to plan event management processes more systematically and to obtain positive outcomes (Cronin et al., 2000). However, both the structure of events and expectations of participants have changed through the development of technology and change of people's lifestyles. In particular, people focused on using their time more efficiently through modernization. Therefore, the number of events taking place on digital platforms is increasing day by day. At first, events organized on Facebook were more of a rendezvous but in time, these events have become the events in which people are active and in which participants invest both their time and money. Correspondingly, many factors affecting satisfaction of participants have appeared. In particular, values like harmonization of technology and humans, time and economy have come to the prominence.

Social media devices like Facebook, YouTube, and Twitter directly affect emotional commitment in the processes of event management and this emotional commitment again directly affects speeches on the event and thus, loyalty between the participant and the event develops organically. Emotional commitment to the event will provide the best opportunity to become different and will support new communication strategies to preserve loyalty between event managers and participants (Hudson et al., 2015). When the contents of "effects of social media" dimension which was determined within the scope of the research are analyzed, especially emotional sharing come to the forefront. Various emotional based factors like participants' sharing with other participants or feeling self-esteem when they see participants on Facebook directly influence the process. According to Lee et al. (2012), emotional sharing via social media influences both potential of participant and satisfaction of participant. Participants keep the interest to the event alive both before the festival and after it through the function of "follow us" on Twitter. Sharings about the event made on various social media devices also create a sustainable story of the event. The concept of multiple intelligence which was put forward by Jenkins (2010) with transmediatic story-telling model appears. Thus, events become platforms which constantly produce information and sharing beyond being a space in which time is spent.

This study aims to gauge and evaluate individual, technological and social effects of digital event world that developed through social media. Within the scope of the aims of this study, factor analyses and regression analyses were made in order to evaluate satisfaction and loyalty of participants in events that were held on social media. As a result of the factor analyses, it is determined that there are four sub-dimensions which are individual-social effect, effects of social media, loyalty, and satisfaction to describe and evaluate social media events. In order to determine the relationship between these four sub-dimensions, regression analyses were made, and it was determined that there are substantial relationships between each of the sub-dimensions. When the contents of these sub-dimensions were analyzed, it was seen that, different from conventional events, factors relating to Facebook and on-line platforms took an important place. This makes re-question of the relationship of events with places from the perspective of digitalization. The dimension of social-individual effects, which was defined within the scope of the study, contains within itself both contents in conventional events and contents relating to Facebook platform. For instance, while the answer to the question "I think this event is beneficial to the society" is sought in conventional events, the event organized on Facebook plays a role in the evaluation of the process. On the other hand, contents like "My self-esteem increases in the environment of the events organized on Facebook" assert a feature of the Facebook platform. Similarly, in addition to the attitudes researched in the studies evaluating conventional events, it was tried to determine attitudes which would reveal effects of social media and on-line platforms in question contents in the other sub-dimension.

Consequently, results of this study suggest that social media-based digital events provide satisfaction to the participants and affect their perceptions in the events. This result supports that this study will be a beneficial source for both sector professionals and individual users.

Further Research and Limitations

Both conventional and virtual events are generally realized by professional companies. However, the events realized on Facebook are at the initial stage in terms of professionalization process since they are quite new. Therefore, there are many difficulties in finding and researching pilot practices. Hence, the study is limited to a single study and the data are only acquired on the online platform.

This study can be a basis for many studies to be conducted. Event management on social webs can be profoundly researched particularly with scale development studies.

Experimental and theoretical studies can be conducted by using new research methods in the studies to be conducted in the future since the topic is quite new and there are not any studies in this area. For instance, events realized on social webs can be profoundly researched by using research methods of netnography and PAR (participatory action research).

REFERENCES

Akay, R. (2014). Etkinlik yönetimi uygulamalarinda yaratici rekabet ve sosyal medyanin entegrasyonu. *The Turkish Online Journal of Design, Art and Communication*, *4*(4), 55–70. doi:10.7456/10404100/005

Argan, T. M., & Yüncü, D. (2015). *Etkinlik Pazarlama Yönetimi*. Istanbul: Detay Yayıncılık.

Arsmtrong, J. S. (2001). *Planning Special Events*. Indiana: Indiana University Center on Philanthrophy.

Carlson, N. (2010). At Last—The Full Story of How Facebook Was Founded. Business Insider. Retrieved from http://www.businessinsider.com/how-facebook-was-founded-2010-3

Cronin, J. J. Jr, Brady, M. K., & Hult, G. T. M. (2000). Assessing the effects of quality, value and customer satisfaction on consumer behavioral intentions in service environments. *Journal of Retailing*, *76*(2), 193–218. doi:10.1016/S0022-4359(00)00028-2

Davis, F. (1989), *A Technology acceptance model for empirically testing new end user information systems: theory and results* [Doctoral Dissertation]. MIT Sloan School of Management.

Delamere, T. A., Wankel, L. M., & Hinch, T. D. (2001). Development of a scale to measure resident attitudes toward the social impacts of community festivals, Part I: Item generation and purification of the measure. *Event Management*, *7*(1), 11–24. doi:10.3727/152599501108751443

Dunteman, G. H. (1989). *Principal component analysis (Quantitative applications in the social sciences)*. Beverly Hills, CA: Sage Publication.

Fishbein, M., & Ajzen, I. (1977). Belief, attitude, intention, and behavior: An introduction to theory and research. *Philosophy & Rhetoric*, *10*(2), 130–132.

Gandhi-Arora, R., & Shaw, R. N. (2002). Visitor loyalty in sport tourism: An empirical investigation. *Current Issues in Tourism*, *5*(1), 45–53. doi:10.1080/13683500208667907

Gedenk, K., & Neslin, S. A. (1999). The role of retail promotion in determining future brand loyalty: Its effect on purchase event feedback. *Journal of Retailing*, *75*(4), 433–459. doi:10.1016/S0022-4359(99)00018-4

Getz, D. (2007). *Event Studies: Theory, Research and Policy for Planned Events.* Oxford: Elseiver Butterworth-Heinemann.

Goldblatt, J. (2010). *Special events: A new generation and the next frontier* (6th ed.). John Wiley & Sons.

Gwinner, K. P., & Eaton, J. (1999). Building brand image through event sponsorship: The role of image transfer. *Journal of Advertising, 28*(4), 47–57. doi:10.1080/00913367.1999.10673595

Halbwachs, M. (1992). *On Collective Memory*. University of Chicago Press.

Huang, C. (2011). Facebook and Twitter key to Arab Spring uprisings: report. *The National* (Vol. 6). Retrieved from http://www.thenational.ae/news/uae-news/facebook-and-twitter-key-to-arab-spring-uprisings-report

Hudson, S., Roth, M. S., Madden, T. J., & Hudson, R. (2015). The effects of social media on emotions, brand relationship quality, and word of mouth: An empirical study of music festival attendees. *Tourism Management, 47*, 68–76. doi:10.1016/j.tourman.2014.09.001

Huyssen, A. (2003). *Present pasts: Urban palimpsests and the politics of memory*. Stanford University Press.

Jenkins, H. (2010). Transmedia storytelling and entertainment: An annotated syllabus. Continuum. *Journal of Media & Cultural Studies, 24*(6), 943–958. doi:10.1080/10304312.2010.510599

Kaplan, A. M., & Haenlein, M. (2010). Users of the world, unite! The challenges and opportunities of Social Media. *Business Horizons, 53*(1), 59–68. doi:10.1016/j.bushor.2009.09.003

Kim, S. S., & Petrick, J. F. (2005). Residents' perceptions on impacts of the FIFA 2002 World Cup: The case of Seoul as a host city. *Tourism Management, 26*(1), 25–38. doi:10.1016/j.tourman.2003.09.013

Kim, W., Jun, H. M., Walker, M., & Drane, D. (2015). Evaluating the perceived social impacts of hosting large-scale sport tourism events: Scale development and validation. *Tourism Management, 48*, 21–32. doi:10.1016/j.tourman.2014.10.015

Kim, Y. G., Suh, B. W., & Eves, A. (2010). The relationships between food-related personality traits, satisfaction, and loyalty among visitors attending food events and festivals. *International Journal of Hospitality Management, 29*(2), 216–226. doi:10.1016/j.ijhm.2009.10.015

Lee, H. S., & Cho, C. H. (2012). Sporting event personality: Scale development and sponsorship implications. *International Journal of Sports Marketing & Sponsorship, 14*(1), 46–63. doi:10.1108/IJSMS-14-01-2012-B005

Lee, W., Xiong, L., & Hu, C. (2012). The effect of Facebook users' arousal and valence on intention to go to the festival: Applying an extension of the technology acceptance model. *International Journal of Hospitality Management, 31*(3), 819–827. doi:10.1016/j.ijhm.2011.09.018

Li, X., & Petrick, J. F. (2006). A review of festival and event motivation studies. *Event Management, 9*(4), 239–245. doi:10.3727/152599506776771526

Liker, J. K., Haddad, G., & Karlin, J. (1999). Perspectives on technology and work organization. *Annual Review of Sociology, 25*(1), 575–596. doi:10.1146/annurev.soc.25.1.575

Ma, W. W. K., Andersson, R., & Streith, K. O. (2005). Examining user acceptance of computer technology: An empirical study of student teachers. *Journal of Computer Assisted Learning*, *21*(6), 387–395.

Majchrzak, A., Rice, R. E., Malhotra, A., King, N., & Ba, S. (2000). Technology adaptation: The case of a computer-supported inter-organizational virtual team. *Management Information Systems Quarterly*, *24*(4), 569–600. doi:10.2307/3250948

Market Research Media. (n.d.). Retrieved from https://www.marketresearchmedia.com/?p=421

McLean, D. J. (2006). *Philosophy and Leisure. In G. Kassing (Eds.), Introduction to Recreation and Leisure*. United States: Human Kinetics.

Nicholson, R., & Pearce, D. G. (2000). Who goes to events: A comparative analysis of the profile characteristics of visitors to four south island events. *Journal of Vacation Marketing*, *6*(3), 236–253. doi:10.1177/135676670000600304

Parasuraman, A., Zeithaml, V. A., & Berry, L. L. (1988). Servqual: A multiple-item scale for measuring consumer perc. *Journal of Retailing*, *64*(1), 12.

Pew Research Center. (2015). Social media update 2014. Retrieved from http://www.pewinternet.org/2015/01/09/social-media-update-2014/

Pierce, R., Stacey, K., & Barkatsas, A. (2007). A scale for monitoring students' attitudes to learning mathematics with technology. *Computers & Education*, *48*(2), 285–300. doi:10.1016/j.compedu.2005.01.006

Pira, A. (2004). *Etkinlik Yönetimi*. Istanbul: MediaCat Yayınları.

Pons, F., Mourali, M., & Nyeck, S. (2006). Consumer orientation toward sporting events scale development and validation. *Journal of Service Research*, *8*(3), 276–287. doi:10.1177/1094670505283931

Rennie, K. M. (1997). Exploratory And Confirmatory Rotation Strategies in Exploratory Factor Analysis. *Paper presented at the Annual Meeting Of The Southwest Educational Research Association*. Academic Press. Retrieved from http://files.eric.ed.gov/fulltext/ED406446.pdf

Roche, M. (1994). Mega-events and urban policy. *Annals of Tourism Research*, *21*(1), 1–19. doi:10.1016/0160-7383(94)90002-7

Rogers, M. (1999), *Boomers And Technogy: An Extended Conversation*. AARP. Retrieved from https://assets.aarp.org/www.aarp.org_/articles/computers/2009_boomers_and_technology_final_report.pdf

Rosen, L. D., Whaling, K., Carrier, L. M., Cheever, N. A., & Rokkum, J. (2013). The media and technology usage and attitudes scale: An empirical investigation. *Computers in Human Behavior*, *29*(6), 2501–2511. doi:10.1016/j.chb.2013.06.006 PMID:25722534

Rothschild, P. C. (2011). Social media use in sports and entertainment venues. *International Journal of Event and Festival Management*, *2*(2), 139–150. doi:10.1108/17582951111136568

Seyfi, M. (2017). *Dijital Etkinlik Yönetimi*. Istanbul: Der Yayınları.

Shone, A., & Parry, B. (2004). *Successful Event Management: A Practical Handbook*. Cengage Learning EMEA.

Stevens, J. P. (2012). *Applied multivariate statistics for the social sciences*. London: Routledge.

Tassiopoulus, D. (2005). *Events-An Introduction. Event Management: A Professional and Developmental Approach* (D. Tassiopoulus, Ed.). Lansdowne: Juta Academic.

Thüring, M., & Mahlke, S. (2007). Usability, aesthetics and emotions in human–technology interaction. *International Journal of Psychology*, *42*(4), 253–264. doi:10.1080/00207590701396674

Woodward, J. (1965). *Industrial Organization: Theory And Practice*. London: Oxford University Press.

Xing, X., & Chalip, L. (2006). Effects of hosting a sport event on destination brand: A test of co-branding and match-up models. *Sport Management Review*, *9*(1), 49–78. doi:10.1016/S1441-3523(06)70019-5

Yürük, P., Akyol, A., & Şimşek, G. G. (2017). Analyzing the effects of social impacts of events on satisfaction and loyalty. *Tourism Management*, *60*, 367–378. doi:10.1016/j.tourman.2016.12.016

KEY TERMS AND DEFINITION

Digital Attitude: Attitudes may change according to the digital platform.

Digital Events: Digital events are classified as Technology-based Events, Virtual-On-line Events, and Transmediatic Events.

Diffusion of Innovation: Innovation is a process that continually renews itself.

Dynamic Structure: Versalite communication platforms.

Event Management: Event management is an applied science in which inter-disciplinary studies are used. It is a format that is continuously renewing and developing its practices through technology that is constantly changing and developing.

Modern-Day Events: Using today's technology in event management.

Social Media: Social media is critical platform in terms of the function it has; that is, it gives the opportunity to reach the individual directly and enables feedback from the individual.

Chapter 11
Co-Creating an Integrative Framework to Enhance the Museum Experience of Blind and Visually Impaired Visitors

Roberto Vaz
University of Porto, Portugal

ABSTRACT

Millions of people with vision impairments worldwide still experience many barriers when visiting museum exhibitions, given the tenacious ocular centricity of these institutions and the lack of assistive technologies that promote multisensory exhibit design. It is urgent to give priority to issues related to the accessibility of those patrons in museums, as well as their social inclusion, considering that visual impairment is a diverse spectrum. This chapter aims to research blind and visually impaired visitors' experiences in these institutions and presents a proposal of an integrative framework, co-created based on semi-structured interviews conducted with 29 individuals from Portugal. It concluded that sensory, intellectual and physical access must be integrated into the pre-visit, on-site visit and post-visit phases of interaction with cultural institutions, in order to promote a sense of autonomy to visually impaired individuals.

INTRODUCTION

According to the World Health Organization (2017), it is estimated that 253 million people worldwide live with visual impairments, of whom about 36 million are blind, and about 217 million have moderate to severe vision impairment (Bourne et al., 2017). This number increases to 442 million when considered the degree of mild visual impairment, which, despite being less severe than blindness, has a substantial effect on the quality of life of nearly 188.5 million people (Bourne et al., 2017; Dandona & Dandona, 2006).

DOI: 10.4018/978-1-7998-1947-9.ch011

In Portugal – where the present research took place – data from the Census of 2011 allowed estimate that among the 17.8% of the population that declared to have any kind of difficulty, 23% (900 thousand) specified the vision as a cause, of which 3.1% (28 thousand) reported blindness (INE, 2012).

About the participation of blind and visually impaired patrons in museums, Dash and Grohall (2016) concluded that only 5.5% visit museums in Europe, which represents a significant potential market for these institutions (Mesquita & Carneiro, 2016). Devile and Kastenholz (2018) corroborate this idea when stressing that the emergence of accessible tourism as a field of academic research, although still being incipient, can be crucial to improve tourism suppliers' capacity of surpassing this people's needs by better understanding their engagement in tourist activities. Consequently, more inclusive approaches to tourism can be provided, alongside promoting individual and social well-being and empowerment.

These facts clearly evidence the need for research in this field, especially when considered the definition of the International Council of Museums (ICOM) for these institutions, which establish the museum's social functions as the very core of their existence, simultaneously with their commitment to safeguard and communicate their collections to the public (Desvallées & Mairesse, 2010). In that sense, it becomes critical to give priority to issues related to social inclusion and accessibility, not only because there are millions of people living with disabilities worldwide, but also because everyone can suffer from temporary impairments during their lifetime, or more permanent ones, like age-related sensory and physical changes.

Among the aspects related to inclusive museums, assistive technologies are highlighted by Freeman et al. (2016, p. 26) as having the "potential to increase access and enable opportunities for disabled persons" since they can promote beneficial impacts on helping museums serve different publics while diminishing the barriers to accessibility (Pillow, Banks, & Reeves, 2015). Besides that, Andrade et al. (2015) expect positive economic effects with the implementation of inclusive technologies in museum exhibitions.

Although the growing concern about increasing accessibility for visitors with visual impairments (Cachia, 2013; Candlin, 2003; European Commission, 2013; IPM, 2004), several authors call attention to the fact that there is a tendency to treat them as a homogeneous group, despite each one of those persons, as individuals, have different reasons to go to museums and galleries (Argyropoulos & Kanari, 2015; Candlin, 2003; Ginley, 2013; Levent & Reich, 2012), and experience varied outcomes from their visits (Argyropoulos & Kanari, 2015; Hayhoe, 2017; Morgan, 2012; Reich et al., 2011; Xicola-Tugas, 2012).

This chapter aims to contribute to the field of accessible tourism and accessibility in museums by providing a framework based on a set of guiding principles to enhance blind and visually impaired patrons' visits to museums. This integrative framework was co-created through interviewing 29 participants with vision impairment, by exploring their habits regarding visits to museums and other cultural sites, positive and negative experiences they faced during past visits, and suggestions on aspects for improvement to make future visits to these places more enjoyable. The chapter begins with a brief discussion of the differences that visually impaired people face regarding their visual deficit, some barriers they experience during visits to museums and how assistive technologies can enhance their experiences by diminishing those constraints. It proceeds with the presentation of the research methods, followed by the findings and discussion. Based on them, a framework is proposed as an integration of the various identified phases of a visit (pre-visit, on-site visit, and post-visit) with the participants' reported necessities regarding accessibility requirements (sensory, intellectual, and physical access) for more autonomous visits. The chapter ends with conclusions and some future research directions.

THE SPECTRUM OF VISUAL IMPAIRMENT AND CONTRIBUTIONS OF ASSISTIVE TECHNOLOGIES TO INCLUSIVE MUSEUM VISITS

The eleventh revision of the International Classification of Diseases systematizes distance vision impairment into groups of severity, taking into account peoples' vision acuity in the better eye (World Health Organization, 2018): mild – visual acuity worse than 6/12 to 6/18; moderate – visual acuity worse than 6/18 to 6/60; severe – visual acuity worse than 6/60 to 3/60; blindness – visual acuity worse than 3/60.

This already denotes that visitors with vision impairments do not constitute a homogeneous group, and it becomes even clear when taken into consideration the moment of sight loss (Burton, 2003): early-blind individuals are those who were born blind or totally lost vision before age 6, and late-blind subjects had sight at birth, learned to read print and lost it after age 12. Cattaneo and Vecchi (2011) highlighted that late-blind benefits from years of normal prior visual experiences over congenitally blind individuals when performing cognitive tasks, and Postma, Zuidhoek, Noordzij and Kappers (2007) also demonstrated differences between early-blind and late-blind participants in haptic spatial-configuration learning and resulting memory traces. Given these reasons, it becomes imperative considering the visitors' different degrees of visual impairments when developing accessible projects in museums.

Devile and Kastenholz (2018), Richards, Pritchard and Morgan (2010), and Small, Darcy and Pacher (2012) suggests that visually impaired people's positive tourism experiences depend on their reliance on other senses – besides vision – to access information, physical space, and other facilities. Nevertheless, there is a literature consensus that museums remain mostly ocular centric oriented, which does not contribute to the engagement with exhibitions through other senses than sight, and also deprives the access to information about art and exhibits (Candlin, 2004; Hayhoe, 2017; Johnson, 2018; Mesquita & Carneiro, 2016; Morgan, 2012).

Digital media can contribute to a more inclusive museum, as suggested by the research of Reich et al. (2011), which revealed that visually impaired patrons consider that technology can support positive experiences during their visit to museums, sometimes without the need for museum staff. Although scarce, there are some examples of projects which use technological solutions as ways of helping improving inclusion in museums.

Audio guides that provide verbal descriptions of selected collection artworks are the most common solution (Osterman, 2017). However, most of the available audio guides merely provide a description of what is seen, which does not allow visually impaired persons to create accurate mental images of the exhibits (Hutchinson & Eardley, 2018; Neves, 2012). Furthermore, Romeo, Chottin, Ancet and Pissaloux (2018) agree that multimodal approaches are necessary to give blind and visually impaired people the opportunity to fully experience museum visits, allowing them to adequately understand the exhibits' characteristics and appreciate their aesthetic qualities, besides benefitting from the emotion inspired by museum objects.

The projects of Cantoni et al. (2018), D'Agnano, Balletti, Guerra and Vernier (2015), and Reichinger et al. (2016), used 3D scanning and printing technologies to provide touch replicas digitally augmented, offering tactile and hearing multisensory experiences that help the blind and visually impaired visitors to formulate mental images of the objects.

Computer-vision systems are also being used to track visitors' activity, facilitating the exploration of specific parts of objects copies in more detail, enriched by contextual audio description. They are particularly useful to provide access to bas-reliefs of 2D original artworks, like paintings, photos and images. In this case, sensory and intellectual accessibility can be improved through the multimodal nature of these systems (Buonamici, Furferi, Governi, & Volpe, 2015; Reichinger, Carrizosa, & Travnicek, 2018; Reichinger, Fuhrmann, Maierhofer, & Purgathofer, 2016).

Apart from helping create new ways for visitors to interactively knowing about museum collections, the projects of Jain, Jain, Paul, Komarika and Balakrishnan (2013), Meliones and Sampson (2018), and Urbas, Pivar and Elesini (2016), also attempt to use technology to provide blind and visually impaired visitors better physical access, indoor navigation and wayfinding in those environments, even though there is a consensus that the issues regarding these activities are not entirely solved yet.

Still about the visit experience – which begins even before entering the museum sites – a United Kingdom survey on best access practices of museums websites concluded that 81% of UK accredited museums provide some kind of information about accessibility online. However, this information is not useful and does not address the access needs of a visually impaired person planning a visit (Cock et al., 2018). On the other hand, results of a study conducted by McMillen and Alter (2017) showed that social media websites like Twitter, Facebook and Instagram, which can be used with an electronic screen reader, can have a positive effect on visually impaired people regarding social inclusion. Furthermore, they have the potential to bridge new connections with this community, as well as previously inform them about the exhibition's accessibility, accessible services and events offered by museums.

Literature review shown that a widespread research is still required to identify what are the blind and visual impaired visitors' needs and expectations from integrating assistive technologies across the continuum of visits to museums – interaction with exhibits, access to contextual information about them, access and orientation inside museum sites and exhibition galleries, online interactions with museums, and so forth. In order to seek to explore new ways of understanding accessible tourism relating with these patrons' experiences in museums, as well as to accomplish a more effective development of future digital accessible solutions, it is essential to give a central role to these people in the research process (Chick, 2017; Devile & Kastenholz, 2018; Levent & Reich, 2012; Richards, Pritchard, & Morgan, 2010). Few studies analyze the blind and visually impaired visitors' positive and negative experiences in museums, but some examples can be found in Spain (Xicola-Tugas, 2012), in Greece (Argyropoulos & Kanari, 2015) and at the United States of America (Asakawa, Guerreiro, Ahmetovic, Kitani & Asakawa, 2018; Hayhoe, 2017). In the Portuguese context, the literature about this subject is scant. The understanding of those dimensions can contribute to help museums develop unified inclusive solutions, which effectively enhance accessible exhibition designs and improve the experience of these patrons when visiting those places.

RESEARCH METHODS

The present study is circumscribed to the North region of Portugal. It was undertaken in four cities – Porto, Viana do Castelo, Vila Real, and Braga, for four months. For the data gathering, it was followed a methodology based on a qualitative research, by using the instrument of semi-structured interview for collecting the data. Nonetheless, some qualitative analyses allowed to generate results of qualitative nature as part of the findings, as well.

This section presents the sampling technique, the research instrument and procedure, and the tools used for data analysis.

Sampling of Participants

In order to randomly sampling participants for the study, the main official institutions for the blind and the visually impaired of the North region of Portugal were identified. A total of 11 institutions (six associations, one school, two formation centers, and two institutions that offer services for these persons) were contacted via e-mail and telephone and asked for collaboration with the present study. The members were then notified by the institutions about the research and 29 people, from 6 institutions (four associations, one service provider, and one formation center), expressed their interest in voluntary collaborating in the study. Finally, the dates for the interviews were agreed upon and scheduled directly with the institutions and, in some cases, with the participants.

Research Instrument and Procedure

The choice of semi-structured interviews as the instrument for data-gathering has to do with the possibility of combining features of unstructured and structured interviews. Moreover, it helps the researchers to probe the interviewees to deepening their answers until no new meaningful information is forthcoming, using a script with preplanned questions, so the same topics are covered with each participant (Preece, Rogers, & Sharp, 2002).

The interview guide included mainly open-ended questions to explore the topics regarding the habits of visiting museums, previous experiences, and suggestions to enhance them; the closed questions were about the demographic information of the participants.

The set of questions prepared for the interview guide were conditioned by the general objectives of the research, and were also based on the conclusions of previous studies (Argyropoulos & Kanari, 2015; Asakawa et al., 2018; Hayhoe, 2017; Xicola-Tugas, 2012), addressing the following topics:

- **Profile of Participants**: Sex, age, category of vision impairment, education level, occupation and area of living.
- **Habits Regarding Visits to Museums and Heritage Sites**: Frequency of visits, who they go with, primary reasons for visiting and access to information.
- Positive experiences of the individuals in visits to museums and heritage sites.
- Negative experiences of the individuals in visits to museums and heritage sites.
- Suggestions on aspects for improvement, in order to make visits to these spaces more enjoyable for these patrons.

The study was conducted between January and April of 2019 at the institutions where interviewees are members, and in seven of the cases at their homes, due impossibilities to move out of their residences. All interviews were performed face-to-face.

A document with the initial protocol, explaining the objectives of the interview, use of data, its anonymity and confidentiality, was read out loud. There was also asked for permission to audio record the data-gathering session. Each participant gave their consent to participate in the study by signing the document.

Tools for Data Analysis

Due to the amount of total audio time (the average duration of each interview was 26 minutes and 37 seconds), tests were performed using the Google Speech to Text API (Google, n.d.) to transcribe the totality of the interviews to text automatically. However, it was concluded that a solution of speech recognition would not work as expected, given the necessity to verify and revise the generated texts. Therefore, it was used the NVivo 12 software 'transcribe mode' tools to transcribe the audio files manually.

Qualitative data analysis was conducted using the same software, where segments of texts containing the interviewees' answers, commentaries, and suggestions were coded, arranged and grouped in order to build theme nodes hierarchies, according to the broad themes being explored. A discovery-focused analytical approach was followed since the focus of the analysis was developing categories that derived inductively from the collected data (Fossey, Harvey, McDermott, & Davidson, 2002).

Regarding the answers related to the profile of the participants, as well as their habits concerning visits to museums and heritage sites, it was possible to convert the qualitative data into quantitative data (Caracelli & Greene, 1993; Gelo, Braakmann, & Benetka, 2008). With this, it was used the SPSS 25 software to perform descriptive statistics and exploratory data analysis about these research topics.

FINDINGS AND DISCUSSION

This section begins with the presentation of the participants' profiles and explores their habits regarding visits to museums and heritage sites. It proceeds with the presentation and discussion of the reasons and factors they reported during their interviews concerning positive visiting experiences. Next, it explores which issues contribute to hindering their accessibility while visiting cultural sites, which were pointed out as promoting negative experiences.

Participants' Profile

The demographic characteristics of the sample are presented in Table 1, grouped accordingly to the 29 participants' different degrees of visual impairments. Of the 17 (58.6%) individuals who reported to be totally blind, 5 were early-blind (EB) and 12 late-blind (LB). The remaining 12 (41.4%) participants had low-vision (6 severe (SVI), 4 moderate (MoVI) and 2 had mild visual impairment (MiVI)).

Since EB individuals correspond to a total of 17.2%, it should be noted that 82.8% of the participants in this research (LB and partially sighted) had visual references – the average age of vision loss of the LB participants is 21 years old ($\sigma = 5.43$).

The group was relatively homogeneous in terms of gender, i.e., 51.7% were male, and 48.3% were female. Note that all EB interviewees were male, and female in the case of the MiVI.

The predominant age group is of 66-73 (20.7%), but this only contemplates LB and SVI participants. The ranges with a more distributed peoples' ages by their degree of visual impairments are 34-41 (13.8%) and 42-49, 50-57 and 58-65, each one representing 17.2%. Interviewees' age ranged from 19 to 77 years old ($\mu = 51.9$, $\sigma = 14.58$).

Table 1. Demographic characteristics of the blind and visually impaired participants

		Blind Participants				Partially Sighted Participants						Total
		Early-blind		Late-blind		Severe Visual Impairment		Moderate Visual Impairment		Mild Visual Impairment		
		n	%	n	%	n	%	n	%	n	%	
Gender	Male	5	17.2%	7	24.1%	1	3.4%	2	6.9%	-	-	51.7%
	Female	-	-	5	17.2%	5	17.2%	2	6.9%	2	6.9%	48.3%
Age Range	18-24	1	3.4%	-	-	-	-	-	-	-	-	3.4%
	25-33	-	-	-	-	2	6.9%	-	-	-	-	6.9%
	34-41	1	3.4%	2	6.9%	-	-	1	3.4%	-	-	13.8%
	42-49	2	6.9%	2	6.9%	-	-	-	-	1	3.4%	17.2%
	50-57	-	-	1	3.4%	1	3.4%	2	6.9%	1	3.4%	17.2%
	58-65	1	3.4%	1	3.4%	2	6.9%	1	3.4%	-	-	17.2%
	66-73	-	-	5	17.2%	1	3.4%	-	-	-	-	20.7%
	74 or older	-	-	1	3.4%	-	-	-	-	-	-	3.4%
Education Level	Basic school	2	6.9%	7	24.1%	3	10.3%	3	10.3%	1	3.4%	55.2%
	High school	3	10.3%	4	13.8%	1	3.4%	1	3.4%	1	3.4%	34.5%
	Higher education	-	-	1	3.4%	2	6.9%	-	-	-	-	10.3%
Occupation	Assistant operator	-	-	3	10.3%	-	-	-	-	-	-	10.3%
	Braille operator	2	6.9%	2	6.9%	-	-	-	-	-	-	13.8%
	Cooking helper	-	-	1	3.4%	-	-	-	-	-	-	3.4%
	Lawyer	-	-	-	-	1	3.4%	-	-	-	-	3.4%
	Masseur	1	3.4%	-	-	-	-	-	-	-	-	3.4%
	Retired	-	-	5	17.2%	3	10.3%	2	6.9%	1	3.4%	37.9%
	Student	1	3.4%	-	-	-	-	-	-	-	-	3.4%
	Unemployed	1	3.4%	1	3.4%	2	6.9%	2	6.9%	1	3.4%	24.1%
Area of Living (by District)	Braga	-	-	1	3.4%	-	-	-	-	-	-	3.4%
	Porto	5	17.2%	6	20.7%	2	6.9%	2	6.9%	-	-	51.7%
	Viana do Castelo	-	-	3	10.3%	3	10.3%	1	3.4%	1	3.4%	27.6%
	Vila Real	-	-	2	6.9%	1	3.4%	1	3.4%	1	3.4%	17.2%
Total of Participants by Degree of Visual Impairment		5	17.2%	12	41.4%	6	20.7%	4	13.8%	2	6.9%	

Source: Author(s) own elaboration.

Regarding their education level, 34.5% concluded high school, 27.6% had the basic first stage and 17.2% had the basic third stage. Three individuals completed higher education: one LB (3.4%) holds a bachelor's degree in social communication and two participants with SVI (6.9%) holds a master's degree, one in social education and the other in law.

The majority of the participants (37.9%) were reported to be retired, 24.1% are unemployed, 13.8% work as Braille operators – two are EB and two are LB, and 10.3%, which are LB, as assistant operators (telephonist). The remaining three individuals work as a lawyer, cooking helper and masseur.

Regarding their area of living, 51.7% of the interviewees live in the District of Porto, 27.6% in the District of Viana do Castelo, 17.2% in the District of Vila Real, and one (3.4%) lives in the District of Braga.

Habits Regarding Visits to Museums and Heritage Sites

Table 2 presents the total of responses about the interviewee's habits regarding their visits to museums and heritage sites.

Concerning the frequency of visits to those places, 62.1% informed they seldom go, 17.2% reported they do it sometimes (one to two times in a year) and two participants (6.9%) use to visit museums more than four times in a year (almost always). Of the 29 blind and visually impaired people, three (10.3%) have never visited a museum at the time of their interviews. Their reasons have to do with the lack of interest, financial reasons, and lack of opportunity. Representative quotations from participants SVI8, LB14, and LB27, respectively, are provided:

- "It is not something that attracts me, but I never had the opportunity to go".
- "I had never had a curiosity to visit museums, even when I was able to see. Also, on a monetary level, it is complicated too, because we cannot see anything and still have to pay for it".
- "I know museums from watching television at the time I could see and have the interest to visit them, but I believe that the experience of visiting a museum as a blind in a wheelchair would be challenging".

Of the total given answers (n=37) – given by the 26 people who frequent museums and heritage sites – to the question on whom they visit museums with, 37.8% mention the associations where they are members. They also stated visits with close family members (27%) and friends (21.6%). All but one interviewee (2.7%) stated they had never visited a museum by themselves.

Regarding their answers about the primary reasons for visiting museums (n=43), it was observed that 39.5% of responses focused on general interest and curiosity, followed by entertainment reasons (25.6%) and socialization (16.3%). Only 7% of the given answers reveal they do so for relaxation and satisfaction, and 11.6% has to do with learning and educational reasons.

Associations are pointed out as essential ways to get information about museums and heritage sites to visit (37.8% of the total answers, n=37), as well as friends, relatives and colleagues (32.4%). Television and radio represent 10.8%, Internet 16.2%, and social media (Facebook) is the less mentioned, with only 2.7% of the total of the answers. It should be noted that the Internet was the most expressive way of getting information for the EB participants (10.8%) but was only cited by one LB and one person with MiVI. Although, they refer to websites as not accessible and its information as being scarce, inadequate and incomplete:

- Interviewee EB11 said, "There is a lack of information on the museums' websites, and they do not have online videos describing the exhibitions".

Table 2. Responses of participants about their habits regarding visits to museums and heritage sites

		Blind Participants				Partially Sighted Participants						Total
		Early-blind		Late-blind		Severe Visual Impairment		Moderate Visual Impairment		Mild Visual Impairment		
		n	%	n	%	n	%	n	%	n	%	
Frequency of visits to museums (n=29)	Almost always	-	-	1	3.4%	-	-	1	3.4%	-	-	6.9%
	Often	-	-	1	3.4%	-	-	-	-	-	-	3.4%
	Sometimes	-	-	3	10.3%	1	3.4%	-	-	1	3.4%	17.2%
	Seldom	5	17.2%	5	17.2%	4	13.8%	3	10.3%	1	3.4%	62.1%
	Never	-	-	2	6.9%	1	3.4%	-	-	-	-	10.3%
Visit museums with (n=37)	Alone	-	-	-	-	-	-	1	2.7%	-	-	2.7%
	Associations	1	2.7%	8	21.6%	3	8.1%	2	5.4%	-	-	37.8%
	Family members	2	5.4%	3	8.1%	3	8.1%	-	-	2	5.4%	27.0%
	Friends	2	5.4%	2	5.4%	3	8.1%	1	2.7%	-	-	21.6%
	Teachers	3	8.1%	-	-	1	2.7%	-	-	-	-	10.8%
Primary reasons for visiting museums (n=43)	Entertainment	1	2.3%	3	7%	3	7%	2	4.7%	2	4.7%	25.6%
	General interest/ Curiosity	5	11.6%	8	18.6%	-	-	2	4.7%	2	4.7%	39.5%
	Learning/ Education	2	4.7%	3	7%	-	-	-	-	-	-	11.6%
	Relaxation/ Satisfaction	-	-	2	4.7%	1	2.3%	-	-	-	-	7%
	Socialize	-	-	2	4.7%	3	7%	2	4.7%	-	-	16.3%
Information about museums to visit (n=37)	Associations	2	5.4%	7	18.9%	3	8.1%	2	5.4%	-	-	37.8%
	Friends/Relatives/ Colleagues	2	5.4%	3	8.1%	3	8.1%	2	5.4%	2	5.4%	32.4%
	Internet	4	10.8%	1	2.7%	-	-	-	-	1	2.7%	16.2%
	Social media	-	-	1	2.7%	-	-	-	-	-	-	2.7%
	Tv/Radio	2	5.4%	1	2.7%	-	-	1	2.7%	-	-	10.8%
Enjoy going to museums (n=26)	Yes	1	3.8%	5	19.2%	1	3.8%	2	7.7%	-	-	34.6%
	More or less	2	7.7%	4	15.4%	3	11.5%	1	3.8%	1	3.8%	42.3%
	No	2	7.7%	1	3.8%	1	3.8%	1	3.8%	1	3.8%	23.1%

- Participant EB17 remarked, "Museum web pages do not follow the rules of web accessibility. That is something that for a while, there was a significant concern, and now it has been neglected again," and explained its importance: "It's essential because if a blind person is minimally documented with what they are going to perceive, they will have something in advance to help to understand the exhibition".
- Interviewee MiVI26 stated, "Often there is a lack of information about the museum subjects. Since we are preparing a visit to their museum, it should have more available information about the things we are going to know".

From the total of the answers obtained to the question about enjoying going to museums and heritage sites (n=26), it was observed that the majority of the participants (42.3%) responded "more or less", 34.6% said "yes" and 23.1% revealed that they do not enjoy visiting those places. The next sections will deeply explore the reasons behind these answers.

Positive Visiting Experiences

When asked about positive experiences in museums and heritage sites, 24 participants shared some reasons, good accessibility practices, and factors that facilitate their visits to those places. From a total of 98 references registered during the interviews, the analysis allowed the classification of those into 13 distinct categories related to positive visit experiences, presented in Table 3.

Guided tours were underlined by 12 interviewees as a crucial factor in accessing the museum themes. Participants agreed that their access to information is facilitated by guided group tours since they can, on the one hand, interact with guides to clear their doubts and, on the other, establish fruitful conversations with other group visitors, share opinions and learn in conviviality and encircling environment:

- According to interviewee EB11, "The usefulness of these visits is that there are always other people asking for things that I do not even know or cannot see, and they end up talking about it. This exchange is what makes them advantageous".
- Participant SVI19 commented, "Most of the positive experiences were in guided visits because they described to us the things we aren't able to see and its history as well".

Table 3. Reasons and factors reported by the blind and visually impaired participants concerning their positive visiting experiences

Reasons and Factors for Positive Visiting Experiences	Participants Who Mentioned Them (n)		
	Blind Participants	Partially Sighted	Total
Museum guides and staff	9	5	14
Touching objects	9	5	14
Guided tours	5	7	12
Braille labels and texts	6	2	8
Participatory activities	4	3	7
Audio description through audio guides	2	4	6
Tactile models (bas-reliefs)	2	2	4
Escorts	3	-	3
Audio guides with navigational indications	-	2	2
Multisensorial experiences	2	-	2
Touching replicas	2	-	2
Circulation route facilitators	-	1	1
Large-sized texts	-	1	1

- Interviewee MiVI29 said, "I have two different experiences going to the same museum: the first time I went the tour was guided, and I found it very interesting, I really enjoyed it.", but the experience was different in the second time: "And then I went again without a guided tour, and it was totally different because nothing means nothing, there isn't anything there to explain".

However, individual guided tours are pointed out by three interviewees as more preferable ones, because they allow, according to LB15, "To interact much more with the guide; the absorption of knowledge is entirely different. In the context of a group visit, there is a rhythm to follow, and often there is no time for everything".

Participants also underlined that the success of guided tours and spontaneous visits is often connected with positive attitudes of museum guides and staff, their knowledge about how to communicate information to blind and visually impaired people, and their attention to visitor's behavior:

- Interviewee EB17 mentioned, "The fact that the tour is guided is not enough by itself; I had to be fortunate to have guides who gave a sufficiently detailed description of the things so that I could have an idea about them".
- Participant LB15 commented, "At the Salamanca Automobile History Museum, I noted that there was an immense concern. A person from the museum asked me if I was blind and if he could accompany me," and during the spontaneous visit, "He had the care to inform me about the things that the people who were with me couldn't convey. He was aware of what was happening all the time, and he was contextualizing the visualization of the objects".
- Interviewee LB4 stated, "At the Calouste Gulbenkian Museum, in Lisbon, a museum worker realized that I was blind and said I could touch the exhibits. It was forbidden, but he opened an exception to me".
- Participant SVI24 observed, "Some museums offer support and have staff available to explain, and this is really good for disabled persons who want to visit a museum without a previous appointment".

Of the total number of interviewees who visited museums and heritage sites, 18 (69.2%) had the opportunity to touch museum objects, replicas or tactile models – even though it happens seldom – while the other eight (30.8%) never had this chance. They reported sensorial access through touch in exhibitions as allowing much-improved experiences, since they are essential to understand what is being described thoroughly and help to create accurate mental images of the samples:

- According to participant LB7, "If there is something touchable to confirm what they are describing to us, the better", because "When people speak, I can get an idea. But some things can only be perceived with the hands, like the size. You can describe the size in centimeters but touching makes all the difference".
- Interviewee LB5 commented, "During one visit they had clothes, and we could touch them. It is crucial for us feeling the texture and sense other things while museum guides are describing the history and the colors", and also clarified, "For me, telling the colors is very important, because as I saw 56 years ago, I still know them".

- Participant LB15 remarked, "By touching, I can perceive. Of course, there are things with small strokes that are difficult to interpret, but as I saw, I have a notion of things", because "The brain automatically associates a visual image to the touch information. For those who have seen, touching is like visually seeing in their minds".
- Interviewee EB18 informed, "At the *Misericórdia do Porto Museum*, the *Fons Vitae* painting was transformed into a bas-relief with several different textures," and said, "I could enjoy the painting and be able to perceive it, given the accessible version they made and the descriptions the museum guide provided".

Furthermore, participants highlighted that hands-on activities could contribute to evoke past memories and to increase the durability of new generated memories. A few representative quotations follow, respectively from interviewees MoVI9, LB21, and EB17:

- "At the Portuguese Toy Museum, it was very interesting to touch toys that existed at the time of my grandparents, and I didn't know about them. Others I knew and had no memory, but then I remembered of them".
- "While the museum guides were explaining the objects, we could touch them. Some of them I knew it from my childhood, and I was able to identify them just by touch".
- "I clearly remember a visit I did 20 years ago to a museum with the customs and traditions of the fishers. But there it is, maybe I memorized this because they had a lot of objects available to touch".

Another positive factor pointed by seven interviewees as providing great museum experiences were participatory activities. Those allowed interactions with the museum themes and objects on a deeper level, putting visitors in the center of the action:

- Participant LB2 said, "I enjoyed a visit I made to the Tramway Car Museum many years ago because we could enter in a tramway and made a mini trip," and commented, "It was only a few meters, but they even talked about the history of asking the driver to open or close the windows because of the wind".
- Interviewee SVI4 remarked, "At the Soares dos Reis National Museum, we did a mini-theater, and since I was a principal character, they allowed me to wear a uniform. It was flawless".
- Participant LB15 commented, "I never forgot the experience I had at the Biscainhos Museum because they let me wear replicas of traditional costumes," and observed, "It is different to touch a thing that someone is wearing or to experience something we are wearing because I had all those accessories, the corsets, and other things," concluding that "By explaining that people could hardly sit down is not as perceivable as experiencing it, when everything is understood without a word. Art is to be experienced, to touch, to feel, and to interact".
- Interviewee SVI19 stated, "I loved going to the Bread Museum in Outeiro where, after the guided tour, we made the bread! I love the parts when we're able to get into the action. Everybody enjoyed that part".

Of all the participants, only one, LB15, mentioned multisensorial experiences in museums: "I once did a visit to blind people where smells, tastes, and sensations were provided, and I never forgot this experience".

Regarding access to information, eight positive comments were registered for the use of Braille labels and texts, one about large-sized texts and six about audio description through audio guides. Interviewees mentioned how they do facilitate access:

- Participant EB11 said, "At the *Futebol Clube do Porto* Museum they gave us a little book with Braille explanations. It was helpful because as the guided tour was going on, I could consult it to guaranty that I didn't forget about anything".
- Interviewee LB7 remarked, "Whenever I go to a museum, I always ask if they have anything in Braille. Sometimes I'm allowed to bring home a volume in Braille and read it calmly. Other times I can only read it there".
- Participant SVI19 commented, "The only time I found large-sized texts was at the Chocolate Factory Museum, and I was able to read everything by myself," and observed, "But I think they were not made for someone with low vision; maybe it had to do with the design of the exhibition".
- Interviewee SVI4 stated, "I did a visit using an audio guide, and it described the pieces in each room. Listening to all the information was a good experience".
- Participant SVI10 mentioned, "I visited an exhibition at the Calouste Gulbenkian Museum, where it was possible to see photographs by using headphones. The audio described everything in the photos".
- Interviewee LB2 said, "I think audio guides play a significant role to communicate the museum information, especially if there aren't museum guides available, as happened when I visited the House of the Prince".

Finally, six commentaries concerning positive mobility and orientation experiences in visits to museums were registered.

Escorts were mentioned by three visitors as facilitators for orientation and mobility inside museum spaces, especially in non-organized visits. Participant LB22 remarked, "I usually take a companion that helps me. I never go to museums with my cane, because it is always a busy hand. With the dog, it would still be worse, because I had to have both hands occupied".

One participant, SVI19, highlighted a good practice for the circulation route inside museum spaces, even though it is not a current practice: "When we go to museums with the association, they prepare the space with a few strings for us to follow," observing that "But if I go there alone with my boyfriend or with my friends, there is nothing there. That was prepared for a group that has already made a prior appointment".

Two interviewees stated using audio guides with navigational instructions during their visits to museums:

- SVI4 commented, "At a museum in Lisbon, I was able to do a visit following the instructions of an audio guide, which gave me directions to move between the rooms and described the exhibits in each space".
- Interviewee MoVI9 remarked, "It was good to be guided, as in a museum we visited and had the headphones describing everything," and concluded "We didn't need an escort because the device was saying: "turn right, turn left, go ahead, get down". But this is rare to find".

Negative Visiting Experiences

Regarding negative experiences in museums, a total of 26 interviewees shared 171 factors, poor accessibility practices, and other reasons that hinder their participation in museum settings. The analysis of those experiences allowed their classification into 19 categories, presented in Table 4.

According to the data, the most reported factor that contributes to negative experiences during museum and heritage sites visits is the unavailability of objects to touch. This was expressed by 80.8% of the participants (13 blind and 8 visually impaired), who also stressed that, by not allowing sensorial access through touch, museums are preventing them from making use of the sense that is in the essence of their understanding and construction of mental images, as well as of the creation of memories about the objects. A few representative quotations follow, respectively from participants EB17, LB7, EB11, LB16, LB5 and SVI24:

- "There are many objects that cannot be touched".
- "During the visits, there was one or another object I was allowed to touch, but in general, there was only a description".

Table 4. Reasons and factors reported by the blind and visually impaired participants concerning their negative visiting experiences

Reasons and Factors for Negative Visiting Experiences	Participants Who Mentioned Them (n)		
	Blind Participants	Partially Sighted	Total
Unavailability of objects to touch	13	8	21
Guided tours	10	5	15
Descriptive language used	7	3	10
Museums architecture and exhibition designs	4	5	9
Museum guides and staff	4	3	7
Lack of Braille labels and texts	6	-	6
Audio description through audio guides	3	1	4
Lack of large-sized texts	-	4	4
Spontaneous visits	1	3	4
Entrance fees and other costs	-	3	3
Gloves to touch exhibits	3	-	3
Inaccessible websites and without useful information	2	1	3
Poor ambient lighting	-	3	3
Unavailability of replicas to touch	2	1	3
Excess of verbal information	1	1	2
Lack of information through audio	-	2	2
Signs do not meet accessibility requirements	-	2	2
Tactile models (bas-reliefs)	2	-	2
Materials of the replicas.	1	-	1

- "No matter how much the persons tell me how the object is, I don't have a good mental representation, because I don't know for sure the outline of things, their textures, how they really are".
- "Touching is essential for us. They [museum guides] can describe us the objects as much as they want, but it is not the same thing as we touch them with our hands. Is different".
- "For our memory, it is very different only listening to the description or touching. If I touched the furniture or the curtains, then I would remind them".
- "In my case, there is the question of not having access to the objects and paintings, of not being able to see nor touching them", mentioned participant SVI24, because "Even as I approach the glass, there is a space between my position and the object that doesn't allow me to observe it. Being able to get close to the exhibits and touch them is crucial".

Notwithstanding recognizing the importance of touch in museum visits, all participants showed to be aware of the constraints of touching original museum pieces and why this is not possible at the majority of times. They agreed that replicas could be an alternative solution to it, although three interviewees reported some negative experiences related to its utilization, and that they still are remarkably rare in museums and other exhibition spaces:

- Participant EB18 mentioned, "I understand that there are objects that cannot be touched in museums, but there are not replicas to touch as well".
- Interviewee LB15 stated, "If we cannot touch, there has to be a smaller size replica as it already exists in other countries. I went to a museum in South Africa where they had several objects on a smaller scale," and commented, "I don't conceive that they have this and museums in Portugal don't. The taste for the art of the blind can only be encouraged if we can touch things".
- Participant LB1 remarked, "I once touched replicas during a visit to a Palace, and it was not useful at all," because, "They didn't let us feel the clothes, just the reproductions of dresses and other garments, made of paper. Touching paper is not the same as touching fabrics".

Of all the participants, only three late-blind reported they used gloves to touch original museum pieces, but their experiences were more negative than positive since it hindered sensing activities:

- Interviewee LB7 said, "I had an unpleasant experience at a museum where they lent us gloves to wear so that we would not damage the sculptures. I understand that, but I didn't enjoy the experience," and commented, "I'd rather just listening to the description than to have such experiences, because the glove, even being very thin, doesn't allow me to feel the things as with direct touch contact".
- Participant LB2 observed, "We live a lot from touch, and with the gloves, we lose a lot of sensitivity. Sometimes we cannot detect the quality of the material, its texture, and so forth".

Even though tactile models were previously pointed as promoting positive experiences, two blind visitors reported that they are not always accessible to people by tactile examination: LB7 said, "They had representations of geometric figures in Braille. I found that a bit confusing and couldn't understand the images. And because I cannot perceive those lines and specks, I find it demotivating and prefer the description instead".

Despite the positive aspects previously expressed by the interviewees about group guided tours, 15 participants (57.7%) also highlighted negative experiences related to them. These have to do with the number of people they usually do visits with, the distraction caused by noises and conversations aside from the museum themes, and the fact that the guide's attention is divided, which makes it impracticable to give attention to all the participants:

- Interviewee MoVI25 said, "In a group, there are always strange noises, mobile phones ringing and people who are not attentive".
- Participant LB1 commented, "During a visit, some persons questions about some things, other comments about other things. It gets to the point of not realizing what the guide is saying".
- According to participant EB17, "When there is something to touch, it's always so fast, because the other colleagues want to touch the objects too. And touching requires calmness and time. It isn't comfortable in these situations".
- Interviewee LB15 remarked, "Group visits are important for socialization but not so much for information".
- Participant MiVI26 noted, "In the Vatican, I did one part where the visit was guided and the other where it wasn't. What I noticed was the hurry of the guides, which I think is not advantageous".

Seven participants also mentioned the lack of awareness and formation of guides and other museum staff in dealing with blind and visually impaired people, as well as attitudes that negatively influence their experiences:

- Interviewee SVI10 stated, "People who are in museums don't have any training nor formation to deal with us".
- Participant EB18 noted, "Sometimes, when the museum workers realize that we are blind, they get all embarrassed and don't know what to do".
- Interviewee EB17 said, "There are museum guides that are different from each other," and exemplified, "I visited the Wax Museum of Fátima twice because I enjoyed the first visit I did. The last time I went there I didn't like anything because of the guide. It seemed that he was doing people a favor".
- Participant EB13 commented, "A guide cannot do a tour to a museum in the same way for all people: if it is for a group of children, it cannot be the same way as for a group of older people," and observed, "It is the same for people with disabilities, mainly depending on the disability: in the case of the blind, it would have to be a precise training so that we could have access to more complete and usefulness information".

The descriptive language used during the visits was pointed out by 38.5% (10) participants as a factor that hinders their access, particularly concerning the quality of the information provided by museum guides:

- Interviewee EB11 said, "They do not indicate the room where we're going to enter and what is in there. Sometimes people forget they are talking to people who are blind or have low vision".
- Participant EB12 remarked, "When they describe that the building is from this height by this breadth, we are never aware of the real dimensions".

- Interviewee LB15 noted, "Not every people can convey what is intended. It's necessary to have a sensitivity of interpretation to communicate to us what sighted people can have access to".
- Participant SVI24 commented, "I went to a museum with several paintings representing spring, summer, autumn, and winter, and that's usually what the guides say. As I couldn't see the details, I didn't know why it represents summer, for example," and observed "They said that was because of the fruits. This information worth's nothing. The "why" stills lacking: what fruit of the time, where are they positioned, it is in the eyes, in the hair, etc.".

Two interviewees, SVI19 and LB22, respectively, added to the issues of the quality of the language used, the quantity of information provided:

- "Sometimes guides talk too much, and it gets to a point where nobody is no longer listening".
- "In general, the information during a visit is too much and provided in a short space of time. It makes it difficult to retain some aspects when everything is so fast and dense".

Six blind participants noted that there is a lack of Braille labels and texts in museums: EB13 said, "Much information in Braille is still missing. They could have labels indicating the name of the pieces, what century they belong to, what they were used for, where they were found, etc.", and concluded, "In short, an abbreviated list, but with enough information for us". They also highlighted that this kind of material in Braille could help retain information: interviewee LB15 remarked, "I don't remember more of the contents not because of the people who transmitted the information, but because there was very little information in Braille".

Four partially sighted participants informed about the lack of large-sized texts and how it was negative during visits they made to museums. A quotation from interviewee SVI19 summarizes this problem: "I recently went to a museum, but I didn't like it. There were posters with lots of text, so I wasn't able to read almost anything besides the titles. My boyfriend read some things to me, but not all", and stated, "I end up telling him to leave because I was enough of it".

Information through audio was pointed by two visually impaired interviewees as facilitating the access to written information. However, MiVI26 observed, "there is often a lack of written information about the subjects, and it's not accessible to me".

Audio guides with audio description can support the lack of written information, as previously discussed, but four participants (three blind and one with mild visual impairment) related negative experiences regarding their use:

- LB7 commented, "There was a time when the description was made for a specific period, and we ended up being late and missing the right time".
- Interviewee MiVI26 said, "I wanted to try to use it by myself, but I had to ask my husband or my daughters every time to help me because I was having trouble working with the keys", and added, "Many times, I even wanted to see a part of the exhibition again, but it was always so hard to work with it that it turned out to be an obstacle".

As expected, negative experiences relating to museums architecture and exhibition designs were expressed by nine participants with all the different types of vision impairments. The main problems registered had to do with stairs, because some people do not feel comfortable with them, and also with obstacles they find while moving inside museum galleries: interviewee SVI24 remarked, "Some museums have display cases in the middle of the route, which in terms of mobility, turns it more difficult".

Three visually impaired participants experienced difficulties regarding poor ambient lighting: MiVI29 commented, "It is always very dark at museums and heritage sites" and MiVI26 said, "I usually notice a lot of the lack of luminosity inside museums, in general. I remember it was very dark in Jerónimos. In Rome, at the Pentagon, there was also darkness", and concluded, "It makes it difficult to move, to read, and also ends up hindering the use of the audio guides". This situation also affects the orientation through signage and graphics, that do not meet accessibility requirements: participant SVI19 mentioned, "It happened I entered the wrong room in Fátima," because "They had tiny arrows, which do not take into account the contrasts necessary for people with low vision. And since a person doesn't know the space and doesn't see the signs, things like this happen".

Even though the majority of the interviewed persons said that they visit museums in contexts of organized tours – that normally require prior scheduling – four participants shared their experiences about spontaneous visits they have made. Parts of two testimonials follow:

- Interviewee LB28 said, "If someone from the museum explained to me the things, maybe I would have like it. Since I had no access to anything, the mental image I have about the objects is nil. And about the information, I didn't have access to anything either," and concluded, "Visiting it was the same as not even entering. I didn't stay outside simply because I didn't want to be alone while my family was doing the visit".
- Participant MiVI29 remarked, "A person who goes by himself with family or friends, without an organized excursion or guided tours, don't have access to objects nor anything, can't see anything, nor knowing the written stories".

The fact that museum websites are inaccessible most of the time and also do not present useful information to someone with visual impairment who is planning a visit was stressed by three people:

- Participant EB17 mentioned, "Online pages should have more detailed descriptions of what museums have to show and should also make use of the accessibility rules for the web," and noted, "This is essential because if a blind person is minimally documented with what is going to attend, even if can't touch the objects, will already have some previous information".
- Participant SVI19 said, "The museum websites I tried to consult do not take into account the contrasts for people with low vision. For example, it uses white on a light blue background, white on a yellow background. Sometimes it drives me crazy".

Finally, three partially sighted interviewees referred entrance fees as hindering access to museums and heritage sites, as well as other costs:

- Participant SVI4 commented, "Nowadays, going to museums is less possible because now museums are paid. Before the visually impaired didn't have to pay, but now we pay for everything".

- Interviewee MiVI26 said, "I usually avoid guided tours because they are more expensive, to be sincere, and it makes them not accessible to everyone," and also mentioned, "At the Colosseum in Rome, for example, we used the audio guides because it was more affordable than the tour with a guide, which was much more expensive".
- Participant MiVI29 observed, "At the Monastery of Batalha, they had audio guides not for free. I don't know how much it was, but it had to be paid to get access to it".

PROPOSAL OF AN INTEGRATIVE FRAMEWORK TO ENHANCE BLIND AND VISUAL IMPAIRED VISITORS' EXPERIENCES IN MUSEUMS

During the interviews, participants were asked to suggest aspects for improvement, based on their previous experiences, in order to enable more enjoyable visits to museums and heritage sites in the future.

Following the above chapter sections, where positive and negative visiting experiences were discussed, and after analyzing the 205 coded interviewees' suggestions for enhancing their experiences in those sites, it was concluded that a framework should support the integration of the various phases identified as relating to the experience of visiting cultural institutions: pre-visit, on-site visit, and post-visit. These findings corroborate with the works of Falk (2013) and Kuflik, Wecker, Lanira, and Stock (2015), despite their researches not including individuals with low vision nor with other impairments.

Moreover, the data analysis performed showed that it is imperative to ensure that the framework also promotes sensory, intellectual, and physical access, so individuals are more likely to feel a sense of autonomy during the various phases of a visit, despite their types of visual impairments. This does not mean that the proposed framework is intended to replace guided tours, but rather to complement them and also allow blind and visually impaired people having access to museums' and exhibitions' contents whenever they wish to make spontaneous visits.

Figure 1 illustrates the integrative framework proposed in this research, which will be presented next.

Pre-visit Planning

The pre-visit phase precedes the actual on-site visit, in which visitors plan their visits and get familiar with the museum. It needs to support full intellectual access to the institution's information and its exhibition contents.

In an era of digital technologies, museum websites commonly constitute the initial impression that individuals have about an institution. With this in mind, access information on museum websites can contribute (or not) to a great first impression of museums, (un)welcoming everyone to visit their spaces.

Intellectual access through this phase is provided by:

- Accessible Museum Website: following the WCGA – Web Content Accessibility Guidelines, developed by the World Wide Web Consortium (2008), in order to provide users with understandable information, text alternatives for non-text content, audio-described multimedia content, and facilities for easy navigation and content finding.
- Informative Museum Website: according to Cock et al. (2018), use inclusive language to provide useful information like contact details, detailed information about getting to the museum, and information about accessible services provided, and exhibitions accessibility.

On-site Visit

It refers to the actual on-site visit, which needs to provide visitors with sensory and intellectual access to the exhibitions, as well as helping them easily find their way in the museum.

When arriving at the museum, visitors can have intellectual access through:

- **Inclusive Leaflet**: With information in Braille and large-sized.
- **Audio Presentation**: Proper descriptive language presents how the museum space is organized and the location and themes of its exhibitions.

During the visit, intellectual access to exhibits is provided by:

- **Labels and Texts**: In Braille and large-size fonts.
- **Audio Information**: Describes the artifacts and provides museum information about them, using inclusive language to the blind and the visually impaired visitors. The quantity of information must be balanced.

Sensory access is also promoted during the visit by:

- **Physical Objects and Tactile Models Digitally Augmented**: Depending on the exhibition, three-dimensional replicas of objects and bas-reliefs of two-dimensional works of art can be sensed by touch and interactively present the audio information.
- Magnified images of the exhibits: to help visitors with moderate and mild visual impairments better observe details, colors, and other visual aspects.

The framework also proposes assistive navigation to enhance visitors' physical access in what concerns to orientation, mobility, and wayfinding:

- **Exhibition Design**: Exhibition spaces are organized, as much as possible, to avoid obstacles on the circulation route.
- **Orientation Signs**: Designed to meet accessibility requirements.
- **Assistive Guides**: Provide audio navigation indications.

Post-visit Phase

The post-visit phase – after visitors leaving the museum – should support reflections and consolidation of the contents learned during the on-site visit.

During the interviews, some participants showed interest in complementing their visits by further exploration of contents not featured in the exhibition. The post-visit phase can also be significant to bridge connections with visitors (McMillen & Alter, 2017) and encourage additional on-site visits.

With this, intellectual access must be granted through:

- **Books or Others**: In Braille and large-size fonts.
- **Museum Website**: With complementary contents about the museum exhibitions.
- **Social Media Platforms**: Information about events and new exhibitions.

CONCLUSION AND FUTURE RESEARCH DIRECTIONS

Museums undertake the hard mission of preserving humanity's heritage, while having to exhibit and to communicate this heritage, in order to provide society with insights and opportunities to experience art, culture, history, sciences, among other subjects. However, given the ocular centricity of those institutions, millions of people with vision impairments worldwide still experience many physical, intellectual, and sensory barriers when visiting museum exhibitions.

Figure 1. Scheme of the proposed framework

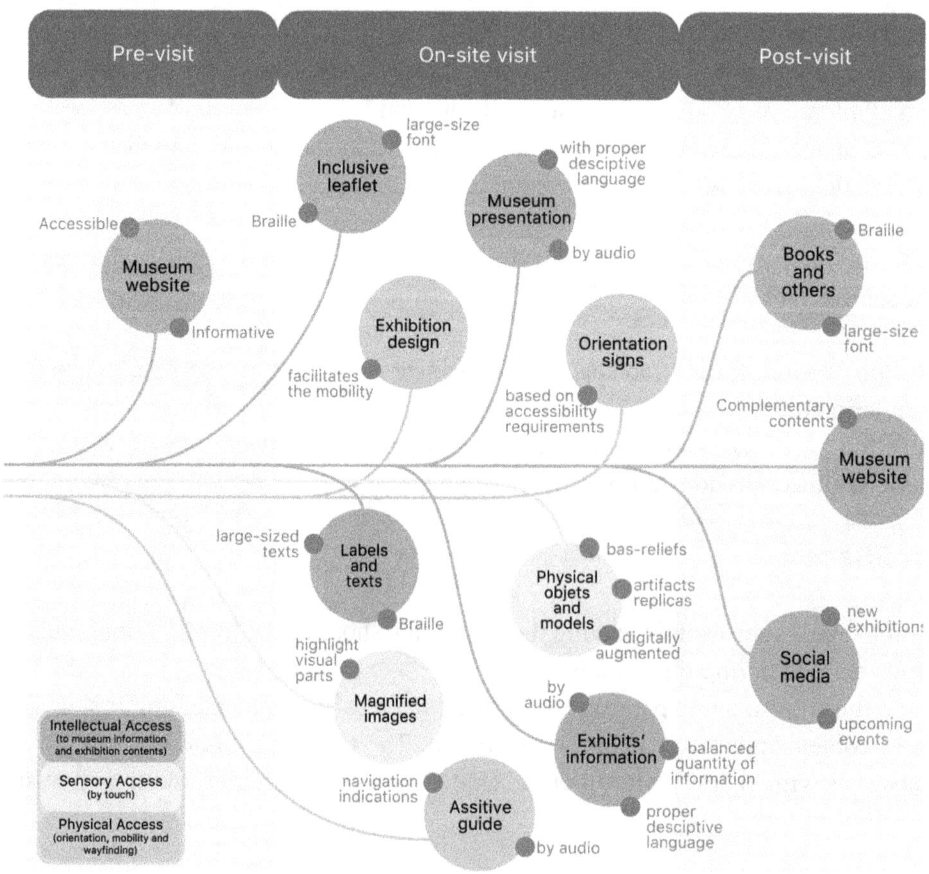

It is urgent to give priority to issues related to the accessibility of those patrons in museums and their social inclusion, taking into account that visual impairment is a very diverse spectrum.

Technology offers the potential of increasing the accessibility of exhibitions and spaces, enhancing blind and visual impaired patrons' experiences when visiting those institutions, by providing opportunities for handling and touching objects, as well as assisting them during their visit. Moreover, websites and social media platforms, when taken into account accessibility concerns, can be used as strategies to bridge new connections with the blind and visually impaired community, contributing to social inclusion in the context of museums and to accessible tourism.

This chapter explored the habits of 29 blind and visually impaired individuals regarding visits to museums and other heritage sites, considering their different degrees of visual impairment. In order to investigate a set of guiding principles to enhance future visits to museums, there were presented and discussed positive experiences they had and also reasons and factors that contributed to negative ones.

According to the performed data analysis, it was possible to conclude that most of the participants visit museums seldom, usually in contexts of visits organized by the associations where they are members, or with close family members and friends. Regarding their primary reasons for visiting museums, general interest and curiosity were highlighted as the usual ones, followed by entertainment reasons. Few participants go to museums for relaxation and satisfaction, and because of learning and educational reasons. When asked if they enjoy going to museums and heritage sites, the majority of the individuals said they like it more or less.

About the ways to get information about museums and heritage sites, it was concluded that associations play an essential role, as well as friends, relatives, and colleagues. The Internet and social media were mentioned as not being accessible and lacking information, which reveals an opportunity to museums magnifying their virtual presence to this community, by addressing issues of digital accessibility.

It was also concluded that participants experienced more negative than positive visits in museums, and those reasons and factors were discussed.

An integrative framework was co-created based on the participants' previous visit experiences, and also on many suggestions on aspects for improvement they made in order to enhance future visits to museum sites and exhibitions. It aims to assist blind and visually impaired people in having access to museum-related content whenever they wish to make spontaneous visits, i.e., without the need for a previous appointment.

The proposed framework links the various phases of visiting cultural institutions (pre-visit, on-site visit and post-visit), and ensures a sense of autonomy of the individuals with different types of visual impairments during all of these, by promoting sensory, intellectual and physical access to museum exhibitions, contents, online sites, and physical spaces.

Since this framework is in its initial development stage, there are not evaluation tests to measure its feasibility and applicability yet. Another limitation is the number of participants in this study and also the sole focus on their viewpoints, rather than also including museum professionals to obtain a more in-depth perception of the explored topics.

Future research will perform interviews with more visually impaired and blind people from other Districts of Portugal, in order to understand if their experiences of visiting museums and heritage sites converge or not with the ones presented in this chapter. It will also be investigated their needs and desires to improve future visits to those institutions, and their expectations from integrating assistive technologies across the continuum of the visits. Besides that, research will be conducted to further understanding how digital media can support the requirements of a framework like this one, following user-centered design principles and accessibility guidelines for digital inclusion of these persons.

ACKNOWLEDGMENT

This work is a result of the project Operation NORTE-08-5369-FSE-000049 supported by Norte Portugal Regional Operational Programme (NORTE 2020), under the PORTUGAL 2020 Partnership Agreement, through the European Social Fund (ESF).

REFERENCES

Andrade, E. C., Delgado, R., Hauptner, B., Kerschbaum, F., List, R., Pobežalová, P., . . . Wögerbauer, S. (2015). *Inclusive technologies in museums: for a better access to culture for blind and visually impaired people.* Vienna, Austria: Project AMBAVis. Retrieved from http://www.ambavis.eu/wp-content/uploads/2016/11/AMBAVis-Broschuere_BF_V01_EN_bf.pdf

Argyropoulos, V. S., & Kanari, C. (2015). Re-imagining the museum through "touch": Reflections of individuals with visual disability on their experience of museum-visiting in Greece. *Alter - European Journal of Disability Research, 9*(2), 130–143. doi:10.1016/j.alter.2014.12.005

Asakawa, S., Guerreiro, J., Ahmetovic, D., Kitani, K., & Asakawa, C. (2018). The present and future of museum accessibility for people with visual impairments. In *Proceedings of the 20th International ACM SIGACCESS Conference on Computers and Accessibility* (pp. 382–384). New York, NY: ACM; doi:10.1145/3234695.3240997

Bourne, R. R., Flaxman, S. R., Braithwaite, T., Cicinelli, M. V., Das, A., Jonas, J. B., ... Taylor, H. (2017). Magnitude, temporal trends, and projections of the global prevalence of blindness and distance and near vision impairment: A systematic review and meta-analysis. *The Lancet. Global Health, 5*(9), 888–897. doi:10.1016/S2214-109X(17)30293-0 PMID:28779882

Buonamici, F., Furferi, R., Governi, L., & Volpe, Y. (2015). Making blind people autonomous in the exploration of tactile models: a feasibility study. In M. Antona & C. Stephanidis (Eds.), *Universal Access in Human-Computer Interaction. Access to Interaction* (pp. 82–93). Cham, Switzerland: Springer International Publishing; doi:10.1007/978-3-319-20681-3_8

Burton, H. (2003). Visual cortex activity in early and late blind people. *The Journal of Neuroscience, 23*(10), 4005–4011. doi:10.1523/JNEUROSCI.23-10-04005.2003 PMID:12764085

Cachia, A. (2013). Talking blind: Disability, access, and the discursive turn. *Disability Studies Quarterly, 33*(3), 1–20. doi:10.18061/dsq.v33i3.3758

Candlin, F. (2003). Blindness, art and exclusion in museums and galleries. *International Journal of Art & Design Education*, *22*(1), 100–110. doi:10.1111/1468-5949.00343

Candlin, F. (2004). Don't touch! Hands off! Art, blindness and the conservation of expertise. *Body & Society*, *10*(1), 71–90. doi:10.1177/1357034X04041761

Cantoni, V., Dondi, P., Lombardi, L., Nugrahaningsih, N., Porta, M., & Setti, A. (2018). A multi-sensory approach to cultural heritage: The Battle of Pavia exhibition. *IOP Conference Series. Materials Science and Engineering*, *364*, 1–8. doi:10.1088/1757-899X/364/1/012039

Caracelli, V. J., & Greene, J. C. (1993). Data analysis strategies for mixed-method evaluation designs. *Educational Evaluation and Policy Analysis*, *15*(2), 195–207. doi:10.3102/01623737015002195

Cattaneo, Z., & Vecchi, T. (2011). *Blind Vision: The Neuroscience of Visual Impairment*. Cambridge, MA: MIT Press. doi:10.7551/mitpress/9780262015035.001.0001

Chick, A. (2017). Co-creating an Accessible, Multisensory Exhibition with the National Centre for Craft & Design and Blind and Partially Sighted Participants. In *REDO: Cumulus International Conference*. Academic Press. Retrieved from http://eprints.linclon.ac.uk/27590

Cock, M., Bretton, M., Fineman, A., France, R., Madge, C., & Sharpe, M. (2018). *State of Museum Access: Does your museum website welcome and inform disabled visitors?* London, UK: Vocal Eyes. Retrieved from https://vocaleyes.co.uk/state-of-museum-access-2018

D'Agnano, F., Balletti, C., Guerra, F., & Vernier, P. (2015). Tooteko: a case study of augmented reality for an accessible cultural heritage. Digitization, 3D printing and sensors for an audio-tactile experience. In *ISPRS - International Archives of the Photogrammetry, Remote Sensing and Spatial Information Sciences* (pp. 207–213). Academic Press. doi:10.5194/isprsarchives-XL-5-W4-207-2015

Dandona, L., & Dandona, R. (2006). Revision of visual impairment definitions in the International Statistical Classification of Diseases. *BMC Medicine*, *4*(1), 7. doi:10.1186/1741-7015-4-7 PMID:16539739

Dash, K., & Grohall, G. (2016). *Economic impact of creating and exhibiting 3D objects for blind and visually impaired people in museums*. Vienna, Austria: AMBAVis Project. Retrieved from http://www.ambavis.eu/wp-content/uploads/2016/09/Economic-Aspects_final-disclaimer.pdf

Desvallées, A., & Mairesse, F. (2010). *Key Concepts of Museology*. Paris, France: Armand Colin.

Devile, E., & Kastenholz, E. (2018). Accessbile tourism experiences: the voice of people with visual disabilities. *Journal of Policy Research in Tourism, Leisure and Events, 10(3),* 265-285. doi:10.1080/19407963.2018.1470183

European Commission. (2013). 19 destinations across Europe win awards for accessible and sustainable tourism. Retrieved from http://europa.eu/rapid/press-release_MEMO-13-972_en.htm

Falk, J. H. (2013). Understanding museum visitors motivations and learning. In I. Lundgaard & J. Jensen (Eds.), *Museums: Social Learning Spaces and Knowledge Producing Processes* (pp. 106–127). Copenhagen, Denmark: Kulturstyrelsen.

Fossey, E., Harvey, C., McDermott, F., & Davidson, L. (2002). Understanding and evaluating qualitative research. *The Australian and New Zealand Journal of Psychiatry*, *36*(6), 717–732. doi:10.1046/j.1440-1614.2002.01100.x PMID:12406114

Freeman, A., Becker, S., Cummins, M., McKelroy, E., Giesinger, C., & Yuhnke, B. (2016). *NMC Horizon Report: 2016 Museum Edition*. Austin, TX: The New Media Consortium; doi:10.1111/j.2151-6952.2010.00064.x

Gelo, O., Braakmann, D., & Benetka, G. (2008). Quantitative and Qualitative Research: Beyond the Debate. *Integrative Psychological & Behavioral Science*, *42*(3), 266–290. doi:10.100712124-008-9078-3 PMID:18795385

Ginley, B. (2013). Museums: A Whole New World for Visually Impaired People. *Disability Studies Quarterly*, *33*(3), 1–15. doi:10.18061/dsq.v33i3.3761

Google. (n.d.). *Cloud Speech-to-Text: Speech-to-text conversion powered by machine learning and available for short-form or long-form audio*. Retrieved from https://cloud.google.com/speech-to-text/

Hayhoe, S. (2017). *Blind Visitor Experiences at Art Museums*. London: Rowman & Littlefield.

Hutchinson, R. S., & Eardley, A. F. (2018). Museum audio description: The problem of textual fidelity. *Perspectives: Studies in Translation Theory and Practice*, *27*(1), 42–57. doi:10.1080/0907676X.2018.1473451

INE. (2012). *Censos 2011 Resultados Definitivos - Portugal*. Lisbon, Portugal: Instituto Nacional de Estatística.

IPM. (2004). *Museus e Acessibilidade. Colecção Temas de Museologia*. Lisbon, Portugal: Instituto Português de Museus.

Jain, D., Jain, A., Paul, R., Komarika, A., & Balakrishnan, M. (2013). A path-guided audio based indoor navigation system for persons with visual impairment. *Proceedings of the 15th International ACM SIGACCESS Conference on Computers and Accessibility* (pp. 33:1-33:2). New York, NY: ACM. 10.1145/2513383.2513410

Johnson, J. (2018). Sensory: Please touch the art. *Art Education*, *71*(1), 12–15. https://doi.or/10.1080/00043125.2018.1389580. doi:10.1080/00043125.2018.1389580

Kuflik, T., Wecker, A. J., Lanir, J., & Stock, O. (2015). An integrative framework for extending the boundaries of the museum visit experience: Linking the pre, during and post visit phases. *Information Technology & Tourism*, *15*(1), 17–47. doi:10.100740558-014-0018-4

Levent, N., & Reich, C. (2012). How can my museum help visitors with vision loss? *Museum. American Association of Museums,* (July-August), 21–22.

McMillen, R., & Alter, F. (2017). Social media, social inclusion, and museum disability access. *Museums & Social Issues*, *12*(2), 115–125. doi:10.1080/15596893.2017.1361689

Meliones, A., & Sampson, D. (2018). Blind MuseumTourer: A system for self-Guided tours in museums and blind indoor navigation. *Technologies*, *6*(1), 4. doi:10.3390/technologies6010004

Mesquita, S., & Carneiro, M. J. (2016). Accessibility of European museums to visitors with visual impairments. *Disability & Society*, *31*(3), 373–388. doi:10.1080/09687599.2016.1167671

Morgan, J. (2012). The multisensory museum. *Glasnik Etnografskog Instituta*, *60*(1), 65–77. doi:10.2298/GEI1201065M

Neves, J. (2012). Multi-sensory approaches to (audio) describing the visual arts. *MonTI. Monografías de Traducción e Interpretación*, (4), 277–293.

Osterman, M. (2017). Accessibility and technology: Developing a virtual access tour. *MW17: Museums and the Web*. Retrieved from https://mw17.mwconf.org/paper/accessibility-and-technology-developing-a-virtual-access-tour/

Pillow, B., Banks, M., & Reeves, D. (2015). *Equal access for all: providing for impaired stakeholders in a museum setting*. Retrieved from http://www.byronpillow.com/uploads/7/6/7/9/7679 7585/disability_access.pdf

Postma, A., Zuidhoek, S., Noordzij, M., & Kappers, A. (2007). Differences between early-blind, late-blind, and blindfolded-sighted people in haptic spatial-configuration learning and resulting memory traces. *Perception*, *36*(8), 1253–1265. doi:10.1068/p5441 PMID:17972487

Preece, J., Rogers, Y., & Sharp, H. (2002). *Interaction design: beyond human-computer interaction*. New York, NY: John Wiley & Sons, Inc.

Reich, C., Lindgren-Streicher, A., Beyer, M., Levent, N., Pursley, J., & Mesiti, L. A. (2011). Speaking out on art and museums : a study on the needs and preferences of adults who are blind or have low vision. Boston, MA: Museum of Science and Art Beyond Sight.

Reichinger, A., Carrizosa, H. G., & Travnicek, C. (2018). Designing an interactive tactile relief of the Meissen table fountain. In K. Miesenberger & G. Kouroupetroglou (Eds.), *Computers Helping People with Special Needs* (pp. 209–216). Cham, Switzerland: Springer International Publishing. doi:10.1007/978-3-319-94274-2_28

Reichinger, A., Fuhrmann, A., Maierhofer, S., & Purgathofer, W. (2016). Gesture-based interactive audio guide on tactile reliefs. *Proceedings of the 18th International ACM SIGACCESS Conference on Computers and Accessibility - ASSETS*, (pp. 91–100). New York, NY: ACM. 10.1145/2982142.2982176

Reichinger, S., S., Löw, C., Sportun, S., Reichl, P., & Purgathofer, W. (2016). Spaghetti, Sink and Sarcophagus: Design Explorations of Tactile Artworks for Visually Impaired People. *Proceedings of the 9th Nordic Conference on Human-Computer Interaction* (pp. 82:1-82:6). New York, NY: ACM. 10.1145/2971485.2996471

Richards, V., Pritchard, A., & Morgan, N. (2010). (Re)Envisioning tourism and visual impairment. *Annals of Tourism Research*, *37*(4), 1097–1116. doi:10.1016/j.annals.2010.04.011

Romeo, K., Chottin, M., Ancet, P., & Pissaloux, E. (2018). Access to Artworks and Its Mediation by and for Visually Impaired Persons. In K. Miesenberger & G. Kouroupetroglou (Eds.), *Computers Helping People with Special Needs* (pp. 233–236). Cham, Switzerland: Springer International Publishing. doi:10.1007/978-3-319-94274-2_32

Small, J., Darcy, S., & Packer, T. (2012). The embodied tourist experiences of people with vision impairment: Management implications beyond the visual gaze. *Tourism Management*, *33*(4), 941–950. doi:10.1016/j.tourman.2011.09.015

Urbas, R., Pivar, M., & Elesini, U. S. (2016). Development of tactile floor plan for the blind and the visually impaired by 3D printing technique. *Journal of Graphic Engineering and Design*, *7*(1), 19–26. doi:10.24867/JGED-2016-1-019

World Health Organization. (2017). *Vision impairment and blindness*. Retrieved from http://www.who.int/mediacentre/factsheets/fs282/en/

World Health Organization. (2018). *International classification of diseases: the global standard for diagnostic health information*. Retrieved from https://icd.who.int/en/

World Wide Web Consortium. (2008). *Web Content Accessibility Guidelines (ACAG) 2.0*. Retrieved from https://www.w3.org/TR/WCAG20/

Xicola-Tugas, H. (2012). Heritage access for the visually impaired: An analysis and assessment of museum adaptations in Catalonia. *The International Journal of the Inclusive Museum*, *4*(4), 59–68. doi:10.18848/1835-2014/CGP/v04i04/44383

ADDITIONAL READING

Classen, C. (2017). The Museum of the Senses: Experiencing art and collections (1st ed.). New York, NY: Bloomsbury Academic.

Eardley, A., Mineiro, C., Neves, J., & Ride, P. (2016). Redefining access: Embracing multimodality, memorability and shared experience in museums. *Curator*, *59*(3), 263–286. doi:10.1111/cura.12163

Hirose, K. (2013). Research on methods of 'touching the world' – the aim of the exhibit area of tactile learning in Japan's national Museum of Ethnology. *Disability Studies Quarterly*, *33*(3). doi:10.18061/dsq.v33i3.3743

Loomis, J. M., Klatzky, R. L., & Golledge, R. (2001). Navigating without vision: Basic and applied research. *Optometry and Vision Science*, *78*(5), 282–289. doi:10.1097/00006324-200105000-00011 PMID:11384005

Oliver, M., & Barnes, C. (2010). Disability studies, disabled people and the struggle for inclusion. *British Journal of Sociology of Education*, *31*(5), 547–560. doi:10.1080/01425692.2010.500088

Rappolt-Schlichtmann, G., & Daley, S. G. (2013). Providing access to engagement in learning: The potential of universal design for learning in museum design. *Curator (New York, N.Y.)*, *56*(3), 307–321. doi:10.1111/cura.12030

Vaz, R., Fernandes, P. O., & Veiga, A. (2018). Interactive technologies in museums: How digital installations and media are enhancing the visitors' experience. In J. M. F. Rodrigues, C. M. Q. Ramos, P. J. S. Cardoso, & C. Henriques (Eds.), *Handbook of Research on Technological Developments for Cultural Heritage and eTourism Applications* (pp. 30–53). Hershey, PA: IGI Global. doi:10.4018/978-1-5225-2927-9.ch002

Wyman, B., Timpson, C., Gillam, S., & Bahram, S. (2016). Inclusive design: From approach to execution. *MW16: Museums and the Web.* Retrieved from https://mw2016.museumsandtheweb.com/paper/inclusive-design-from-approach-to-execution/

KEY TERMS AND DEFINITIONS

Accessibility Standards: Accessibility requirements that can be followed to create high-quality solutions (like websites, interfaces, accessible spaces, and so forth), that allow people to effectively using them.

Accessibility: The user's ability to use products or services, despite their age, disabilities, or special needs, among other human conditions.

Design for All: Refers to approaches, methods, techniques, and tools that can be used to help address the diversity of requirements and needs, to promote equivalent experiences for people with disabilities.

Disability: A permanent condition that limits a person's senses, movements, or activities.

Inclusive Museum: A museum that contributes to social democratization by offering social opportunities to all kinds of visitors, which are perceived as part of the museum sustainability.

Special Needs: The requirements people have according to their actual condition, like age, body function, learning difficulties, emotional problems, among others.

User Experience: The perceptions and responses of the user that result from the use of a product, service or system.

Chapter 12
Accessible@tourism 4.0:
An Exploratory Approach to the Role of Industry 4.0 in Accessible Tourism

Pedro Teixeira
https://orcid.org/0000-0003-4395-4296
University of Aveiro, Portugal

Leonor Teixeira
https://orcid.org/0000-0002-7791-1932
University of Aveiro, Portugal

Celeste Eusébio
https://orcid.org/0000-0002-2220-5483
University of Aveiro, Portugal

ABSTRACT

This chapter describes how Tourism 4.0 is a concept that combines tourism and the fourth industrial revolution, and although the literature in this field is very scarce, this concept has been explored in some research projects, such as the government-sponsored research project in Slovenian tourism. People with various kinds of access requirements represent a combination of challenges and opportunities for the tourism industry. Tourism 4.0 set up the main goals of making tourism accessible to everyone at any time. Therefore, this new phenomenon may have an essential role in the development of accessible tourism. The adoption of technological components in accessible tourism enables the development of a new technological solution that can facilitate access to tourism products for disabled people, contributing to the development of accessible tourism. The new term Accessible@Tourism 4.0 is the answer to the role of the fourth industrial revolution in accessible tourism, emphasizing the effect of Industry 4.0 components in the tourism sector.

DOI: 10.4018/978-1-7998-1947-9.ch012

INTRODUCTION

Currently, there is an ongoing technological revolution that may ultimately change society as a whole. The fourth industrial revolution is evolving at high speed and will lead to a deeply connected world. A significant change in the way that society works and communicates is happening at the moment, and is having impacts on diverse manufacturing and service industries. This new industrial revolution results from a new paradigm called Industry 4.0 (Schwab, 2016:10). The drivers of this revolution came from innovations in different technologies, which are taking the world into a new era of digitalization.

One sector that is being most affected by the new digital era is undoubtedly tourism (Ivanović, Milojica, & Roblek, 2017). The United Nations World Tourism Organization (UNWTO) defined tourism as: "a diverse industry, which is a central economic driver for socio-economic development in many areas/ destinations throughout the world" (UNWTO, 2009). Tourism is one of the economic activities that generates most income worldwide, often having a vital role in the economy of the countries. With the emergence of new technologies, the way of practicing tourism has radically changed.

This new industrial revolution will bring not only benefits but also challenges for tourism. A big challenge is how it can contribute to the social inclusion of people with disabilities (PwD) in tourism practices. Tourism is an activity that enriches people in many dimensions. However, it is difficult for disabled tourists to travel without constraint. People with reduced motor/cognitive skills often end up unable to travel due to various factors such as transport and accommodation. These situations make it impossible for these people to explore and experience tourism.

This chapter intends to explain how technology can improve accessibility in tourism and fight against the social exclusion of PwD. The main objective is to understand the impact of the fourth industrial revolution on accessible tourism, as the literature reveals that research in this field is currently very limited. The study of the connection between tourism and industry 4.0 led to the arrival of a new paradigm on tourism, named Tourism 4.0. With careful exploration of this new concept and its relationship with accessible tourism, it was possible to overcome the gap in the literature on how the fourth industrial revolution is transforming accessible tourism.

The theoretical background provides interesting findings related to accessible tourism and the core concepts related to Industry 4.0 and the fourth industrial revolution. After the theoretical background, the chapter offers insights about the technological impacts of tourism, which led to the appearance of Tourism 4.0. A new concept, Accessible@Tourism 4.0, was created to explain the role of technological innovation in promoting accessible tourism. The chapter ends by discussing the main findings, possible limitations of this study, and recommends targets for future researches.

THEORETICAL BACKGROUND

About Accessible Tourism

Accessibility is a quality concept that may be interpreted differently from person to person (Persson, Åhman, Yngling, & Gulliksen, 2015). Some people think of accessibility as directly serving some of their needs, while others think of it in a more general way. The Convention on the Rights of People with Disabilities states in art. 9 that, in order to enable persons with disabilities to live their life independently and participate in all aspects of life, appropriate measures must be taken. To ensure accessibility conditions for access for PwD, special attention should be given to "the physical environment, transportation, information and communications technologies and systems, and to other facilities and services open or provided to the public, both in urban and in rural areas" (United Nations, 2006). In addition, art. 30 implements the right to participate in cultural life, recreation, leisure, and sport. This convention supported the Manila Declaration on World Tourism signed in September 1980, where tourism was recognized as a fundamental right (UNWTO, 1980).

Accessible tourism encompasses many other concepts. Buhalis and Darcy (2011, pp. 10–11) defined Accessible Tourism as "a form of tourism that involves collaborative processes between stakeholders that enables people with access requirements, including mobility, vision, hearing and cognitive dimensions of access, to function independently and with equity and dignity through the delivery of universally designed tourism products, services, and environments". Accessible tourism seeks to include people with impairments and disabilities in travel and leisure activities (Milicchio, 2016).

It is estimated that by 2020, around 25% of travel and leisure spending will come from PwD (Bekiaris et al., 2018). Demand in the accessible tourism market is increasing, which makes it a very attractive one. However, despite all the potential, the accessibility market is largely ignored (Buhalis & Michopoulou, 2011; Figueiredo, Eusébio, & Kastenholz, 2012). In fact, the accessibility market is not homogenous (Buhalis & Michopoulou, 2011), with different types of disabilities having different demands and requirements. Therefore, it is critical to divide the market into segments with similar needs and wants, in order to create customized offerings and suitable products (Buhalis & Michopoulou, 2011; Figueiredo, Eusébio, & Kastenholz, 2012).

Segmentation can be seen as the key to competitive advantage (McKercher, 2008). According to Darcy and Buhalis (2011: 38), two leading authors in this field of study, seven clusters for people with disabilities/impairments can be identified: mobility; blind or vision impaired; deaf or hearing impaired; speech; cognitive (including mental health/intellectually/learning); hidden impairments and elderly/seniors/boomers.

Accessible tourism has been gaining more exposure in recent times, but there are still many obstacles that need to be conquered. Tourists, especially those with disabilities, face some significant barriers that negatively influence their touristic experiences and impede PwD from traveling and having new experiences. Almost half of all disabled tourists would travel more frequently if there were more barrier-free offers (Pühretmair, 2004). Sometimes, these obstacles are so challenging to conquer that they prevent people with impairments from traveling (Popiel, 2016), making these barriers a big issue. These obstacles exist, as they create "roadblocks" to tourism, threatening the quality of life and independence. Tourism practices recognize the need for information and physical accessibility, but barriers to access are complex and not reducible to the physical environment (Buhalis & Darcy, 2011, p. 50).

In tourism, planning by people with impairments needs more detailed information, focusing on their individual/special requirements. Higher accessibility requirements demand more detailed and more specialized information. However, as found by Waschke (2004), the supply of specific information becomes scarcer for those with higher accessibility requirements.

Technological solutions and more accessible design for all arise as possible solutions to eliminate these obstructions to the tourism industry. Accessible design and the provision of information improve accessibility and usability for PwD, but they also make tourism more approachable for a broader range of the population (Pühretmair, 2004). If the right information is displayed to tourists, correct planning is possible, helping eliminate all other obstacles, including the provision of solutions for physical accessibility. Information is undoubtedly the pillar of accessible tourism, and with the rise of the fourth industrial revolution, connectivity is becoming a significant topic.

About Industry 4.0

Increasing digitalization is changing the everyday lives of people, markets, business models and value chains. Nowadays, companies face various challenges like new corporate strategy, new requirements for the training of personnel, new demands for cybersecurity and radical changes in business processes across all company levels, justifying solutions based on new information and communications technologies (ICT). In fact, emergent ICTs provide the opportunity for companies to build new customer relationships, to invent new business models, and to increase their competitiveness (Parida, Sjödin, & Reim, 2019).

The phenomenon of Industry 4.0, first mentioned in 2011 in Germany (Bundesministerium für Wirtschaft und Energie, 2016) launched the fourth industrial revolution, which is based on a set of concepts enhanced by new technologies. McKinsey, an American management consulting company, studied Industry 4.0 as digitalization of the manufacturing sector which originates from four disruptive technology groups: i) connectivity, data, and computing power; ii) analytics and intelligence; iii) human-machine interaction; and iv) digital-to-physical conversion.

Most of the literature review studies made on this topic rely on the definition provided by Bundesministerium für Wirtschaft und Energie (2016), that define Industry 4.0 as a "new level of organizational control over the entire value chain of the life cycle of products, and it is generated towards increasingly individualized customer requirements". The foundation of the fourth industrial revolution is the availability of all relevant real-time information (Crandall, 2018), which was made possible by focusing on the networking of equipment and processes in the industry through the use of ICT to form an intelligent value chain (Schlüter, Hetterscheid, & Henke, 2017).

Industry 4.0, alongside with digitalization, capitalized on the developments in several technologies. These new technological trends became known as technological drivers, which is a current issue in the literature. Several authors have identified and studied different technological advances that promoted the appearance of Industry 4.0 (Jasperneite, 2013; Kovar et al., 2016; Rüßmann et al., 2015; Saturno, Pertel, Deschamps, De, & Loures, 2018). Cyber-physical systems (CPS), the Internet of things (IoT), big data, cloud technologies, 3D printing, radio frequency identification (RFID), cognitive computing, mobile technologies, machine to machine technologies, virtual reality, and augmented reality represent some of these solutions.

These technological drivers demonstrate that the competitiveness and innovation needed for Industry 4.0 depend on mastering a wide range of ICT skills and competencies (Sedlar et al., 2018). However, the fourth revolution is still in progress, so it is possible that these new technological solutions will be improved and will innovate even further. These technologies have mainly been studied in the production area. However, their utility can be expanded to services industries, like the case of tourism, helping in the transition of tourism into a new digital age.

Table 1 offers a brief description of some Industry 4.0 technological drivers, and some authors that studied them. These technological drivers demonstrate the different areas changing with the fourth industrial revolution. It is almost possible to claim that technological evolution gave rise to this new

Table 1. Industry 4.0 technological drivers

Technological Driver	Author(s)	Description
Cyber-physical-Systems (CPS)	Wolter et al., (2015)	CPS are intelligent systems with integrated physical and computational capabilities, that can interact with humans and are responsible for the interconnectivity of devices, machines and moving objects.
Internet of Things (IoT)	Tobergte and Curtis, (2013) Atzori, Iera, and Morabito, (2010)	IoT refers to the networking of objects with the Internet, providing a worldwide network of interconnected objects, based on standard communication protocols.
Big Data (analysis tools)	Wolter et al., (2015) Schwab, (2016: 135)	Innovative analysis tools (text analytics, machine learning, predictive analytics, and statistics) are tools that have the capacity to distinguish data that should become information from data that has no relevance.
Cloud Technologies	Wolter et al., (2015) IBM, (2015)	These technologies can be perceived as online data centers, where information can be stored and made available for many users.
3D Printing (additive manufacturing processes)	Klein, Avery, Adams, Pollard, and Simske, (2014)	3D printing is a methodology using three-dimensional data sets for producing new physical objects. The replication is made layer by layer, allowing existing objects to be replicated in a three-dimensional manner.
RFID (Radio Frequency Identification)	Pala and Inanç, (2007) Smyth and Crabtree, (2012)	Radio frequency identification is a wireless technology that lets computers or other devices read/identify electronic tags from a given distance, using an object (generally referred to as an RFID tag) for identification and tracking with radio waves.
Cognitive Computing	Esser et al., (2011)	This incorporates digital systems that learn at scale, reason with purpose and interact with humans, in order to achieve big data treatment.
Mobile Technologies	Jazdi, (2014)	Electronic equipment such as mobile phones or small computers that can be used in different places, and the technology connected with them, such as mobile apps.
Virtual Reality (VR) and Augmented Reality (AR)	Nunes, Pereira, and Alves, (2017) Barnes, (2017)	VR can be defined as a computer-generated simulation of a three-dimensional image or environment that can be interacted with using special electronic equipment. AR is a technology that produces a computer-generated image on a user's view of the real world, thus providing an augmented view.

fourth industrial revolution, and the technological drivers prove it. The key technology drivers enable the transformation of processes along the entire value chain, from the development phase up to the use and maintenance phase. Different types of ICT can contribute to Industry 4.0, but some of them have a crucial role for developing products and services, conducting research and promoting innovation activities (Sedlar et al., 2018), making them essential in this new digital era. Industry 4.0 is changing the landscape of industries with a particular impact on the relations between companies and clients and the way they interact (Schwab, 2016: 21).

With technologies dominating markets, clients will become more and more demanding. Services will need to offer quick and customized solutions to clients, which can be obtained with the integration of Industry 4.0 foundations. Adapting services to customer demand and the utilization of new digital technologies is crucial to remain competitive in rapidly changing environments. The application of industry 4.0 principles can promote the digitalization of services, which in turn could provide a source of competitive advantage in the near future. One of the service industries that is quickly changing thanks to new ICTs is tourism. The benefits that the fourth industrial revolution brings to tourism are great and can even revolutionize the sector. In recent years, due to the development of ICTs, tourism has become smarter. These technologies helped improve tourism in many areas. Therefore, it is imperative to understand how the different technological drivers contributed to improving tourism. Of course, with the new industrial revolution ongoing, new technologies have stepped up to develop tourism even further, with particular implications on making tourism more accessible.

THE ROLE OF THE FOURTH INDUSTRIAL REVOLUTION IN TOURISM

Tourism 4.0

In recent times, different types of technologies have become a powerful ally to tourism. Developments became available in ICTs which were responsible for the origins of smart tourism (Gretzel, Sigala, Xiang, & Koo, 2015) and which helped tourism advance to new platforms, reaching more clients and present more quality services. With the fourth industrial revolution, new technologies are taking a step forward to develop tourism even further. The impact of the industrial revolution on the travel industry gave rise to a more advanced paradigm in tourism, named Tourism 4.0.

The concept of Tourism 4.0 first appeared as the largest government-sponsored research project in Slovenian tourism. This research was possible due to the establishment of a partnership between the Arctur company and Slovenia's government. Arctur is a commercial supplier of computing services, which provides innovative and user-friendly digital solutions to businesses, public institutions, and research institutions. Artur's R&D department together with the Slovenian government sought a way to potentialize innovation in the tourism sector. The result of the research was the creation of the Tourism 4.0 consortium. According to the Tourism 4.0 consortium, Tourism 4.0 is defined as "a trend of big data processing, collected from a vast number of travelers, to create personalized traveling experience, based on a variety of modern high-tech computer technologies" (Arctur & Tourism 4.0 Consortium, 2018). Tourism 4.0 follows the slogan: "Green, active, healthy" (Peceny, Urbančič, Mokorel, Kuralt, & Ilijaš, 2019), which demonstrates that the concept encompasses many future trends.

According to the World Economic Forum (World Economic Forum & Accenture, 2017), four subjects are crucial for digital tourism transformation:

- **Living Travel Experience:** According to tourists' habits - optimize and personalize the customer experience by collecting and exchanging data and continuously generating comprehension.
- **Enabling the Travel Ecosystem:** By evolving different stakeholders.
- **Digital Technologies:** That promoted Industry 4.0 - optimizing the real-time use of resources and transforming operations, through innovations such as IoT, virtual reality, and digital platforms will enable flexible working and changes to basic operational processes and will optimize the real-time use of resources.
- **Safety and Security**: Using digital technologies to produce secure environments and enhance the global travel security system (Moavenzadeh & Maar, 2018).

Tourism 4.0 aims to integrate these four topics into the diverse industries connected with tourism. In fact, by applying new technologies in tourism, it is possible to promote personalized solutions to every tourist, while assuring that they are traveling with safely and securely. Tourism 4.0 also promotes sustainable tourism, encouraging a positive environmental, social, and economic impact and collaboration between all stakeholders. The success of Tourism 4.0 depends greatly on the different collaborations in the travel ecosystem.

With the objective of achieving sustainable tourism, Tourism 4.0 sets up a primary goal of making tourism "accessible to everyone at any time," making it tremendously crucial for the development of accessible tourism. Technologies have a crucial role in helping the development of more accessible tourism. The main problem preventing people with different types of disabilities from traveling is mainly related to the shortage of information (Stumbo & Pegg, 2005).

There are several uses of technologies that can help diverse tourism objects display greater accessibility features. With the new industrial revolution, technology is exceeding all expectations, making society evolve in many sectors and providing better life-conditions, especially for those with some type of impairment. Ensuring that tourism is accessible is very important from an economic point of view, due to the potential of the accessible tourism market. Furthermore, Tourism 4.0 has all the tools for achieving more inclusive travel ecosystems and boosting inclusiveness of PwD.

Industry 4.0 was responsible for launching the fourth industrial revolution, and now this revolution is impacting other sectors, like tourism. Industry 4.0 technological drivers are the tools responsible for making travel experiences more accessible, efficient, and affordable for travelers. Innovative information tourism platforms that offer innovative products and services will play a crucial part in the current and future digitalization of the travel industry. However, the question remains: how can the technological impacts of the fourth industrial revolution contribute to a more accessible tourism?

ACCESSIBLE@TOURISM 4.0: HOW THE FOURTH INDUSTRIAL REVOLUTION IS REVOLUTIONIZING ACCESSIBLE TOURISM

The different innovative technologies promoting the fourth industrial revolution became known as technological drivers. As the relationship between new technologies and tourism developed to a very close one, the use of Industry 4.0 technological drivers in touristic activities was mainly responsible for the urge of Tourism 4.0. In smart tourism, technology was already responsible for improving tourism services and products. However, the technologies that Tourism 4.0 presents are one-step forward to the future of tourism. Industry 4.0 technological drivers and other technological innovations have opened the doors to the digitalization of the travel industry. IoT, virtual and augmented reality, big data, RFID, and even 3D printing are technological drivers that were born in the manufacturing world but have the capacity to transform the tourism industry. Other innovative technologies like robotics and virtual assistants are also drivers of the fourth industrial revolution (Schwab, 2016: 13,19) and can have a big impact on many sectors. Tourism 4.0 is a concept that adapts all these technologies to tourism, transporting tourism to a new digital era, stimulating the inclusion of disabled travelers. Accessible@Tourism 4.0 is an overview of different technological innovations developing accessibility conditions in tourism.

This overview will be obtained by illustrating examples of technological innovations that are promoting the fourth industrial revolution and transforming the world of tourism, making it more accessible. The impact of Industry 4.0 technological drivers and other technological innovations in accessible tourism allows the literature gap to be addressed by showing how Tourism 4.0 can help in the development of accessibility conditions for PwD.

A. Mobile Technology

Mobile technologies, especially mobile apps, have become significant players for smart tourism and crucial tools for assuring accessible tourism. Similar to their role in smart tourism, Tourism 4.0 also relies much on mobile apps to take tourism to a new level. The evolution that this type of technology has undergone in recent years made this device turn into a tour guide, a travel agency, a restaurant locator, and a map. Mobile apps and cell phones are side by side with tourists during the entire tourism experience. Services for passengers aligned with the use of social media have made mobile technology indispensable for all phases during a trip: planning (online bookings and places to visit), while traveling (cell phone as the best traveling friend), and post-trip (sharing experiences in social media) (Belén, 2018). Another example is related to apps that allow sharing and matchmaking. Airbnb and Uber became disruptive innovations because technology allows peer-to-peer sharing and matchmaking. (World Travel & Tourism Council, 2015).

B. Virtual Assistants

A virtual assistant is "a computer program or device that is connected to the internet and can understand spoken questions and instructions" (Virtual Assistant, 2019). Voice recognition technology and voice assistant software assistants have experienced growth in recent years. Voice assistant products from Amazon, Apple, and Google enable users to ask questions and issue commands in natural language. Siri and Alexa are examples of voice assistants that can help people in various ways. Some functionalities include: sending and reading text messages, making phone calls, answering basic informational, setting timers, alarms, and calendar entries; controlling media playback and even controlling IoT enabled devices (thermostats, lights, alarms) (Hoy, 2018).

These services on offer can help accessible tourism by making significant contributions to the daily activities of disabled tourists. By helping tourists with these activities, they make traveling more personalized and less troublesome. The main contribution of this type of technology is reducing the information barriers, which disabled tourists' experience, and offering more customized services. The tourism sector is now experimenting with the arrival of virtual assistants specially designed for their atmosphere. IBM launched Watson Assistant, an AI-powered virtual assistant that creates interactive and personalized solutions for tourists. Watson Assistant creates customized experiences, taking into consideration factors like age, gender, and travel interests, forming personalized activity agendas that users can modify (IBM, 2017). The fact that this virtual assistant adapts to users can be fundamental to accessible tourism, due to the particular necessities of disabled tourists.

C. Virtual Reality and Augmented Reality

Virtual reality and augmented reality offer a new set of experiences for tourism. Simulating digital environments can have a significant impact on traveling, leading to big changes in touristic experiences. Virtual reality can help in removing many accessible tourism barriers. It can deliver 3D and immersive experiences to disabled tourists, providing touristic guide services in sign language and even remove physical barriers in museums.

This technology can be even used to access sites that are usually closed or inaccessible (Jung, tom Dieck, Lee, & Chung, 2016). VR can be a platform to capitalize on digital tourism in cultural attractions like museums and archaeological sites.

AR makes use of the traveler's real-time environment, adding it to a 3D environment, and recreates digital tourism experiences (Kounavis, Kasimati, & Zamani, 2012). Augmented reality is also strongly connected to travel mobile apps, which have been revolutionizing tourism.

A practical example of the use of these technological drivers is the Geevor Tin Mine Museum (Jancenelle, Vivien, Storrud-Barnes, Susan, Javalgi, 2017). This museum is a UNESCO world heritage site and uses VR and AR to produce engaging experiences for its main target markets (school groups, families, and elderly visitors). Some parts of this mining museum are difficult to access (underground mines). A VR/AR application is used for pre-visit stage and on-site underground mining experience.

D. The Internet of Things

A network of smart devices enhances businesses by connecting clients and service providers, thus generating a large amount of data that can be stored and computed on available cloud services (Kaur & Kaur, 2016). Tourists trends and patterns can be easily stored. This connection can be useful to offer more personalized tourism offers. The more information known about guests, the better services can be offered, and satisfaction increases.

A practical example is IoT data identifying that a customer has booked a room every year, so it sends a message asking the guest if they would like to make another booking this year (Newman, 2018). IoT is quickly becoming a significant technology for tourism. Some services that IoT has the potential to maximize are: helping to select destinations and searching for suitable travel arrangements; helping tourists during preparation for a trip; providing real-time support to the tourist during the trip, helping managing memories after the trip, for highly personalized onsite support of frequent travelers and personalized services for tourist groups (Balandina, Balandin, Koucheryavy, & Mouromtsev, 2015).

Significant gains that shared data brings are, for example, personalized solutions, as they have a particular influence on accessible tourism. The sharing of information that IoT allows can put an end to the problems of a lack of information. Having all the information available is especially important for disabled people when planning a trip. In addition, having personalized services is crucial for disabled tourists. This particular group has different requirements according to the type of disability, so the offer of rapid and customized solutions is boosting disabled tourism. Examples go from reserving a room with the right requirements to the use of smart devices to enter and explore museums and monuments. IoT can also bring significant improvements to accessible tourism by offering real-time support. Assistance and support for disabled tourists in real time may just unlock all the potential of accessibility tourism markets. It is noteworthy that IoT promotes the quick and efficient flow of information. A better-connected world where information is a crucial aspect is vital for the development of accessible tourism.

E. 3D Printing

3D printing may not be a widespread technology associated with tourism. However, the potential to print perfect replicas can have a significant influence on improving touristic experiences in museums, art galleries, and archaeological sites. 3D printing has particular significance in the context of museums due to the low costs of replicating objects (Allard, Sitchon, Sawatzky, & Hoppa, 2005; Short, 2016).

Another use in museums can be in creating and producing customized 3D souvenirs (Jancenelle, Vivien, Storrud-Barnes, Susan, Javalgi, 2017). In the case of art and design, creating exhibits in art galleries using 3D printing also proves the potential of this technology for tourism (Walters & Davies, 2010). 3D printing has particular importance for accessible tourism because it allows a more immersive experience. This technology helps to make artifacts more tangible and accessible to the public, improving the experience of visiting an art gallery or a museum.

3D printing objects also helps in eliminating barriers for people with cognitive disabilities who may experience learning difficulties during touristic trips to museums or archaeological sites. For this group of disabled tourists, a 3D replica may just be the answer to their particular requirements. A practical example of the application of 3D printing to tourism is the Underwater Maya Project (Mckillop, 2013). The project uses 3D printing to build replicas of ancient Maya wooden buildings. This makes archaeological sites more accessible since exhibiting 3D replicas does not require loans of actual artifacts or extended security measures. The project already led to the opening of exhibits in some remote places, which is also a contribution to accessibility. The main objective is contributing to sustainable archaeological tourism while promoting access to the Internet for "digital museums" and engaging people with archaeological sites.

F. Big Data (Analyze Tools)

With the internet becoming a significant player in tourism, tourists rely on various web-based platforms to plan trips. The electronic trace left by tourists, also known as digital footprint (Girardin, Blat, Calabrese, Dal Fiore, & Ratti, 2008), generates a lot of data. The big data problem is seen once again in Tourism 4.0. Moreover, analysis tools have a critical role in generating the right information and doing this can become a source of competitive advantage.

Data can have different sources. From booking websites to searching for the best places to visit and sharing on social media, the digital footprint can help develop customized solutions in tourism. Li, Xu, Tang, Wang, and Li, (2018) identified three sources of big data generation: user-generated content data (UGC), including online textual data and online photo data; device data (devices), including GPS data, mobile roaming data, Bluetooth data; and transaction data (operations), including web search data, webpage visiting data, and online booking analytics for tourism.

Some practical examples of the use of UGC data come from Meliá hotels and Airbnb. The Meliá hotel chain uses information about its guests to define marketing campaigns. By analyzing their database (amount spent, the reason for the trip, the country of origin), they are able to cross-check information to develop an appropriate customer profile and achieve a higher success rate. The objective is to obtain better market segmentation and optimize their promotional campaigns (Belén, 2018). Airbnb is able to use customer data to identify guests who chose not to book because they were displeased when hosts failed to respond to their expectations (Newman, 2018).

Another example is the Destination Management Information System, Åre, presented by Fuchs, Höpken, and Lexhagen (2014). This information system for the Swedish mountain tourism destination Åre, measures destination performance and customer behaviors by analyzing data from customer-based knowledge sources, like web searches, booking, and feedback. Big data analysis is a vital element of Tourism 4.0, and it is becoming a significant success factor for accessible tourism. Similar to IoT, the main advantage of the analysis of big data is in allowing personalized tourist solutions to be offered. By studying different data sources, it is possible to identify tourists' behaviors and establish patterns.

Through the analysis of data derived from disabled travelers, one can understand the main difficulties and address them in order to optimize touristic experiences. Since disabled tourists needs differ according to the type of disability, much data is generated, making analysis tools indispensable. The data obtained can become valuable knowledge, and a simple understanding of disabled tourist behaviors is essential to promote accessible tourism. Analysis tools should be able to collect and analyze data from accessible tourism knowledge sources in order to deliver a personalized and enlightened service.

G. RFID

Radio frequency identification is an automatic identification method that retrieves and stores data using RFID tags. This technology is being popularized by Industry 4.0, as it is mainly utilized in retail and supply chain management. Nonetheless, while its main application is in manufacturing, RFID can also help enhance services by improving quality, speed, and customer satisfaction.

In the case of tourism, Öztayşi, Baysan, and Akpinar (2009), described RFID applications in four major RFID areas:

- Human tracking and control systems (E-passports, customer loyalty management, tracking children or people with special needs, airport security).
- Assets and valuables tracking systems (luggage tracking and food and beverage management).
- Contactless payment systems (toll collection system, RFID-tagged public transport cards, payments in a hotel, keyless room entry).
- RFID-based information (museums, archaeological sites, art galleries, monuments).

All of the mentioned examples are essential in assuring more accessible tourism. Tracking systems can help disabled tourists with navigation and wayfinding and prevent the loss of specialized equipment like wheelchairs while traveling (Konkel, Leung, Ullmer, & Hu, 2004). Contactless payment systems and digitally activated transport and hotel cards also improve accessibility and inclusiveness by making processes easier for disabled tourists. RFID applications have been studied in maritime cruises (Dias et al., 2016) and even in the organization of cultural events in a tourist destination (Zeni, Kiyavitskaya, Barbera, Oztaysi, & Mich, 2009).

In literature, it is possible to identify some practical examples of RFID use in tourism, which can also contribute to the inclusiveness of disabled tourists. When visiting a capital city, mobility and visually impaired people often find it difficult to cross-busy streets. RFID can allow communication between tourists with disabilities and the crossing equipment to automatically activate a location message and input a request for the green light (Smyth & Crabtree, 2012).

Another practical example is related to navigation at attractions. Disabled tourists can experience difficulties when moving around an attraction, so RFID can propose a route recommendation system that informs tourists which facilities they should visit and in what order. A proposal for this application comes from Tsai and Chung (2012), in the context of theme parks. At the entrance, tourists are provided a wristband embedded with an RFID tag. RFID readers are installed on the entrances and exits of each facility. Whenever a tourist enters the area of an RFID reader, the system records the entrance and time and transfers that information to a route database.

H. Robotics

Automation in tourism attracted the use of robots. Robots are capable of performing automated tasks and delivering value to clients. Their utilization in tourism and hospitality is already a reality. Ivanov, Webster, and Berezina (2017) identified some current uses of robots across the tourism sector. Their findings highlighted six areas where robots are being used to improve services.

Hotels have been evolving for adopting the use of robots in assisting people in their rooms (delivery and cleaning robots) or even working as staff (robot receptionist and porter). One practical example is *Henn Na Hotel* in Japan, which uses technologies in order to maximize excitement, efficiency, and comfort for guests. Amongst other innovations (voice- and face-recognition), the hotel staff consists mainly of robots. The reception is operated by three multi-lingual robots responsible for greeting, checking-in and assisting guests, there is a robotic arm for storing luggage in the cloakroom, and there are porter robots to carry bags to the rooms (Alexis, 2017). Robots also have been used in restaurants (robot servers and even chefs), amusement parks (concierge robots), airports (robot-guide and bag drop robots), at events (delivery and entertainment robots) and galleries and museums (robot guides).

The use of robots in tourism already led to the creation of the concept of R-tourism (robot tourism), proposed by Alexis (2017). R-tourism is deeply connected to Tourism 4.0 since some of the robotic impacts in tourism are already visible. Also, the relation with accessible tourism is quite close since the automation of some services can captivate the interest of disabled tourists. Robot guides represent a significant breakout for tourists to experience museums and art galleries, providing, for example, help to visit inaccessible exhibits. Also, the aid provided in a hotel with cleaning and assistive robots can fulfill some challenging accessibility requirements. In fact, one may recognize that a primary driver of R-tourism is accessibility. Another example of robotics improving accessible tourism is the proposal that a robot could take pictures and records videos from a faraway place and send them to the user in real time. This proposal comes from Cheung, Tsang, and Wong (2017), and aims to build a robot capable of helping disabled people to perform virtual tours of places which they cannot reach because of their mobility impairments.

The following table (Table 2) presents a synopsis of the study conducted. The leading technologies promoting the fourth industrial revolution and impacting accessible tourism were identified. Various projects in the area of these technologies were explored with the help of some authors' work in respective innovative areas. After analyzing the fundamental discoveries, the impact of each technology in accessible tourism was discussed. The table sums up the purpose of Accessible@Tourism 4.0 in describing how Tourism 4.0 has the power to promote more accessible tourism.

FUTURE RESEARCH DIRECTIONS

Although there is a large number of studies in the field of smart tourism, there are not many specially dedicated to accessible tourism. There is an evident lack of research specifying the application of technologies to enhance accessible tourism. The same happens with Industry 4.0 technological drivers in tourism. Many authors describe the potential these technologies can have but never apply it to the case of accessible tourism. The concept of Accessible@Tourism 4.0 was mainly explored by using examples of the application of Industry 4.0 technological drivers in tourism and explaining how they could benefit

Table 2. Technological Drivers Promoting Accessible Tourism

Technological Driver	Author(s)	Impact in Accessible Tourism
Mobile Technology	Emrouzeh, Dewar, Fleet, and Bourgeois, (2017)	• Navigation (wayfinding). • Planning (trips, what to visit, accessible conditions) • Social (sharing, collaborating, and communicating). • Security and emergencies (health monitoring, weather alerts, real-time assistance). • Transactional (buying tickets, making reservations, and shopping). • Entertainment (tour guides, e-readers, etc.). • Information access.
Virtual Assistants	Hoy, (2018)	• Simplifying daily activities of disabled tourists. (setting timers, alarms, and calendar entries; setting reminders, making lists, and doing basic math calculations; controlling media playback; controlling IoT-enabled devices). • Making traveling more personalized and less complicated with fewer information barriers.
Virtual Reality and Augmented Reality	Jung, tom Dieck, Lee, and Chung, (2016) Kounavis, Kasimati, and Zamani, (2012) Jancenelle, Vivien, Storrud-Barnes, Susan, Javalgi, (2017)	• Recreating several points of interest in 3D, like monuments, museums, buildings, and other touristic attractions. • Creating realistic 3D environments for touristic activities and allowing immersive experiences for disabled tourists. • Potentializing better holiday planning functionalities. • Providing touristic services in sign language. • Improving access sites that are generally closed or inaccessible. • Preview of disabled tourism offers. • Helping in travel decisions by providing relevant information about accessibility conditions.
Internet of Things (IoT)	Balandina, Balandin, Koucheryavy, and Mouromtsev, (2015) Kaur and Kaur, (2016)	• Tourists trends and patterns can be easily stored. • More personalized tourism offers (rapid and customized solutions). • The sharing of information can put an end to the problems of lack of information. • Assistant and support to disabled tourists in real-time. • Automatically reserving rooms with the right accessibility requirements. • Allowing the use of smart devices to enter and explore museums and monuments.
3D Printing	Jancenelle, Vivien, Storrud-Barnes, Susan, Javalgi, (2017) Mckillop, (2013)	• Replicating objects offering more immersive touristic experiences. • Artifacts more tangible and accessible to the public. • Eliminating barriers for people with cognitive disabilities who may experience learning difficulties while visiting museums or archaeological sites.
Analysis Tools (Big Data)	Li, Xu, Tang, Wang, and Li, (2018) Fuchs, Höpken, and Lexhagen, (2014)	• Identifying tourists' behaviors and establishing patterns. • Identifying the main difficulties and addressing them in order to optimize touristic experiences. • Collecting and analyzing essential data from different accessible tourism systems can promote interconnection between several digital databases. • Providing more personalized tourism offers.
Radio Frequency Identification (RFID)	Öztayşi, Baysan, and Akpinar, (2009) Tsai and Chung, (2012) Dias et al., (2016)	• Tracking systems help in navigation/wayfinding and prevent the loss of special equipment. • Contactless payment systems. • Digital activated transport. • Autonomous processes in hotels. • RFID in attraction sites can guarantee better access conditions. • Providing digital solutions in tickets. • Informing users by giving information about nearby objects.
Robotics	Cheung, Tsang, and Wong, (2017) Ivanov, Webster, and Berezina, (2017) Alexis (2017)	• Helping disabled people to perform virtual tours of places, which they cannot reach because of their mobility impairments. • Robot guides represent a significant breakout for tourists to enjoy museums and art galleries, providing, for example, help to visit inaccessible exhibits. • Cleaning and assistive robots can help PwD in hotels, restaurants, touristic events, and amusement parks.

and potentialize accessible tourism. Therefore, it is of utmost relevance carry out studies regarding the use of this kind of new technology in the accessible tourism.

Future researches could study the impact of this new revolution in other areas and compare it with the impacts on tourism. Also, studies could be conducted focusing on applying Tourism 4.0 in other types of tourism (educational, religious, medical) and comparing the results. Research in these scopes aiming at clarifying potential opportunities in these particular areas is suggested.

Tourism 4.0 is still a topic that lacks significant literature. The approach to this topic has the goal of stimulating further research and investigation, especially for the importance of enhancing accessible tourism.

CONCLUSION

Tourism 4.0 reflects the integration of technological tools into tourism. This integration has the potential to innovate tourism and make it accessible for everyone, offering disabled tourists the possibility of experiencing the wonders of traveling. Accessible@Tourism 4.0 can be considered the answer to the research gap on the role of Industry 4.0 and new technologies in tourism and their importance in developing more accessible tourism.

Accessible@Tourism 4.0 shows the effect of Industry 4.0 and technologies in the accessible tourism sector, sending shockwaves across tourism and revolutionizing the way of traveling. Tourism 4.0 clearly shows that "Integration of technology makes life easier, especially for disabled people" (Altinay et al., 2016). Technological advantages are finally allowing many disabled tourists to fulfill their dreams of exploring the world.

The adoption of fourth industrial revolution components in accessible tourism (Accessible@Tourism 4.0) enables the development of a new technological solution that can facilitate access to tourism products by disabled people, contributing to the development of accessible tourism. Many of the technologies analyzed demonstrate the significance of connectivity, one of the pillars of the fourth industrial revolution, in the promotion of accessible tourism.

REFERENCES

Alexis, P. (2017). R-Tourism: Introducing the Potential Impact of Robotics and Service Automation in Tourism. *Tourism Management*, *29*(3), 458–468. doi:10.1016/j.tourman.2007.07.014

Allard, T. T., Sitchon, M. L., Sawatzky, R., & Hoppa, R. D. (2005). Use of Hand-held Laser Scanning and 3D Printing for Creation of a Museum Exhibit. *Proceedings of the 6th International Symposium on Virtual Reality, Archaeology and Cultural Heritage*. Academic Press.

Altinay, Z., Saner, T., Bahçelerli, N. M., & Altinay, F. (2016). The role of social media tools: Accessible tourism for disabled citizens. *Journal of Educational Technology & Society*, *19*(1), 89–99.

Arctur, & Tourism 4.0 Consortium. (2018). What is Tourism 4.0? Retrieved from https://www.tourism4-0.org/

Virtual Assistant. (2019) Cambridge dictionary. Retrieved from https://dictionary.cambridge.org/pt/dicionario/ingles/virtual-assistant

Assistant, V. (2019) In Cambridge dictionary. Retrieved from https://dictionary.cambridge.org/pt/dicionario/ingles/virtual-assistant

Atzori, L., Iera, A., & Morabito, G. (2010). The Internet of Things: A survey. *Computer Networks*, *54*(15), 2787–2805. doi:10.1016/j.comnet.2010.05.010

Balandina, E., Balandin, S., Koucheryavy, Y., & Mouromtsev, D. (2015). IoT Use Cases in Healthcare and Tourism. *Proceedings of the 17th IEEE Conference on Business Informatics CBI 2015* (Vol. 2, pp. 37–44). IEEE Press. 10.1109/CBI.2015.16

Barnes, S. (2017). *Understanding Virtual Reality in Marketing: Nature, Implications and Potential*. doi:10.2139srn.2909100

Bassi, A., & Horn, G. (2008). Internet of Things in 2020: A Roadmap for the Future. *European Commission: Information Society and Media*, *22*, 97–114.

Bekiaris, E., Loukea, M., Spanidis, P., Ewing, S., Denninghaus, M., Ambrose, I., . . . Castiglioni, R. C. (2018). Research for TRAN Committee - Transport and tourism for persons with disabilities and persons with reduced mobility. TRAN committee. Retrieved from http://www.europarl.europa.eu/RegData/etudes/STUD/2018/617465/IPOL_STU(2018)617465_EN.pdf

Belén, V. (2018). Tourism and Technology: How Tech is Revolutionizing Travel. Wearemarketing. Retrieved from https://www.wearemarketing.com/blog/tourism-and-technology-how-tech-is-revolutionizing-travel.html

Buhalis, D., & Darcy, S. (Eds.). (2010). *Accessible tourism: Concepts and issues*. Channel View Publications.

Buhalis, D., & Michopoulou, E. (2011). Information-enabled tourism destination marketing: Addressing the accessibility market. *Current Issues in Tourism*, *14*(2), 145–168. doi:10.1080/13683501003653361

Buhalis, D., & Michopoulou, E. (2011). Information-enabled tourism destination marketing: Addressing the accessibility market. *Current Issues in Tourism*, *14*(2), 145–168. doi:10.1080/13683501003653361

Bundesministerium für Wirtschaft und Energie. (2016). Plattform Industrie 4.0 - Was ist Industrie 4.0?

Cheung, C. W., Tsang, T. I., & Wong, K. H. (2017). Robot Avatar: A Virtual Tourism Robot for People with Disabilities. *International Journal of Computer Theory and Engineering*, *9*(3), 229–234. doi:10.7763/ijcte.2017.v9.1143

Clarke, R. H., Twede, D., Tazelaar, J. R., & Boyer, K. K. (2006). Radio frequency identification (RFID) performance: The effect of tag orientation and package contents. *Packaging Technology & Science*, *19*(1), 45–54. doi:10.1002/pts.714

Crandall, R. (2018). Industry 1.0 to 4.0: the Evolution of Smart Factories. APICS Magazine.

Dias, D., Lima, F., & Dias, E. (2016). Tourism management and automation: RFID applications in Brazilian maritime cruises. *International Journal of Internet of Things and Web Services, 1*, 43–49. Retrieved from https://www.iaras.org/iaras/filedownloads/ijitws/2016/022-0006.pdf

Emrouzeh, M. P., Dewar, K., Fleet, G., & Bourgeois, Y. (2017). Implementing ICT for Tourists with Disabilities. *Proceedings of the 2017 International Conference on E-Education, E-Business and E-Technology* (pp. 50–53). Academic Press. 10.1145/3141151.3141152

Esser, S. K., Andreopoulos, A., Appuswamy, R., Datta, P., Barch, D., Amir, A., ... Chandra, S. (2013, August). Cognitive computing systems: Algorithms and applications for networks of neurosynaptic cores. *Proceedings of the 2013 International Joint Conference on Neural Networks (IJCNN)* (pp. 1-10). IEEE.

Figueiredo, E., Eusébio, C., & Kastenholz, E. (2012). How Diverse are Tourists with Disabilities? A Pilot Study on Accessible Leisure Tourism Experiences in Portugal. *International Journal of Tourism Research, 14*(6), 531–550. doi:10.1002/jtr.1913

Fuchs, M., Höpken, W., & Lexhagen, M. (2014). Big data analytics for knowledge generation in tourism destinations - A case from Sweden. *Journal of Destination Marketing & Management, 3*(4), 198–209. doi:10.1016/j.jdmm.2014.08.002

Girardin, F., Blat, J., Calabrese, F., Dal Fiore, F., & Ratti, C. (2008). Digital footprinting: Uncovering tourists with user-generated content. *IEEE Pervasive Computing, 7*(4), 36–43. doi:10.1109/MPRV.2008.71

Gretzel, U., Sigala, M., Xiang, Z., & Koo, C. (2015). Smart tourism: Foundations and developments. *Electronic Markets, 25*(3), 179–188. doi:10.100712525-015-0196-8

Hoy, M. B. (2018). Alexa, Siri, Cortana, and More: An Introduction to Voice Assistants. *Medical Reference Services Quarterly, 37*(1), 81–88. doi:10.1080/02763869.2018.1404391 PMID:29327988

Huang, S. H., Liu, P., Mokasdar, A., & Hou, L. (2013). Additive manufacturing and its societal impact: A literature review. *International Journal of Advanced Manufacturing Technology, 67*(5-8), 1191–1203. doi:10.100700170-012-4558-5

IBM. (2015). What is Cloud Computing? doi:10.1007/978-3-642-13067-0_35

IBM. (2017). Vacation with IBM Watson and discover a whole new world. Retrieved from https://www.ibm.com/blogs/client-voices/vacation-ibm-watson-discover-new-world/

Ivanov, S., Webster, C., & Berezina, K. (2017). Adoption of robots and service automation by tourism and hospitality companies. *Turismo & Desenvolvimento, 27*(28), 1501–1517. Retrieved from https://www.researchgate.net/publication/322635104

Ivanović, S., Milojica, V., & Roblek, V. (2017). A Holistic Approach to Innovations in Tourism. *Proceedings of the International Conference on Economic and Social Studies (ICESoS'16)* (pp. 367–380). Academic Press.

Jancenelle, V., Storrud-Barnes, S., & Javalgi, R. (2017). Augmented reality, virtual reality and 3D printing for the co-creation of value for the visitor experience at cultural heritage places. *Management Research Review, 40*(3), 352–367.

Jasperneite, J. (2013). *Was hinter Begriffen wie Industrie 4.0 steckt*. Internet Und Automation.

Jazdi, N. (2014). Cyber physical systems in the context of Industry 4.0. *Proceedings of the 2014 IEEE International Conference on Automation, Quality and Testing, Robotics, AQTR 2014*. 10.1109/AQTR.2014.6857843

Jung, T., tom Dieck, M. C., Lee, H., & Chung, N. (2016). Effects of Virtual Reality and Augmented Reality on Visitor Experiences in Museum. In Information and Communication Technologies in Tourism 2016. Springer. doi:10.1007/978-3-319-28231-2_45

Kaur, K., & Kaur, R. (2016). Internet of Things to promote Tourism : An insight into Smart Tourism. *International Journal of Recent Trends in Engineering and Research*.

Klein, S., Avery, M., Adams, G., Pollard, S., & Simske, S. (2014). From scan to print: 3D printing as a means for replication. Proceedings of the NIP & Digital Fabrication Conference. Academic Press.

Konkel, M., Leung, V., Ullmer, B., & Hu, C. (2004). Tagaboo: A collaborative children's game based upon wearable RFID technology. *Personal and Ubiquitous Computing*, *8*(5). doi:10.100700779-004-0302-y

Kounavis, C. D., Kasimati, A. E., & Zamani, E. D. (2012). Enhancing the tourism experience through mobile augmented reality: Challenges and prospects. *International Journal of Engineering Business Management*, *4*, 10. doi:10.5772/51644

Kovar, J., Mouralova, K., Ksica, F., Kroupa, J., Andrs, O., & Hadas, Z. (2016). Virtual Reality in Context of Industry 4.0. *Proceedings of the 2016 17th International Conference on Mechatronics - Mechatronika (Me) 2016*. Academic Press.

Li, J., Xu, L., Tang, L., Wang, S., & Li, L. (2018). Big data in tourism research: A literature review. *Tourism Management*, *68*, 301–323. doi:10.1016/j.tourman.2018.03.009

McKercher, B. (2008). A citation analysis of tourism scholars. *Tourism Management*, *29*(6), 1226–1232. doi:10.1016/j.tourman.2008.03.003

Mckillop, H. (2013). Sustainable Archaeological Tourism of the Underwater Maya Project by 3D. *Anthmpology News*, (May/June), 12–13. Retrieved from http://www.ga.lsu.edu/hmckill/publications/2013%20McKillop%20and%20Sills%20Anth%20News.pdf

Milicchio, F., & Prosperi, M. (2016). Accessible tourism for the deaf via mobile apps. *Proceedings of the 9th ACM International Conference on PErvasive Technologies Related to Assistive Environments - PETRA '16* (pp. 1–7). New York: ACM Press. 10.1145/2910674.2910694

Moavenzadeh, J. & Maar, L. (2018). The Known Traveller: Unlocking the potential of digital identity for secure and seamless travel. *World Economic Forum*, (January). Retrieved from http://www3.weforum.org/docs/WEF_The_Known_Traveller_Digital_Identity_Concept.pdf

Newman, D. (2018). Top 6 Digital Transformation Trends in Hospitality and Tourism. Forbes. Retrieved from https://www.forbes.com/sites/danielnewman/2018/01/02/top-6-digital-transformation-trends-in-hospitality-and-tourism/#23f6ddb267df

Nunes, M. L., Pereira, A. C., & Alves, A. C. (2017). Smart products development approaches for Industry 4.0. *Procedia Manufacturing*, *13*, 1215–1222. doi:10.1016/j.promfg.2017.09.035

Öztayşi, B., Baysan, S., & Akpinar, F. (2009). Radio frequency identification (RFID) in hospitality. *Technovation*, *29*(9), 618–624. doi:10.1016/j.technovation.2009.05.014

Pala, Z., & Inanç, N. (2007). Smart parking applications using RFID technology. *2007 1st Annual RFID Eurasia*, (May), 2007–2010. doi:10.1109/RFIDEURASIA.2007.4368108

Parida, V., Sjödin, D., & Reim, W. (2019). Reviewing literature on digitalization, business model innovation, and sustainable industry: Past achievements and future promises. doi:10.3390u11020391

Peceny, U. S., Urbančič, J., Mokorel, S., Kuralt, V., & Ilijaš, T. (2019). Tourism 4.0: Challenges in Marketing a Paradigm Shift. *Intech Open*, *2*, 64. doi:10.5772/32009

Persson, H., Åhman, H., Yngling, A. A., & Gulliksen, J. (2015). Universal design, inclusive design, accessible design, design for all: Different concepts—one goal? On the concept of accessibility—historical, methodological and philosophical aspects. *Universal Access in the Information Society*, *14*(4), 505–526. doi:10.100710209-014-0358-z

Popiel, M. (2016). Barriers in undertaking tourist activity by disabled people. Prace Naukowe Akademii im. *Jana Długosza w Częstochowie Kultura Fizyczna*, *15*(3), 103–110.

Pühretmair, F. (2004, July). It's time to make eTourism accessible. *Proceedings of the International Conference on Computers for Handicapped Persons* (pp. 272-279). Springer. doi:10.1007/978-3-540-27817-7_41

Rüßmann, M., Lorenz, M., Gerbert, P., Waldner, M., Justus, J., Engel, P., & Harnisch, M. (2015). *Industry 4.0: The Future of Productivity and Growth in Manufacturing Industries*. Boston Consulting; doi:10.100712599-014-0334-4

Saturno, M., Pertel, V. M., Deschamps, F., De, F. R., & Loures, E. R. (2018). Proposal of an Automation Solutions Architecture for Industry 4. *0. LogForum*, *14*(2), 185–195. doi:10.17270/J.LOG.266

Schlüter, F. F., Hetterscheid, E., & Henke, M. (2017). A Simulation-Based Evaluation Approach for Digitalization Scenarios in Smart SupplyChain Risk Management. *Journal of Industrial Engineering and Management Science,* (1), 179–206. doi:10.13052/jiems2446-1822.2017.009

Schwab, K. (2016). *The Fourth Industrial Revolution*. World Economic Forum.

Sedlar, U., Kos, A., Pustisek, M., Bester, J., Pogacnik, M., Mali, L., & Stojmenova Duh, E. (2018). Tackling the Challenges of ICT Innovation and Talents for Industry 4.0. *IPSI BGD Transactions on Internet Research, 14*(1).

Short, D. B. (2016). Use of 3D Printing by Museums: Educational Exhibits, Artifact Education, and Artifact Restoration. *3D Printing and Additive Manufacturing*. doi:10.1089/3dp.2015.0030

Smyth, C., & Crabtree, M. (2012). The Use of RFID Technology At Pedestrian Crossings To assist disabled users: on-street trial results.

Stumbo, N. J., & Pegg, S. (2005). Travelers and Tourists With Disabilities : A Matter of Priorities and Loyalties. *Tourism Review International*, *8*(3), 195–209. doi:10.3727/154427205774791537

Tobergte, D. R., & Curtis, S. (2013). *Einführung und Umsetzung von Industrie 4.0 - Grundlagen, Vorgehensmodell und Use Cases aus der Praxis. Journal of Chemical Information and Modeling, 53.* doi:10.1017/CBO9781107415324.004

Tsai, C. Y., & Chung, S. H. (2012). A personalized route recommendation service for theme parks using RFID information and tourist behavior. *Decision Support Systems*, *52*(2), 514–527. doi:10.1016/j.dss.2011.10.013

United Nations. (2006). The UN convention on the rights of persons with disabilities (Vol. 49). Academic Press. doi:10.1057/palgrave.development.1100310

UNWTO. (1980). Manila Declaration on World Tourism. *Proceedings of the World Tourism Conference*. Academic Press.

UNWTO. (2009). Understanding Tourism: Basic Glossary. World Tourism Organization. Academic Press. doi:10.1177/1479164111417471

Walters, P., & Davies, K. (2010). *3D printing for artists: research and creative practice*. Rapport.

Waschke, S. (2004). *Labeling im Barrierefreien Tourismus in Deutschland – Vergleichende Analyse auf Basis Europäischer Beispiele* [Thesis]. Universität Lüneburg.

Wolter, M. I., Mönnig, A., Hummel, M., Weber, E., Zika, G., Helmrich, R., ... Neuber-Pohl, C. (2015). Industrie 4.0 und die Folgen für Arbeitsmarkt und Wirtschaft: Szenario-Rechnungen im Rahmen der BIBB-IAB-Qualifikations- und Berufsfeldprojektionen. *IAB-Forschungsbericht*, *8*, 68. doi:10.100710273-011-1262-2

World Economic Forum, & Accenture. (2017). Digital Transformation Initiative Aviation, Travel and Tourism Industry. Retrieved from http://reports.weforum.org/digital-transformation

World Travel & Tourism Council. (2015). Seven ways technology is changing the travel industry. Retrieved from https://medium.com/@WTTC/seven-ways-technology-is-changing-the-travel-industry-85cff79c1ece

Zeni, N., Kiyavitskaya, N., Barbera, S., Oztaysi, B., & Mich, L. (2009). RFID-Based Action Tracking for Measuring the Impact of Cultural Events on Tourism. In Information and Communication Technologies in Tourism 2009. Springer. doi:10.1007/978-3-211-93971-0_19

ADDITIONAL READING

Avis, A. H., Card, J. A., & Cole, S. T. (2005). Accessibility and Attitudinal Barriers Encountered By Travelers With Physical Disabilities. *Tourism Review International*, *8*(3), 239–248. doi:10.3727/154427205774791591

Chareyron, G., Da-Rugna, J., & Raimbault, T. (2014). Big data: A new challenge for tourism," 2014 IEEE International Conference on Big Data (Big Data), Washington, DC, 2014, pp. 5-7. 10.1109/BigData.2014.7004475

Han, D. I., Jung, T., & Gibson, A. (2014). Implementing Augmented Reality in Tourism. In Information and Communication Technologies in Tourism. Academic Press. doi:10.1007/978-3-319-03973-2_37

Jaremen, D. (2016). Advantages from ICTS usage in hotel industry. Czech Journal of Social Sciences. *Business Economics (Cleveland, Ohio)*, *5*(3), 6–17. doi:10.24984/cjssbe.2016.5.3.1

Koo, C., Gretzel, U., Hunter, W. C., & Chung, N. (2015). The Role of IT in Tourism. *Asia Pacific Journal of Information System*, *25*(1), 99–104.

Li, Y., Hu, C., Huang, C., & Duan, L. (2017). The concept of smart tourism in the context of tourism information services. *Tourism Management*, *58*, 293–300. doi:10.1016/j.tourman.2016.03.014

Michopoulou, E., & Buhalis, D. (2013). Information provision for challenging markets: The case of the accessibility requiring market in the context of tourism. *Information & Management*, *50*(5), 229–239. doi:10.1016/j.im.2013.04.001

Xiang, Z., Magnini, V. P., & Fesenmaier, D. R. (2015). Information technology and consumer behavior in travel and tourism: Insights from travel planning using the internet. *Journal of Retailing and Consumer Services*, *22*, 244–249. doi:10.1016/j.jretconser.2014.08.005

KEY TERMS AND DEFINITIONS

Accessible Tourism: A form of tourism that that enables people with various access requirements, including issues of mobility, vision, hearing and cognitive dimensions to have access to tourism products, services, and environments, without constraints.

Accessible@Tourism 4.0: The contribution of the fourth industrial revolution for improving accessible tourism, resulting in an overview of different technological innovations developing accessibility conditions in tourism.

Industry 4.0: A new level of organizational control over the entire value chain of the life cycle of products, and it is generated towards increasingly individualized customer requirements.

Technological Drivers: Technological trends that capitalize on the fourth industrial revolution and the digitalization era.

Tourism 4.0: The impact of the fourth industrial revolution in tourism, resulting in the creation of more personalized traveling experience, based on a variety of modern high-tech computer technologies.

People with Disabilities: People suffering from any restriction or lack of ability (resulting from an impairment) to perform an activity in the manner or within the range considered normal for a human being.

Chapter 13
Determinants for Value Creation in Mobile Apps Within the Tourist Experience:
An Exploratory Study in a Northern City in Portugal

Ana Margarida Martins
Polytechnic Institute of Cávado and Ave, Portugal

Beatriz Casais
https://orcid.org/0000-0002-7626-0509
School of Economics and Management, University of Minho, Portugal & IPAM, Porto, Portugal & CiTUR, Portugal

ABSTRACT

This study focuses on the use of mobile applications in tourism context (m-tourism). A survey was applied to a hotel in the city of Guimarães, in the north of Portugal, with a sample of 98 tourists. 94 reported to use these devices in tourism context to access information and share experiences in social networks, taking photos and videos, and using GPS. Users have a preference for free applications and look for the creation of itineraries. The use of the app in offline mode is the most valued feature, as well as the availability of updated and useful information. Only 13,7% of the inquired tourists referred the use of the app of the destination Guimarães, where they were staying. This shows that besides the features, it is crucial the promotion of mobile touristic apps, and the establishment of partnerships and rewards to incentivize its use.

DOI: 10.4018/978-1-7998-1947-9.ch013

INTRODUCTION

The constant growth and evolution of technologies and their adoption by consumers have implications in the customer journey and in the orientation of marketing touchpoints (Muzellec & O'Raghallaigh, 2018). Technology is an allied of tourism, with a significant value creation to facilitate and improve the experience through mobile devices, as smartphone, tablets or GPS, in the accomplishment of the most varied tasks in tourism experience (Kim, Park, & Morrison, 2008). With the possibility of accessing the internet anywhere and anytime through smartphones and other mobile devices, the use of mobile apps in tourism experience has been increasing (Sun, Law, Schuckert, Kucukusta, & Guillet, 2017). Tourists tend to become more independent and use the internet for travel planning, with new applications related to hospitality, restaurants and attractions (Tan, Lee, Lin, & Ooi, 2017).

Most tourists, on their trips, use the smartphone or tablet and have easy access to information in real time, with mobile applications, adding value to the tourist experience. In this sense, tourism entities seek to boost their online presence adapted to mobile devices, giving users the possibility to enjoy a more complete and satisfactory experience (Kim et al., 2008). Mobile devices are becoming more important in the tourism sector. Mobile technology is a tool that represents new opportunities for the various entities of the sector to attract consumers. Organizations in the sector should seek to understand this tool in order to develop their services based on the factors that lead to the attraction of consumers, making their services more relevant, effective and attractive (Kim et al., 2008).

Little is known about mobile app features most valued by tourists (Palumbo, Dominici, & Basile, 2013). This study considers the state of the art about consumers' acceptance of mobile technology and look for evidences about the critical features of mobile apps expected by tourists. The research examines tourists' expectations about mobile apps through the analysis of a survey conducted in a hotel in the North of Portugal, Guimarães. This chapter discusses the mobile app features most valued in tourism context and, finally, analyses the use of the official destination app where the respondents were staying at that moment. The results are discussed in the end of the chapter, proposing management recommendations for destination and tourism managers.

BACKGROUND

This section discusses the theoretical background about consumers' adoption of mobile devices and applications. The adoption of technology is related to two fundamental variables - perceived usefulness and perceived ease of use (Davis, 1989). These are subjective perceptions of each individual on performance and effort, respectively, and act according to beliefs. The literature stated that the user may prefer to deal with usage constraints rather than failing in the performance of tasks. However, if the difficulty is high, it may lead to withdrawal. In case the facility is high, it may compensate for its poor utility. The designers and creators of innovations should thus pay great attention to utility, being this strongly linked to the acceptance of innovation, and not only the ease of use in its creation and development.

Technology Acceptance

The technology acceptance model helps to understand the factors that lead to the adoption of new technologies, being value, ease of use, social environment, user resources, hedonic motivations, price/ value, previous experience and habit highly relevant (Doub, Levin, Heath, & LeVangie, 2015).

The intent to continue the use of a technology is related to the perceived satisfaction and utility of the service or product. In the case of mobile applications (apps), users try to use a trial version of the application in order to know the contents and functions before making the purchase. If they are satisfied and feel that there is usefulness in the use, they will consider the purchase. This act of experimenting and then buying is similar to the behavior explained by the Expectation–Confirmation Model (ECM) (Hsu & Lin, 2015). This model gained importance in the understanding of the post-acceptance behavior of technology, having the satisfaction factor and performance perceived as factors that lead to a purchase. The ECM explains the continuity of use through the utility and expectation factors seen as the main determinants of the intended use. It also influences user satisfaction and therefore their loyalty. In most cases, the trial version has the basic features of the application, and the user pays for the possibility of using it more fully. Free app developers will be able to apply advertising strategies or direct in-app purchases as ways to earn revenue.

The perception of the usefulness of the application is based on what the user has to pay for it, that is the perceived value or the understanding about the benefits versus the costs. Perceived value increases when the user believes that one application is more beneficial than another similar and is related to satisfaction, loyalty and intention to use and purchase. Another factor influencing the purchase intention is the classification assigned to the application.

The free alternative or trial version may lead to the purchase and continuity of use of the app since it will be possible for the user to try it before buying and thus reduce the uncertainties. Habit is another factor that influences continuity and loyalty. Users typically prefer free applications or look for alternatives that are free, and substitutes have an impact on user behavior (Hsu & Lin, 2015).

In the tourism sector, there is a behavioral change in travelers, from offline to online, since they no longer feel a dependency on travel agents for booking and searching for information for their trips (Buhalis & Law, 2008). The tourist has become, over the years, more demanding, informed and more willing to provide information in order to obtain a personalized service and satisfy desires and expectations (Palumbo et al., 2013). These changes are also verified at the level of the tourist who has different motivations and desires and, moreover, is increasingly sophisticated due to previous experiences, knowledge of languages and technologies, and always seeking to obtain more value for the cost and time spent. There is also a decrease in interest in organizing groups, but rather planning the trip according to preferences, personalizing experiences through online devices (Buhalis & Law, 2008).

The main function of the tourism-oriented mobile apps is to provide information at various levels and provide advantages over other sources, since the tourist can thus alter and adjust plans according to the obtained information (Dickinson et al., 2014). Mobile applications in tourism have the capacity to make the tourism experience more personalized, more interactive and memorable (Palumbo et al., 2013).

Figure 1 shows categories of mobile applications targeted or used in the context of the tourist experience. Even they may not be created directly for the tourism activity, their use may be made during travel or tourist events (Kennedy-Eden & Gretzel, 2012).

Figure 1. Tourism mobile applications by service provided
Source: (Kennedy-Eden & Gretzel, 2012, p. 48)

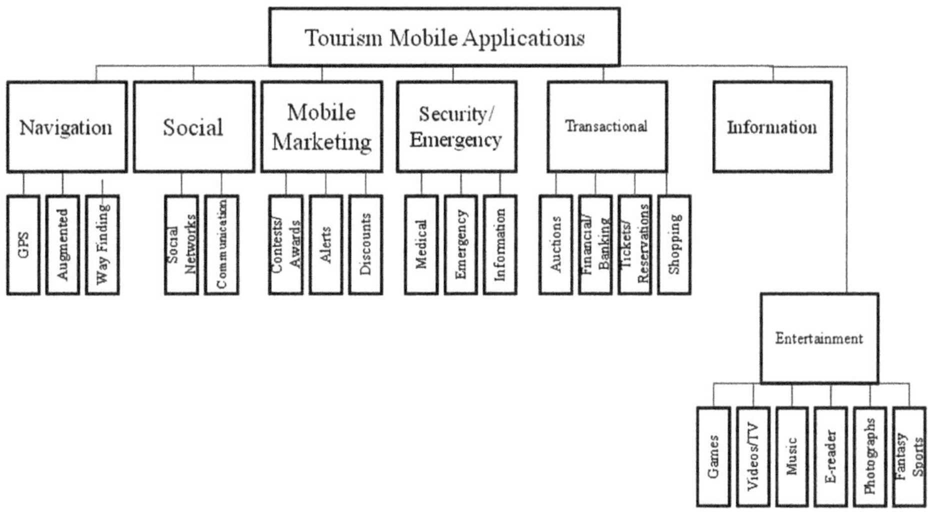

Technological advances, such as mobile applications, can fail if they do not take into account the needs and expectations of the user (Palumbo et al., 2013). Mobile apps can act as a filter of information available, meeting the preferences and profile of the user, not overloading with information that is unnecessary for what the tourist is looking for. In this way, the key to an application's success is to identify consumer needs and reach out to potential customers with vast, customized, and up-to-date products and services to meet those needs (Buhalis & Law, 2008).

Mobile Apps

Mobile tourism applications have as their main success factor the ability to be used by the tourist at all stages of experience, providing information everywhere and in a more dynamic way, for example informing the tourist about the time of the next bus while it moves (Kim et al., 2008). The diversity of mobile devices such as smartphones, PDAs, mobile phones, among others, have different characteristics to which the app needs to be adaptable as displays for different sizes, touchscreen or keypad for mobile phone, or the ease of access to the mobile network. The ability of the application to perform certain functions determines its success. Below are some features that, when integrated in a mobile application, make tourist experience easier, having access to a wide range of information, satisfying needs during the trip: geographical location, weather service, translator, currency converter, time converter, emergency services, taxi service / UBER, bathroom search, social dining, social network integration, reading or writing ratings and reviews, use in offline mode, audiovisual content. Also QR codes and augmented reality make use of mobile device facilities such as camera, GPS and compass to give the user information that completes the visit. This information appears on the image taken by the device's camera.

Electronic Word of Mouth (eWOM) and User Generated Contents (UGC)

With the growing importance given to testimonials, organizations should be more concerned with the so-called eWOM and User Generated Contents. Consumer testimonials are a type of source considered as very credible, having a high persuasion power over potential consumers (Wilson, Murphy, & Fierro, 2012) and is most valued than organization's official information because of its trust in other tourist experiences (Shan, 2015). The consumer then gains the power to create information and share it with potential consumers, thus ending the exclusive power of organizations (Kietzmann, Hermkens, McCarthy, & Silvestre, 2011). These contents created by the consumer are valued, given the difficulty in perceiving the quality of the product or service before consumption as well as the intangibility that characterizes the tourist products (Liu & Park, 2015). This type of content refers to comments, photographs, or videos, created and shared by users. The motivation of the consumer to write a testimony may vary according to the individual, they tend to consist on social interactions, incentives, concerns to help other consumers, or expression of satisfaction, comparisons and contributions to reputation (Henning-Thrurau, Gwinner, Walsh, & Gremler, 2004; Racherla & Friske, 2012).

Testimonials with negative content are usually considered very useful. Information that is perceived as useful will help the potential consumer in the decision-making, leading to evaluate the characteristics of the product and to have greater confidence in the purchase. Thus, the more useful the information on the platforms is considered, the greater its value to consumers (Racherla & Friske, 2012). Providing the writer's identification - name or photo - makes it easy for the message to be perceived as credible to those seeking information, making it more useful and reliable (Shan, 2015). The reputation and expertise of the comment source exert a high influence in the perception of the usefulness of the testimony. Reputation gain is also a motivational factor for writing opinions, that are usually more elaborated in this case (Racherla & Friske, 2012). The reputation factor has greater influence on decision making, than the expertise factor or the fact of the source being a specialist. The rating, or quantitative characteristic, corresponds to the score given by consumers, for example, by using stars or punctuation. The higher the rating, the greater its influence. Consumers are willing to pay more for services or products rated with 5 stars, similarly with the degree of elaboration and detail of the testimonials (Liu & Park, 2015; Shan, 2015). Compared to the positive critics, the negative ones are generally more detailed, hence more credible (Liu & Park, 2015).

Online testimonials are a source of eWOM highly regarded by those seeking information online through the available platforms, due to the quality, convenience, social guarantee and by reducing the degree of uncertainty in the decision (Schuckert, Liu, & Law, 2015). The planning of a trip may now involve the use of the internet, namely for consulting other travelers' testimonies regarding services, searching for maps and directions, consulting online brochures and printing vouchers. Online testimonials play a key role at this stage as they can help the consumer to have ideas, reduce options in case of doubt, learn more about the location or service, and increase confidence in decision making. Frequent readers of these testimonials consider planning the trip somewhat enjoyable, feeling more involved in the process, and use the testimonies at all stages of the journey. For example, in restaurants, online shared pictures are very important for consumers before taking a decision (Oliveira & Casais, 2019). Their decision making is not done impulsively, and they use the internet also for viewing blogs and videos. The most popular channels for sharing experiences are social networks, such as Facebook, and photo and video sharing websites such as Flickr and Youtube, however they vary according to markets, nationality and type of messages (Wilson et al., 2012).

User generated contents have a high degree of influence by the increase of the use of the internet and the time spent online in social media and e-wom sharing platforms, comparing to the official websites of attractions (Wilson et al., 2012). In the tourism sector, there are examples of content sharing platforms as *TripAdvisor, Hotels.com* (Racherla & Friske, 2012), *Yelp, CitySearcher* and *VitualTour* (Liu & Park, 2015). Consumers regularly make a purchase decision after reading positive testimonials regarding hotel and travel. This type of content is thus an important source of information for potential consumers of products and services (Pan & Fesenmaier, 2006; Xiang & Gretzel, 2010; Xiang, Magnini, & Fesenmeier, 2014). In this way, official entities should pay special attention to these contents and realize what leads the consumer to make those shares. It may cause lower costs in marketing actions, since the use of the platforms does not have a high cost. However, there will be an increase in the time spent creating new content and responding to consumer testimonies, seeking to reverse negative shares. As potential consumers are looking for more credible and reliable information from other consumers, organizations should be careful to monitor and control content created. In addition, they should encourage consumers to share their experience, if positive, especially on platforms that make use of rankings such as *TripAdvisor*, where a good score helps in positioning the service (Wilson et al., 2012). The usefulness of the recommendation will lead to the perception of its credibility and the subsequent adoption of the information (Sussman & Siegal, 2003). However, the existence of much information available to the potential consumer makes it difficult to perceive the usefulness of each testimony (Liu & Park, 2015).

Content creation platforms demonstrate their increasing importance in the consumer's decision-making process, since they can access a large number of information easily (Liu & Park, 2015). Regarding the purchase, as mentioned in previous paragraphs, the consumer has more and better access to the information sought and the existing platforms add a greater ease to the traveler, namely for the purchase of tickets online. However, the trust factor is fundamental for the consumer to make purchases via the internet, and for this reason organizations should take special care to e-wom. The dimensions shown in Table 1 are related to the trust in the internet for the purchase (McCole, 2002).

Table 1. Dimensions of trust applied to the internet and online purchase

Availability	Perception of the internet as being always available for use by the potential consumer.
Competence	The internet as a facilitator of the execution of all steps of the purchase, without human intervention. Here is the importance of trust.
Consistency	At this point, it is expected that by using the internet the purchase will follow a normal, easy-to-understand and accomplished process whether it is first used or repeated.
Discretion	This point is about the perception of online security and that the data entered will not be shared with third parties.
Justice	It applies the sense that the consumer hopes to save money given that he/she is making an online purchase for an appropriate price.
Integrity	The consumer hopes that the information they get is reliable and that it meets expectations.
Loyalty	If the service that the consumer buys is satisfactory, he/she can repeat the process through the same platform.
Opening	Perceived that all the information regarding the purchase process is accessible for consumer consultation.
Fulfillment of the expectations	The consumer expects to receive exactly what he/she buys, as described. This point is important for intangible products and services, such as travel - it is only after the purchase that the consumer will come into contact with the product.
Receipt	The effect of the existing information on the willingness to carry out the purchase by the potential consumer.

Source: Adapted from McCole (2002)

THE RESEARCH

Purpose and Unit of Analysis

The focus of the present study was to understand which features or characteristics tourists consider fundamental in a mobile application for tourism use, which they consider adding value and lead to its use. As a unit of analysis, the researchers analyzed the app *Guimarães Mobitur* from Guimarães destination, in the north of Portugal.

The main features of the city app are the map of the city - using GPS, the location of points of interest with their information such as timetables, contacts, photographs. It also allows the application of search filters - personalizing available information, search for events in the city, and sharing in social networks. In order to facilitate access, this is available *online* or *offline*. Free *Wi-Fi* internet access is available in the main public areas and infrastructures of the city, providing comprehensive coverage.

Research Hypothesis

To conduct the research, four hypotheses were built to test the expectations of tourists regarding mobile apps. Table 2 shows the research hypothesis and identify the authors from the literature supporting those hypotheses.

Methodology and Sample

The authors shared a survey to guests of one of the main hotels in Guimarães, between the 1st and 31st of March 2018. It is a four-star hotel and in the month of application it had an average occupancy of 49.46%, with 486 occupied rooms. In 2017, the average annual occupation had been 56%, receiving national and international tourists daily. The reason for choosing a hotel unit to share the survey is the need for the respondents to be considered tourists. The survey was shared in paper, in Portuguese, English and Spanish, seeking to reach the main public of the hotel, and placed in the booked rooms. The questionnaire was built to be simple, for quick response in the attempt to obtain a greater number of responses. A presentation of the research in progress was made, with an email contact provided. The questions used were for single answer to choose only one option from the available multiple choice answers. Other questions gave the respondent the possibility to select more than one answer, and there was also questions using the Likert scale, where the respondent chooses the option which goes according to its

Table 2. Research hypothesis

No. Hypothesis	Description of Hypothesis	Supporting Literature
H1	The price influences the use of mobile applications in a tourism context.	(Hsu & Lin, 2015)
H2	The functionalities integrated in a tourist application determine its adoption.	(Kennedy-Eden & Gretzel, 2012)
H3	The characteristics of the applications determine its use.	(Buhalis & Law, 2008; Palumbo et al., 2013)
H4	For the user, the usability factor of the application is more important than its ease of use.	(Davis, 1989)

opinion of agreement on a scale of 1 - totally agree, to 5 - totally disagree. In the first part, respondents were questioned about their demographic data - gender, age and nationality. The remaining questions were related to the research, regarding the importance given to certain characteristics and functionalities of the mobile applications - where a Likert scale was used. These questions were fundamentally elaborated according to the variables price, functionalities, features, and utility versus facility. Finally, the respondents were questioned about the use of the official application of the city Guimarães, in order to understand its awareness and use by the tourists staying in the city.

98 tourists answered the survey. 55,1% of the respondents are female and 44,9% males. 38,8% of the guests that answered the survey are Portuguese, followed by Spanish (18,1%) and Brazilians (10,2%). Table 3 shows a balanced distribution of the respondents by age.

Only 94 answered that they use mobile devices in the touristic experience, being the sample of the study. The analysis of the data collected through the questionnaire was made using the IBM SPSS Statistics version 23 software, in order to verify the hypotheses posed. This tool was used by performing univariate analyzes in the context of performing the sample characterization, understanding the use or not of mobile devices, frequency analysis of the type of use of mobile device types and the Guimarães Mobitur application, and performing analyzes. Bivariate analysis was conducted using SPSS software, for Cronbach's alpha coefficient and Pearson Correlation analysis.

The data collected through the survey applied made it possible to perceive that the smartphone is the most used device – 91 of the respondents -, perhaps because of its smaller size among other devices and its ability to perform various tasks. The use is mainly for access to social networks – 69 respondents -, taking pictures and video – 64 respondents -, using navigation systems – 57 respondents and access information in the internet – 64 respondents.

Results

After defining the variables under study, the reliability of the scales used was verified, measuring the reliability of the variables. For this purpose the Cronbach's alpha coefficient ranging from 0 to 1 was used, in which the results can be distinguished according to:> 0.9 as excellent,> 0.8 as good,> 0.7 acceptable,> 0.6 questionable,> 0.5 poor and < 0.5 unacceptable. According to Table 4, it is verified that the functionality variable, evaluated through 10 items, presents a Cronbach's alpha of 0.844, which means a reliability of scale defined as good. The features variable obtained 0.760, evaluated from 7 items, being considered as acceptable. The variable price was evaluated with 4 items, obtaining a value of 0.578, considered as questionable. The variable utility vs facility presented a value of 0.438, considered

Table 3. Frequency of respondents by age.

Age	Frequency	%
18-25	12	12,2%
26-36	26	26,5%
37-47	32	32,7%
48-60	22	22,4%
+60	6	6,1%
Total	98	100%

Table 4. Analysis of the measurement ranges of variables

		Average	Standard Deviation	Cronbach Alpha	Items
Functionality	I love it.	2.53	1,334	0.844	10
	Geolocation	1.74	0.961		
	Translation	2.18	1,154		
	Social networks	3.01	1,395		
	Emergency	2.62	1,139		
	Transportation	2.65	1,392		
	Conversion	2.41	1.20		
	Timezone	2.37	1.29		
	Meteorology	2.01	1,159		
	Itineraries	1.68	1.008		
Features	Offline mode	1.64	1.035	0,760	7
	Audio-visual	2.19	1.090		
	Design	2.30	1.076		
	Usable every year	1.65	1,095		
	Custom	1.88	0.890		
	Updated	1.69	1.068		
	ranking	2.27	0.906		
Price	Pay for the perception of utility	2.18	1.087	0.578	4
	More features greater predisposition to pay	2.39	1.090		
	Influence of free applications	1.76	1.034		
	Influence of the trial version	2.84	1,298		
Utility VS Facility	Utility	1.68	1.018	0,438	2
	Facility	2.31	1.098		

Source: Authors

unacceptable. This value can be justified by the reduced number of items under evaluation (Gu, Lee, & 2009) or the heterogeneity of responses according to the means of the 2 items.

Functionality

As regards the presence of platforms that allow the writing and sharing of recommendations, the average value and the value of the standard deviation higher than 0 show very heterogeneous responses. In the case of geolocation, the mean value is between totally agree and agree, with lowest standard deviation. The translation feature presents an average that places it considered important by the respondents since they agree with its integration in applications. However, the answers are very heterogeneous as shown by the value of the standard deviation, above 1.15.

As far as social networking integration is concerned, this is the functionality with the highest average and standard deviation values, showing a great distinction between respondents' responses. This is in line with previous literature that enhance the importance of accessing social media in smartphones (Salehan & Negahban, 2013; Urry, Larsen, & Axhausen, 2007). The same applies to features related to emergency services, transportation services and time zones that revealed similar results. The conversion and weather functionalities obtained high response averages, and a standard deviation a little lower than the results of the previous functionalities.

The route generation functionality has obtained a mean value and standard deviation lower than the previous features, while being aware of the geolocation.

Table 5 analyses the frequency of answers about the most important functionalities in a mobile application for tourists.

In terms of functionalities, the most important for tourists are those that relate to the location, being the possibility of applications to generate routes and itineraries that claim the most important, followed by the geolocation. This result verifies the willingness of tourists to make plans and find locations through their mobile devices, with easy access to maps and *GPS*.

Weather forecasting services, the translation tool and the integration of transport services have similar values and are in 3rd, 4th and 5th place, respectively, showing that they have a similar degree of importance for the tourist. The platforms for recommendations appear in 6th place. Time zone conversion tools, integration of emergency services, currency conversion and integration of social networks were not very important for the respondents.

Features

The possibility of being used in offline mode obtained an average of 1.64. The presence of audiovisual material such as photographs, videos and audio as well as the importance of the design have averages and values of high standard deviation, demonstrating a great discord between the sample.

The possibility of using the application during the whole year obtained a result of 1.65, with most answers between totally agree and agree. Regarding personalized information, the average value is

Table 5. Functionalities by order of importance

#	Functionalities	# Of Answers "1"
1st	Possibility of generating itineraries	55
2nd	Geolocation	48
3rd	Meteorological services	39
4th	Translation tool	34
5th	Transport services such as taxi or *Uber*	33
6th	Platforms of recommendations	30
7th	Time zone conversion	28
8th	Access to emergency services	28
9th	Currency conversion	26
10th	Social Networking Integration	16

1.88. The standard deviation value is the lowest of all characteristics, demonstrating a greater degree of agreement at this point among the respondents. The updated information resulted in a mean of 1.69 and a standard deviation of 1.068, demonstrating that most respondents value this point.

The classification of the application presented relatively high average results, 2.27. The standard deviation shows the second lower heterogeneity of this group, and demonstrate that this item is the second item with the highest degree of agreement among respondents. Table 6 shows the frequency of I totally agree answers in order to respond to the objective of perceiving the features most valued in a mobile application.

Regarding the characteristics, it is verified that the use in offline mode is the most valued by the tourists. This result verifies the importance mentioned in the literature on this point, particularly where it is stated that many tourists may not have access to the internet if there is no *wi-fi*. The second characteristic most mentioned by tourists, very close to the first, is the possibility of being used in all seasons of the year. It is imperative that the application provides relevant information for all times of the year and not just for those considered as high seasons, since tourists travel throughout the year.

As far as personalized information is concerned, it appears in 4th place in the choices of the tourists. This fact verifies the increasing importance of information adjustment to the user, providing information that satisfies their needs and makes their experience unique and memorable (Dickinson et al., 2014; Kim et al., 2008; Palumbo et al., 2013).

Price

Concerning the fact that the user downloads the paid application based on perceived utility, the average concentrates the answers in I agree, with some distinction between the answers, demonstrated by the standard deviation.

The average for the greater willingness to pay for the download in terms of the number of functionalities show a discrepancy between 1 and 3 on the Likert scale.

Next, the negative influence of similar free applications on the download of paid applications is verified, resulting in the average and standard deviation with lower values of this variable, demonstrating that users have preference for free applications that perform functions similar to those that are paid. The value of the standard deviation shows that this question was the one that meet the highest agreement among the respondents.

Table 6. Characteristics in order of importance

#	Features	# Of answers "1"
1st	Use in offline mode	62
2nd	Use at all times of the year	61
3rd	Updated information	56
4th	Custom Information	36
5th	Presence of photographs, audio and video	32
6th	Design as a determining factor in the choice of application	29
7th	The classification influences user intention	20

The positive influence exerted by the trial versions on the purchase of mobile applications is showed by an average value relatively high, which puts the majority of the answers to this question at the level of I do not agree or disagree but with a high degree of heterogeneity in the Likert scale.

Using the SPSS software, and in order to understand if the price influences the use of mobile applications, a correlation was made between the items of the price variable and the question on the use of mobile applications. Pearson correlation coefficient was used. This varies between 1 and -1, with 1 being considered as a perfect positive correlation and -1 a perfect negative correlation.

According to Table 7, the relationship closer to 1 is verified when the question is related to The existence of similar free applications negatively influencing the intention to purchase applications, and the lowest correlation was verified with The existence of the trial version positively influencing the intention to buy the full version, with a value very close to 0.

These results demonstrate the general preference of users for free applications, which may not be the same as the paid application, performing the same functions. By the value of the correlation it is also verified that the user who pay for a *download* feels it useful in relation to the cost - Expectation-Confirmation Model (Hsu & Lin, 2015).

Thus, it is clear that the price variable has an influence on the purchase decision of a mobile application, since the user mainly appreciates the free applications that perform the same or similar functions as the applications that entail acquisition costs. Therefore, the hypothesis 1 is accepted - The price influences the use of mobile applications in a tourism context.

Verification of Hypotheses 2 and 3

In order to evaluate the relationship between the constituent items of the functionalities and usage, the study of the Pearson correlation was carried out. The values obtained are shown in Table 8.

It was verified that the functions related to the location have a higher relation with the predisposition to use mobile applications if the user understands that these will improve the accomplishment of tasks. This result coincides with those obtained in the analysis of the most important functionalities for tourists in the use of an application.

As applications are generated with the aim of providing information of various types of tourists (Dickinson et al., 2014), these results demonstrate, although not applied to all the items under analysis, that the most important functionality considered are geolocation and creation of itineraries, so the tourist will use a certain application if it has these functionalities, more significantly the creation of itineraries

Table 7. Pearson's correlation

	You will use an application if you understand that it improves the performance of a particular task
The existence of the trial version positively influences the intention to purchase the full version	- 0.17
Similar free applications negatively influence the purchase intent of applications	0.662
The download a paid application will be influenced by the realization that this is useful in relation to its cost	0,421
The more features the application has, the greater the predisposition to pay for its download.	0.288

Table 8. Correlation between functionalities and application usage

	"You will use a mobile application if you understand that it improves the performance of a certain task"
Access to recommendations platforms	0.081
Geolocation tool	0,492
Translation tool	0.103
Social Networking Integration	-0.243
Integration of emergency services	-0,126
Integration of transport services	-0.07
Currency Conversion Tool	0.008
Time Zone Conversion Tool	0.075
Integration of meteorological services	0.287
Route generation tool	0.688

(value closer to 1). Here it is verified that the integrated functionalities influence the use of a mobile application in a tourist context - hypothesis 2.

Table 9 shows the results obtained through the Pearson correlation between features and application usage.

As verified, all values obtained are greater than 0, being more significant in the following characteristics: updated information, use in offline mode, personalized information and application use throughout the year, revealing some relation between these items and the use of mobile applications.

Also, the features selected by the tourists as the most important, previously verified in the descriptive analysis by frequency, demonstrate that the characteristics of the app influence the use of mobile applications in the tourist context - hypothesis 3.

Table 9. Correlation between characteristics and application usage

	"You will use a mobile application if you understand that it improves the performance of a certain task"
Use in offline mode	0.625
Presence of photographs, audio and video	0,336
Design as determinant in choosing the application	0.103
Usable all year round	0.541
Custom Information	0.598
Updated information	0,677
Classification influences the intended use	0,139

Utility / Facility

As previously mentioned, the utility/ease of use variable was evaluated through 2 items, the first, the utility, found a significantly low mean response, mostly in point 1 of the Likert scale, and a standard deviation that showed that there was not much dispersion in the answers, since the answers vary between 1 and 2.

Regarding the ease of use of the application, by the responses of the users, a higher average was verified than the one verified in the previous item, mainly in the value 2 of the Likert scale. The standard deviation also reveals more heterogeneity of responses in relation to the previous item, varying between 1 and 3 in the scale.

We analyzed the importance of the usefulness of the application and the ease of use. More than 50% of respondents fully agree with the assertion that the usefulness of the application is the most relevant factor in the use of an application. These results are in agreement with the literature review, where it is suggested that the user prefers to deal with constraints and difficulties of use, but to perform tasks and perform desired functions, rather than to easily use a technology that does not have the desired utility (Davis, 1989), thus confirming the hypothesis 4 - For the user, the usefulness of the application is more important than its ease of use.

Guimarães Mobitur App

The city of Guimarães has a mobile application for a better knowledge of the city. As a unit of analysis, the researchers proceeded to understand the awareness of this application within the tourists in the city. This application uses the user's position and provides tools for a better knowledge of the city. The main functions are the city map - using GPS, the location of points of interest with their information such as times, contacts, photos, among others, also allows the application of search filters - personalizing the available information, the looking for events in the city, sharing on social networks. In order to facilitate access, it is available online or offline. The app, on the Google Play store, has a rating of 3.4 based on 25 reviews.

In addition to the application, free Wi-Fi internet access is available in the city's main public areas and infrastructures, providing comprehensive coverage. Therefore, Guimarães was considered a good example for the application of this study.

As shown in Table 10, 60% of respondents reported not having knowledge about the application and 26.3% did not use it. Only 13.7% said they had used the app.

Table 10. Guimarães Mobitur application analysis

	Frequency	%
Yes	13	13.7%
Not	25	26.3%
Didn't know	57	60.0%
Total	95	100%

Source: Own elaboration based on data generated by *IBM SPSS*

After analyzing the data collected earlier, and in order to understand which nationalities used the application the most, it was crossed with the nationality of tourists, verifying through Table 11 that the market that used the application was mostly the Spanish, followed by Portuguese and French.

In relation to the existing application in Guimarães, it is verified that greater efforts should be made to promote it, since it is still very unknown and little used by tourists, and only 13.7% of the sample reported using the application. Although it is quite complete, its weak adoption should be studied by the surrounding organizations in order to enhance and improve it.

DISCUSSION

As far as the functionalities are at the top of the tourists' concerns and priorities regarding mobile apps, they are related to the location, obtaining the creation of itineraries for the intended destination. The most valued app feature among tourists it the possibility to use in offline mode, since it is not always possible to have online connections without a wi-fi network in the area, then it was verified that the application should be usable all the times of the year. The updated and personalized information is also highly valued by respondents, demonstrating that tourists like to see the information received, in addition to current, adapted to the particularities of their trip and their personal taste and needs. From the correlations between the items corresponding to the functionalities variables and features with the use of mobile applications, it can be stated that if the application has those functionalities or characteristics most valued by tourists they will tend to use the application, thus demonstrating the influence of these variables in the use of mobile applications in the tourism context.

Table 11. Analysis of Guimarães Mobitur application by nationality

	Yes	Not	Didn't know	Total
German	0	0	3	3
American	0	0	2	2
Angolan	0	1	0	1
Argentine	0	0	3	3
Australian	0	0	1	1
Belgian	0	0	1	1
Brazilian	0	2	8	10
Canadian	1	0	0	1
Spanish	5	5	6	16
French	2	2	1	5
English	1	3	4	8
Italian	1	1	3	5
Portuguese	3	11	24	38
Russian	0	0	1	1
TOTAL	13	25	57	95

Source: Own elaboration based on data generated by IBM SPSS

The analysis of the data collected on the variable price showed that there is a preference for free applications that perform the same or similar functions as paid applications and that the existence of a trial version has little to do with the intention to use an application. However, there is some relationship between the download of a paid application according to the features it has, that is, the more the application offers the greater the predisposition of the user to pay for its download, verifying the influence of the price, functionalities and features in the use of applications.

CONCLUSION AND RECOMMENDATIONS

Knowing that an application is a tool that adds value to the tourism offer, its creation can generate thoughts in organizations that have to be easy to use, tourists have revealed that in their understanding the usefulness of the application is more important than the ease of its use, so developers should assure that the application is useful if they want to enhance its adoption.

This study reinforces the greater understanding of the importance of new technologies in the tourism sector, the great use of mobile devices and the activities carried out with them. The contribution of the research is based on the perception of the factors valued by the tourists in the use of a mobile application in their tourist experiences.

In the sense that tourism organizations optimize their mobile applications in a value creation perspective, the study presents a set of recommendations to follow, namely:

- Promotion of the applications in the hotel units since they are the first tourist station that the tourist finds when arriving at the destination. If they have knowledge about the existence of the application and its functionalities and features, the hotel's human resources may suggest downloading it.
- As verified through the research, mobile apps are important for the users, so organizations should promote them.
- The design of the application should be intuitive, easy to use, and adaptable to different mobile devices, without changing page formats or losing relevant information.
- The organization should ensure that the application is downloaded at no cost to the user, making profits, for example, from hotel units, tourist attractions, shops or restaurants that want to see its information promoted in the application.
- Presence of photographs so people relate to an image and create an expectation;
- The organization should seek to ensure that the application has useful information, considering the importance given by tourists to perceived utility.

LIMITATIONS AND FUTURE RESEARCH DIRECTIONS

In relation to the data collection and methodology, one of the main constraints was the temporal space of the application of the surveys. Hotel occupancy was lower at the beginning of the application of the questionnaires and higher at the end. The sample could be larger if the questionnaire was applied in the middle or high season of occupancy in the hotel. The sample is very limited and due to that reason conclusion should be carefully extrapolated.

This study may act as basis for the development of a mobile application directed to tourism according to the preferences shown in the results of the survey.

REFERENCES

Buhalis, D., & Law, R. (2008). Progress in information technology and tourism management: 20 years on and 10 years after the internet – the state of etourism research. *Tourism Management*, *29*(4), 609–623. doi:10.1016/j.tourman.2008.01.005

Davis, F. (1989). Perceived usefulness, perceived ease of use, and user acceptance of information technology. *Management Information Systems Quarterly*, *13*(3), 319–340. doi:10.2307/249008

Dickinson, J. E., Ghali, K., Cherrett, T., Speed, C., Davies, N., & Norgate, S. (2014). Tourism and the smartphone app: Capabilities, emerging practice and scope in the travel domain. *Current Issues in Tourism*, *17*(1), 84–101. doi:10.1080/13683500.2012.718323

Doub, A., Levin, A., Heath, C., & LeVangie, L. (2015). Mobile app-etite: Consumer attitudes towards and use of mobile technology in the context of eating behavior. *Journal of direct, data and digital marketing practice, 17*(2), 114-129.

Gu, J.-C., Lee, S.-C., & Suh, Y.-H. (2009). Y.-H. (2009). Determinants of behavioral intention to mobile banking. *Expert Systems with Applications*, *36*(9), 11605–11616. doi:10.1016/j.eswa.2009.03.024

Henning-Thrurau, T., Gwinner, K., Walsh, G., & Gremler, D. (2004). Electronic word-of-mouth via consumer-opinion platforms: What motivates consumers to articulate themselves on the internet? *Journal of Interactive Marketing*, *18*(1), 38–52. doi:10.1002/dir.10073

Hsu, C., & Lin, J. (2015). What drives purchase intention for paid mobile apps? An expectation confirmation model with perceived value. *Electronic Commerce Research and Applications*, *14*(1), 46–57. doi:10.1016/j.elerap.2014.11.003

Kennedy-Eden, H., & Gretzel, U. (2012). A taxonomy of mobile applications in tourism. *Ereview of Tourism Research*, *10*(2), 47–50.

Kietzmann, J., Hermkens, K., McCarthy, I., & Silvestre, B. (2011). Social media? Get serious! Understanting understanding the functional building blocks of social media. *Business Horizons*, *54*(3), 241–251. doi:10.1016/j.bushor.2011.01.005

Kim, D., Park, J., & Morrison, A. (2008). A model of traveller acceptance of mobile technology. *International Journal of Tourism Research*, *10*(5), 393–407. doi:10.1002/jtr.669

Liu, Z., & Park, S. (2015). What makes a useful online review? Implication for travel product websites. *Tourism Management*, *47*, 140–151. doi:10.1016/j.tourman.2014.09.020

McCole, P. (2002). The role of trust for electronic commerce in services. *International Journal of Contemporary Hospitality Management*, *14*(2), 81–87. doi:10.1108/09596110210419264

Muzellec, L., & O'Raghallaigh, E. (2018). Mobile technology and its impact on the consumer decision-making journey. *Journal of Advertising Research, 58*(1), 12–15. doi:10.2501/JAR-2017-058

Oliveira, B., & Casais, B. (2019). The importance of user-generated photos in restaurant selection. *Journal of Hospitality and Tourism Technology, 10*(1), 2–14. doi:10.1108/JHTT-11-2017-0130

Palumbo, F., Dominici, G., & Basile, G. (2013, June). Designing a mobile app for museums according to the drivers of visitor satisfaction. In *Recent Advances in Business Management and Marketing*. Dubrovnik, Croatia: WSEAS Press.

Pan, B., & Fesenmaier, D. (2006). Online information search: Vacation planning process. *Annals of Tourism Research, 33*(3), 809–832. doi:10.1016/j.annals.2006.03.006

Racherla, P., & Friske, W. (2012). Perceived 'usefulness' of online consumer reviews: An exploratory investigation across three services categories. *Electronic Commerce Research and Applications, 11*(6), 548–559. doi:10.1016/j.elerap.2012.06.003

Salehan, M., & Negahban, A. (2013). Social network on smartphones: When mobile phone become addictive. *Computers in Human Behavior, 29*(6), 2632–2639. doi:10.1016/j.chb.2013.07.003

Schuckert, M., Liu, X., & Law, R. (2015). Insights into suspicious online ratings: Direct evidence from tripadvisor. *Asia Pacific Journal of Tourism Research, 21*(3), 259–272. doi:10.1080/10941665.2015.1029954

Shan, Y. (2015). How credible are online product reviews? The effects of self-generated and system-generated cues on source credibility evaluation. *Computers in Human Behavior, 55*, 633–641. doi:10.1016/j.chb.2015.10.013

Sun, S., Law, R., Schuckert, M., Kucukusta, D., & Guillet, B. D. (2017). From information technology to mobile information technology: Applications in hospitality and tourism. *Ereview of Tourism Research.*

Sussman, S., & Siegal, W. (2003). Informational influence in organizations: An integrated approach to knowledge adoption. *Information Systems Research, 14*(1), 47–65. doi:10.1287/isre.14.1.47.14767

Tan, G. W. H., Lee, V. H., Lin, B., & Ooi, K. B. (2017). Mobile applications in tourism: The future of the tourism industry? *Industrial Management & Data Systems, 117*(3), 560–581. doi:10.1108/IMDS-12-2015-0490

Urry, J., Larsen, J., & Axhausen, K. (2007). Networks and tourism. Mobile social life. *Annals of Tourism Research, 34*(1), 244–262. doi:10.1016/j.annals.2006.08.002

Wilson, A., Murphy, H., & Fierro, J. (2012). Hospitality and travel: The nature and implications of user-generated content. *Cornell Hospitality Quarterly, 53*(3), 220–228. doi:10.1177/1938965512449317

Xiang, Z., & Gretzel, U. (2010). Role of social media in online travel information search. *Tourism Management, 31*(2), 179–188. doi:10.1016/j.tourman.2009.02.016

Xiang, Z., Magnini, V., & Fesenmeier, D. (2014). Information technology and consumer behavior in travel and tourism: Insights from travel planning using the internet. *Journal of Retailing and Consumer Services, 22*, 244–249. doi:10.1016/j.jretconser.2014.08.005

KEY TERMS AND DEFINITIONS

Augmented Reality: It is an application that allows the integration of the reality captured by a camera with virtual contents.

e-WOM: Word-of-mouth made online through the opinions and feedbacks from customers related to the service or product.

Mobile Apps: A software specifically built for mobile devices that can improve the use of mobile services.

Mobile Device: A device characterized primarily by its portability, wireless communications and its ability to perform small tasks. Examples being smartphones, tablets and reading devices.

QR Code: These codes are read from the user's mobile devices and can be used to store texts, hyperlinks, among other types of information, in various characters like numbers, letters, symbols and binary numbers. QR codes are placed on the tours so that, from its reading, it is easier to know the cities and also in several museums this technology is used to facilitate access to all information.

Usefulness / Ease of Use: The perception about the utility and difficulty on the use of a technology.

User Generated Content: Contents generated by online users with information about experiences, such as photographs, texts and opinions, habits and purchases.

Chapter 14
An Evaluation for Long-Haul Low-Cost Carriers Using User-Generated Content:
The Impact of Perceived Service Quality on Value for Money

Mahmut Bakır
https://orcid.org/0000-0002-3898-4987
Anadolu University, Turkey

Sahap Akan
https://orcid.org/0000-0002-4633-7683
Anadolu University, Turkey

Ozlem Atalik
Eskisehir Technical University, Turkey

ABSTRACT

Since the liberalization of the airline industry, the low-cost business model has been developed worldwide and a new business model of long-haul low-cost carriers (LHLCCs) has evolved. This chapter aims to investigate the LHLCC business model from a customer-oriented perspective in terms of service quality and perceived value. For this purpose, the authors investigated the effect of service quality on perceived value for money for LHLCCs. In this chapter, user-generated content was adopted to collect data, and 824 user-generated airline reviews were collected from TripAdvisor.com, the largest tourism-related repository. In order to investigate the relationship, a predictive correlational design was structured and a logistic regression analysis was applied. To contribute to the regression analysis, a receiver operating characteristic (ROC) analysis was performed to measure the classification success. As a result, the logit model describes well the relationship between variables for LHLCCs.

DOI: 10.4018/978-1-7998-1947-9.ch014

INTRODUCTION

The airline industry has undergone many changes and created different business models to meet the needs of passengers. While these business models aim to satisfy consumers, they are also the key to differentiating airlines from the competitors (Wensveen & Leick, 2009). One of these business models is the long-haul low-cost carrier (LHLCC) that emerges from cost leadership strategy (Francis, Dennis, Ison, & Humphreys, 2007). As a rule of thumb, this business model has evolved from the LCC (Low Cost Carrier) business model and has a more sophisticated structure (Wensveen & Leick, 2009). The airlines, which adopt a point-to-point network structure, have also adopted different cabin classes, like full service carriers (FSCs). Contrary to LCCs, however, seat comfort is more critical in this business model due to long-haul operations. The aircraft utilization rate is very high, since flights can take up to 5-6 hours (Francis et al., 2007). Although the single fleet type is used in terms of fleet structure, LHLCCs are not conservative in the use of different aircraft according to their needs. In operations, hub airports are preferred to secondary airports (Morrell, 2008). Similar to FSCs, frequent flyer programs are an important factor in ensuring loyalty in this business model. Since catering services are an important need for long-haul flights, LHLCCs also focus on this attribute. Finally, air cargo has become a pivotal item in this business model to increase ancillary revenues (Francis et al., 2007). The comparison of the LHLCC business model with short-haul LCC operations is detailed in Table 1.

Historically, Long-Haul Low-Cost business model was first adopted by Loftleidir-Icelandic (1944-1979) (Gudmundsson, 2019). Loftleidir started operations from Iceland to the other side of the Atlantic (Hunt & Truong, 2019; Jiang & Zhang, 2016). Loftleidir conducted its operations within the scope of the 6th Freedom Right which allows airlines to fly between two foreign countries via the home state of the carrier. Indeed, the reason why we accept Loftleidir as the first carrier is that it adopted the low fare

Table 1. A quick comparison between short-haul and long-haul LCC operations

Attributes	Short-haul LCC Operations	Long-haul LCC Operations
Aircraft type	Regional jet or typical aircraft with high density seating (Boeing737 or Airbus A320)	Long range, aircrafts require lower operational costs (Airbus A330-300, Boeing 787-900, Airbus A350-900)
Seat Capacity	Approx. 160-189 seat capacity	250-300 seat capacity
Cabin Class	Single class (economy)	Multiple class (generally "pseudo" business class and economy)
Airport	Secondary	Needs major airports but do not require airbridge etc. services.
Cargo	No	An important source of revenue
Connection	Point-to-point services and no baggage transfer	Point-to-point services and no baggage transfer
Distribution	Online channels and direct distribution	Online channels and direct distribution
Frequency	High	Lower because of the length of flight time
Seating	Small pitch	Comfort is more important
Catering	On-board sales	Passengers are likely to pay more attention to this
In-flight entertainment	Charges extra	Passengers are likely to pay more attention to this
Turnaround time	Important	Less important because of the length of flight time

Source: *(Francis, Dennis, Ison, & Humphreys, 2007; Whyte & Lohmann, 2015)*

concept with Canadair CL-44, which was converted from a cargo aircraft to a passenger aircraft on the North-Atlantic route. Another noteworthy carrier was Laker Airways (1966-1982), an international UK carrier (Gudmundsson, 2019). In 1977, Laker Airways pioneered low fares between Gatwick Airport and New York John F. Kennedy International Airport under the "Skytrain" sub-brand. While Skytrain was very successful in the early years, it went bankrupt in the early 1980s due to Laker Airways' managerial mistakes. Similarly, PEOPLExpress (1981-1987) offered low-cost flights with a Boeing 747 aircraft between New York and London and then Brussels in the 1980s. Eventually, People Express stopped these operations due to a number of managerial problems (Perry & Williams, 2015).

Although there were different initiatives such as Zoom Airways and Oasis Airlines in the follow-up process, the desired success was not achieved. The main reasons behind this were mismanagement and the lack of the technology required by the LHLCCs business model (Morrell, 2008; Whyte & Lohmann, 2015). As of 2000, the Long-Haul Low-Cost business model has entered a phase of revival and has been re-implemented in the Asia-Pacific region. This adventure, which started with Jetstar in 2006 and Air Asia X in 2007, has gradually progressed. As of 2015, airlines such as Westjet and Norwegian began to operate in the North Atlantic market and the number of airlines operating in this business model increased over the years (Soyk, Ringbeck, & Spinler, 2017). Today, a large number of LHLCCs continue flight operations. As Rodriguez and O'Connell stated, carriers such as Norwegian, AirAsia X and Jetstar currently offer more than 100 long-haul flights per week and at least one low-cost flight per week is available at more than 70 airports worldwide (Rodríguez & O'Connell, 2017). Moreover, they are expected to increase their market share in existing markets in the near future (CAPA, 2018a).

Today, LHLCCs strive to survive in an intensely competitive environment. At this point, both LHLCCs and FSCs are trying to attract passengers. One of the conditions by which airlines can attract passengers and ensure customer satisfaction is to create higher customer value. While FSCs offer value by providing a superior service, LCCs and LHLCCs adopt a lower price strategy through value for money (Rajaguru, 2016). Additionally, another important factor for the survival of airlines is service quality. This phenomenon, also referred to as an antecedent of value for money (Šebjan, Tominc, & Širec, 2017), has been found to strengthen customer relationship management and the competitive advantage of airlines (Brochado, Rita, Oliveira, & Oliveira, 2019). Therefore, service quality and value for money are noteworthy concepts for airlines. However, when the literature is considered, it is seen that this relationship was discussed in a rare study (Šebjan et al., 2017) and not in the case of up-and-coming LHLCCs. To address these issues, this research aims to analyze the impact of service quality on perceived value for money in LHLCCs. Therefore, this research contributes to the existing literature in different ways. First, the relationship between service quality and value for money in LHLCCs will be better understood. It will also be understood which airline service attributes more motivate consumers to use the LHLCCs. Finally, it will be tested how the proposed logit model can be adapted to the research problem.

Although the survey method is frequently used to assess service quality and satisfaction in the airline industry, the advantages of using User-Generated Content (UGC) in recent years are noticeable (Brochado et al., 2019). UGC refers to the information that individuals with similar interests come together and produce in a virtual way. Studies have also found that UGC platforms (e.g. TripAdvisor.com) are one of the decisive sources that consumers use before purchasing tourism and travel services (Ayeh, Au, & Law, 2013). Therefore, this research reflects the UGC perspective as these social media platforms, which are important conversation tools between consumers and companies in the airline industry, are convenient for the purpose of the research.

The research process description is structured as follows. Section 2 reviews the background framework and hypotheses. Section 3 introduces the data collection process and the proposed methodology of logistic regression analysis and the ROC analysis. Section 4 includes an empirical study of LHLCC passengers. Finally, recommendations and future research directions are shared with readers and the results are discussed in section 5.

BACKGROUND

In the 1980s, the role of the service industry in the overall economy began to increase. With the increasing competitiveness along with liberalization and globalization, service quality became the focus of both scholars and practitioners (Foxall, 2014). Service quality can generally be defined as the comparison of customer expectations and perceptions about services (Parasuraman, Zeithaml, & Berry, 1985). If the perceived performance exceeds the expected performance as a result of comparison, customer satisfaction occurs. On the other hand, there is dissatisfaction if the perceived quality does not match the expected quality. In short, service quality is the unique result of fulfilling consumer demands and expectations (Okumus & Duygun, 2008). As with any service industry, service quality has come to the forefront of the airline industry and has been addressed in many ways. When the existing literature is considered, it is seen that many of the studies have investigated the attributes of airline service quality, analyzed airline service quality through some methods such as Multi-Criteria Decision Making (MCDM) and investigated the relationship between service quality and different variables such as customer satisfaction, and loyalty (Angelova & Zekiri, 2011; Chou, Liu, Huang, Yih, & Han, 2011; Baker, 2013).

Perceived value is another concept which is related to service quality and has various similarities. Zeithaml (1988) defined value as "the consumer's overall assessment of the utility of a product based on the perceptions of what is received and what is given". In terms of similarity, perceived value and service quality concepts are the basis of gaps. In other words, service quality consists of the gap between expected and received performance, while perceived value is the gap between benefit and cost. Additionally, service quality is an important driver of perceived customer value (Hu, Kandampully, & Juwaheer, 2009). Perceived value is a concept developed from two main perspectives: the utilitarian (the value associated with money) and the psychological perception (the value associated with the emotional dimension). The perceived value in the utilitarian perspective implies value for money (Rajaguru, 2016). Value for money is a trade-off between received and given in monetary terms. In recent years, value for money has become one of the key concepts that has attracted the attention of scholars in the airline industry. Moreover, since value for money is a predictor of customer satisfaction (Rajaguru, 2016; Maulisa & Hijrah Hati, 2018) and intention to recommend (Šebjan et al., 2017), the studies seek to provide a deep insight. Therefore, in order to provide a deep insight on LHLCCs, the hypotheses proposed in this research are as follows.

In-flight entertainment systems ensure passengers have a more comfortable and stress-free flight experience, being also an important element of differentiation for airlines and is considered a way of achieving a competitive advantage (Liu, 2007). In-flight entertainment is a must, especially on long-haul flights, as passengers need more comfort (Rodríguez & O'Connell, 2017). As LHLCCs undertake long-haul flights, a practice to enhance the perceived value of passengers in this business model is to expand the in-flight entertainment possibilities (Francis et al., 2007). Based on the above literature, the following research hypothesis was proposed:

H1: In-flight entertainment quality has a significant impact on an LHLCCs passengers' perception of value for money.

Seat comfort is a physical attribute that passengers considerably care. Although consumer needs and expectations may change, seat comfort is seen as a must, and consumers want to travel on comfortable seats (Gupta, 2018). Moreover, seat comfort plays a pivotal role in satisfaction as consumers spend a long time in one seat during long-haul flights (Šebjan et al., 2017). Lim and Tkaczynski (2017) also found in their comprehensive meta-analysis that seat comfort is one of the most important airline service attributes. Therefore, one of the more evident expectations of the consumers who pay the flight is a high seat comfort in return. In other words, comfortable and ergonomically designed seats are expected (Gupta, 2018). Consistent with previous literature, the following hypothesis was formulated:

H2: Seat comfort has a significant impact on an LHLCCs passengers' perception of value for money.

Among the airline tangibles, the quality of in-flight food and beverages is seen as an important attribute of service quality. Indeed, studies have shown that a satisfactory in-flight food and beverages service plays a critical role in the selection of FSCs (Han, Lee, Chua, Lee, & Kim, 2019). Since consumers are keen on quality foods, airlines employ top chefs to create rich food menus for their passengers. (Messner, 2016). Gupta (2018) emphasized that airlines should improve food and beverage quality, especially on long-haul flights. In their study examining the effect of in-flight food quality on perceived value and loyalty, Lee and Ko (2016) found that in-flight meal quality has a major impact on consumers' perceptions of in-flight service quality. Similarly, Han et al. (2019) found that in-flight food and beverage quality affects perceived reasonableness of price, airline image, and satisfaction. Based on the above literature, the proposed hypothesis is the following:

H3: Food and beverage service quality has a significant impact on an LHLCCs passengers' perception of value for money.

As a service business, airlines need several physical environments to serve their passengers (Barber & Scarcelli, 2010). Some factors related to these physical environments have a significant impact on service quality that passengers received; cleanliness stands out among these factors (Yeoh & Chan, 2011). Kankaew (2013) argued that many factors are important in airline selection and that the cleanliness of the cabin affects the preferences of passengers. Atalik and Arslan (2009), in their study focused on customer loyalty in airlines, stated that the cleanliness of the aircraft cabin is important for customers. Yeoh and Chan (2011) also found that one of the most important basic service attributes is interior cabin cleanliness. Consistent with previous literature, the following hypothesis was formulated:

H4: Cleanliness has a significant impact on an LHLCCs passengers' perception of value for money.

In the airline industry, the expectations of passengers are shaped by the moments of truth in which they make direct contact with employees such as the reservations department, airline sales, ticket sales, cabin crew, cabin services, among others (Nejati, Nejati, & Shafaei, 2009). Existing literature suggests that customer service has a positive impact on many service processes, and this leads to long-term customer loyalty (Tsaur, Chang, & Yen, 2002). For example, Chen and Quester (2006) revealed that services provided by airline personnel have a direct effect on the perception of service quality. Moreover, Liou et al. (2011) found that cabin services are the most important service attribute in airlines. Based on the above literature, the following hypothesis was formulated:

H5: Customer Service has a significant impact on an LHLCCs passengers' perception of value for money.

Airlines apply a number of strategies to improve service quality and perceived value. In this direction, one of the critical areas that airlines need to develop strategies on is to make more efficient check-in and boarding processes. Indeed, it has been argued that improving these processes, on which consumers spend a long time, may increase the image of the airline (Chikwendu, Ejem, & Ezenwa, 2012). From the passenger's perspective, this process also affects the perceived service quality, since it includes aspects of the waiting queue and complex tasks to complete during the check-in and boarding process (Wittmer, 2011). In fact, Chumpitaz and Swaen (2002) revealed that the quality of the check in and boarding service affects consumer satisfaction. Similarly, Liou et al. (2011) found that the check-in and boarding processes increase customer satisfaction. Consistent with previous literature, the formulated hypothesis is the following:

H6: Check-in and boarding quality have a significant impact on an LHLCCs passengers' perception of value for money.

METHODOLOGY

In this research, the aim was to empirically test the formulated hypotheses. Therefore, quantitative research methods were adopted. In this context, we used the correlational research design. Correlational research is a research design that allows the examination of the relationship or correlation level between two or more variables (Marczyk, DeMatteo, & Festinger, 2005). Therefore, the dependent variable is explained on the bases of predictor variables and through predictive correlational design. At this point, the variable used in the prediction is called as a predictor variable, and the variable to be predicted is called the criterion/dependent variable (Fraenkel, Wallen, & Hyun, 2012). In this research, predictor variables are in-flight entertainment, seat comfort, food and beverage, cleanliness, customer service, and check-in and boarding. The dependent variable is value for money (Figure 1). This section covers the proposed model, sampling process and proposed research method.

Conceptual Model

The research problem focuses on the relationship between service quality and value for money in LHLCCs. Therefore, the authors aim to gain a deep insight into how the service quality of airlines affects consumers' value for money perceptions. The post-purchase product evaluations of consumers are an important issue, and many structures such as the expectation-disconfirmation paradigm (EDP) and the Means-End theory (MET) have been proposed in this area (Oliver, 1977; Zeithaml, 1988). EDP suggests that consumers compare their pre-purchase expectations with post-purchase sentiments based on some attributes.

Accordingly, these evaluations lead to satisfaction or dissatisfaction and to direct future purchases. On the other hand, the MET argues that consumers are goal-oriented in nature and that product/service attributes are the means to achieve desired results. Thus, the desired results will strengthen the desired end state of consumers (Gardial, Clemons, Woodruff, Schumann, & Burns, 1994). We were inspired by these theories in the research because the perceptions about quality and value for money of LHLCC passengers might affect their future purchases. The proposed research model is shown given in Figure 1.

Sampling Process

This research covers LHLCCs' passengers. We adopted secondary data sources, frequently used in the studies related to the airline industry. In this context, due to physical and economic constraints, we chose TripAdvisor, the largest tourism-related UGC repository. The TripAdvisor website offers travelers information on tourism-related products, such as hotels, flights, restaurants. Travelers share their experiences with other consumers via this website by creating post-purchase reviews and pictures related to the tourism products they experienced (Berezina, Bilgihan, Cobanoglu, & Okumus, 2016). Created user content is published after approval and reaches millions of travelers through this platform. Another advantage of such content is that it is considered highly credible by consumers and it can change their buying behavior (Park, 2007).

For this research, the LHLCC list proposed by the "Center for Aviation" was used in the selection of airlines that make up the population (CAPA, 2018b). From this list, only 17 LHLCCs were included due to the lack of some data. In the data collection process, systematic sampling method, which is one of the random sampling methods, was adopted in the stratified structure (Hair, Wolfinbarger, Ortinau, & Bush, 2010). The sampling interval was 40. In the research, only the reviews that were written between March

Figure 1. Proposed conceptual model

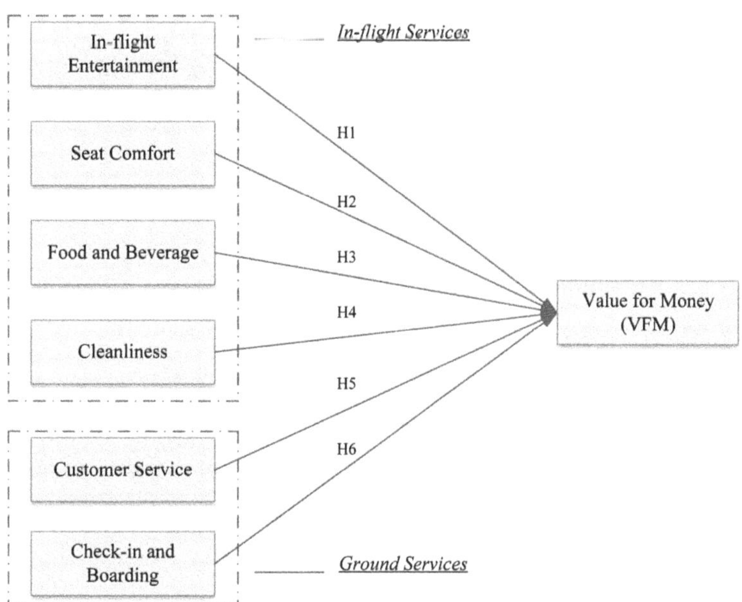

2018 and May 2019 were taken into consideration, and a total of 824 user-generated airline reviews were collected. Our sample contains numeric reviews written in all languages. The predictive variables were measured on a 5-point Likert scale ranging from "1=low quality" to "5=high quality", and the dependent variable was recoded as a dichotomous (0=unsatisfied, 1=satisfied). Detailed information on sampling is given in the Appendix.

An Introduction to the Logistic Regression Analysis

Logistic regression is used to estimate group membership in models containing two or more grouping variables. If the grouping variable is only two categories (dichotomous), binary logistic regression is employed, while multinomial (or polychotomous) logistic regression analysis is employed when there are more than two categories (Brace, Kemp, & Snelgar, 2016). Logistic regression analysis is similar to discriminant analysis and multiple regression analysis according to the study´s rationality, but it offers a number of unique advantages. In contrast to the other two methods, it is not a problem that predictor variables violate the normality assumption in logistic regression. Similarly, the linear relationship between the predictor and the dependent variable, the independence of the errors, and the fact that each group has equal variance are not sought (Field, 2009; Mertler & Reinhart, 2017).

Unlike other regression analyzes, the logistic regression-specific odds ratio indicates the ratio of the probability of occurrence of an event to the probability of not occurring. Let us assume that five runners will compete, in order to clarify the odds ratio specified by Exp (β). If we randomly select one of these runners, the winners' odds will be 0.20 / (1-0.20) = 0.25. This ratio can be any value between 0 and ∞ (Brace et al., 2016). In logistic regression analysis, the goodness of fit of the model is calculated not by using R^2 although by applying different indices. In this context, the overall model fit, labeled as -2 Log likelihood (LL) shows the accuracy of maximum likelihood estimation, while the perfect fit model is zero or close to zero. In addition, using the pseudo R^2, the model's goodness of fit is also calculated. Finally, the significance level of predictors in logistic regression is calculated with Wald Statistics instead of t-test (Mertler & Reinhart, 2017).

The logistic regression equation (or function) is shown as follows (Field, 2009; Aljandali, 2017):

$$P = \frac{1}{1 + \exp - \left(\beta_0 + \beta_1 X\right)} \tag{1}$$

Where P represents the probability of occurrence of the event and

$$\exp - \left(\beta_0 + \beta_1 X\right) = e^{\left(\beta_0 + \beta_1 X\right)}.$$

On the other hand, the value of $\left(1 - P\right)$. which indicates the probability of not occurring, is shown as follows:

$$1 - P = 1 - \frac{1}{1 + \exp - \left(\beta_0 + \beta_1 X\right)} = \frac{1 + \exp - \left(\beta_0 + \beta_1 X\right) - 1}{1 + \exp - \left(\beta_0 + \beta_1 X\right)} = \ldots = \frac{1}{1 + \exp\left(\beta_0 + \beta_1 X\right)} \quad (2)$$

Then, we use Equality (1) and Equality (2):

$$\frac{P}{1 - P} = \frac{1}{1 + \exp - \left(\beta_0 + \beta_1 X\right)} \times \left[1 + \exp - \left(\beta_0 + \beta_1 X\right)\right] = \ldots = \exp\left(\beta_0 + \beta_1 X\right) \quad (3)$$

Then, we take the natural logarithm of the dependent variable:

$$\ln\left(\frac{P}{1 - P}\right) = \ln\left[\exp\left(\beta_0 + \beta_1 X\right)\right] \rightarrow \ln\left(\frac{P}{1 - p}\right) = \left(\beta_0 + \beta_1 X\right) \quad (4)$$

In Equation (4), the left side of the equation shows only the logit, while the whole equation creates the logit model. The logit model can also be expanded for multivariate scenarios as follows.

$$\ln\left(\frac{P}{1 - p}\right) = \beta_0 + \beta_1 X_1 + \beta_2 X_2 + \ldots + \beta_n X_n \quad (5)$$

ANALYSIS AND RESULTS

In this section, the application for LHLCCs and the results are presented. For this purpose, the causal relationship between the variables was investigated by the logistic regression analysis and the success of the separation was tested using the ROC analysis. IBM SPSS Statistics 25 and Jamovi statistical software were employed in the analysis process. A preliminary analysis was first performed to test the logistic regression assumptions (Field, 2009). Firstly, the existence of multicollinearity and outliers were checked.

In this context, Tabachnich and Fidell (2013)'s instructions were followed in the control of outliers. Then, univariate outliers were checked based on a threshold of $\pm$ 3.29 ($p <.001$). Multivariate outliers were also examined by Mahalanobis distance (D2) test and no problems were found. Multicollinearity was examined by checking the Variance Inflated Factor (VIF) and the VIF values of the variables were below the threshold. Secondly, the sample size requirement was checked. Since the logistic regression analysis is suitable for larger sample size, the number of observations is adequate ($n = 824$ and no missing data). In the data analysis, the forced entry method was used.

In this research, 824 observations of passengers that had used LHLCCs were analysed. At this point, the research does not include any descriptive demographic questions. Table 2 presents descriptive statistics and correlation matrix. In the table, it is seen that the attribute that consumers assign the highest ratings to is cleanliness (M = 3.77, SD = 1.119), followed by check-in and boarding (M= 3.59, SD=1.370), customer services (M= 3.54, SD=1.367), seat comfort (M= 3.26, SD=1.104), and food and beverages

Table 2. Descriptive and correlational matrix

Predictors	Mean	Std. Dv.	1	2	3	4	5	6
1 - In-flight Entertainment	2.58	1.443	1					
2 - Check-in and boarding	3.59	1.370	.508*	1				
3 - Seat comfort	3.26	1.104	.517*	.552*	1			
4 - Customer service	3.54	1.367	.546*	.715*	.610*	1		
5 - Cleanliness	3.77	1.119	.474*	.627*	.588*	.647*	1	
6 - Food and Beverage	2.83	1.281	.648*	.559*	.581*	.659*	.520*	1
Value for money	0.538	0.499						

*Significant at the 0.01 level (2-tailed).
Source: *Calculated by authors*

(M= 2.83, SD=1.281), respectively. On the other hand, the lowest rated predictor is in-flight entertainment (M = 2.58, SD = 1.443). To sum up, we can conclude that passengers' perceptions of airline services were considerably positive. For the dependent variable, 53.8% of the passengers were satisfied with the value for money. In addition, all the predictors were significantly correlated with each other.

After the descriptive statistics, a logistic regression analysis was applied. The logit model fit test results of the analysis are given in Table 3. Accordingly, the overall six predictors model is statistically compatible to distinguish between satisfied and dissatisfied passengers (-2 Log Likelihood = 639.655, $\chi^2(6) = 497.982$, $p< .001$). In addition, when the pseudo R^2 is examined, it is reported that the model explains 60.6% of the value for money (Nagelkerke's R^2= 60.6). For the Hosmer-Lemeshow goodness-of-fit statistics represented, the model fits the data ($\chi^2(8)$= 8.194, p= 0.415> 0.05).

Logistic regression analysis provides an indicator of how well the model can predict the correct category for each observation. In the case of this research, the categories are "satisfied" and "unsatisfied". Then, the dependent variable was measured on a scale ranging from "1-poor" to "5-excellent". However, in order to allow for a binary assessment, scores 3 and below were coded as "0-unsatisfied" and above as "1-satisfied". When the classification results in Table 4 is examined, it is observed that the model correctly classified 87.6% of the cases. The overall success rate of the model is also 84.3%.

In the next step, logistic model coefficients, Wald statistics and odds ratios are presented in Table 5. When the table is examined, five of the predictor variables contribute significantly to the model. In

Table 3. Goodness-of-fit tests

	-2 Log Likelihood	R^2_{McF}	R^2_{CS}	R^2_N	Overall Model Test		
Step 0	1137.637				χ^2	df	Sig.
Step 1	639.655	0.438	0.454	0.606	497.982	6	< .001
Hosmer & Lemeshow Test		χ^2	df	Sig.			
		8.194	8	0.415			

Source: *Calculated by authors*

Table 4. Classification table

	Predicted			
Observed	**0**	**1**	**Total**	**% Correct**
0	307	74	**381**	80.6
1	55	388	**443**	87.6
Overall Percentage				84.3

Note. The cut value is set to .50. "0" denotes unsatisfied and "1" denotes satisfied consumers.
Source: *Calculated by authors*

Table 5. Logistic regression results

Predictors	**β**	**S.E.**	**Wald**	**Sig.**	**Exp(β)**	**Δ Odds**	**95% Confidence Interval**	
							L. L.	**U. L.**
In-flight Entertainment	-0.027	0.093	0.082	0.775	0.974	-0.26%	0.811	1.169
Check-in and boarding	0.372	0.102	13.213	< .001	1.450	+45.0%	1.187	1.772
Seat comfort	0.431	0.126	11.741	0.001	1.539	+53.9%	1.203	1.969
Customer service	0.469	0.116	16.398	< .001	1.598	+59.8%	1.274	2.005
Cleanliness	0.719	0.132	29.709	< .001	2.053	+105.3%	1.585	2.659
Food and Beverage	0.397	0.119	11.133	0.001	1.487	+48.7%	1.178	1.877
Constant	-8.041	0.584	189.477	< .001	0			

Note. Estimates denote the log odds of "VFM=1" vs. "VFM=0".
Source: *Calculated by authors*

this context, cleanliness (Wald's χ^2 = 29.709, p =< .001) is the strongest predictor of value for money [Exp(β) = 2.053; 95% CI, 1.585-2.659]. In addition, customer service (Wald's χ^2=16.398, p=< .001), seat comfort (Wald's χ^2 = 11.741, p = .001), food and beverages (Wald's χ^2=11.133, p= .001) and check-in and boarding (Wald's χ^2 = 13.213, p=< .001) are significantly associated with value for money. On the other hands, in-flight entertainment (Wald's χ^2=0.082, p=0.775) is found not to be significantly associated with value for money. Hence, it can be concluded that H1 is rejected, and H2, H3, H4, H5 and H6 are supported.

After examining the effect of predictors on value for money, the ROC analysis is also conducted. As is well known, the success of the model in the logistic regression analysis is associated with several indicators. One of them is the Hit Rate Method, which measures how successfully the predictor variables classify observations into dependent variable categories. This method proposes two different terms, called sensitivity and specificity. Accordingly, sensitivity implies a hit rate that includes a successful classification of observations into the first category, and a specificity to the other category (Huck, 2012). In other words, sensitivity refers to the number of true positives while specificity is the number of true negatives (Peacock, Kerry, & Balise, 2017). To calculate on the basis of Table 4, true positives/total positives= 388/443= 87.6 and true negatives/total negatives= 307/381= 80.6. The ROC curve allows us to visualize sensitivity and specificity. Moreover, the success of classification in the ROC curve is measured by the Area Under the Curve (AUC). Figure 2 shows the ROC curve of the analysis while AUC values of the predictors are given in Table 6 (Sing, Sander, Beerenwinkel, & Lengauer, 2015). The AUC values of the predictors were higher than 0.750 ($p < .001$), indicating that our model was practically and statistically significant. The model's sensitivity is also 87.6% and its specificity 80.6%.

Figure 2. ROC Curve for prediction

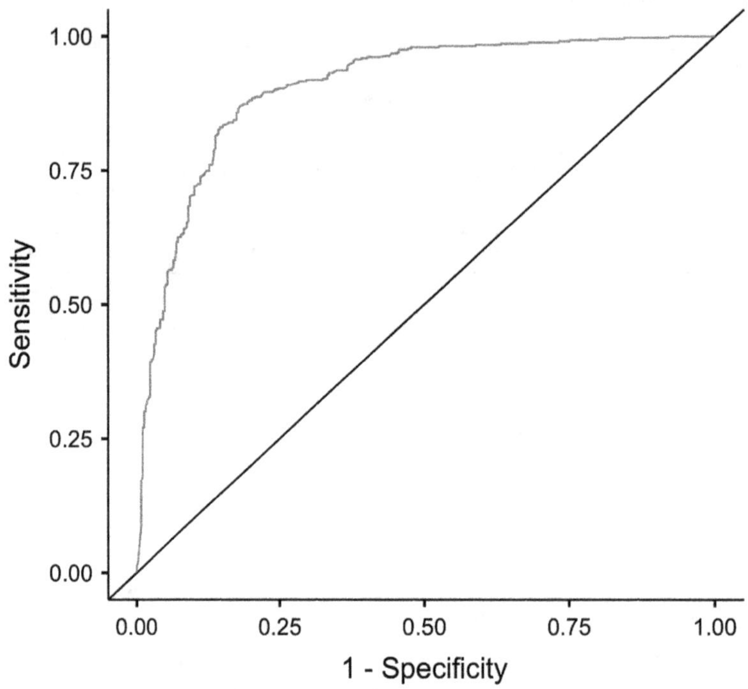

Table 6. Separation success indicators

Predictors	Area	S.E.	Sig.	95% Confidence Interval		Sensitivity	Specificity
				L. Bound	U. Bound		
In-flight Entertainment	0.754	0.017	< .001	0.721	0.787	0.876	0.806
Check-in and boarding	0.824	0.015	< .001	0.795	0.853		
Seat comfort	0.801	0.015	< .001	0.771	0.831		
Customer service	0.845	0.014	< .001	0.818	0.872		
Cleanliness	0.827	0.014	< .001	0.799	0.855		
Food and Beverage	0.808	0.015	< .001	0.778	0.838		

Note. The cut-off value is set to 0.50.

SOLUTIONS AND RECOMMENDATIONS

This research focused on the LHLCC business model, the relationship between service quality and perceived value for money was investigated via User-Generated Content. As a result, service quality attributes were to be associated with value for money that reflects the post-purchase evaluation of consumers. Therefore, consistent with previous studies (Chen & Liu, 2017; Šebjan et al., 2017), the findings support the hypotheses H2, H3, H4, H5 and H6. When the findings are examined, the predictor that explains better value for money is cleanliness. Accordingly, when holding other variables constant, if the level of cleanliness is increased by one unit, the probability of being satisfied with the value for money is about 105.3% higher [Exp (β) = 2.053; 95% CI, 1.585-2.659]. This finding confirms previous findings in the literature (Tsaur et al., 2002; Liou & Tzeng, 2007). Correspondingly, Chou et al. (2011) found that comfort and seat cleanliness are increasingly important service items in recent years.

The second most important determinant of service is customer service. The service staff, which is an important factor in the value co-creation process, has come to the fore in several studies, consistent with our findings (Liou & Tzeng, 2007; Liou et al., 2011). Liou and Tzeng (2007) argued that personnel are critical in managing customer complaints, and that good cabin services are compensating any problem. Similarly, Liou et al. (2011) found that cabin services are the most important service attribute. Considering that the LHLCC business model offers other services at a basic level, this finding is not surprising. Other determinants that affect value for money are seat comfort, food and beverage service and check-in and boarding processes. Seat comfort is an important item that was reported in past studies (Tsaur et al., 2002; Chou et al., 2011). Since LHLCCs are the long-haul version of the LCC business model, one of the main discomforts in long-haul flights is seat comfort and seat pitch (Camilleri, 2018). Therefore, although the business model characteristics force LHLCCs to reduce CASM (Cost per Available Seat Mile), this factor should not be ignored. Just as the quality of food and beverages, Messner (2016) found that the quality of in-flight food is a pivotal attribute in the airlines' service environment. Similarly, Chou et al. (2011) emphasized that the food and beverages factor is an important determinant of service quality. Additionally, Perçin (2018) emphasized the need to address check-in and also baggage delivery problems in order to increase customer satisfaction by preventing customer complaints. In other words, when ticketing, boarding and check-in processes go smoothly, satisfaction increases (Liou et al., 2011). Finally, it is surprisingly reported that in-flight entertainment has not a significant impact on value for money. At this point, airlines can make some arrangements to the in-flight entertainment tools without damaging consumers' perceptions of service quality. Indeed, airlines have implemented more sophisticated applications in recent years. For example, Etihad Airways recently announced the launch of wireless streaming by removing entertainment screens from their narrow-body fleet (Reuters, 2019).

FUTURE RESEARCH DIRECTIONS

In this research, the effect of service quality on value for money perception of consumers in LHLCCs was investigated. In order to carry out the research purpose, online customer-generated airline reviews were adopted, and a logistic regression analysis was conducted. Kotler et al. (2017) stated that consumers are increasingly inspired by f-factors (friends, families, fans, followers) in response to marketing communication bombardment. In other words, despite intensive marketing communication, consumers do not find these tools reliable. Instead, they are influenced by the ideas of their friends/families etc. In this respect, the growth of the f-factor beliefs increases the importance of UGC platforms such as TripAdvisor.com. Therefore, such platforms will attract more consumers and researchers in the future. Similarly, for this business model, it is also necessary to gain a deep understanding of satisfaction, purchase intention, and similar variables in addition to service quality. Finally, it is recommended that future studies adopt the fuzzy sets theory in service quality measurements because of the uncertainty of consumer judgment.

CONCLUSION

The customer-oriented approach requires high-quality service delivery to resist competitive pressure. In this respect, it is very important for airlines to identify and prioritize key service quality factors (Jiang & Zhang, 2016). Moreover, the main advantage of such a prioritization allows the airlines to use their scarce resources most effectively (Kurtulmuşoğlu, Can, & Tolon, 2016). It should be noted that all these actions should be based on the perceptions of consumers. At this point, one of the theories that can be related is the Means-End theory. The Means-End approach addresses the retention level of service information in memory. Accordingly, while service attributes affect service quality, service quality also affects service value, that addresses a cognitive trade-off between received and given (Chou et al., 2011). In this respect, the relationship between service quality and perceived value is an attracting and promising issue for researchers. Therefore, in this research, 17 LHLCCs were examined, based on the service quality-value for money paradigm. To the best of our knowledge, this research is the first study investigating value for money in LHLCCs.

The proposed logit model is also presented as a track that will help airline managers to provide satisfactory service. Thus, cognitive trade-offs can be enhanced by providing high service quality to passengers who pay money for travelling. Therefore, it can be said that this research presents the voice of customer as a key to a sustainable and competitive advantage for airlines. From a theoretical standpoint, the key contribution of this research is to investigate the service quality-perceived value for money paradigm for the first time in the LHLCC business model literature. In addition, service quality factors are prioritized in LHLCCs through the logistic regression.

As with any study, it should be noted that this research has some limitations. Firstly, the data used are based on the assumption that they are reliable and unsuspicious since they are generated voluntarily and TripAdvisor.com has zero tolerance for fake comments. The platform requests reviewers to provide information such as destination, service class and flight number to maintain this policy. Secondly, the data only covers the research period, and consumer perceptions represent only airlines that adopt the LHLCC business model. Thirdly, since the research data contains secondary data, only the evaluation criteria proposed by TripAdvisor.com was examined. In addition, the research does not include any demographic questions because TripAdvisor.com platform does not share information such as age and income of the participants. Moreover, the data are not generalized to other sectors due to the structure of the data and the characteristics of the airline sector. Finally, since the study is a first, the authors hope that the findings will be replicated in further studies.

ACKNOWLEDGMENT

This research received no specific grant from any funding agency in the public, commercial, or not-for-profit sectors.

REFERENCES

Aljandali, A. (2017). *Multivariate Methods and Forecasting with IBM SPSS Statistics*. Cham: Springer International Publishing.

Angelova, B., & Zekiri, J. (2011). Measuring Customer Satisfaction with Service Quality Using American Customer Satisfaction Model (ACSI Model). *International Journal of Academic Research in Business and Social Sciences*, *1*(3), 232–258. doi:10.6007/ijarbss.v1i2.35

Atalık, Ö., & Arslan, M. (2009). A Study to Determine the Effects of Customer Value on Customer Loyalty in Airline Companies Operating: Case of Turkish Air Travellers. *International Journal of Business and Management*, *4*(6), 154–162. doi:10.5539/ijbm.v4n6p154

Ayeh, J. K., Au, N., & Law, R. (2013). "Do we believe in TripAdvisor?" Examining credibility perceptions and online travelers' attitude toward using user-generated content. *Journal of Travel Research*, *52*(4), 437–452. doi:10.1177/0047287512475217

Baker, D. (2013). Service quality and customer satisfaction in the airline industry: A comparison between legacy airlines and low-cost airlines. *American Journal of Tourism Research*, *2*(1), 67–77. doi:10.11634/216837861302317

Barber, N., & Scarcelli, J. M. (2010). Enhancing the assessment of tangible service quality through the creation of a cleanliness measurement scale. *Managing Service Quality: An International Journal*, *20*(1), 70 88. doi:10.1108/09604521011011630

Berezina, K., Bilgihan, A., Cobanoglu, C., & Okumus, F. (2016). Understanding Satisfied and Dissatisfied Hotel Customers: Text Mining of Online Hotel Reviews. *Journal of Hospitality Marketing & Management*, *25*(1), 1–24. doi:10.1080/19368623.2015.983631

Brace, N., Kemp, R., & Snelgar, R. (2016). SPSS for Psychologists (6th ed.). Hampshire: Macmillan International Higher Education.

Brochado, A., Rita, P., Oliveira, C., & Oliveira, F. (2019). Airline passengers' perceptions of service quality: Themes in online reviews. *International Journal of Contemporary Hospitality Management*, *31*(2), 855–873. doi:10.1108/IJCHM-09-2017-0572

Camilleri, M. A. (2018). *Travel marketing, tourism economics and the airline product. An introduction to theory and practice*. Cham: Springer International Publishing.

CAPA. (2018a). *A321neoLR: Ideal for new North Atlantic routes, potentially a global game changer*. Retrieved from https://centreforaviation.com/analysis/airline-leader/a321neolr-ideal-for-new-north-atlantic-routes-potentially-a-global-game-changer-439182

CAPA. (2018b). *Longhaul low cost airlines: World Airways to be US' first.* Retrieved from https://centreforaviation.com/analysis/reports/longhaul-low-cost-airlines-world-airways-to-be-us-first-405559

Chen, C. M., & Liu, H. M. (2017). Exploring the Impact of Airlines Service Quality on Customer Loyalty: Evidence from Taiwan. *International Journal of Business and Management, 12*(5), 36–50. doi:10.5539/ijbm.v12n5p36

Chen, S. C., & Quester, P. G. (2006). Modeling store loyalty: Perceived value in market orientation practice. *Journal of Services Marketing, 20*(3), 188–198. doi:10.1108/08876040610665643

Chikwendu, D., Ejem, E., & Ezenwa, A. (2012). Evaluation of service quality of Nigerian airline using servqual model. *Journal of Hospitality Management and Tourism, 3*(6), 117–125. doi:10.5897/JHMT12.019

Chou, C. C., Liu, L. J., Huang, S. F., Yih, J. M., & Han, T. C. (2011). An evaluation of airline service quality using the fuzzy weighted SERVQUAL method. *Applied Soft Computing, 11*(2), 2117–2128. doi:10.1016/j.asoc.2010.07.010

Chumpitaz, R., & Swaen, V. (2002). Service Quality and Brand Loyalty Relationships. *Proceedings of the 31st Conference of European Marketing Academy*, (pp. 1-7). Academic Press..

Field, A. (2009). Discovering Statistics Using SPSS (3rd ed.). London: SAGE Publications Ltd.

Foxall, G. R. (2014). *Marketing in the Service Industries*. New York: Routledge.

Fraenkel, J. R., Wallen, N. E., & Hyun, H. H. (2012). How to Design and Evaluate Research in Education (8th ed.). New York: Mc-Graw Hill.

Francis, G., Dennis, N., Ison, S., & Humphreys, I. (2007). The transferability of the low-cost model to long-haul airline operations. *Tourism Management, 28*(2), 391–398. doi:10.1016/j.tourman.2006.04.014

Gardial, S. F., Clemons, D. S., Woodruff, R. B., Schumann, D. W., & Burns, M. J. (1994). Comparing consumers' recall of prepurchase and postpurchase product evaluation experiences. *The Journal of Consumer Research, 20*(4), 548–560. doi:10.1086/209369

Gudmundsson, S. (2019). Limits to the Low-Cost Niche? Finding Sustainable Strategies for Low-Cost Long-Haul Airlines. Retrieved from https://www.researchgate.net/profile/Sveinn_Gudmundsson3/publication/278966773_Limits_to_the_low-cost_niche_Sustainable_strategies_for_low-cost_long-haul_airlines/links/5d00a73792851c874c5fcac3/Limits-to-the-low-cost-niche-Sustainable-strategies-for-low-c

Gupta, H. (2018). Evaluating service quality of airline industry using hybrid best worst method and VIKOR. *Journal of Air Transport Management, 68*, 35–47. doi:10.1016/j.jairtraman.2017.06.001

Hair, J. F., Wolfinbarger, M. F., Ortinau, D. J., & Bush, R. P. (2010). Essentials of Marketing Research (2nd ed.). New York: McGraw-Hill/Irwin.

Han, H., Lee, K. S., Chua, B. L., Lee, S., & Kim, W. (2019). Role of airline food quality, price reasonableness, image, satisfaction, and attachment in building re-flying intention. *International Journal of Hospitality Management, 80*, 91–100. doi:10.1016/j.ijhm.2019.01.013

Hu, H. H., Kandampully, J., & Juwaheer, T. D. (2009). Relationships and Impacts of Service Quality, Perceived Value, Customer Satisfaction, and Image: An Empirical Study. *Service Industries Journal, 29*(2), 111–125. doi:10.1080/02642060802292932

Huck, S. W. (2012). Reading Statistics and Research (6th ed.). Boston: Pearson Education.

Hunt, J., & Truong, D. (2019). Low-fare flights across the Atlantic: Impact of low-cost, long-haul trans-Atlantic flights on passenger choice of Carrier. *Journal of Air Transport Management, 75*, 170–184. doi:10.1016/j.jairtraman.2018.12.005

Jiang, H., & Zhang, Y. (2016). An investigation of service quality, customer satisfaction and loyalty in China's airline market. *Journal of Air Transport Management, 57*, 80–88. doi:10.1016/j.jairtraman.2016.07.008

Kankaew, K. (2013). Importance-Performance Analysis in Airlines Service Quality: A Case Study of legacy Airlines in Thailand. *Proceedings of the International Conference on Tourism, Transport and Logistics* (pp. 363-369). Paris: UP Organizer and Publication Co., Ltd.

Kotler, P., Kartajaya, H., & Setiawan, I. (2017). *Marketing 4.0 Moving from Traditional to Digital*. New Jersey: Wiley.

Kurtulmuşoğlu, F. B., Can, G. F., & Tolon, M. (2016). A voice in the skies: Listening to airline passenger preferences. *Journal of Air Transport Management, 57*, 130–137. doi:10.1016/j.jairtraman.2016.07.017

Lee, J., & Ko, S. (2016). Effect of the In-Flight Meal Service Quality on the Customer Value and Loyalty. *Indian Journal of Science and Technology, 9*(26), 1–6. doi:10.17485/ijst/2016/v9i26/97276

Lim, S. S., & Tkaczynski, A. (2017). Origin and money matter: The airline service quality expectations of international students. *Journal of Hospitality and Tourism Management, 31*, 244–252. doi:10.1016/j.jhtm.2017.03.001

Liou, J. J., Hsu, C. C., Yeh, W. C., & Lin, R. H. (2011). Using a modified grey relation method for improving airline service quality. *Tourism Management, 32*(6), 1381–1388. doi:10.1016/j.tourman.2011.01.013

Liou, J. J., & Tzeng, G. H. (2007). A non-additive model for evaluating airline service quality. *Journal of Air Transport Management, 13*(3), 131–138. doi:10.1016/j.jairtraman.2006.12.002

Liu, H. (2007). In-Flight Entertainment System: State of the Art and Research Directions. *Proceedings of the Second International Workshop on Semantic Media Adaptation and Personalization (SMAP 2007)* (pp. 241-244). Uxbridge, UK: IEEE. 10.1109/SMAP.2007.37

Marczyk, G., DeMatteo, D., & Festinger, D. (2005). *Essentials of Research Design and Methodology*. New Jersey: John Wiley & Sons, Inc.

Maulisa, A. R., & Hijrah Hati, S. R. (2018). The Anomaly of Airline Passenger Behavioral Intention: A Comparison Between Low-Cost and Full-Service Airlines. *ASEAN Marketing Journal, 10*(2), 75–90.

Mertler, C. A., & Reinhart, R. V. (2017). Advanced and multivariate statistical methods: Practical application and interpretation (6th ed.). New York: Routledge.

Messner, W. (2016). The impact of an aircraft' s service environment on perceptions of in-flight food quality. *Journal of Air Transport Management*, *53*, 123–130. doi:10.1016/j.jairtraman.2016.02.010

Morrell, P. (2008). Can long-haul low-cost airlines be successful? *Research in Transportation Economics*, *24*(1), 61–67. doi:10.1016/j.retrec.2009.01.003

Nejati, M., Nejati, M., & Shafaei, A. (2009). Ranking airlines' service quality factors using a fuzzy approach: Study of the Iranian society. *International Journal of Quality & Reliability Management*, *36*(3), 247–260. doi:10.1108/02656710910936726

Okumus, A., & Duygun, A. (2008). Service Quality Measurement on Education Services Marketing and Relationship Between Perceived Service Quality and Student Satisfaction. *Anadolu University Journal of Social Sciences*, *8*(2), 17–38.

Oliver, R. L. (1977). Effect of expectation and disconfirmation on postexposure product evaluations: An alternative interpretation. *The Journal of Applied Psychology*, *62*(4), 480–486. doi:10.1037/0021-9010.62.4.480

Parasuraman, A., Zeithaml, V. A., & Berry, L. L. (1985). A Conceptual Model of Service Quality and Its Implications for Future Research. *Journal of Marketing*, *49*(4), 41–50. doi:10.1177/002224298504900403

Park, J. W. (2007). Passenger perceptions of service quality: Korean and Australian case studies. *Journal of Air Transport Management*, *13*(4), 238–242. doi:10.1016/j.jairtraman.2007.04.002

Peacock, J. L., Kerry, S. M., & Balise, R. R. (2017). Presenting Medical Statistics from Proposal to Publication (2nd ed.). Oxford: Oxford University Press.

Perçin, S. (2018). Evaluating airline service quality using a combined fuzzy decision-making approach. *Journal of Air Transport Management*, *68*, 48–60. doi:10.1016/j.jairtraman.2017.07.004

Perry, L., & Williams, C. (2015). *The Viability of Long-Haul Low-Cost Carrier Service*. Leighfisher. Retrieved from http://www.leighfisher.com/sites/default/files/free_files/viabilityoflong-haullow-cost-carrierservice.pdf

Rajaguru, R. (2016). Role of value for money and service quality on behavioural intention: A study of full service and low cost airlines. *Journal of Air Transport Management*, *53*, 114–122. doi:10.1016/j.jairtraman.2016.02.008

Reuters. (2019). *Etihad to offer wireless streaming, remove screens from narrow-body fleet*. Retrieved from https://in.reuters.com/article/gulf-travel-etihad-airways-idINKCN1S40PE

Rodríguez, A. M., & O'Connell, J. F. (2017). Can low-cost long-haul carriers replace Charter airlines in the long-haul market? A European perspective. *Tourism Economics*, *24*(1), 64–78. doi:10.1177/1354816617724017

Šebjan, U., Tominc, P., & Širec, K. (2017). Multiple conceptual modelling of perceived quality of in-flight airline services. *Promet-Traffic & Transportation*, *29*(3), 311–319. doi:10.7307/ptt.v29i3.2195

Sing, T., Sander, O., Beerenwinkel, N., & Lengauer, T. (2015). *ROCR: Visualising the Performance of Scoring Classifiers (R package).* Retrieved from https://cran.r-project.org/package=ROCR

Soyk, C., Ringbeck, J., & Spinler, S. (2017). Long-haul low cost airlines: Characteristics of the business model and sustainability of its cost advantages. *Transportation Research Part A, Policy and Practice*, *106*, 215–234. doi:10.1016/j.tra.2017.09.023

Tabachnick, B. G., & Fidell, L. S. (2013). Using Multivariate Statistics (6th ed.). New Jersey: Pearson Education, Inc.

Tsaur, S. H., Chang, T. Y., & Yen, C. H. (2002). The evaluation of airline service quality by fuzzy MCDM. *Tourism Management*, *23*(2), 107–115. doi:10.1016/S0261-5177(01)00050-4

Wensveen, J. G., & Leick, R. (2009). The long-haul low-cost carrier: A unique business model. *Journal of Air Transport Management*, *15*(3), 127–133. doi:10.1016/j.jairtraman.2008.11.012

Whyte, R., & Lohmann, G. (2015). Low-cost long-haul carriers: A hypothetical analysis of a "Kangaroo route." *Case Studies on Transport Policy*, *3*(2), 159–165. doi:10.1016/j.cstp.2015.01.003

Wittmer, A. (2011). Acceptance of self-service check-in at Zurich airport. *Research in Transportation Business & Management*, *1*(1), 136–143. doi:10.1016/j.rtbm.2011.06.001

Yeoh, E., & Chan, J. K. (2011). Malaysian Low Cost Airlines: Key Influencing Factors on Customers' Repeat Purchase Intention. *World Applied Sciences Journal, 12,* 35-43.

Zeithaml, V. A. (1988). Consumer Perception of Price, Quality, and Value: A Means-End Model and Synthesis of Evidence. *Journal of Marketing*, *52*(3), 2–22. doi:10.1177/002224298805200302

ADDITIONAL READING

Atalık, Ö., Bakır, M., & Akan, Ş. (2019). The Role of In-Flight Service Quality on Value for Money in Business Class: A Logit Model on the Airline Industry. *Administrative Sciences*, *9*(1), 26. doi:10.3390/admsci9010026

Hosmer, D. W., Lemeshow, S., & Sturdivant, R. X. (2013). Applied Logistic Regression (3rd ed.). New Jersey: John Wiley & Sons.

Kashyap, R., & Bojanic, D. C. (2000). A structural analysis of value, quality, and price perceptions of business and leisure travelers. *Journal of Travel Research*, *39*(1), 45–51. doi:10.1177/004728750003900106

Leonnard, S. E. (2018). Perceived service quality, perceived value for money, satisfaction and repurchase intention: An evaluation on private university services. *International Journal of Commerce and Finance*, *4*(1), 40–51.

Waguespack, B. P., & Rhoades, D. L. (2014). Twenty five years of measuring airline service quality or why is airline service quality only good when times are bad? *Research in Transportation Business & Management*, *10*, 33–39. doi:10.1016/j.rtbm.2014.06.001

Woo, M. (2019). Assessing customer citizenship behaviors in the airline industry: Investigation of service quality and value. *Journal of Air Transport Management*, *76*, 40–47. doi:10.1016/j.jairtraman.2019.02.006

KEY TERMS AND DEFINITIONS

Airline Business Model: This is the basic framework that directs how airlines produce value and how they do business.

Customer Oriented Approach: It is the conception that the customer is the key driver in all business activities and that the customer needs and wants are at the core.

Forced Entry Method: It is a widely used variable selection method that allows all predictors to be added to the model at one time.

F-Factor: A reference group that includes friends, families, fans and followers, which is highly influential on consumers' daily decisions and choices.

UGC Repository: These are broad platforms that are mainly associated with crowdsourcing and contain user-generated content.

Seat Pitch: The distance between the seats in the plane.

Systematic Sampling: This is the sampling method that allows the sampling to be taken at fixed intervals in large populations.

Value Co-Creation: It is the economic strategy that supports the joint production process by bringing different parties together to create value-added work.

Voice of Customer: It implies insights reflecting the experiences and expectations of the customers towards the attitude object.

Wireless Streaming: It is the technology that enables traveling information to be delivered to different devices on a server through the wireless network.

LONG-HAUL LOW-COST AIRLINES SAMPLE

Table 7. Long-haul low-cost airlines sample

Airline	Frequency	Percent
Jetstar	113	13.7
Scoot	43	5.2
Azul	130	15.8
Indonesia AirAsiaX	17	2.1
Eurowings	108	13.1
WestJet	138	16.7
Wow Air	70	8.5
Thai AirAsiaX	3	0.4
Lion Air	27	3.3
Beijing Capital	2	0.2
NokScoot	4	0.5
French Bee	17	2.1
Level	31	3.8
Thai Lion Air	16	1.9
Jin Air	5	0.6
Cebu	30	3.6
Air Canada Rouge	70	8.5
Total	824	100.0%

Chapter 15
How Web 3.0 Tourism Students See the 1.0 Higher Education System

José Alexandre Martins
Polytechnic of Guarda, Portugal

Romeu Lopes
Polytechnic of Guarda, Portugal

Vitor Roque
https://orcid.org/0000-0001-6027-3340
Polytechnic of Guarda, Portugal

ABSTRACT

Social media does not seem to easily fit in the higher education system. The Web 3.0 students use social networks on a daily basis, however, 1.0 higher education institutions are not taking advantage of these tools. There are some institutional constraints, pedagogical issues, and cultural resistance for its adoption. A survey directed to Portuguese Higher Education students was held, aiming to identify social networks' use either in a personal and academic context. Results show that social networks are under-exploited in the teaching-learning process, both by students and also their teachers. They are both aware of the platforms and applications but are rarely used for an academic purpose. The study raises important issues and gives practical examples for effective use of social media in education. In order to address these challenges and opportunities wisely, it is necessary an effective engagement in social media by students, professors, and all the academic staff.

DOI: 10.4018/978-1-7998-1947-9.ch015

INTRODUCTION

Tourism is one of the industries that most has been affected and boosted by the development of Information and Communication Technologies (ICTs). The relationship between tourism and technology could be seen as a "win-win" situation. Social media platforms are a good example of it, boosted by the advent of smartphones. New disruptive technologies will arise, with significant impacts on innovation, strategy, management, and competitiveness of the entire sector, which require a different way of teaching.

Social media use for academic purposes is still rare. Pedagogical issues and cultural resistance are some of the obstacles for social media adoption in higher education institutions. Despite an extensive literature regarding the implications of the different technologies in the teaching/learning process, little is known concerning social media use in tourism higher education courses and institutions.

Therefore, the present research intends to explore this gap. Are social media technologies being used properly in tourism higher education courses? This is the main question that supports this research. The paper starts with a review of the current literature regarding tourism and social media. The literature review follows with a discussion surrounding the application of social media technologies in higher education institutions. After presenting the research methodology, we discuss our findings, and present the limitations and future research lines.

TOURISM AND SOCIAL MEDIA

The impact of the world wide web and other ICTs on tourism started to reveal itself in the mid-90s, and completely transformed our daily life, contributing to a growing research effort in the field (Buhalis & Law, 2008; Hjalager, 2013). The role of social media in tourism and hospitality has been a hot topic for research (Leung, Law, van Hoof, & Buhalis, 2013; Xiang & Gretzel, 2010; Zeng & Gerritsen, 2014). Social media, as well as other technologies, enable a more interactive tourism experiences, as well as the co-creation of value by travelers throughout all stages of the process. Nowadays, the social web is "the right spot" to obtain and share information, experiences and preferences among consumers. The online reviews are becoming important and credible information sources for increasingly demanding tourists (Lopes, Abrantes, & Kastenholz, 2013). In terms of tourism destination promotion, the Destination Management Organizations (DMO) are giving a relevant role to social media in their communication activities, putting vital information on these platforms, in order to influence consumer behavior (Roque & Raposo, 2016). Smart tourism is also emerging as a new hot topic for research, given the impact of most recent ICTs over destinations, tourists and business (Koo, Park, & Lee, 2017).

As a matter of fact, ICTs have already provoked a tremendous transformation on tourism, boosted by the advent of smartphones, artificial intelligence, cloud computing or the Internet of Things (Xiang & Fesenmaier, 2017). Nonetheless, according to Buhalis (2019), in the future tourism will be driven by disruptive technologies, that will transform structures, processes and practices, with significant impacts on innovation, strategy, management, and competitiveness of the entire sector. The emergence and impact of those disruptive technologies in all sectors and dimensions of our everyday life, require higher education institutions to look for a different way of teaching (Peres, Moreira, & Mesquita, 2018).

SOCIAL MEDIA TECHNOLOGIES IN HIGHER EDUCATION

Higher education teachers are big users of and believers in social media for their professional careers. This is one of the conclusions of a study carried out in a U.S. higher education institution comprising 1920 questionnaires (Moran, Seaman, & Tinti-Kane, 2011). According to the results, over 90% are using social media not only in their professional careers outside the classroom, but also in courses they are teaching, since they believe that those platforms add value to the teaching process. Nonetheless, some concerns are referred, namely privacy and integrity issues. The study was implemented in 2011 and supported by a learning company that emphasizes the effective use of technology, which may raise some doubts about it.

Apparently, in some European countries, higher education professors do not follow these trends. According to a survey in the context of the higher education subsystems in Portugal (public and private), the majority of the respondents, in a sample of 109 teachers, "say that they use very often the technologies in the classroom" [mostly] "the data projector, the laptop and the computer" (Peres, Moreira, & Mesquita, 2017, p. 4).

Indeed, social media use for academic purposes is still rare and somewhat restricted to some scientific disciplines and curricular units (Manca & Ranieri, 2016a). Institutional constraints, pedagogical issues, and cultural resistance, are some of the obstacles for social media adoption in higher education. According to the Italian academic staff, age and seniority seem to influence the adoption of social media. Moreover, the results show a more favourable attitude in terms of personal sharing and connecting with peers in professional networks, rather than exploring its potential for education purposes (Manca & Ranieri, 2016b).

We have also to bear in mind some concerns related to the private/professional life balance on social media, as well as the addictive characteristic of those social networks (Fuciu, Gorski, & Dumitrescu, 2016). According to Boyd & Ellison (2007, p. 211), online social networks are "web-based services that allow individuals to (1) construct a public or semi-public profile within a bounded system, (2) articulate a list of other users with whom they share a connection, and (3) view and traverse their list of connections and those made by others within the system".

High school students demonstrate positive beliefs and attitudes concerning social media usage in education (Mao, 2014), which is not surprising. "These youngsters have grown up in an immersive computing environment. They come to class armed with smartphones, laptops, and iPods. And this has significant implications for higher education, namely on teaching methodologies" (Peres, Moreira, & Mesquita, 2018, p. 1). Many students, in their wireless and network environments, use different ICTs in the classroom, for parallel tasks, providing a potential problem in terms of learning outcomes (Barry, Murphy, & Drew, 2015). The research performed to determine the frequency of mobile ICT use in the classroom, explores some of the causes for such misuse. According to the students, the main reasons are related to the course content delivery mode, the promotion of passive or active learning, students' nationality (i.e. being domestic or international students), the learning maturity, and the need for supporting course information.

As we know, students depend on social media in their daily lives, not only for social connections and leisure, but also for learning purposes. In certain aspects, the use of these networks, technologies, and devices, leads to better psychosocial and academic outcomes (Barry et al., 2015; Gikas & Grant, 2013; Junco, 2012; Mao, 2014; Moran et al., 2011).

Many authors give evidence for the importance of these tools, as drivers of change for teaching and learning practices. Even Facebook can be used as an effective educational instrument for learning and teaching, with multiple benefits in higher education (Chugh & Ruhi, 2018). According to Sobaih, Moustafa, Ghandforoush, & Khan (2016), social media can provide a set of opportunities to enhance the quality of the learning processes, namely in higher education institutions offering tourism and hotel management degrees.

In this sense, exploring how and why social media technologies are being used in tourism courses can be a fundamental step in order to redesign the classroom activities and prepare students for the job market and challenges ahead. It is crucial to "actively involve those technologies to assist in constructive alignment with the learning outcomes and enhance the student learning experience" (Barry et al., 2015, p. 202). These are the aspects that guide this research and reinforce the relevance and timeliness of the study. The research results and subsequent analyses and reflections may contribute to a better understanding of the thematic.

RESEARCH METHODOLOGY

Are social media technologies being used properly in tourism higher education courses? This is the main question that supports this research, targeted to explore students' perceptions regarding the use of these technologies in the classroom. The main objective is to understand if social media can be used effectively to support the dissemination of content for educational purposes. It is expected to know if there are differences between the use of social media applications in a personal context and its use in an academic context. Moreover, taking into account the market trends, the possibility of students attending e-learning courses is also evaluated.

Beyond other aspects, it is intended to highlight issues such as:

- Time spent on social media (SM).
- Most relevant purposes for SM use.
- Content typology shared on SM platforms.
- Computer software and applications used for academic purposes.
- Contents typology made available by teachers.
- SM applications usage (student vs. teachers).

Moreover, students also indicate what are the most relevant applications to improve the teaching-learning process, according to their point of view.

In order to design the questionnaire properly and accurately, some guidelines and recommendations were followed (Hair, Bush, & Ortinau, 2003; Jennings, 2010; Malhotra, 2010). To avoid ambiguity, the questions were expressed clearly and without technical or sophisticated language. Descriptive words and qualifying phrases were also avoided. Some filter questions were introduced, in order to better characterize the respondents' profile and direct them to the correct section (Foddy, 1996).

The initial questionnaire was composed by thirty-four questions, distributed in three main sections: A- Social Networks´ Use in Personal Context; B- Social Networks' Use in Academic Context; C- Demographic Information. Related subject questions were grouped in the same section and presented in a logical sequence using the funnel strategy (Ghiglione & Matalon, 2001; Oppenheim, 2000), starting with generic aspects, followed by others more specific ones, in order to increase and improve the quality and credibility of the information gathered.

The questionnaire was based on other investigations on the same thematic. For example, in the academic context, we ask which applications students' and teachers' use, according to the classification followed by Peres, Moreira, & Mesquita (2018).

Four experts were invited to assess and validate the questionnaire. First, an analysis in terms of Portuguese language, namely grammar, syntax, adequacy, and logic of the questions (Alumran, Hou, & Hurst, 2012). After some corrections and suggestions, the questionnaire was submitted to content analysis (Yaghmaie, 2003) by three experts: one in the tourism field; another one in computer science; and the third one as an independent expert. As a result, one question was excluded, and three were added. Then, a set of students (n=5) answered the questionnaire (pre-test) for content validation. As a result, there was no need to make any further correction or adjustment.

The final version of the questionnaire is composed by 36 questions, distributed in 3 sections. Figure 1 represents the different phases referred to above.

The questionnaire was built using Google Docs and was available since 20th May 2019, for a 10 days' period. The sample comprises all students enrolled in the Higher School of Tourism and Hospitality (ESTH) of the Polytechnic Institute of Guarda (IPG), in a total of 296 students, distributed by 1 master's degree, 3 undergraduate courses and 1 Higher Professional Technical Course (TeSP), all belonging to the scientific area of Tourism. This research is applied to tourism degrees because the structure of these courses is deficit in terms of curricular units in the area of ICT. According to Roque, Martins, Silva, Rodrigues, & Simões (2010), tourism courses in Portuguese public higher education institutions present only an average of 1,5 curricular units of this scientific area in 2010.

The research follows a non-probabilistic sampling method. This technique implies that the basis for sample selection is a convenience, and the ease to access participants. Convenience samples might offer no guarantee of a representative and unbiased sample. Nonetheless, it is the most employed method in social and behavioral sciences, especially in exploratory research. (Carmo & Ferreira, 2008; Gravetter & Forzano, 2017). This method has been recommended by many scholars as a suitable, fast and economical method to get a significant number of respondents, namely for tourism research (Jennings, 2010).

RESULTS

The sample comprises 111 students, representing 37,5% of the population, consisting of 296 students. In terms of gender, 64,9% are female and 35,1% are male, the majority from Portugal (91%), but there are students also from São Tomé and Príncipe (3,6%), Cape Verde (1,8%), Switzerland (1,8%), Guinea-Bissau (0,9%) and Canada (0,9%).

About 3,6% are master's students, 84,7% are from undergraduate study cycles, and 11,7% are enrolled in a TeSP. About 84,7% have between 18 and 23 years (the mean age is 22,25), the youngest with 18 years old and the oldest student aged with 55 years old. Regarding students' inscription in the institution, 3,6% have "working student" status, but the majority (90,9%) have no special status.

The 3 social networks considered as the most used by students of higher education are Instagram (98,2%), Facebook (82,9%) and WhatsApp (50,5%). According to respondents' perception, Youtube (38,7%) and Twitter (20,7%) have still a significant reference. They usually access to social media networks on a daily basis (92,8%). Nonetheless, among the 111 respondents, 4 of them refer that they do

Figure 1. Development flow diagram of the questionnaire

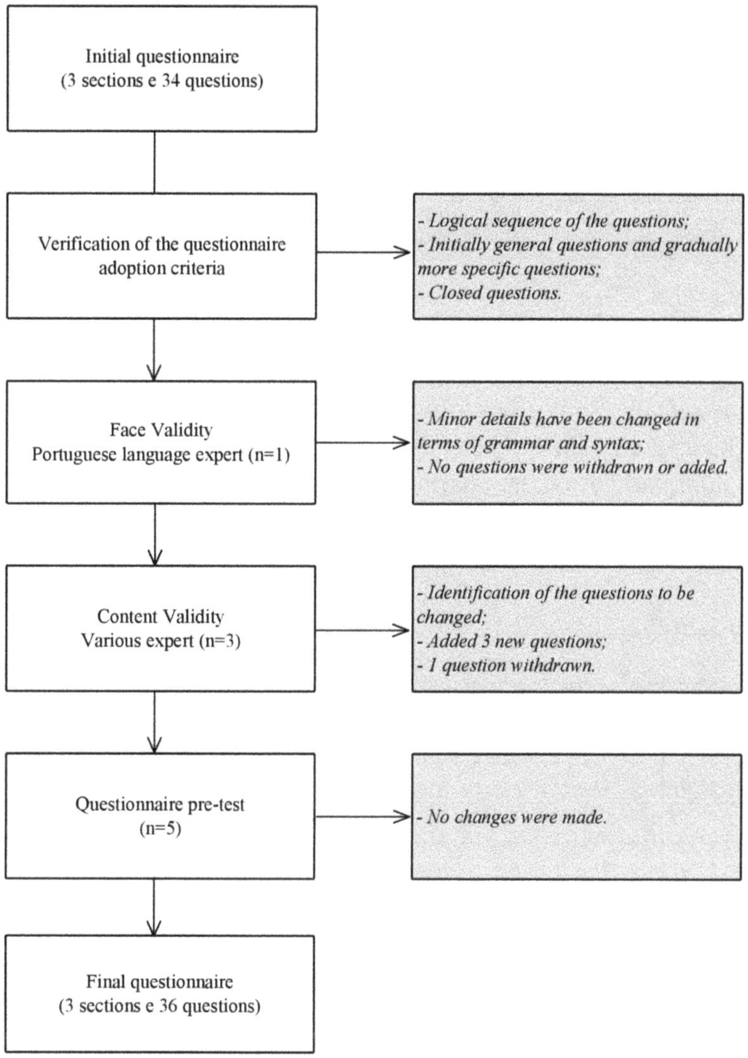

not use social media networks, representing 3,6%. According to the questionnaire structure, they did not answer any further questions of this research.

Therefore, the following statistics are based on 107 respondents. Table 1 resumes some of the most salient aspects that will be analyzed next.

When asked about the year in which they created their first profile/account in a social media network, 80% of the answers are comprised between 2008 and 2013, with the year of 2010 as the most representative (28%). According to the majority of the respondents, Facebook (68,2%) was the 1st social media profile/account created, followed by Hi5 (22,4%). Currently, 87,9% indicated having other profiles/accounts in more social media networks. The most mentioned are Facebook (95,7%), Instagram (93,6%), WhatsApp (88,3%) and Youtube (80,9%). However, curiously Instagram (62,6%) stands out as the most used network, with a considerable distance from Facebook (16,8%), WhatsApp (9,3%), Youtube (7,5%) and Twitter (3,7%).

Table 1. Students main characteristics regarding social media use

1st social media Profile	Facebook	68,20%
Time spent social media	Average (daily)	3,4 hours
Social media access	Period (19:00-24:00)	62,60%
Social media function	Leisure/Entertainment	80,40%
Equipment	Smartphone	89,70%
Connection	Home Wi-Fi	87,90%

Students spent an average of 3.4 hours daily on social media (Table 2), and they prefer to access during the night period, especially between 19:00 to 24:00 (representing 62,6%).

When asked to indicate the motivations for social media use, mostly they refer Leisure and Entertainment (80,4%), the Interaction with Friends/Family (79,4%), the Acquisition of Information/Contents (49,5%), as well as Information/Content Sharing (34,6%). Educational or academic purposes fall short of the aspects referred above, representing only 19,6% of the responses (Figure 2).

Table 2. Time spent on social media

	Absolut Frequency	Relative Frequency	Cumulative Frequency
Less than 1 hour	6	5,60%	5,60%
Between 1 and 2 hours	27	25,20%	30,80%
Between 2 and 3 hours	33	30,80%	61,70%
Between 3 and 6 hours	28	26,20%	87,90%
Between 6 and 9 hours	8	7,50%	95,30%
More than 9 hours	5	4,70%	100,00%
Total	107	100%	

Figure 2. Motivations for social media use

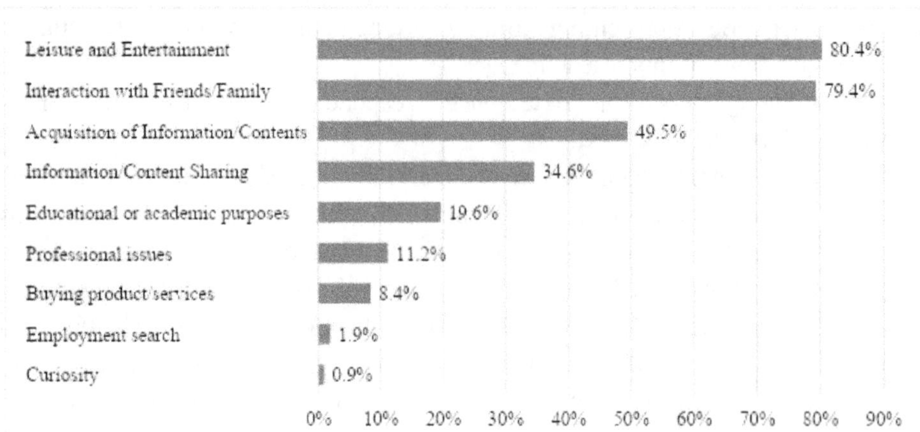

The perceptions regarding content typology available on social media platforms are shown in Figure 3. accordingly, Image is perceived as an important or very important attribute by respectively 35,5% and 45,8% of the respondents (reaching a cumulative percentage of 81,3%). Video is also evaluated as very important (33,6%) exceeding Text and Audio content (which represent 17,8% and 15,9% respectively).

When asked about the typology of content mostly shared by the respondents, Image (71%) is again the most prominent attribute, followed by Text (12,1%) and Web Link (8,4%) format. However, curiously very few prefer to share Video content (3,7%), in contrast to the importance referred above.

The Smartphone is the equipment most used to access social media (89,7%), far surpassing the Laptop (8,4%), and the type of internet connection preferred is home Wi-Fi (87,9%).

The last section of the questionnaire is targeted to analyze the use of social media in an academic context. Table 3 resumes the computer applications used for academic purposes.

According to the results, during the school year and within the educational institution, Smartphone is used on a daily basis by 90,7% of the respondents. Comparatively, the personal laptop is used frequently by 39,3% of the students. On the other hand, Tablet is the least used, with 85% saying they never use this equipment for academic purposes.

In terms of communication with their peers or with professors, for instance, the most frequently used application is Messenger (76.6%). WhatsApp is less widely used (41,1%), as well as Facebook (47,7%), even though, they are important communication tools used with some frequency.

Figure 3. Content typology importance on social media

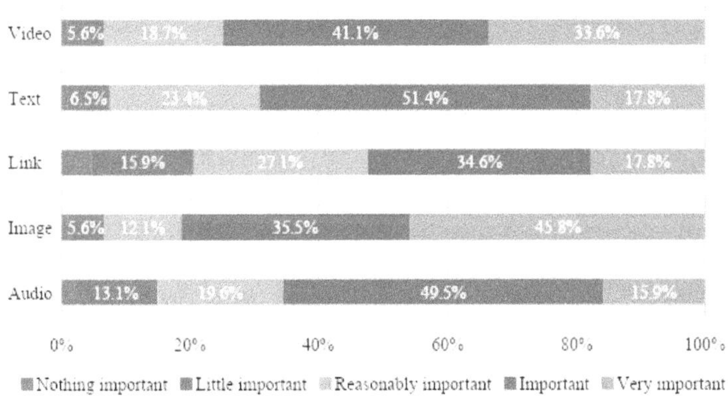

Table 3. Computer applications used for academic purposes

Scope	Computer Application	Frequently
Content production	Office 365	72%
Communication	Messenger	76,60%
Content share	Google Drive	23,40%
Bibliographic references	EndNote	3,70%
Platform	SIGARRA	86,90%

Content sharing does not seem to be widely used by the students, as can be seen in the Figure 4. The less used applications for academic purposes are iCloud and Youtube. Even though, Google Drive stands out as the most used of those presented in the questionnaire.

In the same way, the management of bibliographical references is not used very often. However, EndNote turns out to be the most used, with slight supremacy over other applications, like Mendeley, for instance. According to the results, students' resort to EndNote rarely (16,8%) or sometimes (11,2%). Even so, there are students who indicate frequent use (3,7%).

For academic purposes, the most used platform is Sigarra, which is not surprising since it is the one officially used by the institution (ESTH-IPG) for those purposes. However, 70,1% of the students refer to using Moodle sometimes or frequently, because there are several teachers who use this platform to support the curricular activity of the subjects they teach.

When asked about the content made available by teachers (with a maximum of 2 options), the textual content is the most indicated (73.8%), followed by Web Links (50.5%) and Image (43.9%), as we can see in figure 5.

Figure 4. Content sharing applications for academic purposes

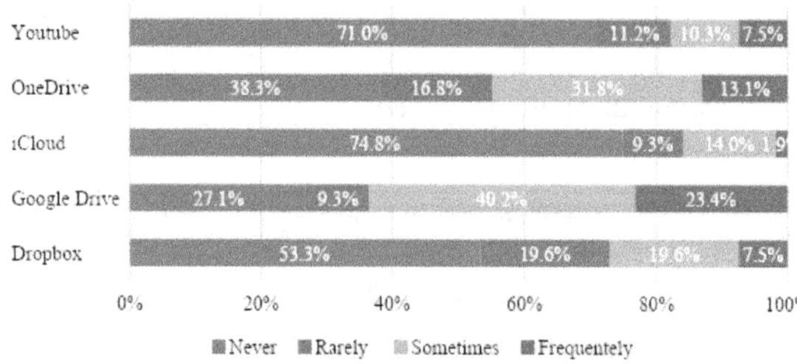

Figure 5. Content made available by teachers

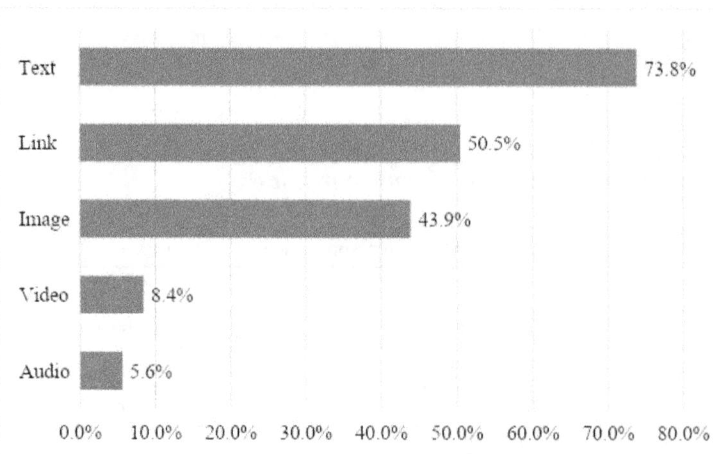

Bearing in mind the importance given by the students in previous questions, it appears that there is a slight discrepancy. In fact, according to students' perceptions, the Image represents a very important attribute, either in a content acquisition or as a content sharing perspective. Nonetheless, for the teachers, it appears in the 3rd position. Video is the type of content less provided by teachers, while students place Video content in the 2nd place in terms of importance. Moreover, teachers provide Web Link content, contradicting once again students' perspective for whom it was perceived as the least important content.

According to the students' perspective, Office365 is the application most used by their teachers for content production, being referred to as frequently used by 57% of the respondents, followed by Google Docs and Prezi with 15,9% and 1,9% respectively.

The Figure 6 represents teachers' use of social networks for communication purposes, according to students' perspective. As shown, Facebook and Messenger stand out, when compared to Skype and WhatsApp.

However, according to the results, this powerful communication tool is still a bit underappreciated by those who teach. Only 13,1% refer to their use by teachers as frequent, in contrast to the 35,5% who report not being a tool used by their teachers to communicate.

Regarding the sharing perspective, the applications less used for these purposes are YouTube and iCloud. Google Drive and OneDrive are used sometimes or frequently, according to 43% and 32,7% of the respondents, respectively.

Moreover, regarding the management of bibliographic references, EndNote and Mendeley have low but similar percentages. Even so, a slightly higher percentage when compared with RefWorks and CiteULike. However, the bibliographic reference is clearly a scope that needs to be improved, both by students and by teachers. This is even more evident when comparing both perspectives, by aggregating the percentage of those who refer its use as frequent or more sporadic (Table 4).

The communication gap is clear when comparing both uses of social media applications. Indeed, students use very often Messenger for communication (91,6%), while teachers seem to prefer to use Facebook, with a much lower frequency of use (39,3%). We must not forget that the results show students' perceptions. Therefore, it would be interesting to replicate this research and applying it to the higher education teaching community, comparing the results reached.

Figure 6. Teachers' use of social networks for communication purposes

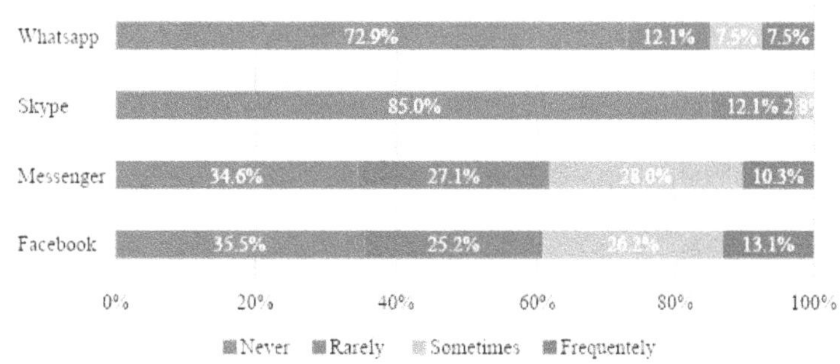

Table 4. Comparison of social media applications usage

Scope	Students Use		Teachers Use	
	Computer Application	Sometimes or Frequently Used	Computer Application	Sometimes or Frequently Used
Content production	Office 365	87%	Office 365	83,20%
Communication	Messenger	91,60%	Facebook	39,30%
Content share	Google Drive	63,60%	Google Drive	43%
Bibliographic reference	Endnote	14,90%	Mendeley	14,90%

The scientific areas where the social media applications are used more frequently are Tourism (63 answers), Hospitality (56 answers) and Information Technology (54 answers). In the opposite direction, Catering and Food Service (26 answers), Languages (18 answers) and Management (13 answers), reveal a lower tendency to the use social media applications for academic purposes. This is clearly an aspect to improve since students are convinced that the use of social media is very important to improve the teaching-learning process. In fact, 43,9% refer to this aspect as Important and 36,4% as Very Important (an average of 4,2 in a 5 points Likert scale).

Figure 7 shows the applications considered as most relevant to the improvement of the teaching-learning process. According to students' perspective, we can stand out: (i) SIGARRA, as the preferred platform; (ii) Office 365, for content production; (iii) Facebook, as a communication tool; (iv) YouTube, to share content. As it is perfectly evident, students still do not recognize the potential of bibliographic reference management tools.

Figure 7. Most relevant applications to improve the teaching-learning process

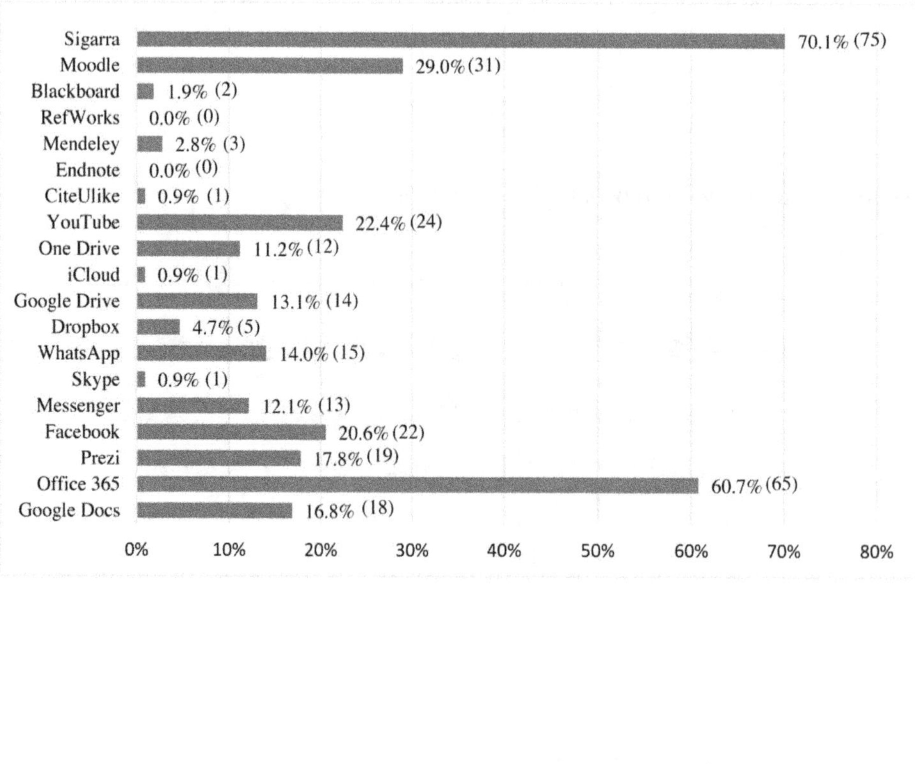

Finally, looking at the future, students were faced with the possibility of attending courses/units/ modules in the format of distance learning. Students prove to be a possibility and an Important (31,8%) or Very Important (30,8%) market tendency.

CONCLUSION

Social media is a reality that we cannot deny, to which higher education institutions and their staff need to adapt. According to Moran et al. (2011), teachers are very familiar with social media, and the level of awareness regarding the most common social media sites does not vary with age or stage in career. On the other hand, students are convinced that social media use is very important to improve the teaching-learning process. Indeed, the web *3.0* students are surrounded by technological environments and use social networks on a daily basis, especially in the personal context. However, *1.0* higher education institutions are not taking advantage of these important tools, which is in line with other researches (Hajli & Lin, 2014; Manca & Ranieri, 2016b; Mao, 2014; Peres et al., 2018).

According to our findings, social networks are under-exploited in the teaching/learning process. As a matter of fact, there are differences between the use of social networks in a personal context and its use in an academic context. The platforms/applications exist but are rarely used for academic purposes. Students have profiles in several social media platforms, especially for content acquisition, more than for sharing purposes. Currently, Instagram stands out as the preferred platform, perhaps by focusing on image and video content. The smartphone is the equipment most used to access social media, far surpassing the laptop or tablet. However, in the long term, the use of reduced graphics media to search and share may have some implications on the visual health of this generation, and we must be attentive to this reality. The smartphone, despite being personal equipment, is available in the classroom and is, sometimes used improperly. Instead of forbidding it, teachers should enhance the use of this tool, promoting interaction in the class, which was also reinforced by other researchers (Barry et al., 2015; Gikas & Grant, 2013). In this sense, teachers may: (i) share audio or video content for group work tasks; (ii) request complementary information that can be found by smartphone and shared immediately with the classmates; (iii) create or answer to different surveys; among other tasks. Those who are responsible for scientific areas such as management or languages, may foment the use of software tools for publishing and managing bibliographies, citations, and references, such as Mendeley or EndNote.

On the other hand, tourism students may act in the classroom like real tourists, finding holiday destinations and flights, simulating hotel reservations, reading restaurant reviews and learn how to answer to them appropriately. Students enrolled in catering and food service disciplines may use their smartphones to photograph the delicacies they have just made, and share those images with their own network on Instagram, for instance, also promoting the educational institution. These aspects would certainly make the classroom more interactive and captivate higher education students for tourism, hospitality and catering courses.

Higher education institutions are concerned with their international brand projection, and are aware of the importance of brand communication through their websites. However, the emotional values are not being part of the equation, which may compromise the overall brand efficiency (Chapleo, Durán, & Díaz, 2011), especially in terms of institutional reputation and individual engagement (Brech, Messer, Vander Schee, Rauschnabel, & Ivens, 2017). Those aspects are crucial for institutions who pretend to recruit foreign students, or reach other stakeholders (Kuzma & Wright, 2013).

According to the findings, the possibility of students attending courses or modules in the e-learning format is an emerging reality, certainly valued by those who have "working student status". This can be part of the strategy of higher education institutions with a peripheral location, particularly those who have difficulty to attract and retain new students due to a process of continuous displacement of the population to the coastal areas (Lopes, Martins, & Roque, 2017). As we know, there are other factors that influence school success in undergraduate programs at Higher Education Institutions (Pimenta, Ribeiro, Sá, & Belfo, 2018). Nonetheless, students' expectations, perceptions and degree of satisfaction, can be determinant for the success of the transition, adaptation, and fixation in the region (Araújo et al., 2015; Soares et al., 2014).

Given the rapid technological evolution, we cannot predict which social media or platform best fits the educational purposes. We are dealing with a social media generation that started with an Hi5 profile, went through Facebook, and is now more interested in sharing content on Instagram. Indeed, to predict the future of social media on education was not the goal of this research, despite all the managerial implications of our findings and the subsequent reflections.

Concerning the limitations, the data was obtained through a convenience sample, which might offer no guarantee of a representative and unbiased sample. Moreover, according to the demographic profile of respondents, the majority are Portuguese students from a tourism school belonging to the Polytechnic subsystem. The number of international students is not significant in order to make any comparisons. Future research should use a more balanced proportion, allowing a comparison with foreign or Erasmus students. In the same sense, students from the university subsystem may bring other important insights. Finally, this research follows a quantitative approach. Despite being subjective and requiring interpretative skills, the qualitative data could complement and contribute to a broader perspective of the phenomenon. Future research should also use other quantitative techniques. It would be interesting to include teacher's perspective in the sample. Other researchers may test if students perceive these limitations in the academic institutions where they develop their learning. A cross cultural research should also contribute for a deeper analysis of the phenomenon. Therefore, our findings and contributions are only a very small piece of the pie.

It is the time to rethink university teaching (Laurillard, 2002), and adapt it to the real "modern times" in which we live. The teaching-learning process needs to be refined and nurtured. In order to improve this bi-univocal relationship, it is necessary an effective engagement in social media, not only by both professors and students but also by their academic staff. The integration of these tools into the teaching and learning environment will provide opportunities for greater flexibility, interactivity, and accessibility for engaging teaching and learning at the individual, group and societal levels, in a globalized digital age (Lawrence & Tar, 2018). Our society needs to address these challenges and opportunities wisely.

ACKNOWLEDGMENT

We would like to thanks to the research centers, Research Unit for Inland Development (UDI/IPG – http://www.ipg.pt/udi/) and Centre for Tourism Research, Development and Innovation (CiTUR – http://citur-tourismresearch.com) for their support in the development of this research.

Due to their work in the development and improvement of the questionnaire we want also to express our gratitude to, Anabela Sardo (PhD) – Portuguese language and culture expert; Cristina Loureiro (BSc) – Tourism professional expert; Marisa Cardoso (BSc) – Computer Science expert and Fátima Roque (PhD) – Independent expert.

REFERENCES

Alumran, A., Hou, X. Y., & Hurst, C. (2012). Validity and reliability of instruments designed to measure factors influencing the overuse of antibiotics. *Journal of Infection and Public Health*, *5*(3), 221–232. doi:10.1016/j.jiph.2012.03.003 PMID:22632596

Araújo, A. M., Almeida, L. S., Costa, A. R., Alfonso, S., Conde, Á., & Deaño, M. (2015). Variáveis pessoais e socioculturais de diferenciação das expectativas académicas: Estudo com alunos do Ensino Superior do Norte de Portugal e Galiza. *Revista Portuguesa de Educação*, *28*(1), 201–220. doi:10.21814/rpe.7057

Barry, S., Murphy, K., & Drew, S. (2015). From deconstructive misalignment to constructive alignment: Exploring student uses of mobile technologies in university classrooms. *Computers & Education*, *81*, 202–210. doi:10.1016/j.compedu.2014.10.014

Boyd, D. M., & Ellison, N. B. (2007). Social network sites: Definition, history, and scholarship. *Journal of Computer-Mediated Communication*, *13*(1), 210–230. doi:10.1111/j.1083-6101.2007.00393.x

Brech, F. M., Messer, U., Vander Schee, B. A., Rauschnabel, P. A., & Ivens, B. S. (2017). Engaging fans and the community in social media: Interaction with institutions of higher education on Facebook. *Journal of Marketing for Higher Education*, *27*(1), 112–130. doi:10.1080/08841241.2016.1219803

Buhalis, D. (2019), Technology in tourism-from information communication technologies to eTourism and smart tourism towards ambient intelligence tourism: a perspective article. *Tourism Review*. doi:10.1108/TR-06-2019-0258

Buhalis, D., & Law, R. (2008). Progress in information technology and tourism management: 20 years on and 10 years after the Internet—The state of eTourism research. *Tourism Management*, *29*(4), 609–623. doi:10.1016/j.tourman.2008.01.005

Carmo, H., & Ferreira, M. M. (2008). *Metodologia da Investigação: Guia para Auto-Aprendizagem (2ª)*. Lisboa: Universidade Aberta.

Chapleo, C., Durán, M. V. C., & Díaz, A. C. (2011). Do UK universities communicate their brands effectively through their websites? *Journal of Marketing for Higher Education*, *21*(1), 25–46. doi:10.1080/08841241.2011.569589

Chugh, R., & Ruhi, U. (2018). Social media in higher education: A literature review of Facebook. *Education and Information Technologies*, *23*(2), 605–616. doi:10.100710639-017-9621-2

Foddy, W. (1996). *Como Perguntar : teoria e prática da construção de perguntas em entrevistas e questionários*. Celta Editora.

Fuciu, M., Gorski, H., & Dumitrescu, L. (2016). Online social networking sites in personal and professional life - A marketing research. *Revista Economica, 68*(3), 57–73. Retrieved from http://economice.ulbsibiu.ro/revista.economica/archive/68305fuciu&gorski&dumitrescu.pdf

Ghiglione, R., & Matalon, B. (2001). *O inquérito: teoria e prática* (4th ed.). Celta Editora. Retrieved from https://books.google.pt/books?id=_1v-SAAACAAJ

Gikas, J., & Grant, M. M. (2013). Mobile computing devices in higher education: Student perspectives on learning with cellphones, smartphones & social media. *The Internet and Higher Education, 19*, 18–26. doi:10.1016/j.iheduc.2013.06.002

Gravetter, F. J., & Forzano, L.-A. B. (2017). *Research Methods for the Behavioral Sciences* (6th ed.). Belmont, CA: Cengage Learning.

Hair, J., Bush, R., & Ortinau, D. (2003). *Marketing Research: Within a Changing Information Environment* (2nd ed.). New York: McGraw-Hill/Irwin.

Hajli, M. N., & Lin, X. (2014). Developing Tourism Education Through Social Media. *Tourism Planning & Development, 11*(4), 405–414. doi:10.1080/21568316.2014.883426

Hjalager, A. M. (2015). 100 innovations that transformed tourism. *Journal of Travel Research, 54*(1), 3–21. doi:10.1177/0047287513516390

Jennings, G. (2010). *Tourism Research* (2nd ed.). John Wiley & Sons.

Junco, R. (2012). The relationship between frequency of Facebook use, participation in Facebook activities, and student engagement. *Computers & Education, 58*(1), 162–171. doi:10.1016/j.compedu.2011.08.004

Koo, C., Park, J., & Lee, J. N. (2017). Smart tourism: Traveler, business, and organizational perspectives. doi:10.1016/j.im.2017.04.005

Kuzma, J. M., & Wright, W. (2013). Using social networks as a catalyst for change in global higher education marketing and recruiting. *International Journal of Continuing Engineering Education and Lifelong Learning, 23*(1), 53–66. doi:10.1504/IJCEELL.2013.051766

Laurillard, D. (2002). *Rethinking University Teaching: A Conversational Framework for the Effective Use of Learning Technologies* (2nd ed.). London: Routledge Falmer. doi:10.4324/9780203160329

Lawrence, J. E., & Tar, U. A. (2018). Factors that influence teachers' adoption and integration of ICT in teaching/learning process. *Educational Media International, 55*(1), 79–105. doi:10.1080/09523987.2018.1439712

Leung, D., Law, R., Van Hoof, H., & Buhalis, D. (2013). Social media in tourism and hospitality: A literature review. *Journal of Travel & Tourism Marketing, 30*(1-2), 3–22. doi:10.1080/10548408.2013.750919

Lopes, R., Abrantes, J. L., & Kastenholz, E. (2013). Online Reviews Credibility: Implications on Traveller's Decision Making. In M. Kosak & N. Kosak (Eds.), *Aspects of Tourist Behavior*. Newcastle: Cambridge Scholars Publishing.

Lopes, R., Martins, J. A., & Roque, V. (2017). Perfil e Impacto Socioeconómico dos Estudantes nas Escolas Politécnicas do Interior de Portugal – O Caso de Estudo da ESTH-IPG. In *Egitânia Sciencia - Edição Especial ISITH - International Symposium on Innovation in Tourism and Hospitality* (pp. 123–143). Academic Press. Retrieved from http://www.egitaniasciencia.ipg.pt/revista.aspx?id=32

Malhotra, N. K. (2010). *Marketing Research: An Applied Orientation* (6th ed.). Pearson.

Manca, S., & Ranieri, M. (2016a). Facebook and the others. Potentials and obstacles of Social Media for teaching in higher education. *Computers & Education*, *95*, 216–230. doi:10.1016/j.compedu.2016.01.012

Manca, S., & Ranieri, M. (2016b). "Yes for sharing, no for teaching!": Social Media in academic practices. *The Internet and Higher Education*, *29*, 63–74. doi:10.1016/j.iheduc.2015.12.004

Mao, J. (2014). Social media for learning: A mixed methods study on high school students' technology affordances and perspectives. *Computers in Human Behavior*, *33*, 213–223. doi:10.1016/j.chb.2014.01.002

Moran, M., Seaman, J., & Tinti-Kane, H. (2011). *Teaching, Learning, and Sharing: How Today's Higher Education Faculty Use Social Media*. Boston. Retrieved from https://files.eric.ed.gov/fulltext/ED535130.pdf

Oppenheim, A. N. (2000). *Questionnaire Design, Interviewing and Attitude Measurement*. Bloomsbury Academic. Retrieved from https://books.google.pt/books?id=6V4GnZS7TO4C

Peres, P., Moreira, F., & Mesquita, A. (2017). Higher education teachers in Portugal use technology in their educational activities: Myth or Reality? *Proceedings of the 12th Iberian Conference on Information Systems and Technologies (CISTI)* (p. 16). Academic Press. 10.23919/CISTI.2017.7975681

Peres, P., Moreira, F., & Mesquita, A. (2018). Are Really Technologies at the Fingers of Teachers? Results from a Higher Education Institution in Portugal. *Journal of Information Systems Engineering & Management*, *3*(1), 1–9. doi:10.20897/jisem.201808

Pimenta, C., Ribeiro, R., Sá, V., & Belfo, F. P. (2018). Factors Influencing School Success in Undergraduate Programs at a Portuguese Higher Education Institution. In *18ª Conferência da Associação Portuguesa de Sistemas de Informação (CAPSI'2018)* (p. 33). Santarém: Associação Portuguesa de Sistemas de Informação. Retrieved from https://aisel.aisnet.org/capsi2018/16

Roque, V., Martins, J. A., Silva, I., Rodrigues, C., & Simões, P. (2010). The ICT Area in High Education Tourism Courses - The Portuguese Case. *Proceedings of the International Congress on Tourism - CIT2010* (pp. 225–232). Porto: Instituto Superior de Ciências Empresariais e do Turismo (ISCET).

Roque, V., & Raposo, R. (2016). Social media as a communication and marketing tool in tourism: An analysis of online activities from international key player DMO. *Anatolia*, *27*(1), 58–70. doi:10.1080/13032917.2015.1083209

Soares, A. B., Francischetto, V., Dutra, B. M., Miranda, J. M., de, Nogueira, C. C. de C., & Leme, V. R. (2014). O impacto das expectativas na adaptação acadêmica dos estudantes no Ensino Superior. *Psico-USF*, *19*(1), 49–60. doi:10.1590/S1413-82712014000100006

Sobaih, A. E. E., Moustafa, M. A., Ghandforoush, P., & Khan, M. (2016). To use or not to use? Social media in higher education in developing countries. *Computers in Human Behavior*, *58*, 296–305. doi:10.1016/j.chb.2016.01.002

Xiang, Z., & Fesenmaier, D. R. (2017). Big Data Analytics, Tourism Design and Smart Tourism. In Z. Xiang & D. R. Fesenmaier (Eds.), *Analytics in Smart Tourism Design* (pp. 299–307). Switzerland: Springer. doi:10.1007/978-3-319-44263-1_17

Xiang, Z., & Gretzel, U. (2010). Role of social media in online travel information search. *Tourism Management*, *31*(2), 179–188. doi:10.1016/j.tourman.2009.02.016

Yaghmaie, F. (2003). Content validity and its estimation. *Journal of Medical Education*, *3*(1), 25–27. doi:10.22037/jme.v3i1.870

Zeng, B., & Gerritsen, R. (2014). What do we know about social media in tourism? A review. *Tourism Management Perspectives*, *10*, 27–36. doi:10.1016/j.tmp.2014.01.001

KEY TERMS AND DEFINITIONS

CiteULike: A web service that allows users to save and share scientific references amongst researchers (service already disabled).

EndNote: A software tool for publishing and managing bibliographies, citations and references.

Google Docs: An application that allows users to create and edit files online while collaborating with other users in real-time.

Google Drive: A cloud-based storage service that enables users to store and access files online.

iCloud: A cloud storage and cloud computing service similar to Dropbox and Google Drive, that enables users to store data (documents, photos, and music) on remote servers, or even to share and send data to other users.

Mendeley: A free reference manager and an academic social network for managing and sharing research papers, discovering research data and collaborating online.

Prezi: A software tool for creating presentations, similar to PowerPoint.

RefWorks: A web-based bibliography and database manager that allows users to create their own personal database by importing references from various sources.

SIGARRA: An integrated information system that facilitates the access to institution relevant information like pedagogical, scientific, technical or administrative information.

Section 3
Social Media and Hospitality

Chapter 16
Social Media Applied to Tourism and Hospitality:
The Case of Hotels in the Porto Metropolitan Area

Luís Pacheco
Portucalense Infante D. Henrique University, Portugal

Fernando Moreira
https://orcid.org/0000-0002-0816-1445
Portucalense Infante D. Henrique University, Portugal

ABSTRACT

Online hotel reviews, ratings, or opinions have gained importance with the growth of social media tools. The objective of this chapter is to study the impact of specific satisfaction attributes on overall satisfaction. It is used a secondary data set obtained from three of the most influential online travel platforms, being analyzed the guests' average ratings for around 130 hotel units, distributed by four quality segments, located in the Porto metropolitan area. The application of this methodology to a large sample of Portuguese hotels has not been done before, been that the main contribution of this study. It is evidenced that the different platforms, while all incorporating consumer reviews as primary social knowledge, are distinct from each other on some aspects. The three platforms present roughly the same supply of hotels, albeit presenting some differences in terms of volume of data. In terms of specific attributes, with the exception of "service," the three platforms present significant differences that may reflect the different user bases on these platforms.

DOI: 10.4018/978-1-7998-1947-9.ch016

INTRODUCTION

Information has not stopped growing exponentially over time, with acceleration in the last decade. According to IBM's calculations, the human being generated, from the beginning of its history until 2003, about five exabytes of information, five billion gigabytes. This value was generated in 2011, every two days (Dans, 2012). A recent IDC study expects that by 2025 163 Zettabytes of data will have been created in the world, a number that will be 10 times larger than the one recorded in 2016. Therefore, it could be the beginning of the Big Bang of data (Lancestremère, 2015). The question that can be asked is: What is the origin of all this data flow? Companies capture large amounts of transactional data about their consumers, suppliers and operations.

According to a recent IDC study for 2020, 65% of large organizations are information-based companies, with data being an important asset that will allow them to focus on relationships, people, and intangible capital, and use data to create value for the business through 4 actions: make better decisions, optimize operations, look for new sources of revenue and improve customer experience (IDC, 2018). Using the full potential of data has more to do with management than with technology (Redman, 2017). McAfee et al. (2012) have published an article entitled "*Big Data: The Management Revolution*" where they argue that such large flows of information can drastically improve a company's performance if the decision is based on such information flows. In a study presented by Franks (2012), it is emphasized the importance of the action of organizations in the incessant search in the capture and analysis of these new sources of data to reach the knowledge and the opportunities that they offer, since it is a new source of advantages competitive for most companies. However, hardware, cloud architectures, and open source software are available to organizations with more limited capabilities. It is however important to note that according to the value of the Iron Mountain and PwC information index of 2015 (Cavanagh, 2015), only 4% of companies are able to extract the full value of their information, showing that most companies have a long way forward.

In the tourism sector, one of the first experiences in the adoption of techniques of mass data analysis to revenue management was in the 1980s, with American Airlines implementing a demand-driven pricing system based on the analysis of a large volume of information. After extensive adoption in the air transport sector, large American hotel chains implemented it between the late 1980s and the early 1990s. In the last decade, the platform economy (referred to as a collaborative economy, but a very controversial designation) developed, with disruptive business models based on big data solutions. According to Song and Liu (2017), Big Data is one of the most important new tools that have affected the global travel industry. Currently, the most well-known examples that are the greatest exponent in maximizing information with the aid of technology are Airbnb and Uber.

Big data tools play an important role in determining the ways in which tourism companies outline their strategies and policies. However, the great difficulty currently encountered in forecasting tourism through the use of the big data is due to the difficulties of capturing, collecting, manipulating and modeling this type of data (structured and unstructured from internal and external repositories), which is usually characterized by its privacy and potential commercial value.

Knowledge about customers, as well as what they think about the destination, facilities or equipments (hotels, hostels, museums, monuments, etc.) remains a key strategic point for those who manage them. However, the vast amount of information that is now available in organizations' internal systems, as well as in social networks or user-generated contents websites (e.g., TripAdvisor) requires the use of new technologies and solutions for their collection, storage, processing and analysis, so that making decisions more effective. In the case of hotels, for example, the role that can be played by large amounts of information available in Customer Relationship Management (CRM) systems, when using emerging Big Data techniques to efficiently trace customer profiles is an asset to those who have to do the planning and management, among others, of rooms' occupancy. Big Data technologies can be extremely useful for analyzing internal data available in the hospitality industry's CRM information systems, as well as information collected from social networks.

Social networks with georeferenced information are essential for the perception of the tourist destination (Chareyron et al., 2014). These authors classify social networks into two types, the first designated by web communities with geotagged data (e.g., Instagram) and the second, web reviews (e.g., TripAdvisor) that allow to monitor reputation and points of interest. Kisilevich et al. (2010) analyze the possibilities of georeferenced photographs and García-Palomares et al. (2015) demonstrate the potential of photo-sharing services to identify and analyze key city attractions by identifying patterns of behavior based on photographs of residents and tourists. Tavakoli and Wijesinghe (2019) argue that the majority of tourism studies have been conducted on Web 1.0, Web 1.5 and Web 2.0. However, there are other platforms using Web 3.0, Web 4.0 and Web 5.0 technologies, but rarely explored. The same authors argue that there are diverse platforms designed beyond Web 2.0 which could represent cyberspaces for tourists to share their experiences and emotions and in this way it is possible to monitor the reputation, points of interest, and to understand and accommodate customers' needs and desires. Moreover, applications that go beyond Web 2.0 (Web 3.0, Web 4.0 and Web 5.0) could provide contexts for more advanced and immersive forms to explore data, but this level of the Web is still underutilized by the tourism sector.

The monitoring of social networks is a source of information of the tourists' opinions and preferences, very valuable for the managers. Through them, it is possible to know which are the points of interest and less interesting points. Reputation is an intangible asset closely related to sustainability as it reflects the perception of the behavior of the destination.

In the last decade Portugal has become a prime destination for leisure and business travelers, with destinations such as Lisbon, the Algarve, Porto or Madeira Island accounting for the large majority of tourists and overnight stays, in parallel with minor volume destinations such as the Douro valley and the Azores Islands, which have recently begun to gain increased international visibility. Currently, Portugal stands at 14th out of 136 countries in the "*Travel & Tourism Competitiveness Index 2017*" and particularly the city of Porto, located in the North of Portugal, has become a success story in terms of a travel destination in the past decade. In recent years the city has been considered by several specialized travel publications and sites as "Best European destination" and one of the most compelling cities to visit in Europe. The city is the center of one of the most dynamic regions in the world in terms of tourism growth, with the number of tourists and hotel stays steadily increasing, as well as the number of hotels and available beds amidst a surge of urban rehabilitation of Porto's "World Heritage" historical center. Data for the entire northern region of Portugal shows that in 2018 the region attracted more than 5.6 million stays in hotels by non-residents (INE, 2019). The city of Porto constitutes the embryo of the great region that is today the Metropolitan Area of Porto (*Área Metropolitana do Porto* – AMP). The AMP embraces a geographic zone currently composed of 17 contiguous municipalities, in an area of approximately 2,040 Km^2 with a resident population of around 1.7 million inhabitants.

Organizations have suffered a large (r)evolution both at social, economic and technological level, with particular predominance in the past ten years. A change of paradigm leads to the younger generation to share, often through mobile devices, personal information on social media tools. Kaplan and Haenlein (2010, p. 61) define social media as a "group of Internet-based applications that build on the ideological and technological foundations of Web 2.0, and that allow the creation and exchange of user-generated content". In this context social media tools impose great changes the way companies in generally, and tourism and hospitality sector, in particular, communicate with the market. In this way tourists tend to look for their destinations and activities through the use of social media tools to make purchasing decisions. For example, on the one hand, 89% of millennials plan travel activities based on content posted by their peers online. On the other hand, from review sites such as TripAdvisor, people are browsing the Internet for travel inspiration and validation from their peers.

Websites displaying hotel reviews, such as TripAdvisor and Booking.com, have become a vital information source for both independent travelers and service providers (Filieri, Alguezaui, and McLeay, 2015). Online reviews, ratings or opinions are part of the electronic word-of-mouth (eWOM) phenomenon and have gained importance with the emergence of new technology and tools. Litvin, Goldsmith, and Pan (2008, p. 461) define eWOM as "all informal communications directed at consumers through Internet-based technology related to the usage or characteristics of particular goods and services, or their sellers" and it has implications for tourism marketing (Morosan, 2013). Cantallops and Salvi (2014, p. 41) state that the main differences between WOM and eWOM can be identified in the reach of the reviews' impact (number of people who can be influenced) and the speed of interaction, facilitating and expanding people's ability to share travel experiences with a wider audience than ever before. In a nutshell, since over 97% of millennials share photos and videos of their travels online, building an influential web of peer-to-peer content that serves to inspire potential guests, social media has altered the landscape of tourism and hospitality industry, because most travelers determine their travel plans based on reviews and social media shares, making online customer service a crucial part of building a positive brand reputation (Chung & Koo, 2015).

The objective of this chapter is to study the impact of specific satisfaction attributes on overall satisfaction. Namely, the relationship between overall satisfaction and different attributes, such as location, service, value (cost-benefit) and cleanliness. The assumption is that guest ratings on different specific attributes largely determines their overall satisfaction. The study focuses on the quantitative ratings of each attribute, disregarding the guest's subjective judgments in written comments (Jeong & Jeon, 2008; Limberger, Anjos, Meira, & Anjos, 2014a; Limberger, Boaria, & Anjos, 2014b).

Filling a research gap pointed by Melián-González, Bulchand-Gidumal, and López-Valcárcel (2013) about comparing hotel reviews on different websites, this chapter also looks for evidence of significant differences between hotel segments and online travel platforms. To that purpose, it is used a secondary data set obtained from three of the most influential online travel platforms – Booking.com, TripAdvisor and Expedia – being analyzed the guests' average ratings for around 130 hotel units, from different quality segments, located in the AMP. The application of this methodology to a large sample of Portuguese hotels has not been done before. This chapter aims to highlight the main attributes responsible for overall satisfaction, also evidencing some specific differences between those attributes' relevance in terms of hotel segment or online platform.

The analysis and comparison of opinions posted on different online travel platforms is something new in the Portuguese context and we expect that our results highlight some significant differences between attributes and online platforms. Results of this chapter could be useful to hotel managers in resource allocation by identifying those factors that have greater impact on guests' overall satisfaction and helping them to better direct their efforts in terms of answer to the growing flow of data gathered trough social media and different online platforms.

The next section presents a literature review and the subsequent sections present the data and methodology and the results' discussion. The chapter ends with some concluding remarks and suggestions to further research.

LITERATURE REVIEW

Web 2.0 technologies have changed the ways users search for and share information, offering new ways of communication that enable users to share their opinions with all other users in an efficient, instantaneous platform without time or geographical limitations. The influence of eWOM is directly applicable to tourism and hospitality since user-generated reviews are an important source of information for prospective travelers searching for independent evaluations. Indeed, its influence is particularly high for experience goods such as in the hospitality and tourism context, because quality is only perceived after consumption (Lopes, Abrantes, & Kastenholz, 2014; Ye et al., 2009). Additionally, people tend to trust information more when it comes directly from other consumers, posted in popular online communities of travelers and supposedly helping them to avoid inadequate choices (Stringam, Gerdes, & Vanleeuwen, 2010; Ye et al., 2011; Casaló et al., 2015). Research by Lu and Stepchenkova (2012) indicates that almost all respondents read online comments while planning their trips and they believe those online comments are more likely to contain updated, detailed and trustworthy information (see also Yoo & Gretzel, 2008; Vermeulen & Seegers, 2009; Xiang & Gretzel, 2010), resulting in an increase in the selling prices of rooms for every extra point in the ratings (Anderson, 2012).

The monitoring and ability to understand the causes of guests' complaints is critical for hotels to improve their service quality, customer satisfaction and revenue (Baka, 2016). Ratings and reviews can have a significant effect on hotel profitability, with several empirical papers evidencing that hotels with better reviews (and also with a higher number of reviews) are capable of obtaining higher revenues per room (e.g., Ye, Law, & Gu, 2009).

A particularly important issue for the hotel industry is that research indicates that people pay greater attention to criticisms or negative reviews than to positive reviews (Lee, Park, & Han, 2008; Lee, Law, & Murphy, 2011; Sparks & Browning, 2011). Due to the increasing use of social media, online review platforms and other user-generated content, there is a danger of consumers feeling flooded with information in an eWOM overload, rendering it difficult for guests and hotel managers to choose the platforms through which they should best concentrate their attention (Park & Lee, 2008; Lee et al., 2011). Another concern is that it is difficult for a consumer to determine the credibility of a message coming from strangers (Mkono, 2018; Filieri et al., 2015; Filieri, 2016). In regard to the fear of user-generated contents being jeopardized by fake comments made under the cover of anonymity, O'Connor (2010), and more recently Martin-Fuentes, Mateu, and Fernandez (2018a), argue that such fear is baseless since research has found little evidence of fake reviews. For this reason, Booking.com and Expedia reviews are only posted by users that made a transaction through the websites and many online intermediaries – including TripAdvisor – require reviewers to register and provide some additional personal information including name, residence, nationality or gender (Jeong & Jeon, 2008; Xie et al., 2011). Besides the adoption of mechanisms to guarantee authenticity, the community of reviewers might also be a competent review monitoring body. Recently, Bandi Tanner, and Hämmerli (2018), performed a comparative quality analysis, revealing that the greater differences between 10 relevant websites are source credibility and argument quality, while the core competence of these websites is the provision of efficient decision-making aids. Nevertheless, as argued by Ruths and Pfeffer (2014), any study using social media data should be aware of several validity problems, such as platform and data availability biases and data authenticity issues.

Several authors have researched users' motives to post reviews (Zheng, Youn & Kincaid, 2009; Huang, Basu, & Hsu, 2010; Wilson, Murphy, & Cambra Fierro, 2012). For instance, Zheng et al. (2009) suggested that negative experiences are more likely to motivate dissatisfied consumers to write reviews, whereas Wilson et al. (2012) argued that people's attitudes motivations and types of online social networks for posting reviews will be affected by their nationalities. Some research argued that there is an 'inverted-U' relationship between customer satisfaction and the quantity of WOM or eWOM, as customers who are neither extremely satisfied nor unsatisfied share their experiences less (Litvin et al., 2008). Cantallops and Salvi (2014) presented a complete survey, reviewing papers on eWOM published in the 2007-2011 period, ending their paper with some suggestions for future avenues of research on the subject. Those authors concluded that previous research can be grouped into two general lines of work: factors related to the generation of comments and the impact those comments have on consumers and on company perspectives.

Some previous studies used TripAdvisor or other sites as a source of data, collecting the data either manually or automatically (e.g., Gerdes & Stringam, 2008; O'Connor, 2010; Ayeh, Au, & Law, 2013; Xiang et al., 2017). For example, Jeong and Jeon (2008) analyzed the consistency of posted reviews with the expected level of service and room rate. Using guests' reviews of hotels in New York City posted on TripAdvisor, their results indicated that "value" was one of the key predictors for guest satisfaction, which leads to return intentions. Regardless of hotel segment and average daily rate (ADR), location appeared to have the highest mean value among seven performance attributes and hotel quality segments (i.e., number of stars) and ADR appears to influence the relationships of selected hotel performance attributes with both overall guest satisfaction and return intentions.

Using the largest travel website in China, Ye et al. (2009) conducted a study to empirically investigate the impact of online consumer-generated reviews on hotel room sales by tallying the number of reviews received on a travel website, which the authors considered to be a proxy of room sales. Their results indicated that positive online consumer reviews can significantly improve the business performance of hotels, with an increase of the variance in customer ratings diminishing hotel room ratings. Using the Expedia website, Stringam et al. (2010) studied the relationship between the overall satisfaction of guests with hotel services, the condition of hotels, cleanliness of rooms and room comfort. Their study looked at over 60,000 traveler ratings, and the majority of all four quality ratings and overall satisfaction were positive. Additionally, nearly 75% of responses indicated willingness to recommend the hotel. These four category ratings had a similarly strong association with travelers' overall rating and willingness to recommend. Partial correlations differ, with hotel service and room comfort being the highest, followed by hotel condition and room cleanliness.

Stringam and Gerdes (2010) evaluated 60,648 consumer ratings and comments from Expedia to determine what factors drive consumer ratings of hotels. Specifically, the paper yielded information about the most frequently used words, as well as the pattern of word usage with either high or low guest ratings. It also explored the hotel guests' choice of words when assigning lower ratings versus higher ratings. Lee et al. (2011) profiled guests who posted helpful reviews on TripAdvisor. Their key finding was that helpful reviewers are those who travel more, actively post reviews, belong to any age and gender group, and give lower hotel ratings. Wilson et al. (2012) examined the influence of nationality both on the motivation to post (or not) and on where and what type of content consumers share in the context of a recent trip. The method used was an online survey of 568 respondents located in Switzerland, the United Kingdom and Spain, and the results revealed differences in national attitudes and motivations for posting. Respondents in Switzerland and the United Kingdom were more likely to post photos or travelogues on Facebook, but respondents from Spain prefer to post on TripAdvisor with the goal of making recommendations to other travelers. Li, Ye, and Law (2013) analyzed the content of 42,668 online reviews from 774 hotels, showing on one hand that transport facilities, food and beverage management, proximity to tourism destinations and amount paid were the most valued factors, independently of the hotel segment. On the other hand, customers were less satisfied with the bed, reception desk services, room decoration and size. In particular, the authors showed that 97% of the satisfaction variance could be explained by only four variables: room, value, cleanliness and check-in/out. The authors concluded that individuals from different countries have diverse reasons to post comments, so hotel managers should consider those behavioral differences.

Considering both the content of reviews and the volume of eWOM in a broad sample of 16,680 hotels, Melián-González et al. (2013) found a relationship between valence (positive vs. negative posts) and volume, in that initial hotel reviews tended to be disproportionately negative. As hotels received more reviews the average ratings improve and the standard deviation declines, which meant there was a greater agreement in ratings and the negative effects are mitigated. One implication of their study was that hotel managers should try to increase the number of reviews they receive to balance the positive and negative opinions about their property. A similar conclusion was reached by Martin-Fuentes, Mateu, and Fernandez (2018b), who found a significant linear relationship between the number of reviews and the score on TripAdvisor but not on Booking.com. Limberger et al. (2014a) also studied the correlation between overall satisfaction and the evaluation criteria used on TripAdvisor, analyzing 660 reviews for overall satisfaction, value, location, sleep quality, rooms, cleanliness and service. Applying a method of multivariate analysis, their results showed a strong correlation of overall satisfaction with the room, service provided and location.

Finally, Jiang, Gretzel, and Law (2014) analyzed English-speaking traveler reviews posted on TripAdvisor about hotels in Mainland China. The authors compared brand image across hotel types and star rating levels, finding that brand image can vary based on specific hotel characteristics, emphasizing different areas of service. For instance, the authors found that a certain star level does not necessarily imply a predetermined type of review, a feature that highlights the importance of distinguishing ratings between star levels.

Although not specifically considered in the present study, different cultural backgrounds lead to different consumption concepts, perceived satisfaction and ratings (Ayoun & Moreo, 2008; Chen, Cheung, & Law, 2012; Wilson et al., 2012; Li et al., 2013; Buzova, Sanz-Blas, & Cervera-Taulet, 2019). There are several studies that focused on the relationship between cultural and nationality factors and personality traits with hotel reviews and ratings (Huang, Huang, & Wu, 1996; Hsieh & Tsai, 2009; Tse & Ho, 2009; Yoo & Gretzel, 2011; Zeng & Gerritsen, 2014; Schuckert, Liu, & Law, 2015a; Pacheco, 2016). The literature also identified significant generational differences, with the younger generation of users attributing more credibility to user-generated contents and being more influenced by them, when compared to older generations, albeit being more likely to publish their opinions on their own social media than on a commercial website (Zeng & Gerritsen, 2014).

In the Portuguese context, the literature about the characterization of customers using online review platforms is scant. Chaves, Gomes, and Pedron (2012) collected 1500 online reviews from 50 small and medium Portuguese hotels from Booking.com and TripAdvisor. Their study showed that room, staff and location were the criteria which appeared more frequently in the reviews, thus highlighting aspects that hotel managers should prioritize. Previously, Carneiro and Costa (2000) looked at how service quality influenced the competitive positioning of five-star hotels. More recently, Pacheco (2017) analyzed reviews for 534 hotels located in the four main Portuguese tourism regions, finding that the "rooms", "service" and "cost-benefit" attributes showed a greater correlation with overall satisfaction than other criteria, and those attributes are common across hotel segments and regions.

In sum, from the analysis of guests' reviews or comments the literature identifies the following set of relevant variables: room, comfort, service, cleanliness, value, location and facilities. There is a growing competition among similar products to draw the attention of web surfers and hotels are not an exception. For instance, Booking.com, TripAdvisor and Expedia present, respectively, reviews for 222, 182 and 162 hotels in the AMP. Given the huge number of alternatives available online and the tendency of consumers to have low degrees of patience, the key question that motivates our study is exactly to find in which way guests' reviews of hotels located in the AMP differ and correlate with specific attributes of the hotels.

So, this exploratory chapter follows and develops the papers from Chaves et al. (2012), Jeong and Jeon (2008), Li, Ye, and Law (2013), Limberger et al. (2014a, 2014b), O'Connor (2010), Stringam et al. (2010); Pacheco (2017); Xiang et al. (2017) and Martin-Fuentes et al. (2018a), employing a sample of around 130 hotels located in the AMP in order to identify the most important determinants of overall satisfaction and considering differences in terms of online platforms and hotel segments. Thus, the three main research questions that this chapter tries to answer in the following sections are:

1. Are there significant differences in average ratings of different hotel quality segments in different online platforms?
2. What are the main determinants of overall satisfaction, specifically in terms of hotel quality segments and online platforms?
3. Is the number and dispersion of guests' reviews related to the overall rating?

DATA AND METHODOLOGY

The data used in this study were retrieved manually from three major online platforms: Booking.com, TripAdvisor and Expedia. These three platforms are very popular and influential, where reviews are incorporated as eWOM, being Tripadvisor a community-based website and Booking.com and Expedia two transaction-based online travel agencies (OTAs). Booking.com is the world leader in booking accommodation online and TripAdvisor is one of the most influential eWOM sources in the hospitality and tourism context (Yen & Tang, 2015; Martin-Fuentes et al., 2018b). According to Bandi Tanner and Hämmerli (2018), Booking.com and TripAdvisor present a high quality index, whereas Expedia presents and intermediate quality index, being primarily oriented towards hotel booking. Booking.com provides a great variety of efficient decision-making aids and has the highest impact in terms of the number of reviews and website traffic, whereas TripAdvisor performs especially well regarding source credibility.

The three considered platforms enable users to input reviews for their hotel stays and share experiences, posting classifications from 1 (very unsatisfied/terrible) to 5 (very satisfied/excellent) – on TripAdvisor and Expedia – or from 2,5 to 10 on Booking.com (see the recent criticisms to this scale in Mellinas, María-Dolores, and García, 2015, 2016). These ratings are a reliable assessment outcome for this study because customers can evaluate their hotel experience and write comments on a set of attributes. On TripAdvisor and Expedia, the user gives an overall score and after can assess other specific attributes individually, but those are not taken into account in the overall score. In order to compute the overall satisfaction score, TripAdvisor and Expedia take in account all the reviews posted by past users, even the oldest ones, whereas Booking.com discards reviews older than 24 months, albeit presenting them on the website. As pointed out by Mellinas et al. (2015; 2016), Booking-com's overall score is a plain average of a set of specific attributes of the hotel, where the customer is not asked to rate each item from 1 to 10, but instead there are four options corresponding to poor (2,5), fair (5), good (7,5) or excellent (10). Bjorkelund, Burnett, and Norvag (2012) compare hotels on TripAdvisor and Booking.com, concluding that Booking.com scores lean toward the higher end of the scale more often than TripAdvisor.

Is spite of these caveats, the scores presented by the considered platforms represent eWOM in a straightforward manner since clients evaluate hotels in terms of overall satisfaction and specific attributes: cleanliness, comfort, location, facilities, staff, value for money and free wi-fi (Booking.com); value, location, service and cleanliness (TripAdvisor); cleanliness, comfort, service and hotel condition (Expedia).

In April 2019 the selected platforms presented reviews for around 200 hotels located in the AMP, distributed by four quality segments. Due to the frequency with which the online platforms are updated, the time frame for collecting the data was set from April 16 to 18, 2019 and only hotels with more than 50 reviews were chosen. After applying these criteria, we obtained a sample of 120 to 147 hotels, depending on the online platform, distributed by different number of stars/ quality segments (Table 1).

Concerning the distribution by segments (according to the number of stars in the official classification system), the sample reflects a higher preponderance of four-star hotels and a relatively lower number of five-star hotels.

This chapter analyzes the overall satisfaction as well the four (seven in the case of Booking.com) specific ratings. Regression analysis is used to understand key predicting attributes of guests' overall satisfaction. In terms of methodology it is employed an OLS regression, assuming a linear relation between the dependent variable and a set of explanatory variables (Gujarati, 2003). Given the fact that Booking.com's overall score represents an average of the specific attributes, the regression is not performed for that platform.

Table 1. Data set

	Booking.com		TripAdvisor		Expedia	
	Nr.	**Nr. Reviews**	**Nr.**	**Nr. Reviews**	**Nr.**	**Nr. Reviews**
5 stars	14	18948	13	18073	14	6082
4 stars	58	85040	48	49308	51	19546
3 stars	38	55432	35	15382	32	9677
2 stars	37	52791	31	13058	23	4365
TOTAL	**147**	**212211**	**127**	**95821**	**120**	**39670**

Source: self-made.

Note: Hotels classified from five to two stars according to the official classification system.

Data retrieved in 16-19 April 2019.

Prior to the regression, some descriptive statistics are computed (means, standard deviations and correlations), and to better understand the main characteristics of the sample statistical tests of differences in mean ratings across online platforms and hotel segments are performed. Finally, it is also tested if the number and dispersion of reviews influences the overall rating.

RESULTS AND DISCUSSION

As stated above, our research focuses on three points: i) whether there are significant differences in average ratings of different hotel quality segments in different online platforms; ii) whether different attributes have different impacts on overall satisfaction; and iii) whether the number and dispersion of guests' reviews is related to the overall rating.

Table 2 presents the means and standard deviations of the different reviews, both in global terms and in terms of hotel segment for the different online platforms.

As can be seen, mean ratings decrease with lower hotel rankings as expected, indicating that the popularity index coincides with the hotel quality segment. This trend appears in all online platforms with the exception of "value" and "free wi-fi" (Booking.com) and "value" and "location" (TripAdvisor) for

Table 2a. Means and standard deviations

	Overall Satisfaction	Cleanliness	Comfort	Location	Facilities	Staff	Value for Money	Free Wi-fi
Booking.com								
5 stars	8.99 (0.38)	9.34 (0.32)	9.21 (0.32)	8.86 (0.84)	8.97 (0.40)	9.19 (0.33)	8.25 (0.32)	8.71 (0.40)
4 stars	8.53 (0.54)	8.86 (0.61)	8.56 (0.66)	8.49 (0.77)	8.37 (0.65)	8.84 (0.49)	7.99 (0.55)	8.38 (0.54)
3 stars	8.12 (0.50)	8.39 (0.58)	7.80 (0.70)	8.36 (0.74)	7.70 (0.60)	8.54 (0.40)	7.79 (0.58)	7.90 (0.81)
2 stars	7.98 (0.53)	8.23 (0.66)	7.48 (0.78)	8.35 (0.75)	7.41 (0.72)	8.50 (0.45)	7.85 (0.59)	8.00 (0.63)

Note: Standard deviations in parenthesis.

Table 2b. Means and standard deviations

TripAdvisor	Overall Satisfaction	Value	Location	Service	Cleanliness
5 stars	4.46 (0.32)	4.27 (0.26)	4.35 (0.69)	4.58 (0.28)	4.69 (0.38)
4 stars	4.13 (0.36)	4.11 (0.31)	4.17 (0.50)	4.29 (0.31)	4.45 (0.30)
3 stars	3.73 (0.44)	3.99 (0.35)	4.03 (0.48)	3.97 (0.36)	4.06 (0.40)
2 stars	3.61 (0.57)	4.10 (0.45)	4.11 (0.48)	3.94 (0,54)	4.02 (0.54)

Note: Standard deviations in parenthesis.

Table 2c. Means and standard deviations

Expedia	Overall Satisfaction	Cleanliness	Comfort	Service	Hotel Condition
5 stars	4.60 (0.18)	4.78 (0.16)	4.61 (0.16)	4.67 (0.15)	4.58 (0.22)
4 stars	4.34 (0.27)	4.53 (0.24)	4.34 (0.28)	4.46 (0.26)	4.34 (0.33)
3 stars	3.98 (0.38)	4.19 (0.38)	3.83 (0.44)	4.16 (0.30)	3.88 (0.50)
2 stars	3.87 (0.26)	4.17 (0.29)	3.66 (0.30)	4.08 (0.23)	3.76 (0.34)

Note: Standard deviations in parenthesis.

the two and three-star hotels, indicating that guests do not perceive differences in those attributes. The apparently higher ratings on Booking.com are possible a result of the "inflated" scoring system (Mellinas et al., 2015). Also, TripAdvisor reviewers tend to post lower ratings in terms of overall satisfaction and several other specific attributes, suggesting some differentiating particularities of that online platform user. Since the majority of ratings increase from lower to superior hotel segments this finding suggests, as expected, that five-star hotel guests enjoy higher satisfaction levels, with satisfaction decreasing in lower segment hotels (two and three stars). This result confirms the positive relationship between star-rating classification and satisfaction measured by scores on user-generated websites such as TripAdvisor and on sales websites such as Booking.com (Martin-Fuentes, 2016).

Confirming the results from Schuckert, Liu, and Law (2015b) the previous table also shows that, in general, lower segment hotels are more likely to generate a greater dispersion in ratings in comparison to the values presented for 5 or 4-star hotels (notice the higher values presented by 3-star hotels, a segment where coexist some traditional and old fashion hotels with design and trendy hotels focused on younger guests. Finally, comparing the dispersion between online platforms the results seem to present a greater cohesion in Expedia ratings, typically presenting lower standard deviations. Notice that not all previous research confirms a positive relationship between the star-rating system and quality. For instance, López Fernández and Serrano Bedia (2004) found a lack of correspondence between guests' expectations, perceptions and hotels' quality segments, suggesting that star-rating criteria should take account of user-generated reviews, thus avoiding regulations that can become outdated with the passage of time.

Table 3 presents the Spearman's rank correlation coefficients between overall satisfaction and the different attributes for the three different online platforms.

In terms of correlations between attributes, the previous table evidences that they are all positive and mostly significant. There are some interesting results, namely lower values for "free wi-fi" (Booking.com) and "location" (Booking.com and TripAdvisor), suggesting that overall satisfaction is less dependent on those two attributes, both presenting a higher dispersion of ratings.

To test the first research question a t-test is performed in order to determine whether there are significant differences in means for the different attributes across different platforms, which comprises a total of three combinations. Table 4 shows the results for the three online platforms, considering the common attributes between platforms and differentiating between attributes and hotel quality segments

Table 3a. Correlations between overall satisfaction and attributes

Booking.com	Cleanliness	Comfort	Location	Facilities	Staff	Value for Money	Free Wi-fi
5 stars	0.972 (***)	0.941 (***)	0.881 (***)	0.905 (***)	0.897 (***)	0.746 (**)	0.846 (***)
4 stars	0.956 (***)	0.948 (***)	0.662 (***)	0.956 (***)	0.902 (***)	0.934 (***)	0.671 (***)
3 stars	0.969 (***)	0.936 (***)	0.455 (**)	0.956 (***)	0.807 (***)	0.881 (***)	0.422 (**)
2 stars	0.968 (***)	0.940 (***)	0.236	0.958 (***)	0.848 (***)	0.935 (***)	0.683 (***)

Note: * $p < 0.05$; ** $p < 0.01$; and *** $p < 0.001$.

Table 3b. Correlations between overall satisfaction and attributes

TripAdvisor	Value	Location	Service	Cleanliness
5 stars	0.637 (*)	0.727 (**)	0.505	0.743 (**)
4 stars	0.803 (***)	0.381 (**)	0.714 (***)	0.751 (***)
3 stars	0.679 (***)	0.380 (*)	0.683 (***)	0.841 (***)
2 stars	0.724 (***)	0.378 (*)	0.853 (***)	0.829 (***)

Note: * $p < 0.05$; ** $p < 0.01$; and *** $p < 0.001$.

Table 3c. Correlations between overall satisfaction and attributes

Expedia	Cleanliness	Comfort	Service	Hotel Condition
5 stars	0.790 (***)	0.699 (**)	0.891 (***)	0.914 (***)
4 stars	0.918 (***)	0.899 (***)	0.828 (***)	0.931 (***)
3 stars	0.976 (***)	0.961 (***)	0.935 (***)	0.973 (***)
2 stars	0.837 (***)	0.945 (***)	0.875 (***)	0.895 (***)

Note: * $p < 0.05$; ** $p < 0.01$; and *** $p < 0.001$.

in each platform. Since the t-test needs to consider an assumption regarding the equality of variances or not, previous to this test it was tested whether the different pairs presented equal variances in order to perform the appropriate test for differences in means. Notice that a parametric test is used, since after computing the Doornik-Hansen test it was found that the large majority of the data series followed a normal distribution.

Table 4a. T-test for differences in average ratings between online platforms and hotel segments

	Overall Satisfaction	Cleanliness	Comfort	Service	Location
Booking.com vs. TripAdvisor	4.74 ***	1.08			1.74
Booking.com vs. Expedia	-0.44	-2.00 *	-0.47		
TripAdvisor vs. Expedia	-4.42 ***	-2.61 **		-3.70 ***	

Note: Ratings were standardized in percentage terms prior to the test. Values for t-tests. * $p < 0.05$; ** $p < 0.01$; and *** $p < 0.001$.

Table 4b. T-test for differences in average ratings between online platforms and hotel segments

Booking.com	Overall Satisfaction	Cleanliness	Comfort	Location	Facilities	Staff	Value for Money	Free Wi-fi
5 stars vs. 4 stars	2.95 ***	2.87 **	3.57 ***	1.57	3.31 **	2.46 *	1.67	2.16 *
5 stars vs. 3 stars	5.89 ***	5.82 ***	7.25 ***	2.06 *	7.33 ***	5.43 ***	2.82 **	3.55 ***
5 stars vs. 2 stars	6.48 ***	6.08 ***	7.97 ***	2.11 *	7.60 ***	5.17 ***	2.38 *	3.90 ***
4 stars vs. 3 stars	3.73 ***	3.78 ***	5.39 ***	0.81	5.08 ***	3.23 **	1.78	3.44 ***
4 stars vs. 2 stars	4.88 ***	4.81 ***	7.20 ***	0.91	6.70 ***	3.42 ***	1.19	3.12 **
3 stars vs. 2 stars	1.22	1.14	1.85	0.10	1.89	0.35	-0.50	-0.56

Note: Ratings were standardized in percentage terms prior to the test. Values for t-tests. * $p < 0.05$; ** $p < 0.01$; and *** $p < 0.001$.

Table 4c. T-test for differences in average ratings between online platforms and hotel segments

TripAdvisor	Overall Satisfaction	Value	Location	Service	Cleanliness
5 stars vs. 4 stars	3.02 ***	1.63	1.06	3.03 ***	2.47 *
5 stars vs. 3 stars	5.45 ***	2.63 *	1.79	5.44 ***	4.96 ***
5 stars vs. 2 stars	5.00 ***	1.28	1.29	4.02 ***	4.09 ***
4 stars vs. 3 stars	4.47 ***	1.75	1.26	4.35 ***	5.13 ***
4 stars vs. 2 stars	4.86 ***	0.21	0.48	3.72 ***	4.59 ***
3 stars vs. 2 stars	0.92	-1.11	-0.71	-0.32	0.35

Note: Ratings were standardized in percentage terms prior to the test. Values for t-tests. * $p < 0.05$; ** $p < 0.01$; and *** $p < 0.001$.

Testing for differences in review ratings the previous table answers the first research question, evidencing that in terms of overall satisfaction there are significant differences in means, except between 3 and 2-star hotels, indicating the guests' lack of perception of significant differences between hotels. Also, TripAdvisor's mean overall rating also seems to differentiate from the other two online platforms. In terms of specific attributes, there seems to exist some significant differences in all dimensions, with the exception of "location" and "value for money/value". Also, differences are less evident between 5

Table 4d. T-test for differences in average ratings between online platforms and hotel segments

Expedia	Overall Satisfaction	Cleanliness	Comfort	Service	Hotel Condition
5 stars vs. 4 stars	3.47 ***	3.65 ***	3.45 ***	2.93 **	2.61 *
5 stars vs. 3 stars	5.84 ***	5.54 ***	6.44 ***	5.99 ***	5.00 ***
5 stars vs. 2 stars	9.13 ***	7.32 ***	10.88 ***	8.55 ***	7.98 ***
4 stars vs. 3 stars	5.08 ***	4.98 ***	6.45 ***	4.78 ***	5.02 ***
4 stars vs. 2 stars	7.02 ***	5.69 ***	9.36 ***	5.96 ***	6.92 ***
3 stars vs. 2 stars	1.18	0.27	1.55	1.03	1.03

Note: Ratings were standardized in percentage terms prior to the test. Values for t-tests. * $p < 0.05$; ** $p < 0.01$; and *** $p < 0.001$.

and 4-star hotels and, in particular, between 3 and 2-star hotels, a result also found by Martin-Fuentes, Mateu, and Fernandez (2018c). Nevertheless, as expected, apart from "location", there are significant differences between hotel segments, either in terms of overall satisfaction or in terms of the specific attributes. As ratings clearly differentiate between hotel segments it is evident that higher star-rated hotels have more favorable guest perceptions than their lower counterparts.

In order to address our second research question, Table 5 below presents the results for the OLS regression for the TripAdvisor and Expedia entire samples and the four hotel quality segments.

As evident in the table, the great majority of explanatory variables are significant and the adjusted R^2 is always between 45 and 97%, meaning that a large part of overall satisfaction variance is explained by the considered set of attributes, thus answering the second research question and also confirming the results of Limberger et al. (2014a, 2014b), Stringam et al. (2010) and Pacheco (2017). Considering the TripAdvisor sample, it is observed that all attributes, with the interesting exception of "location", are significant, albeit with a higher coefficient for "cleanliness". For instance, a one-point upgrade in the "cleanliness" rating has an impact of 0.57 in the global rating. For the 4-star hotels there are only two significant attributes – "value" and "cleanliness". Understandably, the attribute "location" is significant for the 2-star hotels, an explanation for this might be that the traveler already knows where the hotel is located before traveling, thus being that an important decision factor and generating less uncertainty on the trip. Regarding the Expedia sample, the attribute "cleanliness" is never significant and items as "comfort" and "hotel facilities" seem to be more relevant for higher and lower segment hotels. These results highlight the need for hotel managers, independently of the hotel quality segment, to minimize the negative impact of generally small details related to room comfort and general decoration of the hotel. Thus, "cleanliness", "comfort" and "hotel condition" seem to be the key predictor of guests' satisfaction in the AMP. Notice that, possibly due to the smaller sub-samples for 5-star hotels (samples with less than 15 hotels) the results did not allow the identification of significant attributes. Nevertheless, "hotel condition" appears as a significant attribute.

Table 6 summarizes the significant variables influencing overall satisfaction in the different hotel segment and online platform considered.

Finally, in order to test our third research question, Table 7 presents the results obtained for the regression of the overall satisfaction level on the number and standard deviation of reviews for the three online platforms. It is included a control variable to assess if the industry structure (number of stars) can explain ratings.

Table 5. Regression results

	Complete Sample	5 Star	4 Star	3 Star	2 Star
			TripAdvisor		
Constant	-0.948 ***	1.583	-0.851	-1.612 *	-1.145 *
Value	0.110 ***	0.237	0.542 ***	0.139	-0,085
Location	0.108 *	0.173	-0.024	0.157 *	0.312 **
Service	0.374 ***	-0.074	0.207	0.268	0.410 *
Cleanliness	0.570 ***	0.310	0.441 **	0.650 ***	0.551 *
R^2	0.795	0.636	0.754	0.782	0,811
adjusted R^2	0.788	0.453	0.731	0.753	0.782
F	161.32	5.97	41.64	37.08	29,94
			Expedia		
Constant	0.233	0.280	0.012	0.007	0.515 *
Cleanliness	0.081	0.344	0.168	0.393	-0.342
Comfort	0.248 ***	-0.238	0.290 **	0.226	0.611 ***
Service	0.310 **	0.255	0.254 *	0.239	0.386 **
Hotel condition	0.298 ***	0.564 ***	0.272 *	0.121	0.258 *
R^2	0.963	0.904	0.931	0.972	0.944
adjusted R^2	0.962	0.862	0.925	0.968	0.931
F	1,086.91	52.53	236.89	272.22	116.81

Note: Robust standard errors (HC1). VIFs consistent with the absence of multicollinearity.
* $p < 0.05$; ** $p < 0.01$; and *** $p < 0.001$.

Table 6. Most determinant attributes

Hotel Segment/Online Platform	TripAdvisor	Expedia
Total	Service, cleanliness	Comfort, service, hotel condition
5 stars	None	Hotel condition
4 stars	Value	None
3 stars	Cleanliness	None
2 stars	None	Hotel condition

Note: Only significant variables are presented ($p < 0.001$).

The results in the previous table partly confirm the hypothesis of Melián-González et al. (2013) and Martin-Fuentes et al. (2018b), showing that an increase in the number of reviews has a significant impact on the average rating. The non-significant result for Booking.com was also found by Martin-Fuentes et al. (2018b), possibly being explained by the fact that reviews with more than 24 months old aren't considered in the overall satisfaction score. The control variable also evidences that the higher a hotel quality segment is, the more likely it has a higher rating. Considering the result for the TripAdvisor sample, the obtained coefficient implies that 100 more reviews increase the average rating by 0.034. This result that volume counts more than valence is particularly interesting to the lower segment hotels analyzed, since

Table 7. Overall satisfaction explained by the number and standard deviation of reviews

	Booking.com		TripAdvisor		Expedia	
Constant	8.97648 (***)	7.88526 (***)	3.95702 (***)	3.26980 (***)	4.65493 (***)	4.06054 (***)
Nr. of reviews	0.000021	0.000026	0.000340 (***)	0.000020 (***)	0.000043	-0.000024
Stand. dev. of reviews	-1.37279 (***)	-1.00327 (***)	-0.86033 (**)	-0.85328 (***)	-2.79282 (***)	-2.05853 (***)
Star		0.27062 (***)		0.23192 (***)		0.14094 (***)
R^2	0.181	0.355	0.236	0.391	0.503	0.583
adjusted R^2	0.169	0.341	0.223	0.376	0.495	0.572
F	9.247	29.504	19.972	29.205	26.142	64.003

Note: Dependent variable = overall satisfaction.
Robust standard errors (HC1). * $p<0.05$; ** $p<0.01$ and *** $p<0.001$.

their lower average review ratings are also accompanied by a lower number of reviews. Also, and more interestingly, as the variance of review ratings decreases, the average rating tends to be higher, reflecting the higher variance observed in low-quality hotels. This result implies that if a hotel gets a lot of mixed reviews its average rating is substantially lower. So, in order to counteract that effect, hotels managers need to encourage loyal customers to write reviews and at least respond to negative reviews. As stated by Bandi Tanner and Hämmerli (2018), the degree of consensus among reviewers constitutes a validity indicator (Schindler & Bickart, 2005), entailing that website impact, as the number of reviews posted by hotel should foster credibility. While the available information on a specific hotel increases with the number of reviews, the bias due to inauthentic reviews decreases, and the online reputation of a hotel might also be strengthened (Park & Lee, 2008).

These results have implications for hospitality management, since they highlight the most important determinants of overall satisfaction ratings. The "service" and "hotel condition" attributes show a greater correlation with overall satisfaction than other criteria, corroborating the results from Jeong and Jeon (2008), Limberger et al. (2014a, 2014b), Stringam et al. (2010) and Pacheco (2017). For instance, the importance of "service" confirms the findings from Carneiro and Costa (2000) who analyzed Portuguese 5-star hotels, noticing that "service" includes aspects such as service efficiency and staff attitude, which seems relevant to overall satisfaction and deserving of more attention from hotel managers. Thus, the "service" attribute contributes to hotel differentiation and rating. It is also important to note that, globally, "location" does yield significant results, a result also found by Martin-Fuentes et al. (2018a) and Limberger et al. (2014a; 2014b), who stress that location does not seem to be relevant to overall satisfaction, albeit being an influential factor when choosing a hotel. Naturally, a hotel cannot readily change its location, but it can assist guests in using public transport or provide a shuttle service. The "hotel condition" attribute is almost always present, showing that it is independent of the hotel segment and online platform, and room size, decoration and modernity are major determinants of overall satisfaction. The significance of "comfort", "service" and "hotel condition" in lower segment hotels should merit the attention of hotel managers to allocate more resources to this, thus avoiding negative reviews.

CONCLUSION AND FURTHER RESEARCH DIRECTIONS

According to some authors, hotel classification systems have lost credibility as a quality standard because some of the criteria are outdated or because the customer expectations are more related to the quality of services than to the hotel classification (López Fernández & Serrano Bedia, 2004; Torres, Adler, & Behnke, 2014). In the last decade, quality indications derived from user-generated platforms assumed increasing prominence, fading traditional indicators such as stars. Nevertheless, our study confirms that the higher a hotel segment is the higher score it obtains in the three considered online platforms. So, hotels' overall quality can be inferred from their stars in line with Fang et al. (2016).

This study contributes to the hospitality literature by identifying the most significant attributes determining overall satisfaction for a large sample of Portuguese hotels, located in the Porto metropolitan area, distributed by four quality segments and three popular online travel platforms. It is evidenced that Booking.com, TripAdvisor and Expedia, while all incorporating consumer reviews as primary social knowledge, are distinct from each other on some aspects. The three platforms present roughly the same supply of hotels, albeit presenting some differences in terms of volume of data, with Expedia being substantially smaller. In terms of specific attributes, with the exception of "service", the three platforms present differences that may reflect the different user bases on these platforms, a fact that limits the comparability of reviews between platforms.

Only were selected hotels located in the Porto metropolitan area with a sufficient number of reviews, so the results have some limitations in terms of generalization for other regions or countries. This study uniquely attempts to identify how consistent guests' perceptions of hotel performance are related with hotel segment and online platform through their reviews on Booking.com, TripAdvisor and Expedia. Thus, this chapter is of an exploratory nature, presenting some limitations that should be surpassed in future research, namely: (i) since this study evaluates secondary data retrieved from online platforms, issues related to the integrity of guests' reviews and the lack of control in data collection could compromise its conclusions. It is possible that some reviews are biased – coming from the hotel itself, from competitors or simply from a "virtual" guest, so that future research should identify the impact of biased comments (this bias could be particularly relevant for TripAdvisor reviews, which mostly come from "non-verified users", contrary to what happens in the other two platforms, where reviews are made by actual guests); (ii) this study is quantitative – it considers only the general classifications posted by different users without analyzing the words used in each post (a semantic content analysis), which could provide additional relevant information. In their written posts guests may show significant differences in satisfaction concerning those attributes, since each review normally has descriptions that are likely to include valuable information that can give further explanations for a particular rating. A deeper analysis would also allow understanding why some hotels have more customer reviews than others; finally, (iii) the different importance of attributes may depend on the type of guests, their gender, age, cultural background and other variables. Online reviews could vary across different platforms in terms of linguistic characteristics, semantic features, sentiment as well as impact on website's users. For instance, business travelers and leisure travelers may show differences in ratings since their expectations regarding room comfort or service are probably different. The same applies for guests of different ages, so it is important to study that issue in greater depth and characterize the review writers.

The power and importance of information is unquestionable, and for the tourist destination to become intelligent and competitive, it is essential to carry out planning based on reliable information that allows efficient and sustainable management. And for help in managing this information, Big Data software becomes a great ally, allowing for quick and in-depth analysis of the data in it. Current product search engines fail to effectively leverage information created across diverse social media platforms. This requires research on how to design and innovate in IT or website design, to adapt data management to evolving social media development and to share information across different media sources (Ghose, Ipeirotis, & Li, 2012). For instance, it seems worthwhile to exploit the idea of closer integration of hotel classifications and guests' reviews, integrating the latter on future quality classification schemes (Blomberg-Nygard and Anderson, 2016).

The opinions of guests constitute a source of quality segmentation that allows better positioning of each hotel. This study contributes to better management of hotels, identifying which evaluation criteria used by guests are more strongly correlated with (and explain) overall satisfaction. It confirms the results from some previous authors, showing that the high correlation between attributes and overall satisfaction supports previous research in service quality management, which argues that the perceptions of parts of the service delivery are closely related to overall perception. Thus, hotel managers should not ignore any of the attributes. These findings provide hotel managers with an understanding of guest's perceptions of hotel performance by hotel segment and online travel platforms, showing which performance attributes contribute most to guest satisfaction. Some areas are identified where small improvements and investments could make a difference in terms of rating and bring a competitive advantage. On one hand, the greatest gains to overall satisfaction seem to be explained by the "service" and "hotel condition" attributes. Albeit personal preferences and expectations may make it difficult to optimally up-grade rooms in order to satisfy a diverse clientele, regarding "service", hotel managers across all hotel segments should prioritize staff training and service-quality assessment tools in order to guarantee consistent levels of service. On the other hand, constant attention to "cleanliness" (in particular, in the lower-segment hotels) and minor improvements in transportation services to and from the city center (also, in particular regarding lower segment-hotels) could potentially upgrade hotel ratings.

Finally, since this chapter evidences the validity of the different platforms considered, it is advisable for hotel managers to regularly monitor hotel review platforms to glean industry-wide trends from guest voices. A comprehensive marketing strategy that considers reviews of guests' comments or user-generated feedback can help hotel managers to improve service quality from the wealth of data provided, translating that into higher ratings and visibility and, potentially, higher occupancy rates and revenues.

REFERENCES

Anderson, C. K. (2012). The impact of social media on lodging performance. *Cornell Hospitality Report*, *12*(15), 4–11.

Ayeh, J. K., Au, N., & Law, R. (2013). "Do we believe in TripAdvisor?" Examining credibility perceptions and online travelers' attitude toward using user-generated content. *Journal of Travel Research*, *52*(4), 437–452. doi:10.1177/0047287512475217

Ayoun, B., & Moreo, P. (2008). Does national culture affect hotel managers' approach to business strategy. *International Journal of Contemporary Hospitality Management*, *32*(1), 52–65.

Baka, V. (2016). The becoming of user-generated reviews: Looking at the past to understand the future of managing reputation in the travel sector. *Tourism Management*, *53*, 148–162. doi:10.1016/j.tourman.2015.09.004

Bandi Tanner, M., & Hämmerli, S. J. (2018). Reviewing TripAdvisor and Co.: A quality analysis of hotel review sites. *Anatolia*, *29*(4), 518–528. doi:10.1080/13032917.2018.1451350

Bjørkelund, E., Burnett, T. H., & Nørvåg, K. (2012, December). A study of opinion mining and visualization of hotel reviews. *Proceedings of the 14th International Conference on Information Integration and Web-based Applications & Services* (pp. 229-238). ACM. 10.1145/2428736.2428773

Blomberg-Nygard, A., & Anderson, C. K. (2016). United Nations world tourism organization study on online guest reviews and hotel classification systems: An integrated approach. *Service Science*, *8*(2), 139–151. doi:10.1287erv.2016.0139

Buzova, D., Sanz-Blas, S., & Cervera-Taulet, A. (2019). Does culture affect sentiments expressed in cruise tours' eWOM? *Service Industries Journal*, *39*(2), 154–173. doi:10.1080/02642069.2018.1476497

Cantallops, A., & Salvi, F. (2014). New consumer behavior: A review of research on eWOM and hotels. *International Journal of Hospitality Management*, *36*, 41–51. doi:10.1016/j.ijhm.2013.08.007

Carneiro, M. J., & Costa, C. (2000). The influence of service quality on the positioning of five star hotels - the case of the Lisbon area. *Journal of Quality Assurance in Hospitality & Tourism*, *1*(4), 1–19. doi:10.1300/J162v01n04_01

Casaló, L. V., Flavian, C., Guinaliu, M., & Ekinci, Y. (2015). Do online hotel rating schemes influence booking behaviors? *International Journal of Hospitality Management*, *49*, 28–36. doi:10.1016/j.ijhm.2015.05.005

Cavanagh, W. (2015). *Índice del Valor de la Información para ayudar a las empresas a medir su propio rendimento*. Retrieved from https://www.pwc.es/es/sala-prensa/notas-prensa/2015/Seizing-The-Information-Advantage.html

Chareyron, G., Da-Rugna, J., & Raimbault, T. (2014). Big data: A new challenge for tourism. *Proceedings of the 2014 IEEE International Conference on Big Data (Big Data)* (pp. 5-7). IEEE Press. 10.1109/BigData.2014.7004475

Chaves, M., Gomes, R., & Pedron, C. (2012). Analyzing reviews in the web 2.0: Small and medium hotels in Portugal. *Tourism Management*, *33*(5), 1286–1287. doi:10.1016/j.tourman.2011.11.007

Chen, R., Cheung, C., & Law, R. (2012). A review of the literature on culture in hotel management research: What is the future? *International Journal of Hospitality Management*, *31*(1), 52–65. doi:10.1016/j.ijhm.2011.06.010

Chung, N., & Koo, C. (2015). The use of social media in travel information search. *Telematics and Informatics*, *32*(2), 215–229. doi:10.1016/j.tele.2014.08.005

Dans, E. (2012). *Big data: una perspectiva histórica.* Retrieved from https://www.enriquedans.com/2012/05/big-data-una-perspectiva-historica.html

Fang, B., Ye, Q., Kucukusta, D., & Law, R. (2016). Analysis of the perceived value of online tourism reviews: Influence of readability and reviewer characteristics. *Tourism Management, 52*, 498–506. doi:10.1016/j.tourman.2015.07.018

Filieri, R. (2016). What makes an online consumer review trustworthy? *Annals of Tourism Research, 58*, 46–64. doi:10.1016/j.annals.2015.12.019

Filieri, R., Alguezaui, S., & McLeay, F. (2015). Why do travelers trust TripAdvisor? Antecedents of trust towards consumer-generated media and its influence on recommendation adoption and word of mouth. *Tourism Management, 51*, 174–185. doi:10.1016/j.tourman.2015.05.007

Franks, B. (2012). *Taming the Big Data Tidal Wave: Finding Opportunities in Huge Data Streams with Advanced Analytics.* John Wiley & Sons, Inc. doi:10.1002/9781119204275

García-Palomares, J. C., Gutiérrez, J., Mínguez, C., Gutierrez, J., & Mínguez García, C. (2015). Identification of tourist hot spots based on social networks: A comparative analysis of European metropolises using photo-sharing services and GIS. *Applied Geography, 63*, 408–417. doi:10.1016/j.apgeog.2015.08.002

Gerdes, J. Jr, & Stringam, B. (2008). Addressing researchers' quest for hospitality data: Mechanism for collecting data from web resources. *Tourism Analysis, 13*(3), 309–315. doi:10.3727/108354208786094906

Ghose, A., Ipeirotis, P. G., & Li, B. (2012). Designing ranking systems for hotels on travel search engines by mining user-generated and crowdsourced content. *Marketing Science, 31*(3), 493–520. doi:10.1287/mksc.1110.0700

Gujarati, D. (2003). *Basic Econometrics* (4th ed.). Boston: McGraw-Hill.

Hsieh, A., & Tsai, C. (2009). Does national culture really matter? Hotel service perceptions by Taiwan and American tourists. *International Journal of Culture, Tourism and Hospitality Research, 3*(1), 54–69. doi:10.1108/17506180910940342

Huang, J., Huang, C., & Wu, S. (1996). National character and response to unsatisfactory hotel service. *International Journal of Hospitality Management, 15*(3), 229–243. doi:10.1016/S0278-4319(96)00009-6

Huang, Y., Basu, C., & Hsu, M. (2010). Exploring motivations of travel knowledge sharing on social network sites: An empirical investigation of U.S. college students. *Journal of Hospitality Marketing & Management, 19*(7), 717–734. doi:10.1080/19368623.2010.508002

IDC. (2018). *Los retos de la gestión de datos.* Retrieved from https://custom.cvent.com/1E8AD1B771DA4B029B78FF1784749EF5/files/c595061048184ba49482223d066ac565.pdf

INE. (2019). *Destaque: Crescimento nos principais indicadores do setor do alojamento turístico.* Instituto Nacional de Estatística. Retrieved from www.ine.pt

Jeong, M., & Jeon, M. (2008). Customer Reviews of Hotel Experiences through Consumer Generated Media (CGM). *Journal of Hospitality & Leisure Marketing, 17*(1-2), 121–138. doi:10.1080/10507050801978265

Jiang, J., Gretzel, U., & Law, R. (2014). Influence of Star Rating and Ownership Structure on Brand Image of Mainland China Hotels. *Journal of China Tourism Research*, *10*(1), 69–94. doi:10.1080/19388160.2013.870506

Kaplan, A. M., & Haenlein, M. (2010). Users of the world, unite! The challenges and opportunities of Social Media. *Business Horizons*, *53*(1), 59–68. doi:10.1016/j.bushor.2009.09.003

Kisilevich, S., Krstajic, M., Keim, D., Andrienko, N., & Andrienko, G. (2010). Event-based analysis of people's activities and behavior using Flickr and Panoramio geotagged photo collections. *Proceedings of the International Conference on Information Visualisation* (pp. 289-296). Academic Press. 10.1109/IV.2010.94

Lancestremère, S. (2015). *Big Data, Internet de las Cosas y sus aplicaciones prácticas. Fundaciontelefonica.* Retrieved from https://espacio.fundaciontelefonica.com/evento/big-data-internet-de-las-cosas-y-sus-aplicaciones-practicas-2/

Lee, H., Law, R., & Murphy, J. (2011). Helpful reviewers in TripAdvisor, an online travel community. *Journal of Travel & Tourism Marketing*, *28*(7), 675–688. doi:10.1080/10548408.2011.611739

Lee, J., Park, D., & Han, I. (2008). The effect of negative online consumer reviews on product attitude: An information processing view. *Electronic Commerce Research and Applications*, *7*(3), 341–352. doi:10.1016/j.elerap.2007.05.004

Li, H., Ye, Q., & Law, R. (2013). Determinants of customer satisfaction in the hotel industry: An application of online review analysis. *Asia Pacific Journal of Tourism Research*, *18*(7), 784–802. doi:10.1080/10941665.2012.708351

Limberger, P., Anjos, F., Meira, J., & Anjos, S. (2014a). Satisfaction in hospitality on TripAdvisor.com: An analysis of the correlation between evaluation criteria and overall satisfaction. *Tourism & Management Studies*, *10*(1), 59–65.

Limberger, P., Boaria, F., & Anjos, S. (2014b). A relação entre a satisfaçãogeral e as variáveis da satisfação na hotelaria em hotéis de excelência. *Revista Brasileira de Pesquisa em Turismo*, *8*(3), 435–455. doi:10.7784/rbtur.v8i3.803

Litvin, S., Goldsmith, R., & Pan, B. (2008). Electronic word-of-mouth in hospitality and tourism management. *Tourism Management*, *29*(3), 458–468. doi:10.1016/j.tourman.2007.05.011

Lopes, R., Abrantes, J., & Kastenholz, E. (2014). Innovation, tourism and social networks. *Revista Turismo e Desenvolvimento*, *21/22*, 151–154.

López Fernández, M. C., & Serrano Bedia, A. M. (2004). Is the hotel classification system a good indicator of hotel quality? *Tourism Management*, *25*(6), 771–775. doi:10.1016/j.tourman.2004.06.007

Lu, W., & Stepchenkova, S. (2012). Ecotourism experiences reported online: Classification of satisfaction attributes. *Tourism Management*, *33*(3), 702–712. doi:10.1016/j.tourman.2011.08.003

Martin-Fuentes, E. (2016). Are guests of the same opinion as the hotel star-rate classification system? *Journal of Hospitality and Tourism Management*, *29*, 126–134. doi:10.1016/j.jhtm.2016.06.006

Martin-Fuentes, E., Mateu, C., & Fernandez, C. (2018a). Does verifying uses influence rankings? Analyzing Booking. com and TripAdvisor. *Tourism Analysis*, *23*(1), 1–15. doi:10.3727/108354218X15143857349459

Martin-Fuentes, E., Mateu, C., & Fernandez, C. (2018b). The more the merrier? Number of reviews versus score on TripAdvisor and Booking.com. *International Journal of Hospitality & Tourism Administration*, 1–14. doi:10.1080/15256480.2018.1429337

Martin-Fuentes, E., Mateu, C., & Fernandez, C. (2018c). Are users' ratings on TripAdvisor similar to hotel categories in Europe? *Cuadernos de Turismo*, *42*(42), 305–316. doi:10.6018/turismo.42.13

McAfee, A., Brynjolfsson, E., Davenport, T. H., Patil, D. J., & Barton, D. (2012). Big data: The management revolution. *Harvard Business Review*, *90*(10), 60–68. PMID:23074865

Melián-González, S., Bulchand-Gidumal, J., & López-Valcárcel, B. (2013). Online customer reviews of hotels: As participation increases, better evaluation is obtained. *Cornell Hospitality Quarterly*, *54*(3), 274–283. doi:10.1177/1938965513481498

Mellinas, J. P., María-Dolores, S. M. M., & García, J. J. B. (2015). Booking. com: The unexpected scoring system. *Tourism Management*, *49*, 72–74. doi:10.1016/j.tourman.2014.08.019

Mellinas, J. P., María-Dolores, S. M. M., & García, J. J. B. (2016). Effects of the Booking. com scoring system. *Tourism Management*, *57*, 80–83. doi:10.1016/j.tourman.2016.05.015

Mkono, M. (2018). 'Troll alert!': Provocation and harassment in tourism and hospitality social media. *Current Issues in Tourism*, *21*(7), 791–804. doi:10.1080/13683500.2015.1106447

Morosan, C. (2013). The Impact of the Destination's Online Initiatives on Word of Mouth. *Tourism Analysis*, *18*(4), 415–428. doi:10.3727/108354213X13736372325993

O'Connor, P. (2010). Managing a hotel's image on TripAdvisor. *Journal of Hospitality Marketing & Management*, *19*(7), 754–772. doi:10.1080/19368623.2010.508007

Pacheco, L. (2016). An analysis of online reviews by language groups: The case of hotels in Porto, Portugal. *European Journal of Tourism Research*, *14*, 66–74.

Pacheco, L. (2017). Customer satisfaction in Portuguese hotels: Evidence for different regions and hotel segments. *Tourism Analysis*, *22*(3), 337–347. doi:10.3727/108354217X14955605216087

Park, D., & Lee, J. (2008). eWOM overload and its effect on consumer behavioral intention depending on consumer involvement. *Electronic Commerce Research and Applications*, *7*(4), 386–398. doi:10.1016/j.elerap.2007.11.004

Redman, T. C. (2017). Does Your Company Know What to Do with All Its Data? *Harvard Business Review*. Retrieved from https://hbr.org/2017/06/does-your-company-know-what-to-do-with-all-its-data

Ruths, D., & Pfeffer, J. (2014). Social media for large studies of behavior. *Science*, *346*(6213), 1063–1064. doi:10.1126cience.346.6213.1063 PMID:25430759

Schindler, R. M., & Bickart, B. (2005). Published word of mouth: Referable, consumer-generated information on the Internet. *Online Consumer Psychology: Understanding and influencing consumer behavior in the virtual world, 32*, 35-61.

Schuckert, M., Liu, X., & Law, R. (2015a). Hospitality and tourism online reviews: Recent trends and future directions. *Journal of Travel & Tourism Marketing, 32*(5), 608–621. doi:10.1080/10548408.2014.933154

Schuckert, M., Liu, X., & Law, R. (2015b). A segmentation of online reviews by language groups: How English and non-English speakers rate hotels differently. *International Journal of Hospitality Management, 48*, 143–149. doi:10.1016/j.ijhm.2014.12.007

Song, H., & Liu, H. (2017). Predicting tourist demand using big data. In *Analytics in smart tourism design* (pp. 13–29). Cham: Springer International Publishing. doi:10.1007/978-3-319-44263-1_2

Sparks, B., & Browning, V. (2011). The impact of online reviews on hotel booking intentions and perception of trust. *Tourism Management, 32*(6), 1310–1323. doi:10.1016/j.tourman.2010.12.011

Stringam, B., & Gerdes, J. Jr. (2010). An Analysis of word-of-mouse ratings and guest comments of online hotel distribution sites. *Journal of Hospitality Marketing & Management, 19*(7), 773–796. doi:10.1080/19368623.2010.508009

Stringam, B., Gerdes, J. Jr, & Vanleeuwen, D. (2010). Assessing the importance and relationships of ratings on user-generated traveler reviews. *Journal of Quality Assurance in Hospitality & Tourism, 11*(2), 73–92. doi:10.1080/1528008X.2010.482000

Tavakoli, R., & Wijesinghe, S. N. (2019). Tourism management perspectives. *Tourism Management, 29*, 48–55.

Torres, E. N., Adler, H., & Behnke, C. (2014). Stars, diamonds, and other shiny things: The use of expert and consumer feedback in the hotel industry. *Journal of Hospitality and Tourism Management, 21*, 34–43. doi:10.1016/j.jhtm.2014.04.001

Tse, E., & Ho, S. (2009). Service quality in the hotel industry when cultural contexts matter. *Cornell Hospitality Quarterly, 50*(4), 460–474. doi:10.1177/1938965509338453

Vermeulen, I., & Seegers, D. (2009). Tried and tested: The impact of online hotel reviews on consumer consideration. *Tourism Management, 30*(1), 123–127. doi:10.1016/j.tourman.2008.04.008

Wilson, A., Murphy, H., & Cambra Fierro, J. (2012). Hospitality and travel: The nature and implications of user-generated content. *Cornell Hospitality Quarterly, 53*(3), 220–228. doi:10.1177/1938965512449317

Xiang, Z., Du, Q., Ma, Y., & Fan, W. (2017). A comparative analysis of major online review platforms: Implications for social media analytics in hospitality and tourism. *Tourism Management, 58*, 51–65. doi:10.1016/j.tourman.2016.10.001

Xiang, Z., & Gretzel, U. (2010). Role of social media in online travel information search. *Tourism Management, 31*(2), 179–188. doi:10.1016/j.tourman.2009.02.016

Xie, H., Miao, L., Kuo, P.-J., & Lee, B.-Y. (2011). Consumers' responses to ambivalent online hotel reviews: The role of perceived source credibility and pre-decisional disposition. *International Journal of Hospitality Management*, *30*, 178–183. doi:10.1016/j.ijhm.2010.04.008

Ye, Q., Law, R., & Gu, B. (2009). The impact of online user reviews on hotel room sales. *International Journal of Hospitality Management*, *28*(1), 180–182. doi:10.1016/j.ijhm.2008.06.011

Ye, Q., Law, R., Gu, B., & Chen, W. (2011). The influence of user-generated content on traveler behavior: An empirical investigation on the effects of eword-of-mouth to hotel online bookings. *Computers in Human Behavior*, *27*(2), 634–639. doi:10.1016/j.chb.2010.04.014

Yen, C. L. A., & Tang, C. H. H. (2015). Hotel attribute performance, eWOM motivations, and media choice. *International Journal of Hospitality Management*, *46*, 79–88. doi:10.1016/j.ijhm.2015.01.003

Yoo, K.-H., & Gretzel, U. (2008). Use and impact of online travel reviews. In P. O'Connor, W. Hopken, & U. Gretzel (Eds.), *Information and Communication Technologies in Tourism 2008* (pp. 35–46). Springer.

Yoo, K.-H., & Gretzel, U. (2011). Influence of personality on travel-related consumer-generated media creation. *Computers in Human Behavior*, *27*(2), 609–621. doi:10.1016/j.chb.2010.05.002

Zeng, B., & Gerritsen, R. (2014). What do we know about social media in tourism? A review. *Tourism Management Perspectives*, *10*, 27–36. doi:10.1016/j.tmp.2014.01.001

Zheng, T., Youn, H., & Kincaid, C. (2009). An analysis of customers' e-complaints for luxury resort properties. *Journal of Hospitality Marketing & Management*, *18*(7), 718–729. doi:10.1080/19368620903170240

ADDITIONAL READING

De Mauro, A., Greco, M., & Grimaldi, M. (2015). What is big data? A consensual definition and a review of key research topics. *AIP Conference Proceedings*, *1644*(1), 97–104. doi:10.1063/1.4907823

Gretzel, U., Sigala, M., Xiang, Z., & Koo, C. (2015). Smart tourism: Foundations and developments. *Electronic Markets*, *25*(3), 179–188. doi:10.100712525-015-0196-8

Gretzel, U., Werthner, H., Koo, C., & Lamsfus, C. (2015). Conceptual foundations for understanding smart tourism ecosystems. *Computers in Human Behavior*, *50*, 558–563. doi:10.1016/j.chb.2015.03.043

Jani, D., Jang, J. H., & Hwang, Y. H. (2014). Big five factors of personality and tourists' internet search behavior. *Asia Pacific Journal of Tourism Research*, *19*(5), 600–615. doi:10.1080/10941665.2013.773922

Marko, K. (2015). Using big data and machine learning to enrich customer experiences. Forbes. Retrieved from http://www.forbes.com/sites/kurtmarko/2015/04/08/big-data-machine-learning_customer-experience/

Shi, Y. (2014). Big data: History, current status, and challenges going forward. *Bridge*, *44*(4), 6–11.

Snijders, C., Matzat, U., & Reips, U.-D. (2012). Big data: Big gaps of knowledge in the field of internet science. *International Journal of Internet Science*, *7*(1), 1–5.

Turner, E. (2014). How big data is changing the travel industry. Retrieved from http://blog.sumall.com/journal/big-data-changing-travel-industry.html

Varian, H. R. (2014). Big data: New tricks for econometrics. *The Journal of Economic Perspectives*, *28*(2), 3–27. doi:10.1257/jep.28.2.3

Xiang, Z., & Fesenmaier, D. R. (2017). *Big Data Analytics in Smart Tourism Design. In Analytics in Smart Tourism Design* (pp. 299–307). Cham: Springer International Publishing. doi:10.1007/978-3-319-44263-1_17

KEY TERMS AND DEFINITIONS

Big Data: Extremely large data sets that can be analyzed in order to reveal patterns, trends and associations.

Electronic Word-of-Mouth (eWOM): Is a form of online recommendation or buzz marketing and it can become viral if the message is persuasive or funny enough.

Hospitality: Friendly and welcoming reception and entertainment of guests or visitors.

Hotel: A commercial establishment providing lodging, meals, and other guest services.

Online Travel Agency (OTA): Travel website specialized in the sale of travel products to consumers.

Porto Metropolitan Area: Geographic zone currently composed of 17 contiguous municipalities centered on the city of Porto, in Portugal.

Social Media: Websites and applications that enable users to create content or to participate in social networking.

Tourism: Comprises all the activities of persons travelling to and staying in places outside their usual environment, for leisure, business and other purposes.

User-Generated Content (UGC): Digital content that is produced and shared by end users of an online service or website (also known as consumer-generated media).

Chapter 17
eWOW of Guests Regarding Their Hotel Experience: Sentiment Analysis of TripAdvisor Reviews

Zelia Breda
https://orcid.org/0000-0002-5882-063X
GOVCOPP, University of Aveiro, Portugal

Rui Costa
GOVCOPP, University of Aveiro, Portugal

Gorete Dinis
GOVCOPP, Polytechnic Institute of Portalegre, Portugal

Amandine Angie Martins
University of Aveiro, Portugal

ABSTRACT

Online comments are increasingly mentioned as an important source of information, simplifying consumers' buying decisions. Online user-generated content has become one of the main sources of information for tourists, who themselves become creators of their own online content. This chapter focuses on sentiment analysis of comments made on TripAdvisor regarding one resort located in the Algarve region, in Portugal. The resort has good reviews, which means that the eWOM is positive. The highest scores relate to the resort's cleanliness, location and quality of sleep, and those that were less relevant were the value for money, the rooms and the service. The most dominant emotion is joy, followed by an analytical response. Negative emotions, such as sadness and anger, were not found very often in the online reviews. These results could be explained by the quality of the service, the kindness of the staff, the facilities for children, the entertainment, and the location, attributes that were often highlighted in the comments.

DOI: 10.4018/978-1-7998-1947-9.ch017

INTRODUCTION

The Internet has become an essential tool for any user, either on a personal or on a professional level. The rapid evolution of information and communication technologies (ICT) has given rise to Web 2.0, mainly characterized by user-generated content (UGC), or electronic word of mouth (eWOM). Companies used to control the information they wanted to share, now users determine the information they want to see and to consume (Limberger, Anjos, Meira, & Anjos, 2014).

In tourism, social media predominates in terms of knowledge and information sharing, being recognized as an innovative knowledge sharing tool (e.g. product reviews, market trends, restaurant and hotel reviews, and travel blogs), allowing users to interconnect, share the desired information and interact with other users (Gaál, Szabó, Obermayer-Kovács, & Csepregi, 2015). This type of user-generated content allows to reach a greater number of users in a shorter time, eliminating restrictions of time, location and speed during the exchange of information. On the other hand, it can affect purchasing decisions, as well as consumer perceptions of product quality, since most of the comments on online platforms serve to recommend or discourage a product (Limberger et al., 2014).

Sentiment analysis can be applied to user-generated content, such as online reviews, survey responses and social media, aiming to identify ideas and opinions about a particular subject or event, classifying and quantifying them as positive, negative or neutral (Liu, 2015). Pang and Lee (2008) state that there are numerous companies in the world, especially in the United States, that provide sentiment analysis services on the Internet. It is considered that through sentimental analysis, e-commerce benefits can be realized, and companies tend to promote their products more on a website, blog or social networks. Organizations will have other benefits such as ease of designing marketing strategies, improved security policies, accurate perception of a product / service by a customer, among others.

Studies on sentiment analysis are still relatively scarce (e.g. Freitas & Vieira, 2015; Neethu & Rajasree, 2013; Pak & Paroubek, 2010; Serrano-Guerrero, Olivas, Romero, & Herrera-Viedma, 2015; Valdivia, Luzón, & Herrera, 2017), thus, this study aims to provide a contribution in the tourism sector. This chapter focuses on sentiment analysis of comments made on TripAdvisor regarding one resort located in the Algarve region, in Portugal. In general, this study aims to identify: (i) the profile of the visitor; (ii) the general and specific evaluation for each category of the resort; and (iii) the emotional intonations of the visitors' comments. The methodology of the study is based on qualitative and quantitative analysis of secondary sources, i.e. comments posted by customers on TripAdvisor. A sample of 294 comments was used to perform the sentiment analysis, using IBM Watson Tone Analyzer. After this analysis, results were imported on IBM SPSS v. 25, where univariate and bivariate analyses were performed.

The chapter is organized as follows: first, the literature on social media and tourism is reviewed, following which user-generated content and eWOM are addressed. Subsequently, the methodology used for the empirical study is described in more detail. Finally, research findings are reported and discussed, and their implications for further research and applications are highlighted.

LITERATURE REVIEW

Social Media and Tourism

The availability of the Internet through various platforms allows users to access, in real time, a large amount of information, such as maps, information about a particular destination, attraction or service, among others. It is inevitable that people's behaviour has influenced the form of communication (Afonso & Borges, 2013).

In Portugal, the number of residents aged 15 and above accessing social networks increased significantly from 1,417 thousand (17.1%) in 2008 to 5,064 thousand (59.1%) in 2017 (Marktest, 2017). According to the same study, the proportion of users in the 15-24 age group is not much higher than that of users between 45 and 64 years old. However, the reason why young people are connected is mainly because entertainment and leisure, while adults, in addition to this, use more social networks to search for health information, opinions on articles, among others. Facebook (95.3%), WhatsApp (74.2%) and Instagram (67.9%) are the most used social networks in Portugal (Marktest, 2019).

Technological advances influence the way companies and consumers relate to each other, undermining traditional communication management models of transactional marketing processes. Interest in social networking systems is growing as industries have to consider new forms of planning, marketing of tourism products and services (Di Pietro, Di Virgilio, & Pantano, 2012).

Social networks have become increasingly relevant and are generally used to: (i) improve product / service and brand awareness, thereby managing reputation; (ii) increase customer loyalty, that is, the customer is more exposed to publications and these may affect the feeling of loyalty to the product or service; (iii) simplify market research through the creation of new ideas, also helping in the projection of new products or services; and (iv) emphasize eWOM, stimulating users' trust among themselves (Berthon, Pitt, Plangger, & Shapiro, 2012; Vermeulen & Seegers, 2009; Vlachvei & Notta, 2014).

For marketers, social networks have advantages such as: (i) simple and instant information update; (ii) greater freedom and efficacy in measuring the results of a marketing campaign; (iii) marketing strategies aimed at triggering a viral result; (iv) plurilateral and participatory communication in order to create greater contact with the public; (v) user interaction and learning about the particularities of products, services, news; (vi) reduced marketing costs; (vii) insight into customer needs to take into account in future strategies; (viii) direct traffic to the company's website, blog, articles, among others; (ix) campaigns aimed at a certain audience; (x) real time market perception; (xi) input of ideas by users; and (xii) improved customer service through closer relationships (Afonso & Borges, 2013; Hays, Page, & Buhalis, 2013; Leung, Law, Van Hoof, & Buhalis, 2013; Sigala, Christou, & Gretzel, 2012). However, social networks have also some disadvantages that should be taken into account when the manager or marketer decides whether or not to be present on social networks. These usually refer to the visibility and reach of the social network, which, in case of negative comments generated by users, it will contribute to a pejorative image of the company; and the need to have employees with availability and with specific knowledge for effective social network management (Afonso & Borges, 2013; Hays, Page, & Buhalis, 2013; Leung, Law, Van Hoof, & Buhalis, 2013; Sigala, Christou, & Gretzel, 2012).

Regarding the organization and fruition of travel, the desire to use social networks is directly related to the user's perception of their benefits (Wang, Yu, & Fesenmaier, 2002). However, the process of identifying the benefits of using social networks is complex, since there are so many travellers and not everyone is looking for the same information, which also depends on their ability to use it. According to Munar and Jacobsen (2014), altruistic and community-related motivations are most relevant for information sharing, differing depending on the type of content and type of social media. Other authors (Hsu & Lin, 2008; Wang & Fesenmaier, 2004; Wang, Yu, & Fesenmaier, 2002) also report that tourists can enjoy positive dynamics in using social networks depending on the functional, social, psychological and hedonic benefits of their use.

In the case of tourism, in terms of functional benefits, the use of social networks is effective because it allows travellers to keep up to date on tourist sites they want to visit and the activities of interest (Yoo & Gretzel, 2008). Since the price of travel arrangements is one of the motivations for seeking online reviews (Goldsmith & Horowitz, 2006), active Internet collaboration certainly helps tourists save money; and the exchange of information between social network users provides mutual benefits (Wang & Fesenmaier, 2004).

With regard to social benefits, social networks allow: (i) visitors to always share and maintain contact with others who share the same interests in travel; (ii) increase interpersonal relationships between people who have the same motivations for travel, further emphasizing interest; (iii) develop a strong feeling in the users that they belong to a group with the same interest (Leung et al., 2013; Wang & Fesenmaier, 2004).

Regarding psychological and hedonic benefits, online travel communities, in addition to meeting functional requirements and social needs, also meet basic psychological needs (Wang & Fesenmaier, 2004). The psychological benefits are related to users' trust in social networks and belonging to a particular group. These benefits can be achieved through persistent communication and participatory activity in a fully collaborative environment, meeting social benefits (Bressler & Grantham, 2000, cited by Chung & Buhalis, 2008). On the other hand, community members want not only functional, social and psychological benefits, but also fun and enjoyment, which fit the hedonic benefits. Using social media is fun and enjoyable, and travellers who participate in online communities to voice their opinions and share their tourist experiences are proud to do so (Wang & Fesenmaier, 2004).

TripAdvisor does not offer any kind of reward but a recognition for its loyal users. This form of gamification of the interaction with the platform becomes, in a way, a motivation for its use, since, when giving their opinion, users accumulate more points and "climb up the hierarchy", thus achieving the desired merit and notoriety among other members (Mkono, 2012; Yoo & Gretzel, 2010). Similar to the motivations for using social networks, the drive comes from the need for social recognition and is also a psychological incentive (Chen, Fay, & Wang 2011; Lecture, Fehr & Falk, 2002). Thus, the motivation results in the search for this social status that has as its principle to achieve more economic and social power (externally) and is also stimulated by psychological and emotional reasons (internally) (Perretti & Negro, 2006). However, the essence and main objective of this platform is to spread the experience of each user as their social status increases, which is important for virtual communities (Chen et al., 2011). The form of expressing feelings or something that may be happening through the publication of comments posted on other users' pages or on their own profile is also pointed out by Brandtzaeg and Heim (2009) as a reason for using social networks.

User-generated Content and eWOM

Today, WOM has moved from an informal conversation between people who get the word out about a product or service to sharing opinions / information on online platforms, such as social networks, blogs, review websites, discussion boards, among others (Cheung, Chiu, & Lee, 2012; Litvin, Goldsmith, &

Pan, 2018). A retrospective view of electronic word-of-mouth in hospitality and tourism management. International Journal of Contemporary Hospitality Management, 30(1), 313-325.). Online comments are increasingly mentioned as an important source of information, simplifying consumers' buying decisions (Zhang et al., 2014). Online user-generated content becomes one of the main sources of information for tourists, who themselves become creators and users of their own online content (Marchiori & Cantoni, 2015). In addition, this type of communication allows to reach a larger number of users in a shorter time, eliminating time, location and speed constraints during information exchange (Chaves, Gomes, & Pedron, n.d.).

According to a Skift.com report, overall, 89% of travellers and 64% of the hotel industry believe online reviews have an influence on reservations (Skift, 2014). In addition, 95% of travellers report they would rather read online reviews before making a reservation as they became a deciding factor in hotel selection (Ady & Quadri-Felitti, 2015). Thus, online content not only highlights opinions, but encompasses legitimacy and power of influence for other consumers, whether positively or negatively (Sales, Carvalho, Arruda, & Albuquerque, 2015). The online opinions of experienced and trusted users have a significant influence on the purchasing decisions of other travel consumers (Litvin, Goldsmith, & Pan, 2008, 2018). Today, online content influences other consumers' opinions, and numerical ratings are an important tool that can measure the value of comments themselves without having to read them in full (Godes & Mayzlin, 2004).

Thus, eWOM may, according to Stringam and Gerdes (2010), affect consumers' purchasing decisions and product quality perceptions, as most comments on online platforms provide information about the recommendation or discouragement of a product to other consumers. Cantallops and Salvi (2013) argue that eWOM includes various forms of online communication, such as opinions, recommendations and comments. This means that the type of information is both informative, providing additional information as a way to complement and guide the consumer's decision, as well as advisory, where the popularity of the product is indicated by a positive or negative score (Jalilvand, Esfahani, & Samiei, 2011).

There are many online platforms that allow users to interact and provide reviews on tourism services, namely TravBuddy, Travellerspoint, WAYN, Woophy, Wheretostay, Zoomandgo, Passportstamp, and TripAdvisor (Jeong & Jeong, 2008; Miguéns, Baggio, & Costa, 2008). The latter is probably the most recognized online review platform in the tourism and hospitality industry. It provides reviews (both quantitative and qualitative) from millions of users and a wide range of destination choices and planning features with built-in links to booking tools that check hundreds of websites to find the best prices. Studies addressing user-generated content and eWOM have already focused on TripAdvisor as a social media platform in travel content (e.g. Barcala, Díaz, & Rodriguez, 2009; Huang, Basu, & Hsu, 2010; Jeong & Jeon, 2008; Lei, & Law, 2015; Limberger et al., 2014; O'Connor, 2010; Sparks & Browning, 2010; Weilin & Svetlana, 2012).

METHODOLOGY

The main purpose of this study is to analyse the opinions shared by guests of a resort on the TripAdvisor online platform, in order to identify the emotion associated with these comments and therefore the experience gained during their stay. The specific objectives are to identify: (i) the profile of the visitor; (ii) the general and specific evaluation made for each category; and (iii) the emotional intonations of the visitors' comments. Thus, it is intended to qualify the comments separately by categories and to identify the type of traveller, and to quantify and analyse the emotion present in each one.

The study was applied to a resort located in the Algarve region, which is the only development of a renowned French hotel group existing in Portugal. This hotel is made up of 389 rooms, divided into 3 buildings, and has various activities available for its guests (tennis, trapeze, archery, fitness, golf, ping pong and mini golf). Being a family-oriented resort, it is suitable for children from 4 months to 17 years old, with special food services for babies and various activities for all children. In this resort there are two restaurants and three bars, as well as a spa service with different packages for its guests. It should also be noted that the resort is not open all year round, it closes from November to mid-February.

The data used in this study was collected on TripAdvisor. TripAdvisor is, in part, similar to a social network, a virtual community, and a blog, being hard to categorize its type (O'Connor, 2010). It is, therefore, considered a multifaceted platform, recognized for providing information and user feedback, where it is possible to respond to each user comment (not being able to remove or edit comments). This platform was chosen for this study because it is the world's largest travel website that allows easy interaction with travellers, and with more user-generated content, including comments, ratings, photos and videos (TripAdvisor, 2017).

The study collected a sample of 294 comments from September 1, 2017 to April 15, 2018, to provide a sample of comments from visitors who had stayed until Easter. Data extraction from guest posts on TripAdvisor was possible using filters, namely the language filter. Since most reviews for resort are available in French (2,719 reviews), which may be due to the nationality of the hotel chain, and since 75% of French people use TripAdvisor reviews for the choice of destination (Wesgro, 2017), it was decided to choose to analyse only the comments of the francophone public.

The comments were collected individually through manual extraction into an Excel file. Along with the comment, elements that allow a characterization of the visitor (gender, city and country of residence) and the trip (month and year, and type of trip) were also collected. In addition, the overall rating and rating by category (value for money, location, sleep quality, rooms, cleanliness and service) were collected, being the TripAdvisor rating on a scale of 1 to 5.

After data collection, the comments were manually entered individually into an artificial intelligence software, the IBM Watson Tone Analyzer. This tool analyses feelings automatically, based on algorithms designed to extract feelings in a similar way to humans. That is, the software "questions / identifies" the feeling(s) present in each comment, regarding terms and topics, making a classification taking into account its polarity (positive, neutral and negative). This software, in its context, identifies emotional intonations (anger, fear, joy, sadness, analytical, confident and tentative), which are classified by a score. The score may be less than 0.5 (none or weak intonation), between 0.5 and 0.75 (median intonation) and greater than 0.75 (strong intonation). The program assigns a colour to sentences according to the punctuation. Usually if there is more than one emotion extracted, the score shown will be the strongest. Emotion results were also integrated into the Excel database, which later were imported and analysed on the IBM SPSS Statistics v.25 statistical program. In this study, univariate and bivariate analysis were performed.

RESULTS

Profile of the Guests

From the analysis of the reviews, it was found that the profile of guests was mostly female (46%), 35% were male and 19% did not specify their gender. Guests live predominantly in France (44.9%), Belgium (10.9%) and Switzerland (3.1%). Other countries represent only 2.7% of the comments (Luxembourg, Portugal, Canada and Russia). From 38.4% of the comments it was not possible to identify the users' country of origin. Regarding the cities of origin, there is a large dispersion. It is found that, generally, guests come from the main cities of each country, however, users did not insert their city of residence in 39.1% of the comments. It was also found that, from the French guests, 44.7% are residents of the Île-de-France, 13.6% of the Provence-Alpes-Côte d'Azur region, 10.6% of the Hautes de France, and 7.6% from Auvergne-Rhône-Alpes, with the other regions having a reduced expression.

Regarding the characteristics of the trip, it was found that 69.7% of the guests travelled in 2017, having commented mostly on September (30.6%), October (20.7%) and August 2017 (13.6%), and 30.3% in 2018, mainly in March (14.3%). Most guests (55.7%) travel with their families, 26.2% travel as a couple or with friends (11.8%), 5.9% travel alone and only 0.5% travelled for business reasons. People from the Île-de-France and Hautes-de-France regions travelled mostly with their families (57.9% and 64.3%, respectively), while people from Provence-Alpes-Côte d'Azur travelled mostly as a couple (50%) and in family (38.9%).

Review of the Guests' Ratings

The resort scored an average rating of 4.3 out of 5, with a standard deviation of 0.896 (Table 1); 53.7% of the guests rate it as excellent, 32.7% give it a very good rating and 1.7% give it a terrible rating.

In this context, Cantallops and Salvi (2013) argue that the type of information is both informative, helping to complement and guide the consumer's decision, as well as advisory, where the popularity of the product is indicated by a positive or negative score (Jalilvand et al., 2011). The comments in which tourists rated the resort are mostly positive (86.4%), showing their relevance.

Table 1. Descriptive frequency of the 'general classification'

Classification	Frequency	%	Average	Standard Deviation
Terrible	5	1,7	4,3	0,896
Weak	10	3,4		
Average	25	8,5		
Good	96	32,7		
Excellent	158	53,7		
Total	294	100		

Source: Own construction

Regarding the classification by parameter, measured on a 5-point scale, it was found that the highest emphasis goes to the cleanliness and location with an average of 4.30, while the least rated attribute was the value for money, with an average of 4.0, which, although with a lower rating, is a good reference for the quality / price ratio (Figure 1).

There was a high frequency of comments regarding the service (mentioned 159 times), with a strong correlation coefficient regarding the total rating (> 0.8). Other frequently mentioned categories are rooms (109 times) and location (106 times), which show a moderate (> 0.6) and strong (> 0.7) correlation, respectively (Table 2).

Figure 1. Ratings by category
Source: Own construction

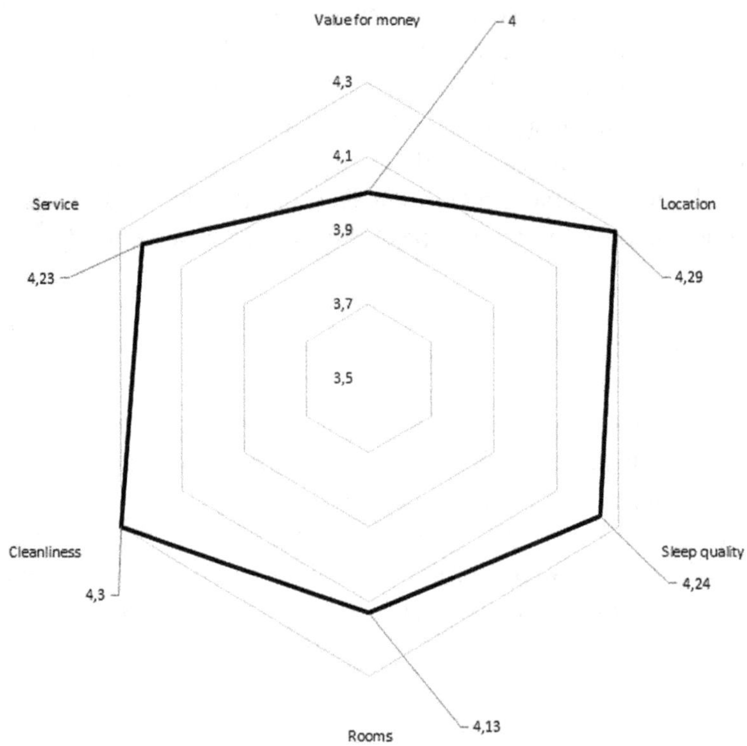

Table 2. Pearson correlation

Specific Classifications	Coefficients	Number of Comments	p-Value
Value for money	0.701	95	0.000
Location	0.722	106	
Sleep quality	0.564	106	
Bedrooms	0.656	109	
Cleanliness	0.635	102	
Service	0.828	159	

Source: Own construction

Overall, the guests greatly appreciate the strategic positioning of the resort and rate the location as one of the service's key points. They also refer to the kindness of staff, both in the bar and restaurant, as well as in the personalized attention and care of children. The results obtained are in line with other studies, namely Ong (2012), who argue that the location and distance from the hotel to the most well-known attractions, the cleanliness of the room and the value for money are important attributes as well. According to Ögüt and Taş (2012), the most mentioned attributes were the location, service, price, room and features of the hotel, safety, reputation and category. In the case of the resort under study, the most relevant attributes were the service, the cleanliness, the location and the sleep quality.

Sentiment Analysis of the Comments

Regarding the sentiment analysis of the comments, it was found that "joy" is the most relevant emotion, with a strong intonation of 73.8% (Table 3). Joy can be explained by the many services and features that the resort offers, such as: (i) the all-inclusive regime; the kindness of the staff, since they have the duty to interact with customers and always be available; (ii) the childcare setting where parents can rest while enjoying the pool or spa because their children are busy with activities with a team of certified educators; (iii) the entertainment in which thematic shows are presented every night; and (iv) prime location.

Following is "analytical", with 23.1% of medium intonation and 14.3% with strong intonation. However, "confident" and "tentative" also stood out with 7.1% and 5.8% of strong intonation, and 5.8% and 3.1% with medium intonation. Finally, the emotions "sadness" and "anger" had 14.3% and 11.9% medium intonation, and 3.4% and 2.0% referred to strong intonation.

This resort is characterized as being enjoyable for holidays, the emotions that stand out mainly are those related to joy and analytical, as it may mean that people feel happiness when staying at the resort.

Table 3. Descriptive frequency of the emotions

	Analytical	**Confident**	**Joy**	**Sadness**	**Tentative**	**Anger**
No emotions	184 62.6%	256 87.1%	18 6.1%	242 82.3%	268 91.2%	253 86.1%
Weak	-	-	1 0.3%	-	-	
Medium	68 23.1%	17 5.8%	58 19.7%	42 14.3%	9 3.1%	35 11.9%
Strong	42 14.3%	21 7.1%	217 73.8%	10 3.4%	17 5.8%	6 2%
Mean	0.263	0.096	0.743	0.114	0.067	0.085
Median	0.00	0.00	0.80	0.00	0.00	0.00
Standard deviation	0.349	0.255	0.211	0.251	0.219	0.216

Source: Own construction

CONCLUSION

Technological advances in recent years have had a significant impact on the tourism sector. Social media has become a powerful tool to leverage services for the tourism industry and, therefore, caused a change in the role of traditional intermediaries (Buhalis & Law, 2008). The online environment enables the effective exchange of information between users, allowing them to access information faster, easier and more conveniently (Wang & Fesenmaier, 2004). Today, Internet users contribute significantly to the generation of online content and information. They share their experiences, describing their levels of satisfaction with various services in order to inform and help other visitors. It is the consumer who manages the information, sharing it easily in an informal way and influencing the choice of other consumers. Therefore, user-generated content has been identified as an important source of information, influencing consumer choice and decision.

Overall, this study allowed to identify a resort's rating assigned by tourists on TripAdvisor and the emotions associated with the comments. Regarding the profile of visitors to the resort under analysis, they are mostly women, French, and from the Île-de-France region. The date of the trip and the comments on TripAdvisor were mainly September and October 2017. Concerning the travel typology, most visitors are accompanied by their family or travel as a couple. In addition, the results show that the resort has a strongly positive rating from guests who visited it, meaning that eWOW and influence on other potential consumers is positive. The most relevant criteria for guests were the cleanliness, location and sleep quality, and the ones that obtained lower ratings, although they may be considered good, were the value for money, the rooms and the service. As far as emotions were concerned, the ones that were most evident were joy and the analytical, and the less strong were sadness and anger.

This study, while focusing on a case study in the hospitality industry, allowed to reflect on how user-generated content on online platforms, such as TripAdvisor, can provide relevant data and information for tourism companies, which should be seen as a competitive advantage and incorporated into their online marketing strategies. It showed that it is possible to group different opinions and understand the consumers' needs through the emotional intonations of each comment and, therefore, be able to improve the service quality. For companies starting their business or those looking to embrace digital technologies, given that many of the tourists' experiences are shared on online social networks and online booking platforms, this kind of information becomes increasingly important. It is, therefore, relevant to realize the potential of this data, which may impact on several areas, such as customer loyalty, making predictions in order to study the popularity or introduction of a new product or service, and understanding the profile of customers who express opinions about the company. Thus, eWOM must be seen as an opportunity rather than a threat. As a result, tourism companies increasingly have to adapt, create and / or streamline user-generated content-based strategies.

The study presents some limitations, namely the limited time period for data collection, as well as the fact that certain comments did not have complete information about the guests' profile, and the lack of classifications in the different categories. In future research, it is suggested that the study integrates comments published at other times of the year, in order to assess the impact of seasonality, as well as comments from guests from other countries, on the other hand, it would be interesting to make a comparative study with other similar hotels of the same group.

REFERENCES

Ady, M., & Quadri-Felitti, D. (2015). Consumer research identifies how to present travel review content for more bookings. Retrieved from https://static1.squarespace.com/static/56dd934137013bc69c80ae95/t/56e6c50462cd94987f7ad826/1457964296558/Consumer+Research+Identifies+How+To+Present+Travel+Review+Content+for+More+Bookings.pdf

Afonso, C., & Borges, L. (2013). *Social target: Da estratégia à implementação – Como tirar partido das redes sociais e potenciar o negócio*. Top Books.

Barcala, M. F., Díaz, M. G., & Rodriguez, J. P. (2009). Factors influencing guests' hotel quality appraisals. *European Journal of Tourism Research*, *2*(1), 25–40.

Berthon, P. R., Pitt, L. F., Plangger, K., & Shapiro, D. (2012). Marketing meets Web 2.0, social media, and creative consumers: Implications for international marketing strategy. *Business Horizons*, *55*(3), 261–271. doi:10.1016/j.bushor.2012.01.007

Brandtzaeg, P., & Heim, J. (2009). Why people use social networking sites. In A. A. Ozok & P. Zaphiris (Eds.), *Online Communities and Social Computing (OCSC 2009)*. Berlin: Springer. doi:10.1007/978-3-642-02774-1_16

Buhalis, D., & Law, R. (2008). Progress in tourism management progress in information technology and tourism management: 20 years on and 10 years after the Internet – The state of eTourism research. *Tourism Management*, *29*(4), 609–623. doi:10.1016/j.tourman.2008.01.005

Cantallops, A. S., & Salvi, F. (2014). New consumer behavior: A review of research on eWOM and hotels. *International Journal of Hospitality Management*, *36*, 41–51. doi:10.1016/j.ijhm.2013.08.007

Chaves, M. S., Gomes, R., & Pedron, C. (n.d.). Decision-making based on online reviews: The small and medium hotels management. Retrieved from https://repositorio-cientifico.essatla.pt/bitstream/20.500.12253/639/1/ecis-2012-mchaves-etal-cr.pdf

Chen, Y., Fay, S., & Wang, Q. (2011). The role of marketing in social media: How online consumer reviews evolve. *Journal of Interactive Marketing*, *25*(2), 85–94. doi:10.1016/j.intmar.2011.01.003

Cheung, C. M. K., Chiu, P.-Y., & Lee, M. K. O. (2011). Online social networks: Why do students use Facebook? *Computers in Human Behavior*, *27*(4), 1337–1343. doi:10.1016/j.chb.2010.07.028

Chung, J. Y., & Buhalis, D. (2008). Information needs in online social networks. *Information Technology & Tourism*, *10*(4), 267–281. doi:10.3727/109830508788403123

Di Pietro, L., Di Virgilio, F., & Pantano, E. (2012). Social network for the choice of tourist destination: Attitude and behavioural intention. *Journal of Hospitality and Tourism Technology*, *3*(1), 60–76. doi:10.1108/17579881211206543

Freitas, L., & Vieira, R. (2015). Exploring resources for sentiment analysis in Portuguese Language. *Brazilian Conference on Intelligent Systems*. Retrieved from https://www.researchgate.net/publication/300409116_Exploring_Resources_for_Sentiment_Analysis_in_Portuguese_Language

Gaál, Z., Szabó, L., Obermayer-Kovács, N., & Csepregi, A. (2015). Exploring the role of social media in knowledge sharing. *Electronic Journal of Knowledge Management, 13*(3), 185–197.

Gallegos, J. A. (August 23, 2016). What is user generated content (and why you should be using it). Tintup. Retrieved from https://www.tintup.com/blog/user-generated-content-definition/

Godes, D., & Mayzlin, D. (2004). Using online conversations to study word-of-mouth communication. *Marketing Science, 23*(4), 545–560. doi:10.1287/mksc.1040.0071

Goldsmith, R. E., & Horowitz, D. (2006). Measuring motivations for online opinion seeking. *Journal of Interactive Advertising, 6*(2), 2–14. doi:10.1080/15252019.2006.10722114

Hays, S., Page, S. J., & Buhalis, D. (2013). Social media as a destination marketing tool: Its use by national tourism organisations. *Current Issues in Tourism, 16*(3), 211–239. doi:10.1080/13683500.2012.662215

Hsu, C.-L., & Lin, J. C.-C. (2008). Acceptance of blog usage: The roles of technology acceptance, social influence and knowledge sharing motivation. *Information & Management, 45*(1), 65–74. doi:10.1016/j.im.2007.11.001

Huang, Y., Basu, C., & Hsu, M. K. (2010). Exploring motivations of travel knowledge sharing on social network sites: An empirical investigation of U.S. college students. *Journal of Hospitality Marketing & Management, 19*(7), 717–734. doi:10.1080/19368623.2010.508002

Jalilvand, M. R., Esfahani, S. S., & Samiei, N. (2011). Electronic word-of-mouth: Challenges and opportunities. *Procedia Computer Science, 3*, 42–46. doi:10.1016/j.procs.2010.12.008

Lecture, J. S., Fehr, E., & Falk, A. (2002). Psychological foundations of incentives. *European Economic Review, 46*(4-5), 687–724. doi:10.1016/S0014-2921(01)00208-2

Lei, S., & Law, R. (2015). Content Analysis of Tripadvisor Reviews on Restaurants: A Case Study of Macau. *Journal of tourism, 16*(1), 17–28.

Leung, D., Law, R., Van Hoof, H., & Buhalis, D. (2013). Social media in tourism and hospitality: A literature review. *Journal of Travel & Tourism Marketing, 30*(1-2), 3–22. doi:10.1080/10548408.2013.750919

Limberger, P. F., Anjos, F. A., Meira, J. V., & Anjos, S. J. (2014). Satisfaction in hospitality on TripAdvisor.com: An analysis of the correlation between evaluation criteria and overall satisfaction. *Tourism & Management Studies, 10*(1), 59–65.

Litvin, S. W., Goldsmith, R. E., & Pan, B. (2008). Electronic word-of-mouth in hospitality and tourism management. *Tourism Management, 29*(3), 458–468. doi:10.1016/j.tourman.2007.05.011

Litvin, S. W., Goldsmith, R. E., & Pan, B. (2018). A retrospective view of electronic word-of-mouth in hospitality and tourism management. *International Journal of Contemporary Hospitality Management, 30*(1), 313–325. doi:10.1108/IJCHM-08-2016-0461

Liu, B. (2015). *Sentiment analysis: Mining opinions, sentiments, and emotions*. New York: Cambridge University Press. doi:10.1017/CBO9781139084789

Marchiori, E., & Cantoni, L. (2015). The role of prior experience in the perception of a tourism destination in user-generated content. *Journal of Destination Marketing & Management*, *4*(3), 194–201. doi:10.1016/j.jdmm.2015.06.001

Marktest. (2017). Os portugueses e as redes sociais 2017. Retrieved from http://www.marktest.com/wap/a/grp/p~96.aspx

Marktest. (2019). Os portugueses e as redes sociais 2019. Retrieved from http://www.marktest.com/wap/a/grp/p~96.aspx

Miguéns, J., Baggio, R., & Costa, C. (2008). Social media and tourism destinations: TripAdvisor case study. *Advances in tourism research, 26*(28), 1–6.

Mkono, M. (2012). Netnographic tourist research: The Internet as a virtual fieldwork site. *Tourism Analysis*, *17*(4), 553–555. doi:10.3727/108354212X13473157390966

Munar, A. M., & Jacobsen, J. K. S. (2014). Motivations for sharing tourism experiences through social media. *Tourism Management*, *43*, 46–54. doi:10.1016/j.tourman.2014.01.012

Neethu, M. S., & Rajasree, R. (2013). Sentiment analysis in Twitter using machine learning techniques. *Proceedings of the 2013 Fourth International Conference on Computing, Communications and Networking Technologies (ICCCNT)* (pp. 1–5). IEEE. 10.1109/ICCCNT.2013.6726818

O'Connor, P. (2010). Managing a hotel's image on TripAdvisor. *Journal of Hospitality Marketing & Management*, *19*(7), 754–772. doi:10.1080/19368623.2010.508007

Öğüt, H., & Onur Taş, B. K. (2012). The influence of Internet customer reviews on the online sales and prices in hotel industry. *Service Industries Journal*, *32*(2), 197–214. doi:10.1080/02642069.2010.529436

Pak, A., & Paroubek, P. (n.d.). Twitter as a corpus for sentiment analysis and opinion mining. *LREC*, 1320–1326. Retrieved from https://pdfs.semanticscholar.org/ad8a/7f620a57478ff70045f97abc7aec9687ccbd.pdf

Pang, B., & Lee, L. (2008). Opinion mining and sentiment analysis. *Foundations and Trends in Information Retrieval*, *2*(12), 1–135. doi:10.1561/1500000011

Perretti, F., & Negro, G. (2006). Filling empty seats: How status and organizational hierarchies affect exploration versus exploitation in team design. *Academy of Management Journal*, *49*(4), 759–777. doi:10.5465/amj.2006.22083032

Sales, R., Carvalho, C., Arruda, D., & Albuquerque, J. (2015). Clique aqui para decidir: As recomendações online na decisão de compra dos serviços hoteleiros. *Revista de Turismo y Património Cultural*, *13*(3), 541–550. doi:10.25145/j.pasos.2015.13.038

Serrano-Guerrero, J., Olivas, J. A., Romero, F. P., & Herrera-Viedma, E. (2015). Sentiment analysis: A review and comparative analysis of web services. *Information Sciences*, *311*, 18–38. doi:10.1016/j.ins.2015.03.040

Sigala, M., Christou, E., & Gretzel, U. (Eds.). (2012). *Social media in travel, tourism and hospitality: Theory, practice and cases*. Ashgate Publishing, Ltd.

Skift.com. (2014). State of travel: 2014. Retrieved from http://skift.com/wp-content/uploads/2014/08/Skift-2014-State-of-Travel-Report-Full.pdf

Sparks, B. A., & Browning, V. (2010). Complaining in cyberspace: The motive and forms of hotel guests' complaints online. *Journal of Hospitality Marketing & Management*, *19*(7), 797–818. doi:10.1080/19368623.2010.508010

Stringam, B. B., & Gerdes, J. Jr. (2010). An analysis of word-of-mouse ratings and guest comments of online hotel distribution sites. *Journal of Hospitality Marketing & Management*, *19*(7), 773–796. doi:10.1080/19368623.2010.508009

TripAdvisor. (2017). PT Media Center | About TripAdvisor. Retrieved from https://tripadvisor.mediaroom.com/pt-about-us

Valdivia, A., Luzón, M. V., & Herrera, F. (2017). Sentiment analysis on TripAdvisor: Are there inconsistencies in user reviews? In F. Martínez de Pisón, R. Urraca, H. Quintián, & E. Corchado (Eds.), *Hybrid artificial intelligent systems*. Cham: Springer. doi:10.1007/978-3-319-59650-1_2

Vermeulen, I. E., & Seegers, D. (2009). Tried and tested: The impact of online hotel reviews on consumer consideration. *Tourism Management*, *30*(1), 123–127. doi:10.1016/j.tourman.2008.04.008

Vlachvei, A., & Notta, O. (2014). Social Media adoption and managers' perceptions. *International Journal of Strategic Innovative Marketing*, *1*(2), 61–73. doi:10.15556/IJSIM.01.02.001

Wang, Y., & Fesenmaier, D. R. (2004). Modeling participation in an online travel community. *Journal of Travel Research*, *42*(3), 261–270. doi:10.1177/0047287503258824

Wang, Y., Yu, Q., & Fesenmaier, D. R. (2002). Defining the virtual tourist community: Implications for tourism marketing. *Tourism Management*, *23*(4), 407–417. doi:10.1016/S0261-5177(01)00093-0

Wang, Y., Yu, Q., & Fesenmaier, D. R. (2002). Defining the virtual tourist community: Implications for tourism marketing. *Tourism Management*, *23*(4), 407–417. doi:10.1016/S0261-5177(01)00093-0

Weilin, L., & Svetlana, S. (2012). Ecoturism experiences reported online: Classification of satisfaction attributes. *Tourism Management*, *33*(3), 702–712. doi:10.1016/j.tourman.2011.08.003

Wesgro. (2017). France: Tourism Market Insights 2017. Retrieved from http://www.wesgro.co.za/pdf_repository/Tourism Market Insights_France.pdf

Yoo, K. H., & Gretzel, U. (2008). The influence of perceived credibility on preferences for recommender systems as sources of advice. *Information Technology & Tourism*, *10*(2), 133–146. doi:10.3727/109830508784913059

Yoo, K. H., Gretzel, U., Yoo, K.-H., Lee, Y., & Fesenmaier, D. R. (2010). Antecedents and impacts of trust in travel-related consumer-generated media. *Journal of Information Technology and Tourism*, *12*(2), 139–152. doi:10.3727/109830510X12887971002701

Zhang, K. Z. K., Zhao, S. J., Cheung, C. M. K., & Lee, M. K. O. (2014). Examining the influence of online reviews on consumers' decision-making: A heuristic–systematic model. *Decision Support Systems*, *67*, 78–89. doi:10.1016/j.dss.2014.08.005

Chapter 18
The Importance of Social Media for the Improvement of the Tourist Experience Offered by Rural Tourism Enterprises:
The Case of the Alentejo

Marta Isabel Amaral
Department of Management, Polytechnic Institute of Beja, Portugal

Ana Isabel Rodrigues
Department of Management, Polytechnic Institute of Beja, Portugal

ABSTRACT

Social networks and the use of social media have been gaining more and more importance in recent years and have had a very significant impact on the tourism sector. The way this influences activity can be seen in two ways. On the one hand, from the point of view of the consumer himself, especially the way in which tourists access information. On the other hand, social networks are used by tourism companies as a means to support their marketing activities, such as the promotion of tourism products. This chapter explores how rural tourism companies use social media as a way to improve the tourist consumer experience. By assessing managers of the rural tourism businesses in the Alentejo Region, Portugal, the aim is to identify entrepreneurs' perceptions regarding the factors that influence the tourism experience, the use of social media and their relationship with the improvement of the tourist experience and customer loyalty. It is possible to conclude that rural tourism establishments are no longer ignoring the important role of social media in promoting their tourism experiences.

DOI: 10.4018/978-1-7998-1947-9.ch018

INTRODUCTION

It is clear from the authors consulted that price and quality are no longer the only points for competitiveness (Shaw & Ivens, 2005, cited by Veríssimo & Menezes, 2016). Today, the focus must be on a differential customer experience. It is important to highlight that tourist consumption goes beyond material goods; when the emotional aspects involved are considered, then it must focus on affective and cognitive aspects. This is the case even when tourist services have a clearly functional value for customers, such as hotels and transportation, because experiences can directly influence the process of evaluation of services and consequently the tourist experience (Veríssimo & Menezes, 2016).

Social media are changing the way society consumes and contributes to the creation of information. Technology now allows individuals to easily contribute their thoughts, opinions and creations to the Internet (Hays, Page & Buhalis, 2013). Their use has been gaining more and more importance in recent years and has had a very significant impact on the tourism sector. It has brought more interactivity to internet and forged the relationship between customer and firms (Veríssimo & Menezes, 2016). The way it influences activity can be viewed in two ways. On the one hand, from the point of view of the consumer himself, especially the way in which tourists access information, in particular to decide and plan their trip and to share their experiences. On the other hand, and for those who are interested in investigating, social networks are used tourism companies as a means to support their marketing activities, such as the promotion of tourism products (Danhil, Marzuki, Langgat & Fabeil, 2014; Zeng & Gerritsen, 2014).

In a study called "Digital Channels in Travel", conducted by Deloitte and Facebook, it was concluded that social networks have an important role in the choice of travel and that is why companies in the industry should take advantage of this business opportunity (https://www2.deloitte.com/en/pages/consumer-business/articles/influencia-dos-social-media.html). Social media could be used by the hospitality industry as a tool to enhance marketing experience and to connect to customers, because thus it is possible to understand their needs and wishes, offering differentiated services, creating brand value and getting a competitive advantage over their competitors (Veríssimo & Menezes, 2016, p. 24).

So, the use of social media and on-line channels are new challenges for tourism enterprises in rural areas, particularly for hotel managers. These information and communications technologies (ICT) are a means by which tourism subsector companies can communicate with customers more easily; they make it possible to perceive consumer behaviour and to estimate the choices that the consumer may make in terms of accommodation. In this way, it has become indispensable to maintain a constant presence in these online channels, not only to take notice of what consumers are talking about in terms of the hotel accommodation in question, but also to determine the best communications strategy to undertake and which information should be made available, taking into account what the consumer is looking for.

This is the situation. However, many of these small and medium-sized hotel companies have been slow to adopt technologies due to lack of money and/or time, as well as a lack of knowledge about how they work (Oliveira, Correia & Fernandes, 2017).

In view of the above, the main objective of this article is to seek to understand how rural tourism accommodation enterprises (small and medium-sized enterprises – SMEs) use social media as a way to improve the tourist consumer experience. The specific objectives of this article are to indicate which components of the experience are most used by enterprises in social networks; to determine which social networks are the most used; and to highlight the practical terms of how entrepreneurs direct their attention.

In order to accomplish the purposes in this paper, first a review of the background theory is given, followed by the methodology, main results and limitations and implications for future researches.

LITERATURE REVIEW

Experience as a New Framework in Tourism

Most of the growth sectors of the global economy are related to the consumption of experiences. The tourism industry – not only because of its nature as an economic, social, cultural and human activity profoundly connected with the act of experiencing, but also because it is considered to be the "industry of dreams" – is framed by the experiential age, where "work is theatre and every business is a stage" (Pine & Gilmore, 1999). Tofler (1970), a visionary, in his 1970s book "Future Shock" mentions an upcoming "experiential industry", where people in the future would be willing to allocate high percentages of their salaries to have amazing, emotional experiences. As he advocated in this milestone book, in future societies the idea of change will not be only necessary to live, it is life itself. People need to live within a constant idea of change. Also because of that, Jensen (2001) argues that we live in a new society in which business, communities, and people as individuals will thrive on the basis of their stories, not just on data and information. Jensen stated that consumers prefer products with a better story and because of that people are moving from an information-based market to one of dreams; from a market of tangible products to a market of feelings; from pragmatism to values; from physical comfort to spiritual comfort; from rational intelligence to emotional intelligence. Based on Pine and Gilmore (1999, cited by Oh, Fiore & Jeoung, 2007),

in the emerging experience economy, consumers seek unique experiences beyond merely consuming products and services because consistent, high level of product and service quality can no longer be used to differentiate choices for consumers. This new demand for unique and memorable experiences requires firms to develop a distinct value-added provision for products and services that have already achieved a consistent, high level of functional quality. For instance, a cooking demonstration or a themed guestroom at a rural bed-and breakfast facility offers guest educational, entertaining, and aesthetic experiences, or forms of experiential value added to its highly personalized, professionally managed lodging facility (p. 119).

As is well known, tourism products are a composite of elements, tangible and intangible, based on an activity at a destination (Middleton & Clarke, 2004). Tourism products are genuinely based on tourism experiences. Intangibility, heterogeneity, and inseparability are the main characteristics of service products (Cooper, Fletcher, Gilbert, & Wanhill, 1998; Holloway, 1995; Seaton, 1996). Intangibility means that the tourist's travel decision is mostly based on impressions, perceptions and ideas. What consumers really buy are invisible elements of the product or destination and not the product itself. At the same time, inseparability and heterogeneity indicate that there is great subjectivity in providing tourism services. The producer (service provider) and consumer (tourist) not only a determinant participant in the service, but they are the service itself. In sum, the tourism product is underpinned by impressions, interpretations, perceptions, sensations, and meanings. More simply put, the tourism product is grounded on images and perceptions rooted in experiences (Rodrigues, 2004, 2015). Tourism has been at the forefront of staging experiences.

Because of this 'new reality', increasing attention has been paid to consumers' experiences and their components over the past few decades, not only by academia, but also by the industry i.e. practitioners, marketers and enterprises. As argued by Stamboulis and Skayannis (2003), tourism is undergoing a significant change and faces new challenges, and thus new perspectives are called for. According to the authors, at least two dimensions of this change can be identified: (i) new forms of tourism are arising, characterized by the tendency to depart from mass tourism and, (ii) there is a diffusion of information and communications technologies, with a pervasive effect on the creation, production and consumption of the tourist product.

Quan and Wang (2004) perceived two broad, academic perspectives in tourism experience studies (Volo, 2007):

1. A social science approach, more related to the tourist perspective and that focuses on travel experience in contrast to daily experience; the tourist experience is conceived as the "pure" or "peak" experience usually derived from the attractions that can be added to the "supporting consumer experiences", such as accommodation or transport, which are taken for granted (this means a "demand side perspective").
2. A marketing/management approach, more related to understanding attractions, services and facilities as products that need to be conceived, promoted and sold in the tourism market as consumer experiences (this means a "supply side perspective"). This study examines tourism experiences based on this twofold perspective, although framed by the supply angle.

For that reason, Table 1 exhibits a definitional approach grounded on these two insights. The aim is to contribute to a deeper understanding of the nature, dimensions, modes and components of tourism experience as a concept.

In a demand-side approach, ideas such as "multi-sensory, fantasy, emotions and feelings", "high levels of emotional intensity", "constant flow of thoughts and feelings", "consuming is engaging in tourism experiences", "primarily seek and consume at destinations is engaging experiences", "long-term memory", and "experience interrupts daily lives" have been used to define a tourism experience. These elements incorporate the nature of the tourism experience, a definitely "complicated psychological process" (Cutler & Carmichael, 2010, p. 3) since it circumscribes a complexity of various elements. As stated by O'Dell "experiences are argued to be subjective, intangible, continuous and highly personal phenomena" (2007, cited by Cutler & Carmichael, 2010, p. 3). These are considered the main features of tourism experience. In a supply-side approach, ideas such as "distinct economic offering", "different elements of a context created by a service provider", "a combination of factors that shape the tourist's feelings", "experiences usually derived from attractions" and "relationships between the host and guest" should be taken into consideration by a tourist service provider (e.g. accommodation owners or marketers) during the process of conceiving a tourism experience around accommodation as the main service. It is important to highlight the idea that a tourism experience is a result of an aggregation of multiple elements (supporting elements, such as accommodation of transport, but also attraction elements, such as exploring and enhancing the natural and cultural resources of a specific destination). In addition, in a management and marketing approach, when conceiving a tourism experience, different types and modes of experiences might be considered, some of them involving the consumer more emotionally, physically, intellectually or even spiritually.

Table 1. A definitional approach to tourism experience grounded on two main perspectives (demand and supply side)

Perspectives/ Approaches	Definitions/Author/Date
Demand (tourist)-side approach	Those facets of consumer behaviour that relate to the **multi-sensory, fantasy and emotive** nature of one's experience with products. (Hirschman & Holbrook, 1982). Extraordinary experiences are those characterized by **high levels of emotional intensity.** (Arnold & Price, 1993) An experience can be defined as a **constant flow of thoughts and feelings** that occur during moments of consciousness. (Carlson, 1997) "**Different people** may engage in **different experiences".** (Uriely, 2005, pp. 205-206) In essence, what tourists **primarily seek and consume at destinations is engaging experiences** accompanied by the goods and/or service components of the destinations. (Oh et al., 2007) A past travel-related event which was **significant enough in long-term memory** (Larsen, 2007) **Experiences interrupt people from their lives** and expectations to provide something that demands attention; experiences themselves are **incredibly involving.** (Ray, 2008)
Management/ marketing (supply)-side approach	A **distinct economic offering** that is as different from services as services are from goods (Pine & Gilmore, 1998,1999) An experience occurs when a customer has any sensation or knowledge acquisition resulting from some level of interaction with **different elements of a context created by a service provider.** (Gupta & Vajic, 1999). The goal of experience design is to **orchestrate experiences** that are functional, purposeful, engaging, compelling, and memorable. (McLellan, 2000) A **combination of factors that shape the tourist's feelings** and attitudes towards his or her visit (Page et al., 2001, p.412-413) An **interaction between tourists and destinations**, with the destinations being the site of the experience and tourists being the actors of the experience. (Stamboulis & Skayannis, 2003) "The tourist experience is thus understood as the "pure", "net" or "peak" experience, **usually derived from attractions, rather than "mixed", "gross" or "supporting" experiences** such as eating, sleeping and so on." (Quan & Wang, 2004, p. 297) Tourists' experiences must be organically whole, in which peak (art, culture and heritage) and experiences (accommodation, transportations, shopping, food establishments) **complement each other**. (Quan & Wang, 2004) The **tourist experience is proposed** as the moment when tourism consumption and tourism production meet. (Andersson, 2007) A **blend of many elements coming together** and involving the consumer emotionally, physically, intellectually and spiritually. (Mossberg, 2007) Creates hospitable **relationships between the host and guest**; these experiences engage emotions, which is essential to create a memory (Lashley, 2008)

Source: Elaborated by the authors, adapted from Walls et al. (2011), among other sources.

Rural Tourism, Rurality Experiences, Consumer Loyalty

Rural tourism is considered a useful tool for development, that optimize benefits for both tourists and areas they visit (Fotiadis, 2009). The reality of tourism is to link tourism with development, in particular in peripheral, rural or developing countries, and this justifies the various resources involved and the investments made by communities, governments, and development associations, all over the world.

Tourism is seen as an important agent for regeneration and the economic redevelopment of certain rural and isolated regions of attraction that these present, constituting a means of sustenance and diversification for rural economies and societies (Jackson & Murphy, 2006; Sharpley & Telfer, 2002; Fotiadis, 2009). In this sense, all efforts should be based on the development of innovative and differentiated products, for the reinforcement of strategic partnerships and for preservation of the environmental balance and the enhancement of cultural heritage.

The dynamism of tourist demands a quantitative point of view, both of a permanent diversification of its profiles and also the appreciation of these for tourist products, especially those associated with the environmental (Silva & Silva, 2003). Therefore, tourism should be seen beside its economic dimension, adding other dimensions (social, cultural, environmental, institutional framework). It can thus become an alternative strategy for preserving and valuing key aspects such as traditions, to rationalize the use of natural resources and also to take advantage of local human capacities (Sampaio, 2003, as cited by Pellin, 2005; Silva & Perna, 2002).

The forms of tourism that emerge in a context of rural development are associated with some changes in society itself, namely the increase in levels of education, the greater interest in natural and artificial heritage and the development of transport and communication. In these circumstances, rural areas can hardly be considered totally remote and isolated and constitute privileges to meet ever more pressing needs: health, aesthetics and well-being (OECD, 1994). This is what consumers demand when they buy rural tourism products.

Rural tourism and rurality are intertwined, and the term "rurality" is used to capture generalized rural features; these features are then reproduced for tourist consumption (Lane, 1994, cited in Zhou, 2014, p. 227). According to the same author, consumption is the "imagined rural idyll." This means that the rural concept is an imaginative place with a significant range of cultural meanings (Cloke, 2006; Woods, 2011, p. 92, cited by Zhou, 2014). Furthermore, tourism has the capacity to revitalize the culture and heritage of rural communities, and represent local people, societies, and lifestyles (Garrod, Wornell, & Youell, 2006). Then, the role of tourism in making, promoting, and representing rurality needs to be further understood (Pritchard & Morgan, 2001, cited in Zhou, 2014).

It is a fact that not all regions offer the same conditions for tourism, considering the immense fragilities of the rural territories and of the interior and, therefore, a region with potential for development should, in fact, have some characteristics in the cultural, physical-natural and social domains that define its regional identity (Cunha & Cunha, 2005).

The package has to be sufficient to attract and keep tourists, offering conditions for spending (Gannon, 1994). It is also crucial for development and the organization of rural tourism that a significant investment is required, which is not always available (Fleisher & Felenstein, 2000). In addition, sometimes local communities and entrepreneurs may find it difficult to adapt to the new role of 'servility' (Fleischer & Piozam, 1997). In addition to the above, the territories that promote rural tourism should have quality products and services that meet the demands and expectations of tourists and the demand of local experiences. Finally, however, people often lack the skills and resources for effective marketing (Sharpley & Telfer, 2002). Usually these rural tourism destinations are characterized by a small private sector, mostly family members, who may benefit greatly from cooperative marketing.

The "economy of experience", presents the "new tourist" as a more demanding "tourist of experiences and emotions", who seeks greater differentiation, a made-to-measure product and high correspondence to his expectations. In this new model of tourism, the tourist has the central role, in which his motivations and needs undertake the command to live a set of emotions (Cupeto, Figueiredo, Silva, Santos, Figueiredo, Carvalho & Preguiça, 2014).

So, considering those facts, it is important for managers of rural tourism enterprises to try to attend to customers' needs with activities that (1) highlight the value contained in legends, myths and even stories of the people of a locality that can be incorporated into the products offered by the enterprise;

(2) create rewarding experiences in the region in which they are inserted and that make it possible to feel the culture and the peculiarities of the place; (3) add elements that increase the use of the senses, providing sensations that add value and create real and meaningful emotional connections, and that can reach an audience willing to pay even more for the product; and (4) implement actions that reduce energy consumption, water and solid waste generation to charm customers, who today are more connected to the care of the environment and the planet (SEBRAE, Boletim Turismo de Experiências). The specific characteristics of rural tourism make it necessary to look for the definition of loyalty because the multiplicity of destinations and lodgings that it offers is so wide and varied that the probability of visiting may be low (Cerro, Mogollón, & Alves, 2010).

It is important to realize the capacity of rural tourism to retain clients through the destination or accommodation or both and to understand the factors that contribute to this (Cerro et al., 2010). The research into tourism customer loyalty has not been thoroughly studied, and this is based on the fact that rural tourism encompasses an integrated concept, as rural tourism involves visiting a rural area with the use of its accommodation infrastructure and the activities that the surroundings offer. The study of loyalty in rural tourism should be based on loyalty to a destination, since in most cases, and with rare exceptions, there are no strong brands associated with accommodation that can generate loyalty to the brand itself, as is characteristic of major hotel brands (Cerro et al., 2010).

Although there are some explanatory models of loyalty to tourist destinations, the study of Chi and Qu (2008, cited by Cerro et al., 2010, p. 234) emphasizes that destination image directly influences satisfaction, and both are direct antecedents of global satisfaction, which has a positive impact on loyalty to the destination. The indicators applied to rural tourism are: (1) image of the destination (e.g., environmental – cleanness, natural and cultural resources, events, infrastructures, accessibility); (2) satisfaction with the characteristics of the destination (e.g., purchases and quality of typical products, accommodation, variety of tourism resources); (3) overall satisfaction with the travel experience; (4) loyalty to the destination (intention to return and recommend).

According to the study developed by Cerro et al. (2010), a group of experts explains their claims about (1) the concept of loyalty; (2) rural tourism customer characteristics/profile and (3) tourism behaviour pattern in rural tourism fidelity. First of all, for the same authors (p. 235).

loyalty in rural tourism is seen in the repetition or recommendation of a particular destination or accommodation of this typology, influenced by high levels of satisfaction that will depend on specific factors such as: personalized, close and family-like treatment that the owners of the enterprises offer, as well such as the hospitality of the resident community. It is important to complement this with the authenticity of the heritage (natural and cultural) and the singularity of the place. Secondly, the pattern of customer loyalty in rural tourism is based on the valorisation of quality and customer service.

Considering the subject of this chapter, it is important to ally all these characteristics with the possibility of having experiences, making the tourist become the protagonist of the activity and of the destination. The customer wants to be surprised and excited looking to (1) create links with the place visited; (2) get more complex and complete experiences and (3) experience something unusual, completely different from their current reality. When it comes to the experience itself, the proposal is to make the client immerse himself in the customs of a region and go beyond traditional visits. (SEBRAE, Boletim Turismo de Experiências).

Tourist Consumer Experience and Business Competitivity

The tools and strategies for communicating with customers have changed significantly with the emergence of the phenomenon known as social media, also referred to as consumer-generated media (Mangold & Faulds, 2009, p. 357). Social media are an important tool for the analysis of tourists' attitudes, and this is confirmed by the increased purchases and recommendations to other users. Building a successful service-based brand in tourism means that each offer should be a unique value proposition based on the customer experience (Živković, Gajic & Brdar, 2014).

First, it is important to clarify concepts. The definition of social media marketing includes using social media platforms and their related technologies and features to help achieve marketing objectives in conjunction with other marketing communication tools. So, in this paper, the terms social media marketing and Internet marketing will be taken to include the activities of e-marketing (Dahnil, Langgat, Mizal & Fabeil, 2014). For Brake and Safko (2009, cited Hays, 2013, p. 212), social media refers to activities, practices, and behaviours among communities of people who gather online to share information, knowledge, and opinions using conversational media. This form of media "describes a variety of new sources of online information that are created, initiated, circulated and used by consumers intent on educating each other about products, brands, services, personalities, and issues" (Blackshaw & Nazzaro, 2004, p. 2, cited by Mangold & Faulds, 2009, p. 358).

According to new information technology trends, consumers have started being more adaptive and flexible, and a new consumer profile has emerged – the digital user (Buhalis, 2011 cited by Živković et al., 2014). So, it is important to notice that the "new" type of consumer leads to new experiences and with new technologies being developed, the main interest of tourism subsequently lies in exploring the potential of ICTs, and particularly social networking, as strategic instruments for positive enhancement of tourism experiences (Neuhofer, Buhalis & Ladkin, 2013).

To reduce doubts and uncertainty and create some expectations of what can be found at a destination, potential tourism consumers need to know reliable sources and how they can obtain the necessary information in order to make the right decision concerning travel. Today it is understandable that modern tourists have more trust in other travellers' opinions using social media than in official marketing advices (Živković et al., 2014, p. 759).

In fact, the rise of new media channels enables customers to take a more active role as market players and reach (and be reached by) almost everyone anywhere and anytime (Thurau, Malthouse, Friege, Gensler, Lobschat, Rangaswamy, Skiera, 2010, p.312). Buhalis and Law (2008) explain that the Internet has fundamentally transformed the way tourism-related information is distributed and the way tourists plan for and consume travel. The tourism industry is currently in need of technology-based integrated value-added services, which are highly dynamic and offer interactivity and entertainment (Garcia-Crespo et al., cited by Kounavis, Kasimati & Zamani, 2012). One such integrated value-added service is social media, which has allowed the Internet to evolve from a broadcasting medium to a participatory platform, enabling people themselves to become the "media" for collaborating and sharing information (Leung, Buhalis & Hoof, 2013). There is no doubt of the growing importance of social media in the online tourism domain (e.g. Xiang & Gretzel, 2010). For these authors, two "mega trends" have emerged on the Internet, "underscoring changes that can significantly impact the tourism system" (2010, p. 179): (1) on one hand various forms of consumer-generated content (CGC) such as blogs, virtual communities, wikis, social networks, collaborative tagging, and media files shared on sites like YouTube and Flickr, the so-called social media websites, (2) on the other hand, the use of searching as an increasingly dominant mode in travellers' use of the Internet (e.g. the use of search engines for travel planning, which has become a powerful interface, serving as the "gateway" to travel-related information). This is the domain of consumer-generated media (CGM), online content created primary by internet users themselves, through sharing images of tourism experiences (Tussyadiah & Fesenmaier, 2009). The act of sharing all these images helps tourists in the recollection process and the remembrance of the past experiences after returning home.

The classification of social media types cannot be consensual [e.g. online, word-of-mouth forums (blogs, chat rooms, consumer-to-consumer e-mail, consumer product or service ratings websites and forums, Internet discussion boards and forums, moblogs), sites containing digital audio, images, movies, or photographs and social networking websites] and the numbers of social media are changeable, but their most important role is to encourage users and travellers to post and share their travel experiences, comments and opinions, by having them serve as a source of information for other users (Friedman & Friedman, 2008; Mangold & Faulds, 2009).

Social media play an important role in information search and decision-making processes (Zeng & Gerritsen, 2014, cited in Lopes, Abrantes & Kastenholz, 2014). Indeed, many companies are taking advantage of social networks to inform tourists and enrich their experiences (Lopes et al., 2014). According to Dellarocas (2003, cited by Leung et al., 2013) social media provide tourism companies with the opportunity to understand and respond to consumer preferences. Through analysing the comments on online communities (e.g. TripAdvisor), hotels and other travel-related companies can better understand what their guests like and dislike about them and their competitors. So, considering the rapid growth of this trend and also the social media benefits already described, tourism businesses have been integrating social media applications into their websites to enhance customers' experience in searching for travel information (Leung et al., 2013).

Sophisticated target groups and modern ICT environments are a serious challenge for the tourism industry and it is of great importance to send the right messages through the proper media channels (Edelman, 2010 cited by Živković et al., 2014). The use of social media and the Internet are considered as fundamental tools for business competitivity (Hays et al., 2013). This means that entrepreneurial SMEs are competing in highly dynamic environments (Chan, Teoh, Yeow & Pan, 2018, cited by Cenamor, Paridab & Wincent, 2019) and in order to response to competitive pressure, many SMEs use digital platforms to improve their business strategy (Li, Liu, Belitski, Ghobadian & O'Regan, 2016, cited by Cenamor et al., 2019).

As social media rise in popularity, users gain more power as the authority of marketers and institutions declines (Thevenot, 2007, cited by Hays et al., 2013). Particularly in tourism, marketers and institutions no longer have ultimate control over their destination image or product and Web 2.0 facilitates the integration of five properties: information representation, collaboration, communication, interactivity, and transactions (Gretzel, Fesenmaier, Formica & O'Leary, 2006, p. 147, cited by Hays, 2013, p. 212).

Driven by user-generated content – including the ability to share photos, videos and comments with other readers – academic scholars have also noted the capacity of social media in helping tourism and hospitality companies to engage potential guests, increase their online presence and thereby lead to greater online revenues (Leung, Buhalis & Hoof, 2013, p. 4). In fact, the use of social networks has brought many benefits to tourism companies (and in general), especially at the level of communication. Their use is an opportunity to get closer to the client through the possibility of direct contact in real time, as well as a challenge, encouraging the finding of solutions to monetize business success. In addition, the costs associated with the use of these tools are low compared to the traditional means of communication also used by companies (Oliveira, Correia & Fernandes, 2017). As authors say, on the one hand, consumers are changing the way they access information using it to plan their travel and share their travel experiences; companies, on the other hand, use social networks as a means to support their marketing activities, namely to identify new needs, to evaluate the degree of satisfaction with a given service, to implement improvements, and to promote their offer (Lopes et al., 2014).

Rural tourism is a special form of tourism, characterized by a natural environment and distance from urban areas (Clarke, Denmanb, Hickmanc & Slovak, 2001 cited by Vučetić, 2017). The deployment of ICT is crucial for the competitiveness of rural tourism businesses, so these companies have embraced the technological advances that have made a positive contribution to their rapid expansion in recent years (Ruiz-Molina, M.E., Gil-Saura, I. & Moliner-Velazquez, B., 2015; Polo Peña & Frías Jamilena, 2009 cited by Vučetić, 2017). But for this type of enterprises (SMEs) the challenges for business development are further accentuated by the typical characteristics of rural destinations such as seasonality and low occupancy rates. Many companies are family owned, and do not have enough knowledge and management skills to implement strategic plans in order to reduce the negative effects of seasonality (North & Smallbone, 1996 and Irvine & Anderson, 2004 cited by Oliveira et al., 2017).

Summing up, there is no doubt that nowadays the Internet mediates tourism experiences since tourists use social media sites to describe, recreate and recall their travels. In fact, images, videos, photos and films available through the various social media channels provide various messages that represent destinations and serve as mediators of tourist experiences (Tussyadiah & Fesenmaier, 2009). Related to this, Urry states that "there is no evidence that virtual and imaginative travel is replacing corporeal travel, but there are complex intersections between these different modes of travel that are increasingly de-differentiated from one another" (2001, p. 2). According to Urry, there is a distinction between (a) "virtual travel" through the Internet; (b) imaginative travel through phone, radio and TV, and (c) corporeal travel along the infrastructures of the global travel industry. Mobile technologies have changed the nature of vision (Urry, 2001). The 'static' forms of the tourist gaze have been transformed into a variety of tourist glances, due to different forms of capturing and sharing of sights. As stated by Urry, "globalisation has thus ushered in some momentous reconfigurations of the tourist gaze" (2001, p. 7).

RESEARCH METHODOLOGY

The methodology used was based on a "quali-quantitative" exploratory research with the purpose of identifying rural tourism accommodation enterprises. The selection of the companies (mainly small and medium enterprises – SME) was based on the national registration of accommodation in Portugal (vide https://rnt.turismodeportugal.pt/RNT/ConsultaAoRegisto.aspx), selected by rural tourism enterprises in Baixo Alentejo. Since one of the difficulties was the lack of updated data of the enterprises (some no longer existed; others were closed; others had recently opened), the researchers try to increase and complete the sample by updating more tourism SME according to some criteria, such as having: a WEB page; a page on a social network (e.g. Facebook); e-mail; reviews on the WEB or on TripAdvisor.

The empirical study was based on a questionnaire. It was designed based on a review of the literature and considering dimensions of the survey such as: differentiated factors for competition, namely local tourism resources; expected tourist experiences; cultural and natural experiences offered; hospitality concept. After selection of the companies, data collection was carried out through an online questionnaire (between March and May 2019), structured with closed and open questions with the entrepreneurs/

managers of the enterprises in order to identify their perceptions regarding the use of the networks and their relationship with the improvement of the tourism experience, customer loyalty and the guarantee of profitability of the business. The results were subject to a content analysis.

The article focuses on the suggestions for entrepreneurs to currently implement the use of social networks and ICT to improve the experience of their tourist consumers and invest in the use of these networks for the promotion and marketing of their companies.

Data analysis was performed through quantitative data (with Excel software) and content analysis with a Computer-Assisted Qualitative Data Analysis Software (in this case with webQDA®). This study is essentially explorative and a more "deductive procedure" was undertaken since a previously formulated theoretical schema was used (Mayring, 2000). The textual data was content analysed using webQDA software as previously mentioned, which carries out qualitative data analysis individually or collaboratively, synchronously or asynchronously (Souza, Costa & Moreira, 2016). In terms of coding procedure, a direct approach was undertaken where "the researcher uses existing theory or prior research to develop the initial coding scheme prior to beginning to analyse the data" (Hsieh & Shannon, 2005, p. 1286). The qualitative step of these categorizations and analysis consists of a methodological, controlled assignment of the category to a passage of text. A descriptive method that "summarizes in a word or short phrase – most often as a noun – the basic topic of a passage …" (Saldaña, 2009, p. 70) was adopted. In some cases, it was used a pre-determined categorization ("tree nodes") validate by theoretical models from the literature review, in other cases a free coding procedure in order to extract the main ideas ("free codes").

Other difficulty of the study was directly reported by the owners of the enterprises: most of these businesses are family-owned and are also SME in which the owner carries out several roles. It is thus difficult to motivate these owners to participate in this kind of research. Besides that, the researchers achieved a sample of 76 entrepreneurs characterized mainly by female managers (55.4% - 41 respondents), with a age average of 45 years old. Most of the respondents have undergraduate qualification (64% - 47 respondents) and they are de owners of the enterprises (71.6% - 53 respondents), having it as a principal activity (68.9% - 51 respondents). Considering the region, we had a quite balanced distribution of the subregions: 36.5% of the rural enterprises are from Baixo Alentejo; 24.3% are from Alentejo litoral; 20.3% from Alto Alentejo and the last, from Alentejo central.

4. RESULTS

Type of Experiences Offered in Rural Tourism Enterprises

In order to better define the components of the experience that are most often used by enterprises on social networks, it is important firstly to explore and understand the type of experiences offered to tourists in this region. To this end, a two-side approach was considered, i.e. not only knowing what tourists want to experience in this region (demand-side approach), but also to explore the type of experiences that in fact are accessible to tourists, prepared and organized by rural accommodation (supply-side approach). It will be interesting to explore if there is a match or a disparity between these two perspectives of the tourism experience, as discussed in the literature review section. Therefore, the research question that guided this analysis was the following:

RQ1: *"What type of tourism experiences would tourists like to have in this region related to the type of tourism experiences that are actually made available for tourists by the companies themselves?"*

First, a 'classical' content analysis based on counting words and phrases (manifest content) was undertaken as an explorative analysis. Figure 1 displays a word cloud for the demand-side perspective.

Words such as "nature", "tranquility", "rest", "peace", "landscapes", "culture", "sun" and "relax" emerge from the analysis. These means that tourists are willing to "consume" more functional and objective tourism experiences, mainly related to the region's natural and cultural heritage. Local resources such as "landscape", "history" or even "sun" seem to be central to them. Additionally, it should be noted that a more affective and psychological side of experiencing the destination is also mentioned through the words "tranquility", "rest", "relax", "peace" or even "beauty". In a general sense, these results show that tourists experience the destination not only based on their objective attributes (such as "heritage", "landscape" or even "sun"), but also based on their affective attributes, i.e. sensations and emotions that tourists feel when they are "consuming" a destination. This is totally in line with the concepts of tourism experiences presented and discussed previously. From this standpoint, a tourism experience should be planned and conceived strongly based on the feelings and sensations that tourists might have when experiencing it. This seems to be an important issue to consider.

Moving now to the supply-side perspective, Figure 2 presents a word cloud with the most frequent words related to the type of tourism experiences provided by rural accommodation.

Words such as "nature", "cultural", "gastronomy", sports", "walking", "tours", "traditional" or "contact" emerge from the analysis. However, in this case more information is provided about the specificity of the tourism experience, i.e. about the type of experience that need a certain level of organization (e.g. worlds such as "tours", "trails", "hiking", "walking", "pedestrian" or "cycling"), since this is an answer from the standpoint of who organizes it. In this case it is possible to find more diversity of tourism experiences corroborating the idea expressed in some definitions of tourism experiences, which requires

Figure 1. What tourists want to experience in this region, according to the perception of the rural tourism accommodation enterprises (demand-side perspective)
Source: Generated by WordItOut. Retrieved from https://worditout.com/

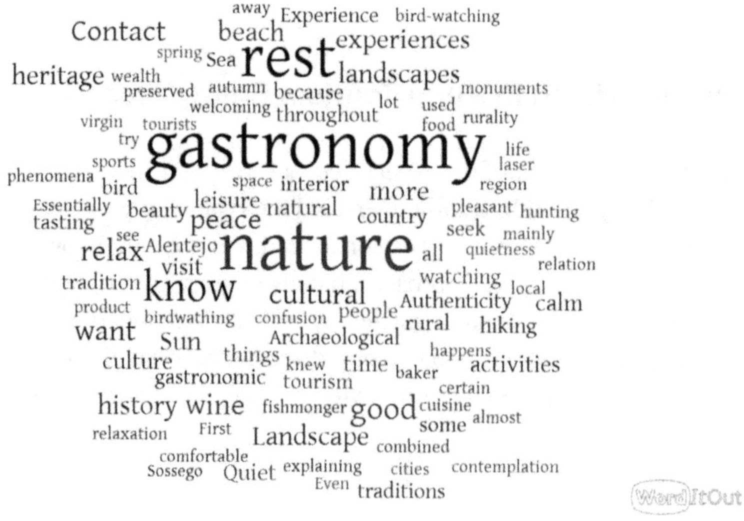

Figure 2. Type of tourism experiences that can the tourists enjoy in rural tourism accommodation (supply-side perspective)
Source: Generated by WordItOut. Retrieved from https://worditout.com/

a level of interaction with different elements of a context created by a service provider. It is also worth mentioning how sports and outdoor activities are highlighted in the supply perspective, corroborating the idea that tourism experiences are a blend of many elements coming together, involving the tourist not only emotionally, but also physically (and even intellectually or spiritually).

Table 2 displays and compares the top 25 most frequent words for the demand and supply side, aiming to give a preliminary analysis of both perspectives, corroborating the previous considerations.

With the aim of classifying tourism experience, according to Pine and Gilmore's (1999) model adapted to tourism (Oh, Fiore & Jeoung, 2007), based on four realms or dimensions of experiences (entertainment, educational, aesthetics and escapism), a descriptive codification (Saldaña, 2009, p. 70) was also adopted. The choice of items to depict was determined beforehand (through the tree nodes option from webQDA) as part of the study, which means that the codebook relied on four codes/categories from the above-mentioned Pine & Gilmore model. Table 3 exhibits the number of segments (references) coded per category (node), within a comparative analysis between the demand and supply perspectives. On both sides, "educational realm" and "escapism realm" present the large number of coded references.

Due to the lack of space, some supportive qualitative data examples about tourism experiences are presented here, in order to better understand the opinions from both sides:

1. **From the demand-side**: entertainment category e.g. "*have a good time*"; educational category, e.g. "to *know the landscapes*"; "*to know traditions and rural activities*"; "*they seek gastronomy and winery experiences a lot*", "*visitors/tourists like to talk to them, things, people, what is done .... Even more mundane things*", "*Know the region's genuineness*"; aesthetics category, e.g. "*Enjoy the beauty of country life*", "*preserved and natural landscape quality (escape large cities*", "*see*

Table 2. Comparison of the most repeated words between demand and supply-side about tourism experiences

Demand-Side		Supply-Side	
Word	**Repetition**	**Word**	**Repetition**
nature	19	nature	15
gastronomy	18	cultural	12
tranquillity	13	visit	12
landscapes	6	tours	10
contact	5	gastronomy	9
cultural	5	contact	7
peace	5	sports	7
relax	5	traditional	7
beach	4	walking	7
experiences	4	animals	6
heritage	4	beach	5
history	4	cycling	5
activities	3	hiking	5
culture	3	region	5
leisure	3	rural	5
natural	3	museum	4
quiet	3	through	4
visit	3	trails	4
watching	3	trips	4
archaeological	2	visits	4
Alentejo	2	where	4
authenticity	2	Alentejo	3
because	2	festival	3
beauty	2	information	3
country	2	nearby	3

Source: Generated by webQDA software.

Table 3. Different realms of experience based on demand and supply-side approach: nº of references coded with webQDA software

Approach/Perspective	Entertainment	Educational	Aesthetics	Escapism
Demand-Side	5	54	38	50
Supply-Side	11	45	21	40

Source: Generated by webQDA software

the landscapes of the coast and also more and more of the interior zone"; Escapism category, e.g. "*rest in protected nature*", "*The calm and the peace of the Alentejo*", "*pleasant and comfortable space and peaceful quietness*".

2. **From the supply-side**: entertainment category e.g. "*You can enjoy the traditional festivals and fairs in the nearby villages*", "*participate in traditional games*", "*visit the theme parks*"; educational category, e.g. "*visit megalithic monuments, archaeology, astronomical observations*", "*Visiting wineries*", "*Classes of Ceramics – wine tasting at Herdade do Gamito, snacks and dishes of the region's cuisine with chef José Julio Vintém*", "*Experience a typical rural Alentejo house, contact with traditional agricultural and cultural practices that identify the region and its people*"; aesthetics category, e.g. "*You can enjoy all the surroundings and the beautiful landscape*", "*tourists will have at their disposal varied information about the Rota Vicentina for sightseeing*"; "Escapism category, e.g. "*hiking, cycling or car tours, surfing*", "*Walking trails through the Montado*", "*canoeing, boat trip on the dam, sport fishing ride*", "*river beach with water sports, boat trips, free fall and parachute jump (Évora), balloon trips*", "*Walking or cycling (free of charge)*".

These results demonstrate that tourism experiences more related to the "educational real" and "escapism realm" should be taken more into consideration by the service provider i.e. rural accommodation. To truly create an educational experience, a tourist must increase his knowledge or skills through educational tourism experiences that actively engage the mind, causing the feeling that the tourists are truly learning something about the destination during their visit. Similarly, escapist experiences require tourists to really engage with the destination, mentally or physically. Tourists, when participating in this type of experiences, participate in activities worthy of their time, engaging with the destination, diverging to a new self (Oh et al., 2007).

Marketing and Social Media - Means Used to Publicize the Enterprise

Considering the perception of managers of rural tourism enterprises regarding the importance of social media for the communication and dissemination of the experiential activities that the company sells (makes available to the client), managers emphasize how the most important and most used in their enterprise are various means together. Of the answers obtained three are the means: (1) company website; (2) personal contact with the customer and (3) social networks. In fact, although social media plays an important role, managers greatly value direct contact with the customer upon check-in. This result reinforces the realization that the experiences of the tourist with rurality implies a family relationship and authenticity with the managers of the rural tourism units.

Based on the question about the main channels used in the dissemination of a business, it is noted that social media (e.g. Facebook, Instagram) means the most commonly used media (89% – 65 respondents) closely followed by the WEB pages (84% – 61 respondents). More distanced, digital platforms also appear with 59% of responses (43 respondents). The conclusion is for the use of more modern and current information channels.

The Internet is gaining increasing importance as a means of communication, sharing and dissemination of information, as it provides a wealth of tourist information and more and more tourists use online sources to meet their information needs (Nunes, 2014). However, traditional media are not excluded and are also highly valued by 17 of the managers, namely word of mouth and support from other public entities, tourist intermediation (e.g. tour operators, travel agencies and destination management companies – DMC).

In addition, almost all tourist enterprises surveyed have a WEB page (90.3% – 65 respondents) that has images and their use is directed to the provision of tourist information related to the characteristics of the rooms of the development (98.4% – 63 responses), from their location (95.3% – 61 responses), to get contacts to constitute a mailing list (93.8% – 60 responses) and to disclose prices (90.6% – 58 responses).

The main objectives for the use of social networks in their enterprises are the majority of responses: 82.6% (57 responses) to promote the company, followed by the objective of sharing photographs (76.8% – 53 responses), maintaining contact with clients (49 replies – 71%) and also reaching the market faster (60.9% – 42 responses).

Despite the fact that information sharing is mainly related to textual and narrative data, sharing experiences specifically enables creation and sharing of audio-visual content online (Munar, 2014). The main reasons for using images (visual content, specifically photos) on web pages stressed by the respondents were the need to advertise and show the rural accommodation as a way of connecting with the tourist (e.g. "make known what type of accommodation I have to offer" or "make our space known as much as possible"). For them, images (photos in this case) are a perfect way of marketing their houses through their website, since this category presents the highest number of coded references (n = 14). Many respondents highlighted the role of image when promoting their houses through statements such as "the eyes are the first to eat", "an image is worth a thousand words" or even "no photos, no guests". This seems to mean that they are perfectly aware of the importance of visual content for marketing purposes. Another reason mentioned by the respondents is related to the idea of attracting and engaging with the tourist through photos (n = 10). This is in line with the concepts of tourism experience, which stress the fact that what tourists primarily seek and consume at destinations is engaging experiences and because of that, photos seem to be a perfect way to achieve this goal because of their visual effect. It is also worth mentioning, although with fewer coded references (n = 7), that for the respondents, images might also inform (e.g. "inform the customer visually"), not only attract and engage. In fact, this could be achieved if images are conceived as a complement to textual information. Additionally, the respondents mention an interesting reason for using images: they not only attract, engage and inform, but also might lead to closing a sale. That is, images, if appropriately used, inform and even might contribute to selling the service (e.g. "images disclose, inform and book"). To conclude, the respondents also pointed out that images might promote not only the rural accommodation itself, but the whole surrounding area. Moreover, this might be even an excellent opportunity to better position the accommodation within marketing purposes, since photos can highlight differentiating elements about the accommodation and also the region where it is located (e.g. "show the best and differentiate the location"; "advertise the houses and the region".

Regarding the question "how can social networks reach the target audience and attract potential customers" and based on content analysis (free coding) it was possible to extract several ways for social media to reach the target audience, according to the accommodation suppliers as Figure 5 shows. Based on the process of tying each key idea to a theme (thematizing), three main themes were extracted related to the use of social media, from the supplier's perspective: (i) Promotion & Information: social media for reaching a global target audience and disseminating promotional information about the company; (ii) Sharing & Engaging: social media for sharing experiences, sensations and emotions and as a tool for interacting, not only with the tourists, but also with the locals, based on an authentic approach; (iii) Positioning & Segmentation: social media for marketing purposes, within a pre-defined strategy, related to targeting, positioning and branding. There is an absence of responses related to social media within a distribution outlook and how it facilitates online bookings. Regarding the total number of references in each idea/topic, "a tool for communicating efficiently", "sharing experiences and emotions" and "through photos and reviews" are the most referenced key ideas, meaning that the use of social media for the accommodation suppliers seems to be strongly rooted in purposes of promotion & information. It will be interesting to see if companies explore more other aspirations and contributions of social media, according to their marketing strategies.

The perception of the majority of respondents on whether the use of social networks changed the way of communicating/promoting their business (61.8% – 42 respondents) means that a large percentage of managers of rural tourism companies in the Alentejo considered that they have not changed.

Considering that social networks are a powerful vehicle for tourists (clients) to use to comment on their stay in the region and in the enterprise, the majority of entrepreneurs (84.1% – 58) are concerned with responding to the comments published. Another point of interest of this study was to explore how the entrepreneurs react when they have negative comments about their houses from their clients on social networks. Free coding was also adopted with the aim of extracting ideas about this topic freely. Based on this procedure, a total of 11 items that express the reactions of the entrepreneurs were elicited from the answers, showing that in general they have a very positive attitude. Here are some examples of the extracted items: (1) "answer with diplomacy, respect and consideration" (9 references); (2) "Negative comments help us to improve" (8 references; (3) "Accept, apologize and respond nicely" (7 references); (4) "Explain and correct the situation" (6 references). It is interesting to note that the item "no negative comments" also had also a significant number of references (a total of 7), meaning that there are no significant problems among the entrepreneurs about negative opinions. Although it was less often

Table 4. How social media reach the target market according to the accommodation suppliers

Ways/Topics/Themes	Descriptors: Some Examples
Information about tourism experiences	"especially giving tips on tours and activities"
Sharing experiences and emotions	"by sharing their experience on social networks, friends, family"
Use of a specific social media channel	"communication channel par excellence"
No impact/passive impact	"does not reach the public"; "Not very important"
A tool for communicating efficiently	"allows us to reach only the public interested in our product"
Through photos and reviews	"photos, reviews and shares";"Regular publication, mainly images"
Global impact of communication	"through the possibility of reaching many people and quickly"
Allows an efficiently market segmentation	"greater customer targeting";"Through filters of interest, location"
A tool for interacting with the locals	"great tool to promote our services to the local population"
Freedom to choose what to visualize	"customers look for what they want and use the comments"
Fast and direct tool for communicating	"faster and more efficiently"; "a comprehensive and fast way"
Direct interaction with the client	"very fast and direct connection with the customer"
Integration of social media channels	"complement the exposure provided by traditional platforms"
Allows to share authenticity	"active in the publications, be authentic"

Source: Generated by webQDA software analysis

mentioned, it seems important to highlight the fact that 3 references were coded in the item "Completely ignore the comments", which requires some consideration and debate.

When questioned about whether the fact that the company is present in social networks and if this can influence the results of companies, it was concluded that the opinion of the business entrepreneurs was affirmative (83.3% – 55 respondents).

About whether the enterprise is inserted in digital reservation platforms in some association or co-operation network, the conclusion is that the answers are more balanced although more entrepreneurs responded negatively to this question (56.9% – 41 respondents). Of the 43.1% who they claim to belong, the vast majority of the responses indicate the "Heranças do Alentejo Association".

5. CONCLUSION

Given their important role, the use of social media and online channels is a new challenge for tourism enterprises in rural areas, particularly for hotel managers. There is no doubt that these digital platforms are a means by which companies in tourism subsectors can communicate with customers more effectively.

This study aimed to understand how rural tourism rural accommodation enterprises use social media as a way of improving the tourist consumer experience. Tourism products are genuinely based on tourism experiences and for that reason it is of utmost importance to comprehend how the use of it can help accommodation SMEs in rural areas to become more efficient and competitive.

Tools and strategies for communicating with customers have changed significantly with the emergence of social media, also referred to as consumer-generated media, and because of that, these companies are undergoing significant change and face new challenges. New perspectives are thus called for.

The age of "economy of experience" presents the "new tourist" as a more demanding "tourist of experiences and emotions", who seeks greater differentiation, a made-to-measure product and high degree of correspondence to his expectations. This is an interesting opportunity for rural tourism enterprises to attend to customers' needs with activities that creating rewarding experiences in the region in which they operate and that make it possible to feel the culture and the peculiarities of the place. This is a challenge and an opportunity.

When trying to better understand the type of tourism experiences that tourists appreciate in these rural areas, the findings reveal interesting information based on approaches from two sides: demand and supply. From a demand-side approach (even though based on the entrepreneur's perspective) the findings reveal that tourists experience the destination, based not only on its objective attributes (e.g. "heritage", "landscape" or even "sun"), but also on its affective attributes, i.e. sensations and emotions that tourists feel when they are "consuming" a destination. These are important elements that need to be considered when conceiving tourism experiences. It is important to highlight that the tourism experience is a multidimensional takeaway one, a constant flow of thoughts and feelings that occur during moments of consciousness (Carlson, 1997), and because of that, rural places have high value resources at their disposal, strongly linked to the cultural and natural heritage of this territories that can be of value.

From a supply-side approach, the findings reveal that it is possible to find some diversity in tourism experiences, which requires a level of interaction with different elements of a context created by a service provider. Sports and outdoor activities are highlighted by the entrepreneurs as a type of experience that they offer to their clients, suggesting that they explore their natural and cultural resources, putting the tourists in contact with the surrounding area. No longer are tourists mere inert consumers but rather co-producers who actually are engaging more and more with the surrounding area, providers, hosts, etc. These experiences involve the tourist not only emotionally, but also physically (and even intellectually or spiritually).

In this sense, all efforts should be based on the development of innovative and differentiated products in a context of rural development. This is an opportunity for rural tourism, as a special form of tourism, to create unique concepts and marketing strategies that differentiate products and experiences from their main competitors.

Regarding the use of social media by these companies, the findings show that these digital tools play an important role, although managers greatly value direct contact with the customer upon check-in. This result reinforces the idea that, despite all the developments in technology, tourism experiences offered by rural accommodation are strongly intertwined with a close and family-like relationship and authenticity with the rural tourism entrepreneurs. Nevertheless, Facebook or Instagram are the most commonly used media closely followed by the WEB pages.

There is no doubt that nowadays the Internet mediates tourism experiences and images, videos, photos and films available through the various social media channels, which provide various messages that represent destinations and serve as mediators of tourist experiences (Tussyadiah & Fesenmaier, 2009). The findings reveal that the main objectives for using social media are for promotion purposes, as a channel for sharing photographs and a way of keeping in contact with clients. Moreover, the results show that the entrepreneurs are perfectly aware of the advantages in using visual content, such as photos, to promote their houses. For them, it is "no photos, no guests." Undoubtedly, rural accommodation suppliers are no longer ignoring the role of social media in promoting and even distributing their products. Given that, they need to understand even better the technological dynamics in order to reach out, contact their clients and promote their products. More efforts should be taken in this field, such as providing training for rural accommodation entrepreneurs, but profoundly adapted to the reality of the region and the SME itself.

REFERENCES

Andersson, T. (2007). The tourist in the experience economy. *Scandinavian Journal of Hospitality and Tourism*, *7*(1), 46–58. doi:10.1080/15022250701224035

Arnould, E. J., & Price, J. J. (1993). River magic: Extraordinary experience and the extended service encounter. *The Journal of Consumer Research*, *20*(1), 24–45. doi:10.1086/209331

Buhalis, D., & Law, R. (2008). Progress in information technology and tourism management: 20 years on and 10 years after the internet: The state of eTourism research. *Tourism Management*, *29*(4), 609–623. doi:10.1016/j.tourman.2008.01.005

Carlson, R. (1997). *Experience Cognition*. New York: Lawrence Eribaum Association. doi:10.4324/9781410603265

Cenamor, J., Paridab, V., & Wincent, J. (2019). How entrepreneurial SMEs compete through digital platforms: The roles of digital platform capability, network capability and ambidexterity. *Journal of Business Research, 100*, 196–206. doi:10.1016/j.jbusres.2019.03.035

Cerro, A., Mogollon, J., & Alves, H. (2010). Marketing relacional e Turismo: A fidelização de clients no turismo rural, proposta de um modelo para o seu estudo. *Revista Turismo & Desenvolvimento, 13/14*, 231–241.

Cloke, P. (2006). Conceptualizing Rurality. In P. Clocke, T. Marsden, & P. Mooney (Eds.), *The Handbook of Rural Studies.*, doi:10.4135/9781848608016.n2

Cooper, C., Fletcher, J., Gilbert, D., & Wanhill, S. (1998). *Tourism principles and practice*. London: Addison Wesley Longman.

Cunha, S., & Cunha, J. (2005). Competitividade e sustentabilidade de um cluster de turismo: Uma proposta de modelo sistémico de medida do impacto do turismo no desenvolvimento local. *Revista de Administração Contemporânea, 9*(2), 63–79. doi:10.1590/S1415-65552005000600006

Cupeto, C., Figueiredo, M., Silva, V., Santos, S., Carvalho, C., & Preguiça, C. (2014). *Empreendedorismo no turismo de experiência e emoção. In 4ª Conferência Iberica de Emprendimiento, CIEM 2014, Actas de la Conferencia: Jornada Científica*. Universidad de Vigo.

Cutler, S., & Carmichael, B. (2010). The dimensions of the tourist experience. In M. Morgan, P. Lugosi, & J. R. Brent Ritchie (Eds.), *The tourism and leisure experience: Consumer and managerial perspectives* (pp. 3–27). Bristol: Channel View Publications.

Dahnil, M. I., Marzuki, K. M., Langgat, J., & Fabeil, N. F. (2014). Factors Influencing SMEs Adoption of Social Media Marketing. *Procedia: Social and Behavioral Sciences, 148*, 119–126. doi:10.1016/j.sbspro.2014.07.025

Fleischer, A., & Felsenstein, D. (2000). Support for rural tourism: Does it make a difference? *Annals of Tourism Research, 27*(4), 1007–1024. doi:10.1016/S0160-7383(99)00126-7

Fleischer, A., & Pizam, A. (1997). Rural tourism in Israel. *Tourism Management, 18*(6), 367–372. doi:10.1016/S0261-5177(97)00034-4

Fotiadis, A. (2009). The role of tourism in rural development through a comparative analysis of a Greek and a Hungarian rural tourism area [PhD Thesis]. University of Pécs.

Friedman, L., & Friedman, H. (2008) The New Media Technologies: Overview and Research Framework. doi:10.2139srn.1116771

Gannon, A. (1994). Rural tourism as a factor in rural community economic development for economies in transition. *Journal of Sustainable Tourism, 2*(1-2). 51-60

Garrod, B., Wornell, R., & Youell, R. (2006). Re-conceptualising rural resources as countryside capital: The case of rural tourism. *Journal of Rural Studies, 22*(1), 117–128. doi:10.1016/j.jrurstud.2005.08.001

Gupta, S., & Vajic, M. (1999). The contextual and dialectical nature of experiences. In J. Fitzsimmons & M. Fitzsimmons (Eds.), New Service Development (pp. 33-51). Thousand Oaks, CA: Sage.

Hays, S., Page, S., & Buhalis, D. (2013). Social media as a destination marketing tool: Its use by national tourism organisations. *Current Issues in Tourism*, *16*(3), 211–239. doi:10.1080/13683500.2012.662215

Hirschman, E. C., & Holbrook, M. B. (1982). Hedonic consumption: Emerging concepts, methods and propositions. *Journal of Marketing*, *48*(3), 92–101. doi:10.1177/002224298204600314

Holloway, C. (1995). *Marketing for tourism*. London: Addison Wesley Longman.

Kounavis, C., Kasimati, A., & Zamani, E. (2012). Enhancing the tourism experience through mobile augmented reality: Challenges and prospects. *International Journal of Engineering Business Management*, *4*(1-6). doi:10.5772/51644

Jackson, J., & Murphy, P. (2006). Clusters in regional tourism – an Australian case. *Annals of Tourism Research*, *33*(4), 1018–1035. doi:10.1016/j.annals.2006.04.005

Jensen, R. (2001). *The dream society*. London: McGraw-Hill.

Larsen, S. (2007). Aspects of a psychology of the tourist experience. *Scandinavian Journal of Hospitality and Tourism*, *7*(1), 7–18. doi:10.1080/15022250701226014

Lashley, C. (2008). Marketing hospitality and tourism experiences. In H. Oh & A. Pizam (Eds.), *Handobook of hospitality marketing management*. Oxford, UK: Butterworth-Heineman. doi:10.1016/B978-0-08-045080-3.50004-4

Leung, D., Buhalis, D., & Hoof, H. (2013). Social media in tourism and hospitality: A Literature Review. *Journal of Travel & Tourism Marketing*, *30*(1-2), 3–22. doi:10.1080/10548408.2013.750919

Lopes, R., Abrantes, J., & Kastenholz, E. (2014). Innovation, tourism and social networks. *Revista Turismo e Desenvolvimento*, *21/22*, 151–154.

Mangold, W., & Faulds, D. (2009). Social media: The new hybrid element of the promotion mix. *Business Horizons*, *52*(4), 357–365. doi:10.1016/j.bushor.2009.03.002

Mayring, P. (2000). Qualitative Content Analysis. *Forum Qualitative Sozialforschung / Forum: Qualitative. Social Research*, *1*(2), 20. Retrieved from http://nbn-resolving.de/urn:nbn:de:0114-fqs0002204

McLellan, H. (2000). Experience design. *Cyberpsychology & Behavior*, *3*(1), 59–69. doi:10.1089/109493100316238

Middleton, V., & Clarke, J. (2004). *Marketing in travel and tourism*. Oxford: Butterworth-Heinemann.

Mossberg, L. (2007). A marketing approach to the tourist experience. *Scandinavian Journal of Hospitality and Tourism*, *7*(1), 59–74. doi:10.1080/15022250701231915

Munar, A. M., & Jacobsen, J. K. S. (2014). Motivations for sharing tourism experiences through social media. *Tourism Management*, *43*, 46–54. doi:10.1016/j.tourman.2014.01.012

Neuhofer, B., Buhalis, D., & Ladkin, A. (2013). Experiences, co-creation and technology: a conceptual approach to enhance tourism experiences. In *Tourism and Global Change: on the edge of something big. CAUTHE 2013. Conference Proceedings* (pp. 546-555).

Nunes, A. (2014). *A importância e utilização da internet como fonte de informação para o enoturista que visita regiões portuguesas*. Unpublished MSc thesis, Polytechnic Institute of Porto, Porto.

OECD. (1994). *Tourism strategies and rural development*. Paris: OECD.

Oh, H., Fiore, A. M., & Jeoung, M. (2007). Measuring experience economy concepts: Tourism applications. *Journal of Travel Research*, *46*(2), 119–132. doi:10.1177/0047287507304039

Oliveira, A., Correia, A., & Fernandes, C. (2017). A utilização das redes sociais na perspetiva das empresas de turismo em espaço rural. *Revista Pasos*, *15*(4), 807–822. doi:10.25145/j.pasos.2017.15.055

Page, S. J., Brunt, P., Busby, G., & Connel, J. (2001). *Tourism: a modern synthesis*. London: Thomson Learning.

Pellin, V. (2005). O turismo no espaço rural como alternativa para o desenvolvimento local sustentável: o caso do Município de Rio dos Cedros. *Caderno Virtual de Turismo*, *5*(1). Retrieved from http://www.inclusaodejovens.org.br/Documentos/BIBLIOTECA/Desenvolvimento/turismo_c omo_alternativa.pdf

Pine, B. J., & Gilmore, J. H. (1999). *The experience economy: work is a theatre and every business a stage*. Boston: Harvard Business School Press.

Quan, S., & Wang, N. (2004). Towards a structural model of tourist experience: An illustration from food experiences in tourism. *Tourism Management*, *25*(3), 297–305. doi:10.1016/S0261-5177(03)00130-4

Ray, A. (2008). *Experiential art: Marketing imitating art imitating life*. Retrieved from https://www.experiencetheblog.com/2008/05/experiential-art-marketing-imitating.html

Rodrigues, A. (2004). *The contribution of marketing communication on destination image formation process: the case of the Alentejo region*. Unpublished MSc thesis, Lisbon University Institute (ISCTE-IUL).

Rodrigues, A. (2015). *Lake-destination image assessment: the case of the Alqueva Lake*. Unpublished doctoral dissertation, University of Algarve.

Saldaña. J. (2009). The coding manual for qualitative researchers. London: Sage.

Seaton, A. V. (1996). The Marketing Concept in Tourism. In A. V. Seaton & M. M. Bennet (Eds.), *Marketing tourism products* (pp. 350–376). London: International Thomson Business Press.

Sharpley, R., & Telfer, D. (2002). *Tourism and development – concepts and issues. Channel View Publication*. Clevedon: Library of Congress Cataloging in Publication Data.

Silva, J., & Silva, J. (2003). Inserção territorial das actividades turísticas em Portugal: Uma tipologia de caracterização. *Revista Portuguesa de Estudos Regionais.*, *1*, 53–73.

Silva, J., & Perna, F. (2002). Turismo e desenvolvimento auto-sustentado. In J. Costa (Ed.), Compêndio de Economia Regional (pp. 451-474). Coimbra: APDR.

de Souza, F. N., Costa, A. P., & Moreira, A. (2016). webQDA - Qualitative Data Analysis (versão 3.0). Aveiro: Micro IO e Universidade de Aveiro. Retrieved from www.webqda.net

Stamboulis, Y., & Skayannis, P. (2003). Innovation strategies and technology for experience-based tourism. *Tourism Management*, *24*(1), 35–43. doi:10.1016/S0261-5177(02)00047-X

Toffler, A. (1970). *The future shock*. USA: Bantman Books.

Thurau, T; Malthouse, E.; Friege, C.; Gensler, S.; Lobschat, L.; Rangaswamy, A.; & Skiera, B., (2010). The impact of new media on customer relationships. *Journal of Service Research, 13*, 311-330. doi:. doi:10.1177/1094670510375460

Tussyadiah, I., & Fesenmaier, D. R. (2009). Mediating tourist experiences: Access to places via shared videos. *Annals of Tourism Research*, *36*(1), 24–40. doi:10.1016/j.annals.2008.10.001

Uriely, N. (2005). The tourist experience. *Annals of Tourism Research*, *32*(1), 199–216. doi:10.1016/j.annals.2004.07.008

Urry, J. (2001). *Globalizing the tourist gaze,* published by the Department of Sociology. Lancaster University. Retrieved from https://www.lancaster.ac.uk/fass/resources/sociology-online-papers/papers/urry-globalising-the-tourist-gaze.pdf

Veríssimo, M., & Menezes, N. (2016). Social media as a tool to enhance customer experience in hospitality industry. *Revista Portuguesa de Marketing*, *38*(4), 23–30.

Volo, S. (2004). Conceptualizing experience: A tourist-based approach. *Journal of Hospitality Marketing & Management*, *18*(2), 111–126.

Vučetić, S. (2017). The Importance of Using ICT in the Rural Tourism of the Zadar County. *DIEM – 4th Dubrovnik International Economic Meeting*. Academic Press.

Walls, A. R., Okumus, F., Wang, Y. R., & Kwun, D. J. (2007). An epistemological view of consumer experiences. *International Journal of Hospitality Management*, *30*(1), 10–21. doi:10.1016/j.ijhm.2010.03.008

Xiang, Z., & Gretzel, U. (2010). Role of social media in online travel information search. *Tourism Management*, *31*(2), 179–188. doi:10.1016/j.tourman.2009.02.016

Zeng, B., & Gerritsen, R. (2014). What do we know about social media in tourism? A review. *Tourism Management Perspectives*, *10*, 27–36. doi:10.1016/j.tmp.2014.01.001

Zhou, L. (2014). Online rural destination images: Tourism and rurality. *Journal of Destination Marketing & Management*, *3*(4), 227–240. doi:10.1016/j.jdmm.2014.03.002

Živković, R., Gajić, J., & Brdar, I. (2014). *The impact of social media on tourism. Sinteza 2014 - Impact of the Internet on Business Activities in Serbia and Worldwide*. Retrieved from; doi:10.15308inteza-2014-758-761

ADDITIONAL READING

Amaral, M. (2013). *Cooperação entre os atores sociais para o desenvolvimento turístico – o caso da subregião do Baixo Alentejo*. Unpublished doctoral dissertation, University of Évora, Évora.

Bec, A., Moyle, B., Timms, K., Schaffer, V., Skavronskaya, L., & Little, C. (2019). Management of immersive heritage tourism experiences: A conceptual model. *Tourism Management*, *72*, 117–120. doi:10.1016/j.tourman.2018.10.033

Binkhorst, E., & Dekker, T. D. (2009). Agenda for co-creation tourism experience research. *Journal of Hospitality Marketing & Management*, *18*(2-3), 311–327. doi:10.1080/19368620802594193

Kelliher, F., Reinl, L., Johnson, T., & Joppe, M. (2018). The role of trust in building rural tourism micro firm network engagement: A multi-case study. *Tourism Management*, *68*, 1–12. doi:10.1016/j.tourman.2018.02.014

Kiráľová, A., & Pavlíčeka, A. (2015). Development of social media strategies in tourism destination. *Procedia: Social and Behavioral Sciences*, *175*(12), 358–366. doi:10.1016/j.sbspro.2015.01.1211

Li, C., Guo, S., Wang, C., & Zhang, J. (2019). Veni, vidi, vici: The impact of social media on virtual acculturation in tourism context. *Technological Forecasting and Social Change*, *145*, 513–522. doi:10.1016/j.techfore.2019.01.013

Morgan, M., Lugosi, P., & Ritchie, J. R. (Eds.). (2010). *The tourism and leisure experience*. UK: Channel View Publications.

Ritchie, J., Tung, V., & Ritchie, R. (2011). Tourism experience management research. *International Journal of Contemporary Hospitality Management*, *23*(4), 419–438. doi:10.1108/09596111111129968

Sugathan, P., & Ranjan, K. R. (2019). Co-creating the tourism experience. *Journal of Business Research*, *100*, 207–217. doi:10.1016/j.jbusres.2019.03.032

Vada, S., Prentice, C., & Hsiao, A. (2019). The influence of tourism experience and well-being on place attachment. *Journal of Retailing and Consumer Services*, *47*, 322–330. doi:10.1016/j.jretconser.2018.12.007

Zatori, A., Smith, M. K., & Puczko, L. (2018). Experience-involvement, memorability and authenticity: The service provider's effect on tourist experience. *Tourism Management*, *67*, 111–126. doi:10.1016/j.tourman.2017.12.013

KEY TERMS AND DEFINITIONS

Consumer-Generated Content: Online content from blogs, virtual communities, wikis, social networks, collaborative tagging, and media files shared on sites like YouTube and Flickr, the so-called social media.

Creative Tourism: Extension of cultural tourism, in which consumers seek interactive and dynamic experiences.

Experience Economy: Age where business must orchestrate memorable events for their customers, and that memory itself is the product, the experience.

Rural Tourism: Form of tourism develops in rural areas (as opposed to the urban environment); supposes a reduced tourist offer; uses its own resources in the rural territory (natural, heritage and cultural) with a vision of sustainability.

Social Media: A variety of new sources of online information used by consumers to promote products, brands, services or even personalities.

Tourism Experience: Tourist behaviour related to the multi-sensory, fantasy and emotive nature of tourism products.

Tourism Product: A combination of elements underpinned by impressions, interpretations, perceptions, sensations, and mainly by experiences that shape the tourist's feelings and attitudes towards his or her visit.

Chapter 19
Local Accommodation Establishments' Strategies in Social Media

Rui Costa
University of Aveiro, Portugal

Gorete Dinis
GOVCOPP, Polytechnic Institute of Portalegre, Portugal

Raquel Seabra
University of Aveiro, Portugal

ABSTRACT

Tourism is a phenomenon of contemporary society, being one of the economic sectors where information has an essential function and where the impact of social media had more evidence. Digital technologies are increasingly present in life and have radically changed the way people think, make decisions and interact. Consumers are more demanding and informed, supporting their buying decisions in the online world. These changes also have implications in the tourism sector and in the promotion of tourist destinations. Given this new paradigm, the local accommodation units, as the most recent accommodation modality, have registered a remarkable growth and due to the characteristics of the activity itself, have to accompany and adapt to this new reality. The present work intends to analyze how Local Accommodation units in Aveiro use the Internet, and in particular the applications of social media in their communication strategies, as well as the importance they give to this reality of the digital world.

DOI: 10.4018/978-1-7998-1947-9.ch019

INTRODUCTION

Tourism is a phenomenon of contemporary society, being one of the economic sectors where information has an essential function and where the impact of social media had more evidence.

The digital world where we live today has transformed corporate communication as well as its strategies. Digital technologies are increasingly present in life (social, professional and personal) and have radically changed the way people think, make decisions and interact. Consumer/ tourist behaviour has become more flexible (more spontaneous, hybrids, technology fans), more experienced (informed, active), more independent (more adventurous and involved), have new values (seek unique, more real experiences and authentic), new lifestyles (more free time, travel more, make more short breaks) and have undergone demographic changes (more singles, smaller families). In this way, consumers are more demanding and informed, supporting their buying decisions more and more in the online world.

These changes also have implications in the tourism sector and in the promotion of tourist destinations, giving rise to the emergence of new intermediaries and a redefinition of the organizations' strategies of (Buhalis & Licata, 2002). Traditionally, the promotion of tourism was done through conventional media. Currently, this trend has been declining, since consumers / tourists prefer to get the information they need online, using two major forms (Xiang & Gretzel, 2010): social media websites and search engines rather than using traditional tour operators or travel agencies.

Due to the increasing use of the Internet, mainly due to the exponential growth in the use of social media platforms and the influence of the content and opinions generated by their users, tourism organizations must define a strategy and select the appropriate digital marketing channels to each phase of the trip and target audience.

In recent years we have witnessing an exponential growth of accommodation, legally qualified in Portugal as Local Accommodation (LA] units. According to the National Registry of Local Accommodation (RNAL), between 2012 and 2016, as occurred an exponential growth of 637.6%, in the number of LA units.

One of the main factors contributing to the proliferation and success of the LA model is the easy contact between citizens around the world brought by the internet. Today, any property owner that wants to monetize it through temporary lease can use the internet to promote it, resorting to the various portals of LA, such as Airbnb, the FlipKey (owned by TripAdvisor), HomeAway or Roomorama. Portugal was the sixth most requested European country on Airbnb in the summer of 2017, accounting for 1,113 million reservations on platform accommodations between June and August (Macário, 2017). If, prior to globalization, this phenomenon represented only a marginal fringe of tourist flow, now is seen as a real competitor to traditional hospitality, with a growing weight in the sector as a whole.

Although it is recognized that marketing strategy of tourism organizations, and especially of these small accommodation units, should favour digital channels, scientific studies addressing the online presence of tourism organizations and especially LA units are incipient. Considering all the presented questions, the main objective of the present study is to analyse LA units online strategies, moreover, understand how they use online channels and, in particular the social media in their communication strategies.

To this end, was prepared an observation grid consisting of a set of parameters that were defined from the literature review, proceeded through direct observation to the analysis of the online presence of 239 LA units from the city of Aveiro[1], the observation period occurred from October 7 to 21, 2018. The chapter is structured in three main sections. (i) the literature review, which focuses in the relevance digital marketing for tourism and hospitality, the social networks in tourism and hospitality from a business perspective; (ii) the methodology used in the study; and (iii) the conclusion and discussion of the results.

LITERATURE REVIEW

Relevance of Digital Marketing for Tourism and Hospitality

The development of the internet has had a significant impact on the way communication is performed today. The internet has facilitated access to information and the development of relationships between users. Furthermore, new media have changed and revolutionized the design and conduct of marketing strategies for a particular business. In this sense, because of the emergence of new digital media, the relationships between consumers and brands have suffered strong impacts and changes, becoming significantly connected through social media channels. (Edelman, 2010)

Currently, consumers are constantly online, using the internet at all stages of the decision-making process to purchase a tourism product or service. As stated by Kotler and Armstrong (2007), companies have to consider the possibility also starting to work in the digital world, and should be present at least in the most popular marketing channels (Kotler, Kartajaya, & Setiawan, 2017). These changes on social media and technological mobility as led to changes in communication, consumer attraction and product differentiation, requiring new marketing approaches from tourism organizations. Marketing is, also, in the context of destination retaining policies, one of the most important dimensions of quality and performance of a destination, which depends, among others, on the originality and diversity of tourism products and services (Moura, Mónico, & Mira, 2019)

For the success of a business, it is crucial to understand the functioning of the media present on the internet, to obtain a competitive advantage over its competition (Yannopoulos, 2011). According to Cintra (2010), marketers have a growing need to find new ways to communicate, to be innovative, thus getting the attention of their target audiences, creating a relationship with them so that they are loyal to the brand. Thus, from the internet, it is possible to apply a variety of brand promotion strategies, identify the target audience and provide information that is more directed to consumers' needs and to capture potential customers.

Social media is one of the most important promotional tools for a company, given that if they are properly used (i.e. based on a specific strategy and marketing plan), they can meet the needs of online customers and transform them on potential customers. According to Digital Marketing (2015), social networks are one of the most powerful tools of today's tourism. The use of these tools for travel-related purposes has grown considerably, with Facebook being one of the most popular in terms of monthly visitors. According to StatCounter (2019), the social network Facebook, in July 2019, dominated 72, 1% of the social networking market worldwide.

Social networks as a means of communication have enabled users to communicate more easily and effectively (Yusuf, Al-banawi & Al-iman, 2014). Thus, companies can use social media to promote their suggestions and innovative activities to the public (Danias & Kavoura, 2013). On the other hand, presence in social networks presents itself as a key way for brands / companies to increase their visibility, as well as to develop a relationship with consumers (Harris & Rae, 2009), all at a low cost, which will increase revenue and stimulate innovation in brand communication.

Definitely, Social networks have come to revolutionize business and the way people and brands communicate with their targets, which suggests that the way a business can grow rapidly will be boosted by its presence on social networks. Indeed, as Edosomwan, Prakasan, Kouame, Watson & Seymour (2011) point out, the widespread use of social networks by companies has seen profound and striking transformations in the conduct of business, publicity made by the brands, as well as in the products supply.

Regarding tourism and the hospitality, the UNWTO World Tourism Organization (2014) cited in Leite and Azevedo (2017) highlights that the internet and new technologies, such as email marketing, websites, blogs, social networks, together with mobile applications and online booking platforms are fundamental tools in the marketing strategies of tourism and hotel sector. According to Blanco (2011), tourism activity is increasingly being shaped by technologies combination, which reinforces the need for companies to understand these changes that also led to the emergence of new opportunities and triggered the development of new business models (UNWTO, 2011). Tourism and hospitality organizations should monitor these changes and use Internet data to understand consumer profile and behaviour, to provide timely information to make informed decisions appropriate to the type of consumer and depending on the phase of travel.

Thus, hotel managers should strategically plan, build and update frequently the online presence throughout the entire decision-making process. (Leite & Azevedo, 2017). González-Soria and Moreno de la Santa refer that travel marketer have been under-served the early stages of dreaming/researching and the experiencing stages, and this stage represents a major opportunity to influence tourist decision and build loyalty (UNWTO, 2011).

Social Networks in Tourism and Hospitality, from a Business Perspective

Since the mid-1990s, businesses / activities in the tourism and hospitality sector have started to use the internet as a distribution channel and as a marketing or commercial tool (Xiang, Magnini & Fesenmaier, 2015). For example, online travel agencies such as TravelBids, Expedia, PreviewTravel and, in 2000, TripAdvisor, started using the internet as a distribution channel and allowed direct access to the travel market (Xiang et al., 2015).

Kavoura and Stavrianea (2015) describe that tourists are eager to be part of an online travel community that will allow them to share experiences and ideas with other online members. However, these online communities cannot only be found on social media, but also on various other interactive websites providing information required by tourists. TripAdvisor is an example indicative of a travel community where you can find information for your destination, but also comment and exchange information with other users (Miguéns, 2008).

Information and communications technologies has strongly influenced and altered the tourism sector (Buhalis & Law, 2008). This technological age has developed means of communication that enable contact with the world (people, moments and places), eliminating the barriers that existed in communication (Riu, 2015). Therefore, a positive development in performance and quality of service is expected, as well as the optimization of operational efficiency in order to reduce costs (Buhalis & Law, 2008).

With a dedicated and increasing competition in the tourism and hospitality sector, it is crucial for operators in this field to recognize the importance of designing marketing strategies, with a perspective to public visibility and to position themselves appropriately against other competitors (Ip, Leung &

Law, 2011), something that digital tools can deliver faster and more effectively. Hotels can from social networks, run online campaigns through paid advertising, promoting their services and managing their reputation (Milović, 2012).

Thus, social networks are a distribution channel for a hotel unit, as well as a challenge and an opportunity. According to Brusha (2013), some opportunities should be considered when using social networks in tourism (see Table 1).

Thus, it is essential to study, reflect and analyse strategic and operational objectives when creating a profile in a social network. It is important to define which segments, the target, the objectives, and the message to transmit and have control over it. A strategy defined by a marketer is needed to identify where potential customers are and how best to communicate with them (Afonso & Borges, 2013).

In addition to all the strategies that needs to be defined to be present in the digital world, an action plan must be outlined to achieve the desired results, through marketing tactics, editorial calendar, campaigns, policy making, procedures and budgets, and, devices to identify, evaluate and select the most appropriate and cost-effective social platforms (Balegno, 2010).

Social media makes it possible to reach more communities, with greater online involvement and a greater relationship and sharing of content production between the company and the consumer. However, it is not enough to be present in social networks; companies must measure and monitor the actions they perform. According to Silva (2012), the metrics in digital social media can be described, more broadly, dividing them into groups.

Table 1. Opportunities to be considered by tourism organizations when using social networks

Opportunities or Advantages	Description
New customers	Using new channels other than usual increases the likelihood of reaching new people and may also gain new customers
Effective communication	By knowing what they say about a company on social networks, and participating in communication, the company attracts and enhances its relationship with current and potential customers through an image of credibility and trust;
WOM (word of mouth)	The so-called "word of mouth", a powerful source of advertising in tourism. People on travel-related social networks are eager for relevant and original information, which when they get it, easily share it on the web, making it viral;
Online reputation management	Social networks are a way for companies to constantly manage their brand image, given the ease of monitoring and control they provide;
Dynamic Technology	There is a panoply of online tools and applications that can help you to create innovative and unique content that engages users to attract potential customers, such as videos, online events, and contests;
Partnerships	A good way to reach more users while making the customer happier is to create partnerships with complementary brands, i.e. if a business has accommodation and does not serve breakfast, you can collaborate with a restaurant / pastry shop. Thus the partner, complementing the presence and offer on social networks; also performs online communication;
Publicize the region where it operates	In the tourism sector, any business is located in a particular region that people want to visit and know, so it is important to arouse the curiosity of the visitor. Social networks should also be used to promote and publicize the region, stimulating the curiosity of the internet user to visit it.

Source: Brusha (2013)

Table 2. Metric groups and their definitions

Metrics Groups	Definition
Reach (or visibility)	degree of disclosure that a particular content has
Influence	degree of attention and mobilization that one profile can generate in others
Commitment	degree of participation and involvement of a particular profile or group of people in relation to a theme or subject
Adequacy	degree of relevance / proximity between the actions performed and their values, proposals and characteristics, taking into account the users

Social networks, such as Facebook Insights, provide metrics and statistics that can, and should be used, by companies to improve their marketing strategies. However, these can be complemented with metrics provided by other online monitoring service companies (e.g. Radian6, Omniture, Coremetrics, Webtrends, Google Analytics). These companies provide different data and information; the company can and should use more than one, according to the phase of the web analytics strategy.

From the literature review it was found that there are some studies that address the presence or strategy of online tourism organizations (Boaria, Anjos & Raye, 2014; Abreu & Costa, 2000; Leite & Azevedo, 2017), but we found no references to studies whose scope was local business, namely LA units. Boaria, Anjos and Raye (2014) analysed the main e-marketing strategies used by national and international hotel chains established in Brazil and concluded that, in general, hotel chains are not aware of technological changes and innovations. Abreu and Costa (2000) conducted a feasibility study on the use of digital marketing in the hotel industry in Maceió, and concluded that, overall, hotel entrepreneurs in Maceió were not using the full potential of the internet, and one of the reasons pointing to the lack of online establishment is lack of experienced human resources. Leite and Azevedo (2017) analysed the online presence of four and five-star establishments of Porto (Portugal) and verified that, although there is a large online presence, managers and marketers are not fully aware of all the advantages that digital marketing can bring to their business. Most hotels need to improve the use of digital marketing as marketing and communication tool, in order to take advantage of all its potential.

METHODOLOGY

The main objective of these changes is to analyse Local Accommodation Establishments' strategies in Social Media, moreover, understand how they use online channels, in particular the applications of social media in their communication strategies. To this end, direct observation was used as a methodology in this study.

According to Quivy & Van Campenhoudt (2008), "direct observation is that in which the researcher proceeds directly to collect information, without the intervention of the observed subjects. It focuses on all relevant indicators and it is supported by an observation grid that is based on indicators and designates the behaviours to observe." Thus, the data collected about the use of the internet and social media by local accommodation units was made directly by the authors and recorded in an observation grid developed for this purpose.

The observation grid is formed by a set of indicators, as shown in scheme 1, which resulted from the literature review and which, in this context, should be understood as the behaviours to be observed.

Scheme 1: Observation Grid

The main objective of this grid is to analyse the relations between the internet and the social media and the Local Accommodation units of the city of Aveiro. With the collected data, it was possible in the end:

1. Know which applications of social media and social networks are used most by LA in the city of Aveiro.
2. Know the type of content published and which are the most frequent.
3. Compare practices of using different social media applications among the LA analysed.

Information regarding the previous indicators was obtained through the Google search engine, and the website was always checked, as well as social networks, to confirm that they were the officers of the LA unit and only after confirmation; data were collected to fill in the observation grid. Research was performed, first, by the name of the RNAL, followed by the number of the RNAL. This sequence helps to solve the problem of not finding the local accommodation unit in the first search, since many local accommodation units do not use in their social media communication the same name by which they are registered.

The study focused on local accommodation units. In Portugal, local accommodation units, in 2015, had a significant growth, around 65% (Marques, 2015). Between May and July 2017, there were 8,650 new official LA registrations, which double the number for the same period of 2016. The phenomenon is explained not by the large-scale entry of houses in the tourist rental market, but rather by the legalization of existing units (Villalobos, 2017). In addition, according to Airbnb, Portugal was the sixth most requested European country on Airbnb in the summer of 2017, accounting for 1,113 million bookings on the hosting platform between June and August (Macário, 2017).

At the end of 2017, according to the National Register of Local Accommodation (RNAL), there are 55,653 registered units, of which 36,499 are apartments (65.6%), 15,311 are houses (27.5%), 3,843 are accommodation establishments (6.9%), of which 510 are registered as hostels. The Aveiro district is the eighth most registered district in LA (RNAL, 31/12 2017). Despite this exponential growth in the number of LA in the country, there are still illegal operating housing, so it is almost impossible to ascertain the exact number of existing LA.

The sample to study coincides with the total population considering that exists 239 LA units in the city of Aveiro. Of these, 170 are apartments, 28 are villas and 41 lodging establishments (7 of these are hostels). Also, through the RNAL, it is possible to verify that of the 239 units, 135 are registered as a sole proprietorship and 104 as a collective proprietorship. Looking at the tax identification numbers, it can be concluded that the 104 registrations as a collective proprietorship have a repeated Tax Identification Number. This is explained by the fact that there are entrepreneurs with more than one unit, more precisely 41 entrepreneurs. Thus, we have 176 entrepreneurs in the city of Aveiro. The city of Aveiro was chosen as the geographical area of the study, mainly for its growing tourist importance at national level.

Figure 1. Scheme 1: observation grid
Source: Own elaboration

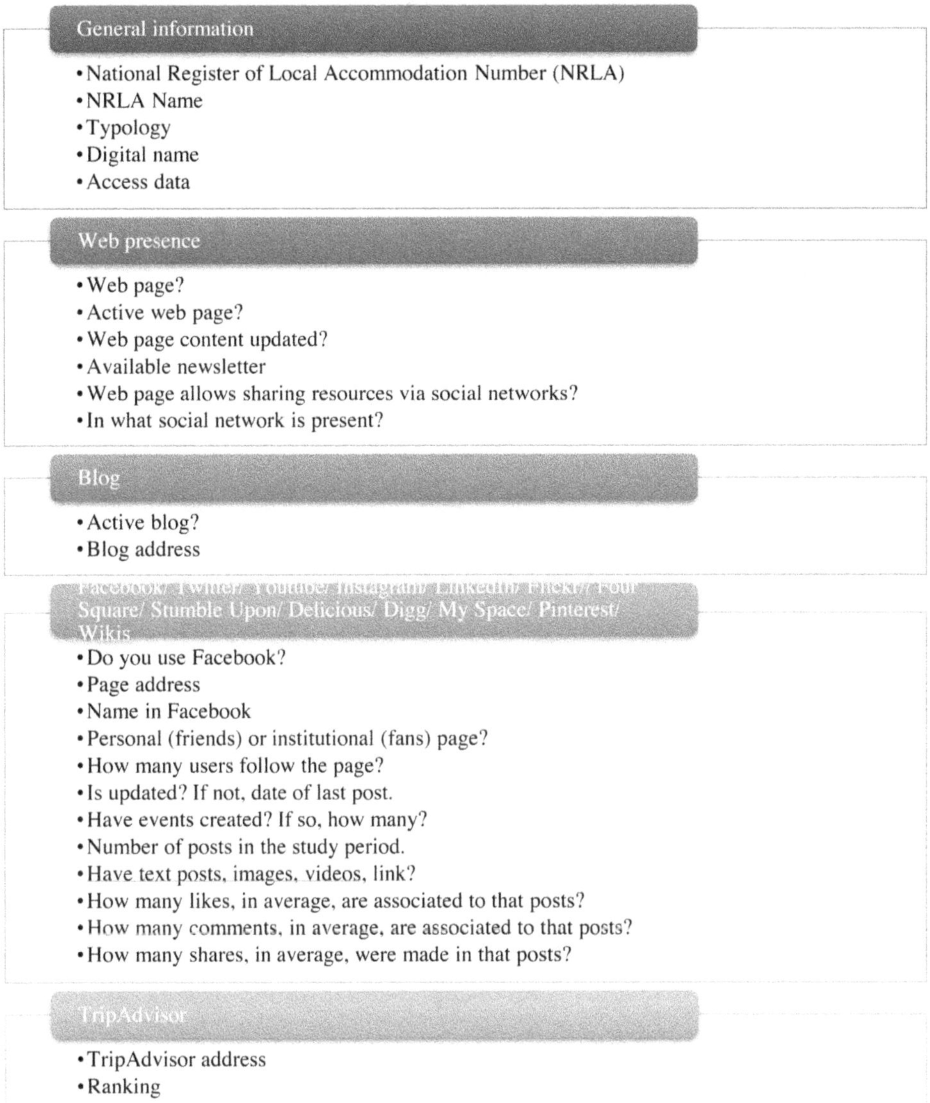

Data were collected on October 22, 2018, but the observation period was from October 7 to 21, 2018. The two-week period was chosen to understand whether LA units have a defined content strategy, such as a fixed editorial calendar, that is, days and times when the engagement with the audience is greater. After collecting all data, the database was processed through Excel.

RESULTS

Regarding web presence, it was concluded that only 63 LA units have a website, which corresponds to a 26.36% ratio, however all existing pages are active and up to date. Only 39 of these units have a newsletter and 45 share buttons for their social networks. It was also possible to conclude that of the 63 LA units with website in the city, 49 (78%) are apartments, 8 (12%) lodging establishments, 3 (3.5%) hostels and 3 (3.5%) villas, as shown in figure 2. It was also possible to conclude that 48 of the units of LA with Website are registered as a collective proprietorship, i.e. they are more likely to have more LA units in Aveiro or other cities and as such invest more in communicating and promoting their units.

From the analysis carried out it was found that only 1 LA unit has a Blog, but by evaluating the specific case, it was concluded that this LA unit uses the Blog as a Website, having all the information one would expect to find on a Website. Probably they may have chosen to use a blog instead of creating a website for the inherent cost savings.

Regarding the use of social networks, it was possible to analyse that LA units in the city of Aveiro are only present on Facebook, TripAdvisor and Instagram. It was concluded that of the 239 LA units analysed; only 108 are on Facebook (45.18%), 48 on TripAdvisor (20.08%) and only 6 on Instagram (2.5%).

Starting by analysing the most used social network in the city of Aveiro, by the LA units, it was concluded that of the 108 LA units with Facebook in the city of Aveiro, 75 (69%) are apartments, 16 (15%) lodging establishments, 6 (6%) hostels and 11 (10%) villas, as it can be observed in Figure 3. It was concluded that 100% of them use the institutional page, with an average of 556.8 followers. Moreover, it was concluded that 78.70% of LA units with Facebook page have not updated it, and there have been no publications in some cases since 2015. As there were no publications in the period under study, it was considered that despite these units have actually Facebook page, have not updated it, finalizing the study of these LA units in this parameter.

Only 23 (21.30% of the units with Facebook) are updated, the average post in the study period is 3.13, and the average of "likes" in the posts is 7. The average of comments is 0.4 and ultimately the average share is 2. Despite the small number of posts over this period, it was possible to conclude that there are text, image and link sharing posts (Table 3).

Figure 2. LA typologies with website in Aveiro city
Source: Own elaboration

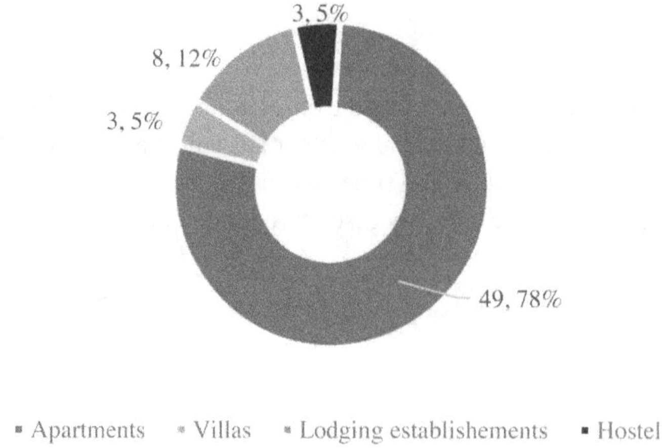

Figure 3. LA typologies with Facebook in Aveiro city
Source: Own elaboration

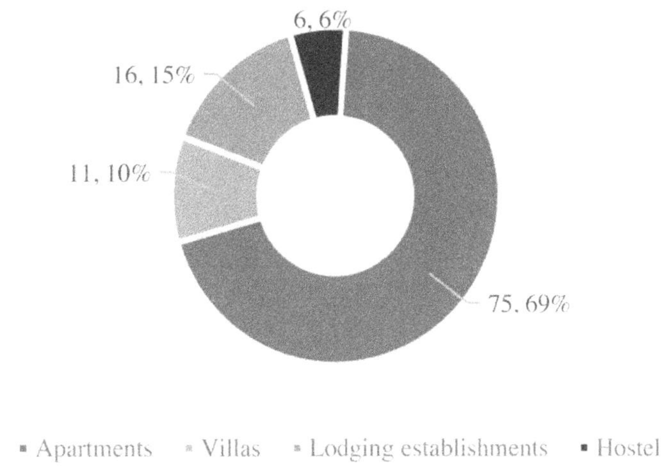

Table 3. Observation summary of the Aveiro City AL Facebook pages

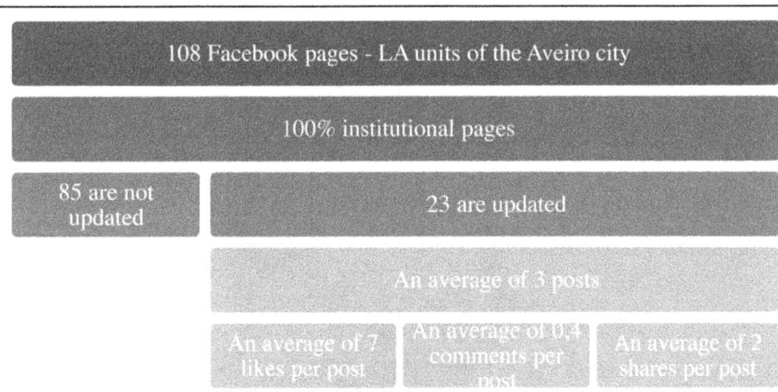

Regarding TripAdvisor, it was concluded that only 48 LA units are present on TripAdvisor (20.08%) with an average rating of around 4 out of 5. It was also possible to conclude that out of 48 LA units of Aveiro city on TripAdvisor, 14 (29%) are apartments, 23 (48%) lodging establishments, 5 810%) hostels and 6 (13%) villas, as shown in Figure 4.

Finally, the Instagram social network has considerably less users from LA units, respectively 6 units (2.51%). It was also possible to conclude that of the 6 LA units on Instagram, 4 (67%) are apartments and 2 (33%) villas, as shown in Figure 5.

The average of followers is around 390.83. Of these LA units, only 50% have updated this social network, i.e., with post, in the period under study. It was possible to determine from the 3 active units that the average posts in the period under study was 4.67, which obtained an average of 45.2 likes and 0.78 comments. All posts were photographs and text (Table 4).

From the analyses performed, it can be verified that only 2 LA units in the city of Aveiro have Website, Facebook and Instagram pages and are on TripAdvisor. The LA typology more frequent are Apartments and the type of holder is sole proprietorship; the typology of LA with higher incidence in

Figure 4. LA typologies on TripAdvisor in Aveiro city
Source: Own elaboration

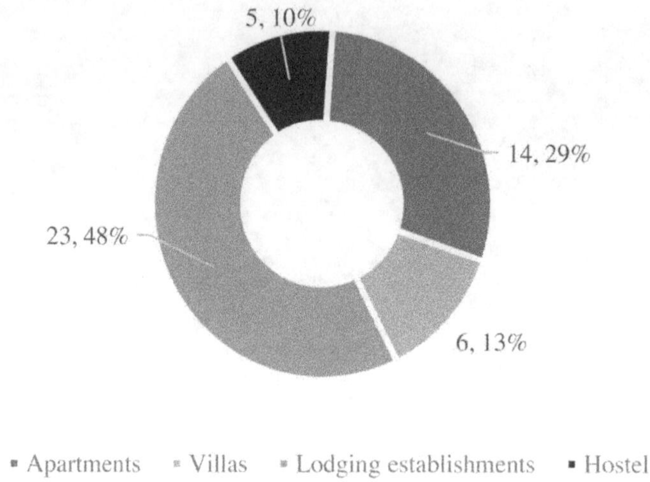

Figure 5. LA Typologies in Instagram in the Aveiro city
Source: Own elaboration

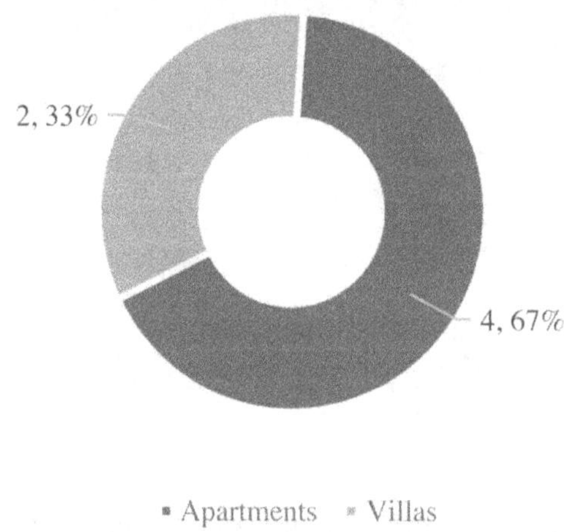

Table 4. Observation summary of the Aveiro City LA Instagram pages

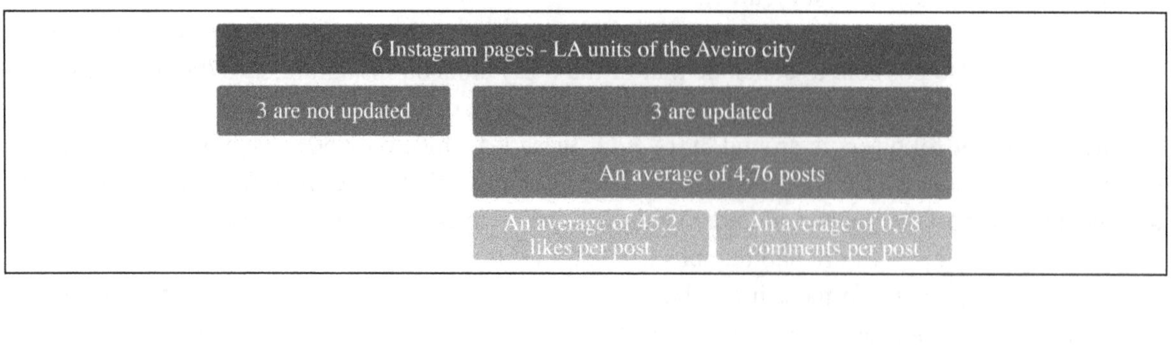

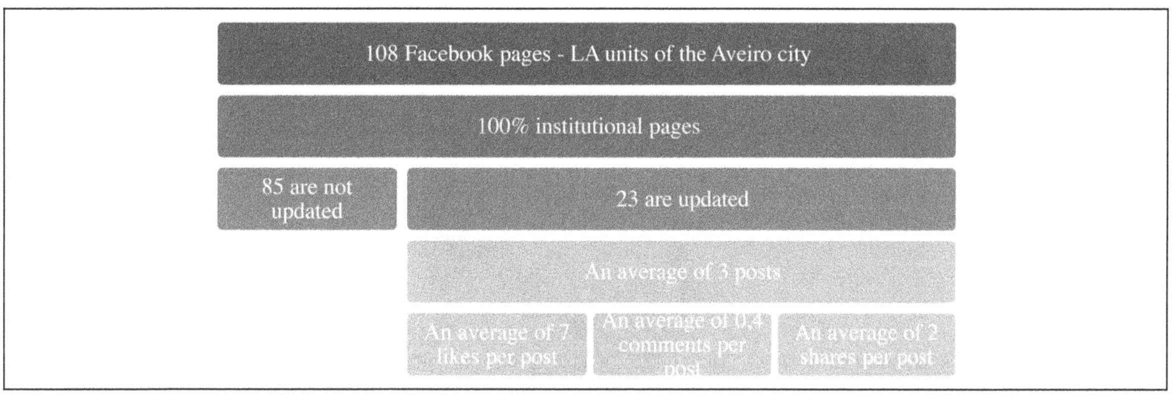

social networks is Apartments. Of the 239 LA units in the city of Aveiro, 118 do not have a website or blog, nor are they on any social network, so it was possible to find out that 109 units are only promoted on online sales platforms such as Booking and Airbnb.

In synthesis, and by type of LA unit, most part of all typologies have an active online presence and are present in social networks, mainly Facebook and TripAdvisor. The only typology to quote Instagram was the Apartment. The online presence and use of social networks, is in line with what happens in previous analyses, with no difference in relation to the type of holder. Both have mostly an active online presence and use social networks, Facebook and TripAdvisor, also having presence on Instagram.

Therefore, it can be concluded that few companies are using the digital world to analyse demand, building a closer relationship with the customer, to increase trust / brand awareness. The purpose of most of them is just to sell, which reinforces the results observed in the observation grid in which 109 of the units analysed are exclusively investing in online selling platforms and not in ways to create relationship with the consumer.

Moreover, most part of the LA units don't have a defined strategy, which is in line with what was analysed on social networks, through the observation grid, since of the 108 Facebook pages of LA units in the city of Aveiro, only 23 were up to date.

CONCLUSION

In recent years, social networks have played a crucial role in users' travel, from trip planning to experience sharing. In the past, the holiday destination was chosen through the tour operator's catalogues and reliance on travel agents, but nowadays people search the internet for all the information about the place or places they want to visit before traveling and often look for suggestions of places to go on holiday, thus creating an holiday package that really satisfies them.

When companies realize the potential of social media to boost the market, they adhere to these tools. Social media is a tool that, when used based on a specific marketing strategy and plan, can meet consumer needs and notably social media, is a useful, practical and cost-effective tool for businesses. (Danias & Kavoura, 2013)

The present study focused on Local Accommodation units of the Aveiro city, and was intended to assess to what extent they already use the internet and social media for the development of their promotion and communication activity. For this, 239 LA from Aveiro city were observed through a grid, in order to know which applications of social media and social networks are most used by the respective LA.

The results show that only 26.36% had a website, only one has a blog. The most used social network is Facebook with only 108 LA units, and only 23 were up to date.

In summary, it can be concluded that the LA units of Aveiro city does not have a clearly defined online strategy. Online presence through its own website is small and communication with the customer through social networks, is not consistent, which makes customer engagement more difficult. Thus, LA entrepreneurs in the Aveiro city should give more importance to the online world, as competition in the sector is increasingly strong, and the active presence in social networks can contribute to the company's strategic positioning.

The study has a limitation considering the period of observation (only 15 days), because data were obtained by direct observation, without option to any tool, in a period of only 15 days. In addition, it was not possible to analyse in depth the content of the publications. In future research, it is suggested that the study should be extended to other local accommodations in other locations, making it possible to carry out comparative analysis.

REFERENCES

Abreu, N., & Costa, E. (2000). Estudo da viabilidade da utilização de marketing na internet no setor hoteleiro de Maceió. Retrieved from https://bit.ly/2B6HmzG

Afonso, C. & Borges, L. (2013). *Social Target: Da estratégia à implementação – Como tirar partido das redes sociais e potenciar o negócio*. Editora Top Books.

Balegno, S. (2010). *2010 Social Media Marketing Benchmark Report Warren*, MarketingSherpa. Retrieved from https://bit.ly/2yspj5B

Boaria, F., Anjos, S., & Raye, R. (2014). A aplicação do e-marketing nas redes hoteleiras do Brasil. *TMStudies*, *10*(2), 116–122.

Brusha, P. (2013). Turismo e redes sociais: porque devem andar juntos? Retrieved from https://bit.ly/2OK4m0G

Buhalis, D., & Law, R. (2008). Progress in information technology and tourism management: 20 years on and 10 years after the Internet-The state of eTourism research. *Tourism Management*, *29*(4), 609–623. doi:10.1016/j.tourman.2008.01.005

Buhalis, D., & Licata, M. C. (2002). The future eTourism Intermediarie. *Tourism Management*, *23*(3), 207–220. doi:10.1016/S0261-5177(01)00085-1

Cintra, F. C. (2010). Marketing Digital: A era da tecnologia on-line. *Investigação*, *10*(1), 6–12.

Danias, K., & Kavoura, A. (2013). The role of social media as a tool of a company's innovative communication activities. *Zeszyty Naukowe Małopolskiej Wyższej Szkoły Ekonomicznej w Tarnowie*, *23*(2), 75–83.

Edelman, D. C. (2010). Branding in the Digital Age. *Harvard Business Review*, *88*(12), 6269.

Edosomwan, S., Prakasan, S. K., Kouame, D., Watson, J., & Seymour, T. (2011). The History of social media and its impact on business. *The Journal of Applied Management and Entrepreneurship*, *16*(3), 79–91.

Grönroos, C. (1994). From marketing mix to relationship marketing: Towards a paradigm shift in marketing. *Management Decision*, *32*(2), 4–20. doi:10.1108/00251749410054774

Harris, L., & Rae, A. (2009). Social networks: The future of marketing for small business. *The Journal of Business Strategy*, *30*(5), 24–31. doi:10.1108/02756660910987581

Ip, C., Leung, R., & Law, R. (2011). Progress and development of information and communication technologies in hospitality. *International Journal of Contemporary Hospitality Management, 23*(4), 533–551. doi:10.1108/09596111111130029

Kavoura, A. (2014). Advertizing activities in social media and the creation of a community belonging in the digital era. *Zeszyty Naukowe Małopolskiej Wyższej Szkoły Ekonomicznej w Tarnowie, 25*(2), 97–106.

Kavoura, A., & Stavrianea, A. (2015). Following and Belonging to an Online Travel Community in Social Media, its Shared Characteristics and Gender Differences. *Procedia: Social and Behavioral Sciences, 175*, 515–521. doi:10.1016/j.sbspro.2015.01.1231

Kotler, P., & Armstrong, G. (2007). *Princípios de Marketing (12a Edição)*. São Paulo: Pearson Prentice Hall.

Kotler, P., Kartajaya, H., & Setiawan, I. (2017). *Marketing 4.0: Moving from Traditional to Digital*. New Jersey: John Wiley & Son, Inc.

Leite, R. & Azevedo, A. (2017). The Role of Digital Marketing: A perspective from Porto Hotels' managers. *International Journal of Marketing, Communication and New Media,* (2), 88-105.

Macário, J. (2017). Portugal em sexto entre os países europeus mais requisitados na Airbnb. Retrieved from https://bit.ly/2KoUK6a

Marques, N. (2015). Crise ajudou à expansão do alojamento local. Retrieved from https://bit.ly/2OBO7mD

Miguéns, J., Baggio, R., & Costa, C. (2008). Social media and tourism destinations: TripAdvisor case study. *Advances in Tourism Research, 26*(28), 1–6.

Milović, B. (2012). Social Media and eCRM as a Prerequisite for Hotel Success. *Management Information Systems, 7*(3), 26–31.

Moura, A., Mónico, L., & Mira, M. (2019). Measuring quality regarding destination marketing: Perceptions from local public stakeholders in Portugal. *Tourism & Management Studies, 15*(1), 44–53. doi:10.18089/tms.2019.150104

Quivy, R., & Campenhoudt, L. (1995). *Manual de Investigação em Ciências Sociais* (2nd ed.). Lisboa: Gradiva.

Silva, T. (2012). Para Entender o Monitoramento de Mídias Sociais. Editora Bookess Online. Retrieved from https://bit.ly/2SY6aBY

StatCounter. (2019) Social Media Stats Worldwide. Retrieved from https://bit.ly/2zNzuTr

TrueSocialMetrics. (2018). Glossário. Retrieved from https://bit.ly/2ku1FP2

UNWTO. (2011). World Tourism Organization (UNWTO) Affiliate Members: AM-reports – Technology in Tourism (Vol. 1). UNWTO. Retrieved from https://bit.ly/2MyXNKi

Villalobos, L. (2017). Novas regras fazem duplicar registos de alojamento local. Retrieved from https://bit.ly/2Ka1xlm

Xiang, Z., & Gretzel, U. (2010). Role of social media in online travel information search. *Tourism Management, 31*(2), 179–188. doi:10.1016/j.tourman.2009.02.016

Xiang, Z., Magnini, V. P., & Fesenmaier, D. R. (2015). Information technology and consumer behavior in travel and tourism: *Insights from travel planning using the Internet. Journal of Retailing and Consumer Services, 22*, 244–249. doi:10.1016/j.jretconser.2014.08.005

Yannopoulos, P. (2011). Impact of the Internet on Marketing Strategy Formulation. *International Journal of Business and Social Science, 2*(18).

Yusuf, N., Al-Banawi, N., & Al-imam, H. A. R. (2014). The Social Media As Echo Chamber: The Digital Impact. *Journal of Business & Economics Research, 12*(1), 1–8. doi:10.19030/jber.v12i1.8369

ADDITIONAL READING

Chaffey, D., Ellis-Chadwick, F., Mayer, R., & Johnston, K. (2009). *Internet marketing: strategy, implementation and practice*. Pearson Education.

Edelman, D. C. (2010). Branding in the Digital Age. *Harvard Business Review, 88*(12), 62–69.

Harris, L., & Rae, A. (2009). Social networks: The future of marketing for small business. *The Journal of Business Strategy, 30*(5), 24–31. doi:10.1108/02756660910987581

KEY TERMS AND DEFINITIONS

Apartment: Establishment of local accommodation whose accommodation unit consists of an autonomous fraction of a building or part of an urban building that can be used independently.

Lodging Establishments: Local accommodation establishment whose accommodation units consist of rooms, integrated in an autonomous fraction, in an urban building or part of an urban building that can be used independently. Lodging establishments may use the name "hostel" when the predominant accommodation unit is a dormitory [i.e. when the number of dormitory users is greater than the number of guests in a room], and if they comply with the other requirements laid down for the effect.

Metrics: Measures, normally quantitative, used for assessing, comparing, and tracking performance of an organization or activity.

Rooms: Holding of local accommodation made at the holder's residence - corresponding to his tax domicile - when the accommodation unit is no more than three rooms.

Villas: Establishment of local accommodation whose accommodation unit consists of an autonomous, single-family building.

ENDNOTES

1 City located in the Centre region of Portugal

Chapter 20
Online Reputation of Porto Hotel Supply, Northern Region of Portugal

Noelia Araújo Vila
https://orcid.org/0000-0002-3395-8536
University of Vigo, Spain

Pablo de Carlos Villamarín
University of Vigo, Spain

Alexandra Matos Pereira
https://orcid.org/0000-0001-6928-2040
ISLA – Instituto Politécnico de Gestão e Tecnologia, Portugal

ABSTRACT

Recent advances in ICT has led to changes in the tourism sector, mainly in information dissemination and online reputation. Information no longer flows from suppliers to tourists, being the tourist, simultaneously an information consumer and an information creator or generator based on his tourism experience. Through networks, searches, and metasearch engines tourists can review and rate products and services and can access other users' information. They can access online content before, during, and after the trip. This information serves as a reference to choose the next destination to travel to, being much more influential than that generated by the tourism provider or the destination management office. As a case study, this chapter analyses online reputation of the hotel offer in the city of Porto (Portugal), the second most important city in the country, using information from Booking.com. Known as one of the most well-known search engines worldwide, Booking.com allows the guests to assess the hotel units they have stayed in, using seven different items, from staff to free Wi-Fi.

DOI: 10.4018/978-1-7998-1947-9.ch020

INTRODUCTION

The internet and social media have changed the way individuals communicate and take decisions (Manaman, Jamali, & AleAhmad, 2016). More and more companies are using social media to interact with their clients and selling their services (He, Zha, & Li, 2013). That explains why companies, from various sectors, use these online media to manage and evaluate their reputation. Reputation is understood as "the perceptual representation of a company's past actions and future prospects that describe the firm's overall appeal to all its key elements when compared to other leading rivals" (Fombrun, 1996, p. 72). Therefore, they act as safety nets, and by developing their reputation, companies can reduce the stakeholders' uncertainty about their capacity to create value (Rindova, Petkova, & Kotha, 2007). In other words, "reputation appears to be a fundamental extension of the organizational credibility whereby stakeholders evaluate past actions and records of behaviour" (Baka, 2016, p. 149).

All information spread through the internet is powerful, which justifies the need to design and use tools that can analyse and manage it. Besides, consumers usually review other consumers' comments before making a purchase, and the level of confidence in these assessments is usually high (Manaman et al., 2016). It is not an isolated evaluation, since a single valuation can be biased, but rather thousands or millions of opinions about a certain product or service, where it is unlikely that the knowledge obtained from these social media is biased (Mostafa, 2013).

Until the rise of online markets, it was difficult for researchers to study reputation systems' feedback. They were aware of the relevance of reputation and its influence on sales, hence the dangers of hidden information. Also, with internet transparency, understanding online reputation and feedback mechanisms became easier (Tadelis, 2016).

Reviews are at the core of online markets' reputation systems. Products and services offered online, but also sellers, are reviewed by effective buyers enabling potential buyers to use this information to choose which products to buy, and with which sellers to interact. Quality incentives are also created, as buyers and sellers' behaviour is made public (Luca, 2017).

Tourism is one of the sectors that has been highly affected by this framework. In fact, nowadays, no tourist or traveller makes a reservation or decides to travel to a tourism destination without reviewing other visitors' opinions. It has become a common practice to check reviews and buy online tourism products and services, making search engines such as Booking.com or Trivago popular worldwide. With these tools, and similar ones, you can read thousands of comments on hotels, restaurants, vacation rentals or vacation packages. Other web portals, such as Minube, allow reading recommendations from other travelers as what to do, where to eat, where to stay or what to visit in a tourist destination.

Given the growing popularity of Porto as a tourism destination and consequential rising of online reviews and comments impacting significantly on the hotel market of Northern Portugal, this paper analyses the online comments of hotels in Porto, Portugal's second biggest city and capital of the Northern region. The city is in the estuary of the Douro River in Northern Portugal. Being the largest city in the region, Porto is considered the economic and cultural heart of the whole region.

In this regard, we start by a theoretical review of the online reputation in the tourism industry and user generated content applied to this sector, since they are key variables within the online reputation. Then, we explain the goal of the study and the methodology used to develop the empirical work of the study. The aim of this work is to measure online hotel market reputation in the city of Porto. For this purpose, we have collected quantitative data in the evaluations by Booking.com of the hotels in the region concerned and analysed them. Finally, we have drawn the main findings of the research.

ONLINE REPUTATION IN TOURISM

The evolution and growth of the so-called web 2.0 has made it possible for consumers to react actively to everything that companies do and disseminate. No one depends on anyone, as in real time, and just with a smartphone anyone can reach a large audience, even worldwide (Cormode & Krishnamurthy, 2008). Information searches are no longer made offline (Sillence, Briggs, Harris, & Fishwick, 2007) but through the internet, using online word-of-mouth, best known as eWOM (electronic worth-of-mouth). Social media or review sites are very suitable online environments for these processes. The offline and more conventional communication strategy is losing effectiveness, compared to online content generation that is gaining more and more power (Gligorijevic & Luck, 2012).

Companies are aware of this change in society and consumer behaviour, so they have actively changed their communication strategy. Being now aware of the role of consumers and online users, companies lead their consumers to participate in online discussions about their products and services, using social media such as Twitter and Facebook. The Fortune Global 100's companies are active on at least one social media platform (87% of them), mainly Twitter (82%) and Facebook (74%) (Marsteller, 2012). Among major reasons for the use of these strategies are the drive to reliability, attitude towards the brand, and commitment to customers (Weinberg & Pehlivan, 2011). In addition, a strong reason why companies engage on this type of online activities is the assumption they are beneficial to their reputation (Dijkmans, Kerkhof, & Beukebook, 2015).

The concept of reputation is important for various reasons. It is a key variable for potential clients in the process of selecting companies (and their respective products and services) (Walsh, Mitchell, Jackson, & Beatty, 2009). That is, the better the company reputation, the more likely is to be chosen by a potential customer who can even be willing to pay more for its products or services (Graham & Bansal, 2007). In addition, a good reputation can assure a company a good position in the market, creating entry barriers to other competitors. The benefits are substantial, such as encouraging customer loyalty or attracting more customers. It can help a company in a crisis context (Shamma, 2012) or call for more investors. In short, many of the factors that ensure a company's success depend on its reputation, the image it projects on its customers and potential customers.

In the tourism industry it is also more and more common to access information based on online data generated by users. Tourists play a dual role, of co-creators and online content users about tourism experiences, before, during, and after the tourist experience.

Nowadays, tourists and travellers use platforms such as TripAdvisor to exchange information about tourism companies. This interaction allows users to reveal their experiences to tourism entities through the network. All kinds of information are shared, including their emotions about brands, services, destinations and suppliers (Horster & Gottschalk, 2012).

This explains why such online information not only affects tourists, but also suppliers and tourism destinations (Neuhofer, Buhalis, & Ladkin, 2014). To be aware and to understand tourist generated information is part of the digital management functions, as well as assuring the organization's online presence and reputation, since the perception of other travellers depend on it.

This flow of information has been driven by the evolution of ICTs (information and communication technologies), having access to it through new devices. Smartphones and social media have been some of the means to enhance the effectiveness of this phenomenon. You can access comments, images, and record, share or publish data, all in real time (Marchiori & Cantoni, 2015).

Throughout the pre-travel experience, information may be a determining factor. Prior knowledge may influence information search behaviour and affect the credibility of the consulted sources. In the current online context, experience and prior knowledge of a destination can influence the kind of content that is searched, and the time consumed to find it (Lehto, Kim, & Morrison, 2006). According to Cheung et al. (2009, p. 17), "users can detect the level of confirmation/ disconfirmation between the received information and their prior beliefs about the reviewed product/service through various direct or indirect experiences." Cheung, Luo, Sia & Chen (2009) underlined how, in the online context, the message receiver is more likely to rely on information that confirms his/her prior beliefs but perceives this as a belief change.

In addition, it is very likely that a satisfied consumer will recommend a product/service to another consumer (Chiu, Wang, Fang, & Huang, 2014).

Until this century, the consumer had a passive role, but later he/she becomes a content producer made available to the general market. This trend makes the tourism companies advertisements increasingly doubtful. The image a company or tourism destination wants to convey through its communication strategy will not have any impact or expected effect if it does not coincide with the users' reviews posted in the internet. This turn has meant a revolution in marketing, where information used to flow from service providers and has now been replaced by consumers who take the role of issuing information (Horster & Gottschalk, 2012).

The tourism industry as an economic segment is largely underpinned on the dissemination of information, as booking a trip is taken as a risk. The experience will depend on the climate, the food or the staff quality, in other words, on unpredictable heterogeneous factors. For this reason, to compare tourism products has become very difficult, so a great part of the decision making depends on trust. The more information the tourist has access to, the less uncertain he/she will be. Therefore, the search for information is a key issue, and the decision taken will depend significantly on what is found. Suppliers can issue information promoting their tourism products and services advantages, yet their views will be complemented by another users' information. That is why a high online reputation is a valuable tool to increase customer confidence (Chu, Choi, & Song, 2005).

Managing online reputation in the tourism industry involves determining who influences the providers' reputation, in what areas and through which factors, as well as ascertain the impact it will have on tourists' behaviour. Several models emerged, such as the one by Inversini et al. (2010) 'Destination Online Reputation Model' (DORM), which offers an overview of important issues that must be considered when assessing a tourism destination online reputation. This analysis allows evaluating which reputation factors UGC reveals and how they can be interpreted (positive, indifferent or negative). The findings give way to identify important issues related to a tourism destination online reputation, although they need to be further analysed.

The consequences of online reputation are also important, particularly in products that rely heavily on service provision, such as vacation travel. A good reputation corresponds to a greater client-supplier confidence, resulting in the client´s greater level of support towards the tourism product provider (Liehr, Peters, & Zerfaß, 2009).

USER GENERATED CONTENT (UGC) IN TOURISM

Online product reviews are increasingly important for consumer decisions (Goes, Lin & Au, 2014). This means that internet generated content is also a source of important and real information for suppliers (Benwell & Stokoe, 2006) who need to be aware of everything that is published online about their brand, products, and services. In other words, information that must be monitored (Sparks & Browning, 2011). A dissatisfied customer can spread his/her dissatisfaction or disappointment by personal word of mouth, but it has become more common to use the internet for it (Grégoire, Laufer & Tripp, 2010). Therefore, understanding and knowing what leads a client to generate online information, especially in a negative way, is vital for companies that need to anticipate outcomes and avoid possible crisis. Several researchers such as Yoo and Gretzel (2008) or Henning-Thurau, Gwinner, Walsh, and Gremler (2004) have conducted investigations on this field but the diffusion of negative experiences through the internet is still an open research field.

Focusing on the tourism industry, online travel purchases have several advantages: convenience, financial benefits, time saving, variety of products and comparability (Jensen, 2009; Wong & Law, 2005). In addition, according to Lee, Qu, and Kim (2007) behaviour, attitudes and predispositions to buy online increases online travel purchase intentions. In other words, consumers who are used to shop online also follow this consumption pattern with tourism products. Comparability stands out as the most advantageous feature, as in tourism is very relevant to know the opinion of other counterparts to compare different services, being them accommodation, restaurants, attractions or even destinations. Based on hotel industry literature, online reputation plays an important role when deciding on booking a hotel (Tsao, Hsieh, Shih, & Lin, 2015; Anderson & Anderson, 2014).

Yet, online purchases were and are complex processes, especially for certain population segments (Morosan & Jeong, 2008). Some people, faced with so many possibilities, feel overwhelmed when making vacation reservations through internet (Rheem, 2012). That is why social media such as Facebook, Twitter, Instagram, etc. and travel search websites such as Booking.com, TripAdvisor or Expedia are so popular as a mean to post trips, accommodations and attractions reviews and get information. These online channels are more reliable than family and friends' opinions (Ayeh, Au & Law, 2013).

The importance of personal referrals within the tourism industry has long been referenced in academic studies and is known in the marketing literature as WOM (Jeacle & Carter, 2011; Carl, 2006; Arndt, 1967). These recommendations, hosted by the website www architecture, are called eWOM (already mentioned in the previous section) being the Tripadvisor website a prominent example of an eWOM platform within the tourism industry (Jeacle & Carter, 2011; Vermeulen & Seegers, 2009; Buhalis & Licata, 2002). Consequently, online reputation management (ORM) has become a key factor in the present and future of marketing and feedback (TripAdvisor, 2014).

Any individual who has taken a trip can share their experience on these social media and become a reviewer (Baka, 2016). That is why User Generated Content (UGC) is defined as any type of content that has been created and put out there by unpaid contributors or, to use a better term, fans. It can refer to pictures, videos, testimonials, tweets, blog posts, and everything in between and is the act of users promoting a brand rather than the brand itself (Gallegos, 2016). UGC has become an important part of travel planning, as travellers evaluate travel products based on previous reviews.

The UGC concept allows consumers to read, learn about something and share their experiences with others. This term, together with consumer-generated media (CGM), has been used indistinctly in scientific literature. UGC is understood as the content created by users on the internet (Moens et al., 2014), such as social tags, reviews, blogs and tweets. CGM is defined as "media impressions created by consumers, typically informed by relevant experience and archived or shared online for easy access by other impressionable consumers" (Gretzel, Kang, & Lee, 2008, p.100).

UGC is used for different purposes, such as evaluating the quality and price of a service, searching for the best attractions, food, and destinations (Lee, Xiong, & Hu, 2012). Others do it as social acceptances, enjoyment, a communal feeling and involvement (Ukpabi & Karjaluoto, 2018).

The emergence of the UCG concept, whose rapid growth makes online reputation vital to an organization and daily management of a tourism company, is receiving growing attention from tourism researchers, who are studying how UGC comments intensifies the contingency of organizational reputation in the travel industry (Van Norel et al., 2014; Williams & Buttle, 2011). In such a fast-paced, constantly changing online landscape, content needs to be quick and on-trend. Instead of spending months and millions coming up with an advertising campaign that might be out of date by the time it finally airs, UGC allows brands to stay on the ball and stay current with their customers (Gallegos, 2016).

Thus, UGC data analysis has become an approach of great interest, since it can identify the characteristics of a disclosed destination image and its resources, via online users (Marine-Roig & Clavé, 2015). Perceptions, experiences, and tourist behaviours can be analysed through this information, which allows, at the same time, to investigate the tourism destination reputation, brand, and image (Lu & Stepchenkova, 2015). In addition, UGC can be linked directly to the eWOM effect, previously referred, resulting in an even more interesting effect, that of conveying the destination image of some tourists (online users) to others (Koltringer & Dickinger, 2015). With the eWOM, tourists and users can influence not only other tourists or potential tourists, but also companies' marketing and communication programs (Groeger & Buttle, 2014). This propagated online image is considered influential when tourists make decisions about which destination to visit (Yoo & Gretzel, 2010), as they trust on it (Leung et al., 2013). Just because these comments and assessments are from other tourists in person, and not from companies or organizations with economic interests.

On the other hand, sociodemographic variables influence the perception of UGC. Younger users and travellers often review a large number of sources before deciding, while the senior tourists' searches are more limited (Ayeh et al., 2013). Besides, those users who enjoy reviewing evaluations and comments from other users are more likely to accept their opinions as credibility is greater. The kind of information and the way it is shared also influence acceptance. Sparks and Browning (2011) found that consumers are particularly influenced by negative reviews when such information is negatively valued.

There is a large body of literature dealing on the UGC in the hotel sector, which is the focus of this study. On their review of papers related to online reviews in tourism and hospitality published in academic journals between 2004 and 2013, Schuckert, Liu, and Law (2015) found that more than half of the 50 relevant papers identified are focused on the hotel sector. From very early contributions, the analysis has focused on the comments and scores given by these online reviews, be it separately or jointly. Therefore, Crotts, Mason and Davis (2009) apply the technique of the stance-shift analysis to the comments in TripAdvisor.com published by the guests of three high-end hotels in New York City, identifying the different features in the hospitality service scored with the best and the worst marks. The evaluation in

Tripadvisor.com of a set of features in a sample of hotels in New York City is also the object of study of an early contribution by Jeong and Jeon (2008), even though in this case the analysis is focused on the numerical evaluations published by the guests. Finally, one of the most recent examples is the combined analysis of comments and scores by Stringam and Gerdes (2010), which analyses the words used in the comments published in Expedia.com by guests within a sample of hotels in the major U.S.A. cities, depending on the high or low scores given. Anyway, as noted by Sparks and Browning (2011), within a context marked by an excess of information, the numerical evaluations that are the basis of this work are a type of information that can be easily processed and those who search for a hotel are more prone to take them from online reviews. In this respect, it should be mentioned that Lee, Law and Murphy (2011) show that the helpful reviewers are those who, among other things, give lower hotel ratings.

In recent years, the literature on this topic has continued to increase, with contributions focused on the influence of WOM on the occupancy ratios (Viglia, Minazzi, & Buhalis, 2016), WOM of hotel performance (Yang, Spark & Hu, 2018), the impacts of Booking.com in rating system (Mariani & Borghi, 2048) or the role of eWOM in reducing information of online hotel booking (Manes & Tchetchik, 2018).

OBJECTIVE AND METHODOLOGY

The aim of this work is to measure online hotel market reputation in the city of Porto. For such, one review corpus has been examined based on the overall distribution of Booking.com property ratings in "Porto/Northern Region/Portugal".

Accommodation estates listed on Booking.com do not use a bilateral reviewing system which provides modest corroborating evidence for bias observed in this study. Prior works revealed that bilateral reviewing systems, as used in Airbnb, inflate ratings by incentivizing hosts to provide positive feedback so they are positively judged in return (Zervas et al., 2015; Fradkin et al., 2014; Bolton et al., 2013; Dellarocas & Wood, 2008). This suggests that Booking.com reviewers have factual and faithful opinions when ranking and rating accommodations as they are based on real experiences. "Only customers who have stayed at the Accommodation will be invited by Booking.com to comment on their stay at the relevant accommodations and to provide a score for certain aspects of their stay or may receive a rating request during their stay" (Booking.com, 2019a). The reason why we have selected this platform to perform the analysis is its reliability in respect of evaluations and because they have the highest number of items evaluated. On the other hand, we have selected accommodation because it is one of the most important subsectors within the tourism sector; being indispensable to create tourism. Each and every tourist will use any type of accommodation. The data has been collected in March 2019.

The completed guest review may be uploaded onto the relevant Trip Provider's information page on Booking.com platform rating 7 items on a scale of 1 to 10: staff, facilities and services, cleanliness, comfort, value for money, location and free Wi-Fi for "the sole purpose of informing (future) customers of your opinion of the service (level) and quality of the Trip Provider" (Booking.com, 2019a).

This observational study will attempt to examine this dataset, ranking Porto online hotel scores, finding possible clusters and determining valuable differences between factors. Two statistical techniques have been used: descriptive analysis, by computing the mean values of the scores given by the guests, and correlation analysis.

The population for sample analysis is composed by 3,033 hotel accommodations in Porto listed at Booking.com. As a preliminary study, a sample of 99 hotels (Table 1) registered at the Platform was analysed according to the star ratings (sampling error for a finite population of ± 9.7%, for a confidence level of 95%). Hotels are pooled by the number of stars, a system that ranks them in Portugal, being the 5-star hotels those which provide the best services and premises. The statistical program used to analyse the data was SPSS.

RESULTS

Porto, the second most important city in Portugal (after its capital, Lisbon), is one of the oldest destinations in Europe. It stands out for its wine, artistic heritage, cultural life, and wide spaces dedicated to leisure. It exhibits rich landscapes, by the river or the sea, as well as charming small places. The city conformed to a conservative set of values is associated with a creative and contemporary vibe, with cultural sites, iconic buildings, religious blue tile façades, but also shopping streets, urban leisure spaces, and unique terraces facing the River Douro or the Atlantic Ocean. Indeed, it's those co-existing opposites that underpin the whole destination. The Historical Centre of Porto has been Cultural Heritage of Humanity since 1996, and its environment and street winding layout give it a mesmerizing beauty. It is considered the city of Port Wine and has survived several invasions throughout history, regenerating and reinventing itself continuously (visitporto.travel, 2019).

In 2018 the city received 2.39 million tourists, and that same year it entered the top 100 most visited cities in the world, ranking 96, according to Euromonitor International (2018). But in 2017 the city has been awarded European Best Destination (europeanbestdestinations.com, 2017). According to this European organisation based in Brussels, "Porto is exceptional in more ways than one and the 'Cidade Invicta' (Invictus/Unconquered city), its history, architecture, culture, gastronomy, trade, encounters and discoveries, will seduce you". After receiving the vote of travellers from 174 countries, the city won this well-deserved award. But this was not the first time, as it had been awarded the same distinction in 2012 and 2014.

Aware of its importance as a tourist attraction, the city offers a wide range of tourism services, including accommodation. When compared to the Lisbon region that offers 5,908 accommodations, in Porto 57% of the accommodation venues are in the city centre, with 3,033 places, placing it as the second city with the largest accommodation capacity in the country (Booking.com, 2019a).

Table 1. Sample

Categories	Sample
1*	2
2*	26
3*	22
4*	37
5*	12
TOTAL	99

So many tourists have visited Porto that the number of hotel reviews in digital platforms is huge. That is also the case in Booking.com, the search engine used as data source for this study. Sample data shows that 1,541 is the average number of reviews per accommodation, reaching the maximum of 5,597, number that corresponds to the 5-star hotel HP Ipanema Park. Comparing figures between different hotel categories, using the European hotel star rating system, we came to the conclusion that 3-star hotels got the highest number of reviews (see Figure 1).

Stepping further on the analysis, an overall assessment of Booking.com customer review scores of Porto hotel supply was made, revealing an average of 8.34 or 'Very Good', a very positive value according to Booking.com rating system, making it the second-best score, as the maximum review score is between 9 and 10 or 'Superb.'

Analysing the sample per individual categories, hotels in Porto stand out for Staff, Cleanliness, and Location, items that score above the mean. As for the remaining items (Comfort, Facilities, Value, and Wi-fi), the results prove well with a score range of 8-8.3, which corresponds to the 'Very Good' category. It is also worth mentioning that the lowest score was for Value for Money, though it should not be considered excessively low (see Figure 2).

If the same comparative analysis is made, but taking into account the hotel star classification, 5-star units are those with the highest overall rating, 8.99, practically a 9, which puts them at the top qualitative score of 'Superb'.

Figure 1. Reviews' number (average) according to European hotel star rating system

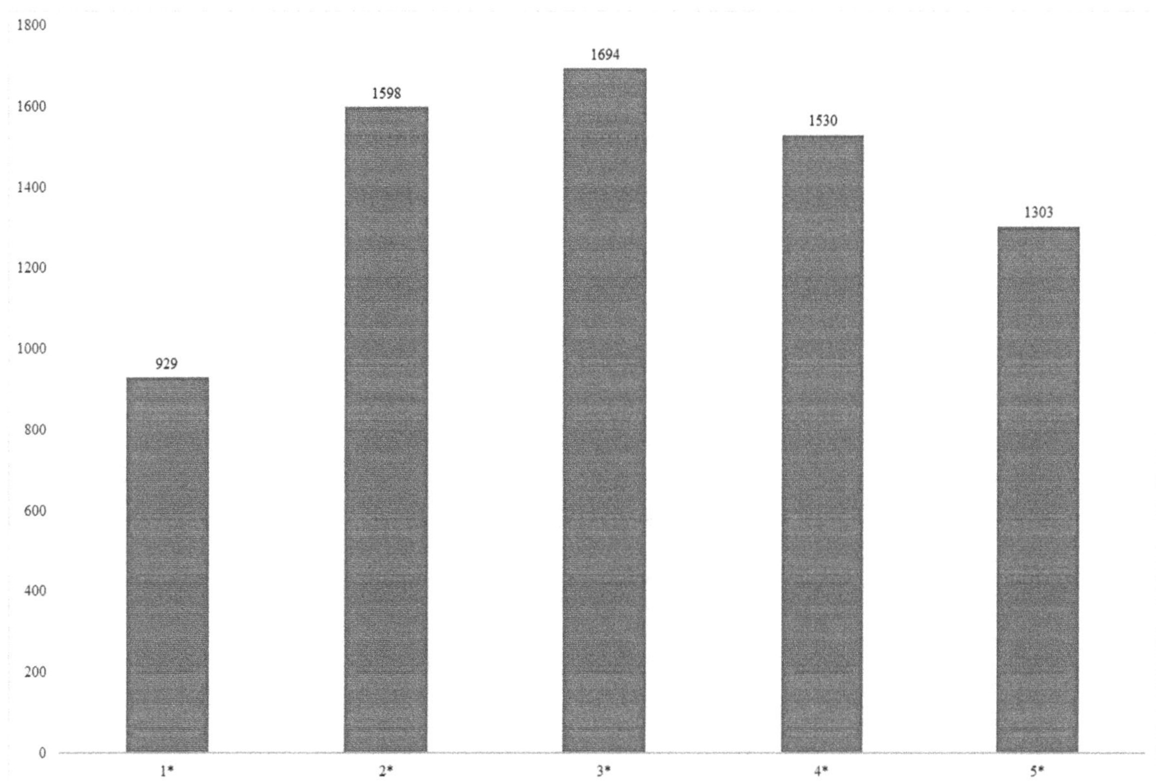

Figure 2. Average score of the 7 review items in Booking.com

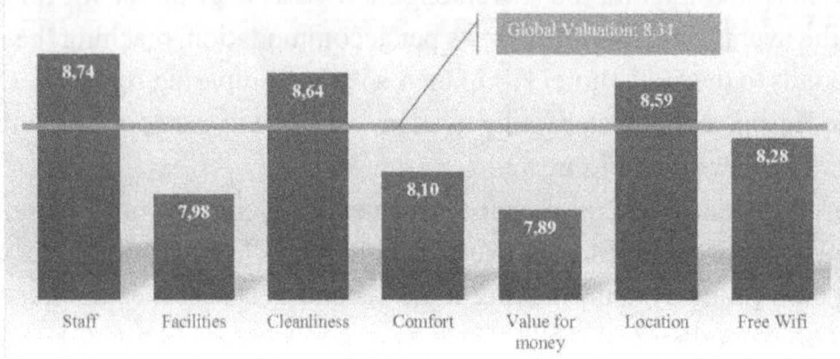

Analysing the relationship between hotel star classification and the seven items customer review scores on Booking.com, it is clear that 5-star hotel units receive the highest scores, exception made for Value for Money ratio, which scores best in 1-star hotels, and Location that scores 9 (Superb) in 2-star hotels (see Table 2).

Looking for the hotel unit in the city that stands out and scores the best in guest review items, it is interesting to verify that Maison Albar Hotel, a 5-star unit recently inaugurated in the very heart of the city centre, is highly ranked by its Staff (9.6), Facilities (9.7), Comfort (9.8), Cleanliness (9.9), and Free Wi-Fi (10), thus being the best appraised hotel in the city.

Nevertheless, three other hotels in Porto (3 and 4 stars) are equally well ranked in Staff. The Value for Money item is led by the 4-star hotel Design Palacio Flats, and the 2-star hotel Moov Hotel Porto Norte. In relation to Location, Design Palacio Flats, near the Cathedral, scores the highest, ex aequo with the 4-star Porto River Aparthotel, in Ribeira (see Table 3).

Finally, possible relationships between guest review items are examined creating a correlation matrix (Table 4). The correlation coefficients for the Facilities and Cleanliness variables show a high correlation (near to 1) as well as with Comfort and Facilities. In other words, there is a large positive relationship because as the variable Facilities increases, the other variables Cleanliness or Comfort also increase, and vice versa.

Table 2. Average scores according to hotel classification (European star rating system)

1*		2*		3*		4*		5*	
Global Scores	8.35	Global Scores	7.89	Global Scores	8.11	Global Scores	8.59	Global Scores	8.99
Staff	8.80	Staff	8.30	Staff	8.57	Staff	8.90	Staff	9.23
Facilities	7.70	Facilities	7.25	Facilities	7.67	Facilities	8.39	Facilities	8.98
Cleanliness	8.75	Cleanliness	8.05	Cleanliness	8.37	Cleanliness	8.96	Cleanliness	9.35
Comfort	7.80	Comfort	7.25	Comfort	7.72	Comfort	8.57	Comfort	9.22
Value for money	8.35	Value for money	7.80	Value for money	7.67	Value for money	7.96	Value for money	8.29
Location	8.55	Location	9.00	Location	8.54	Location	8.69	Location	8.84
Free Wi-fi	8.15	Free Wi-fi	8.30	Free Wi-fi	8.02	Free Wi-fi	8.48	Free Wi-fi	8.81

Table 3. Hotel units with the best scores according to guest review items

Item	Hotel Unit	Classification	Score
Global Score	Maison Albar Hotel	5*	9.6
Staff	Maison Albar Hotel	5*	9.6
	Torel Avantgarde	5*	
	Infante Sagres	5*	
	Design Palacio Flats	4*	
	Porto Charming Hotel	3*	
Facilities	Maison Albar Hotel	5*	9.7
Cleanliness	Maison Albar Hotel	5*	9.9
Comfort	Maison Albar Hotel	5*	9.8
Value for money	Design Palacio Flats	4*	8.9
	Moov Hotel Porto Norte	2*	
Location	Porto River	4*	9.8
	Design Palacio Flats	4*	
Free Wi-fi	Maison Albar Hotel	5*	10

Table 4. Correlation Matrix

	Staff	Facilities	Cleanliness	Comfort	Value for Money	Location	Free Wi-fi
Staff	1	0.860	0.884	0.835	0.809	0.426	0.652
Facilities	0.860	1	0.970	0.989	0.834	0.329	0.708
Cleanliness	0.884	0.970	1	0.961	0.866	0.395	0.729
Comfort	0.835	0.989	0.961	1	0.828	0.287	0.700
Value for money	0.809	0.834	0.866	0.828	1	0.312	0.625
Location	0.426	0.329	0.395	0.287	0.312	1	0.365
Free Wi-fi	0.652	0.708	0.729	0.700	0.625	0.365	1

There are notable links (value close to 1) between the scores of two pairs of items: Facilities and Cleanliness, thus linking the hotel facilities to their cleanliness and comfort and facilities, that is, the greater the evaluation of the variable Facilities, the better the variable Comfort is valued, and vice versa.

CONCLUSION

In the new technologies and information age, the tourism industry communication model is no longer the same. Information is much less provided by companies and tourism destinations than by tourists who generate content and consume content from other tourists. Smart devices, social media and tourist search engines have contributed greatly to this circumstance. A destination or tourism company online reputation does not depend entirely on the image it projects, but rather on the matching with the image tourists perceive of that same reality, and also on the way tourists generate information on the internet. Before making a trip, any tourist looks for online information from other tourists, and the lower the age the more the access. Previous studies have already shown that this kind of information creates more confidence in travellers than that disclosed by the tourism provider. This justifies why online reputation or information dissemination through internet about any tourism company is vital for its survival.

Referring to accommodation supply, the use of metasearch engines and search engines as a common online consultation practice is becoming popular. Among the most well-known are Booking.com, Trivago or TripAdvisor. Booking.com has been the agency business model selected for this study, as it obliges the guest to stay one night in the estate to review it, giving an up-to-date and hopefully unbiased selection of comments. The sample was taken from Porto hotel supply, chosen for its world's recognition and European Best Destination awards (2012, 2014, and 2017). It is widely recognised that a city value and online reputation is directly related to its accommodation supply.

Having analysed a sample of 99 hotels in the city of Porto, the overall score is high or "Very Good", using Booking.com review ratings. The large number of ratings per estate (more than 1,500 from tourists of all nationalities) assures unbiased data. Items such as Staff, Cleanliness or Location are the Booking.com rating items that ranked the best. As for the hotels, not only the 5-star hotels, something to be expected by their ranking in the European star rating system, but also 1-, 2-, and 3-star hotels showed ratings higher than 8 (or around it) in a 10 scale. Last but not least, the 2-star hotels are the best rated by tourists, when evaluating their Location in the city of Porto.

Regarding the implications for the sector, findings prove that the most important features when it comes to invest on a hotel estate are Staff, Cleanliness, and Location, with evaluations above the average for this region. These are the features that must be emphasised if a hotel wants to be competitive in this market. On the other hand, it is true that 5-star hotels are those that get the best scores in general, but some features are better evaluated among hotels with a lowest number of stars. This means that a market exists where different categories make sense because they have demand. Such information is useful for those who want to start a business, with no need of high investment in respect of the highest end.

Regarding the constraints of this study, the results achieved have been taken from 99 hotels sample whose evaluations have been collected in a specific moment, March 2019. The number of comments in Booking.com increases every day, and changes may occur over time. Therefore, as a future research topic data will be collected from a new sample, within a medium term, to analyze all the new evaluations and compare them with the present results to understand the evolution of hotel supply in Porto and its tendency.

REFERENCES

Anderson, M., & Anderson, M. (2014). 88% of consumers trust online reviews as much as personal recommendations. *Searchengineland.* Retrieved from [REMOVED HYPERLINK FIELD]http://searchengineland.com/88-consumers-trust-online-reviews-muchpersonal-recommendations-195803

Arndt, J. (1967). Role of product-related conversations in the diffusion of a new product. *JMR, Journal of Marketing Research, 4*(3), 291–295. doi:10.1177/002224376700400308

Ayeh, J. K., Au, N., & Law, R. (2013). "Do we believe in TripAdvisor?" Examining credibility perceptions and online travelers' attitude toward using user-generated content. *Journal of Travel Research*, *52*(4), 437–452. doi:10.1177/0047287512475217

Baka, V. (2016). The becoming of user-generated reviews: Looking at the past to understand the future of managing reputation in the travel sector. *Tourism Management*, *53*, 148–162. doi:10.1016/j.tourman.2015.09.004

Benwell, B., & Stokoe, E. (2006). *Discurso e identidad*. Editorial de la Universidad de Edimburgo.

Bolton, G., Greiner, B., & Ockenfels, A. (2013). Engineering trust: Reciprocity in the production of reputation information. *Management Science*, *59*(2), 265–285. doi:10.1287/mnsc.1120.1609

Booking.com. (2019a). Ranking, Preferred Programme, Stars and Guest Reviews. Retrieved from https://www.booking.com/content/terms.en-gb.html?aid=304085&label=Hoteis-iQ35xiFDNh3r-Jo4ReZuiSgS14356879461%3Apl%3Ata%3Ap1%3Ap2230.000%3Aac%3Aap1t1%3Aneg&sid=ff86ec28cdd97bcd214ff482d9f74277&tmpl=docs%2Fterms-and-conditions&lang=en-gb&soz=1&lang_click=top;cdl=pt-pt;lang_changed=1

Booking.com. (2019b). Retrieved from https://www.booking.com/

Buhalis, D., & Licata, M. (2002). The future eTourism intermediaries. *Tourism Management*, *23*(3), 207–220. doi:10.1016/S0261-5177(01)00085-1

Carl, W. J. (2006). What's all the buzz about? Everyday communication and the relational basis of word-of-mouth and buzz marketing practices. *Management Communication Quarterly*, *19*(4), 601–634. doi:10.1177/0893318905284763

Cheung, M. Y., Luo, C., Sia, C. L., & Chen, H. (2009). Credibility of electronic word-of-mouth: Informational and normative determinants of on-line consumer recommendations. *International Journal of Electronic Commerce*, *13*(4), 9–38. doi:10.2753/JEC1086-4415130402

Chiu, C. M., Wang, E. T., Fang, Y. H., & Huang, H. Y. (2014). Understanding customers' repeat purchase intentions in B2C e-commerce: The roles of utilitarian value, hedonic value and perceived risk. *Information Systems Journal*, *24*(1), 85–114. doi:10.1111/j.1365-2575.2012.00407.x

Chu, W., Choi, B., & Song, M. R. (2005). The role of on-line retailer brand and infomediary reputation in increasing consumer purchase intention. *International Journal of Electronic Commerce*, *9*(3), 115–127. doi:10.1080/10864415.2005.11044336

Cormode, G., & Krishnamurthy, B. (2008). Key differences between Web 1.0 and Web 2.0. *First Monday*, *13*(6). doi:10.5210/fm.v13i6.2125

Crotts, J. C., Mason, P. R., & Davis, B. (2009). Measuring guest satisfaction and competitive position in the hospitality and tourism industry: An application of stance-shift analysis to travel blog narratives. *Journal of Travel Research*, *48*(2), 139–151. doi:10.1177/0047287508328795

Dellarocas, C., & Wood, C. A. (2008). The sound of silence in online feedback: Estimating trading risks in the presence of reporting bias. *Management Science*, *54*(3), 460–476. doi:10.1287/mnsc.1070.0747

Dijkmans, C., Kerkhof, P., & Beukeboom, C. J. (2015). A stage to engage: Social media use and corporate reputation. *Tourism Management*, *47*, 58–67. doi:10.1016/j.tourman.2014.09.005

Euromonitor International. (2018). *Top 100 City Destinations 2018.*

European Best Destinations. (2017). European Best Destinations 2017. Retrieved from https://www.europeanbestdestinations.com/best-of-europe/european-best-destinations-2017/

Fombrun, C. J. (1996). Reputation: Realizing value from the corporate image.

Fradkin, A., Grewal, E., Holtz, D., & Pearson, M. (2014). Reporting bias and reciprocity in online reviews: Evidence from field experiments on Airbnb.

Gallegos, J. A. (2016). *What is User Generated Content (and Why You Should Be Using it).* Tintup. Retrieved from https://www.tintup.com/blog/user-generated-content-definition/

Gligorijevic, B., & Luck, E. (2012). Engaging social customers–Influencing new marketing strategies for social media information sources. In *Contemporary research on e-business technology and strategy* (pp. 25–40). Berlin: Springer. doi:10.1007/978-3-642-34447-3_3

Goes, P. B., Lin, M., & Au Yeung, C. M. (2014). "Popularity effect" in user-generated content: Evidence from online product reviews. *Information Systems Research*, *25*(2), 222–238. doi:10.1287/isre.2013.0512

Graham, M. E., & Bansal, P. (2007). Consumers' willingness to pay for corporate reputation: The context of airline companies. *Corporate Reputation Review*, *10*(3), 189–200. doi:10.1057/palgrave.crr.1550052

Grégoire, Y., Laufer, D., & Tripp, T. M. (2010). A comprehensive model of customer direct and indirect revenge: Understanding the effects of perceived greed and customer power. *Journal of the Academy of Marketing Science*, *38*(6), 738–758. doi:10.100711747-009-0186-5

Gretzel, U., Kang, M., & Lee, W. (2008). Differences in consumer-generated media adoption and use: A cross-national perspective. *Journal of Hospitality & Leisure Marketing*, *17*(1-2), 99–120. doi:10.1080/10507050801978240

Groeger, L., & Buttle, F. (2014). Word-of-mouth marketing influence on offline and online communications: Evidence from case study research. *Journal of Marketing*.

He, W., Zha, S., & Li, L. (2013). Social media competitive analysis and text mining: A case study in the pizza industry. *International Journal of Information Management*, *33*(3), 464–472. doi:10.1016/j.ijinfomgt.2013.01.001

Henning-Thurau, T., Gwinner, K.P., Walsh, G., & Gremler, D.D. (2004). El boca a boca electrónico a través de las plataformas de opinión de los consumidores: ¿qué motiva a los consumidores a articularse en internet? *Revista de marketing interactivo, 18*(1), 38-52.

Horster, E., & Gottschalk, C. (2012). Computer-assisted webnography: A new approach to online reputation management in tourism. *Journal of Vacation Marketing*, *18*(3), 229–238. doi:10.1177/1356766712449369

Inversini, A., Marchiori, E., Dedekind, C., & Cantoni, L. (2010). Applying a conceptual framework to analyze online reputation of tourism destinations. *Information and communication technologies in tourism*, 321-332.

Jeacle, I., & Carter, C. (2011). In TripAdvisor we trust: Rankings, calculative regimes and abstract systems. *Accounting, Organizations and Society*, *36*(4-5), 293–309. doi:10.1016/j.aos.2011.04.002

Jensen, J. M. (2009). Travellers' intentions to purchase travel products online: The role of shopping orientation. In Advances in tourism economics (pp. 203-215). Physica-Verlag HD.

Jeong, M., & Mindy Jeon, M. (2008). Customer reviews of hotel experiences through consumer generated media (CGM). *Journal of Hospitality & Leisure Marketing*, *17*(1-2), 121–138. doi:10.1080/10507050801978265

Költringer, C., & Dickinger, A. (2015). Analyzing destination branding and image from online sources: A web content mining approach. *Journal of Business Research*, *68*(9), 1836–1843. doi:10.1016/j.jbusres.2015.01.011

Lee, H. A., Law, R., & Murphy, J. (2011). Helpful reviewers in TripAdvisor, an online travel community. *Journal of Travel & Tourism Marketing*, *28*(7), 675–688. doi:10.1080/10548408.2011.611739

Lee, H. Y., Qu, H., & Kim, Y. S. (2007). A study of the impact of personal innovativeness on online travel shopping behavior—A case study of Korean travelers. *Tourism Management*, *28*(3), 886–897. doi:10.1016/j.tourman.2006.04.013

Lee, W., Xiong, L., & Hu, C. (2012). The effect of Facebook users' arousal and valence on intention to go to the festival: Applying an extension of the technology acceptance model. *International Journal of Hospitality Management*, *31*(3), 819–827. doi:10.1016/j.ijhm.2011.09.018

Lehto, X. Y., Kim, D. Y., & Morrison, A. M. (2006). The effect of prior destination experience on online information search behaviour. *Tourism and Hospitality Research*, *6*(2), 160–178. doi:10.1057/palgrave.thr.6040053

Leung, D., Law, R., Van Hoof, H., & Buhalis, D. (2013). Social media in tourism and hospitality: A literature review. *Journal of Travel & Tourism Marketing*, *30*(1-2), 3–22. doi:10.1080/10548408.2013.750919

Liehr, K., Peters, P., & Zerfaß, A. (2009). *Reputationsmessung: grundlagen und verfahren*. Leipzig: DPRG.

Lu, W., & Stepchenkova, S. (2015). User-generated content as a research mode in tourism and hospitality applications: Topics, methods, and software. *Journal of Hospitality Marketing & Management*, *24*(2), 119–154. doi:10.1080/19368623.2014.907758

Luca, M. (2017). Designing online marketplaces: Trust and reputation mechanisms. *Innovation Policy and the Economy*, *17*(1), 77–93. doi:10.1086/688845

Manaman, H. S., & Jamali, S. (2016). Online reputation measurement of companies based on user-generated content in online social networks. *Computers in Human Behavior*, *54*, 94–100. doi:10.1016/j.chb.2015.07.061

Manes, E., & Tchetchik, A. (2018). The role of electronic word of mouth in reducing information asymmetry: An empirical investigation of online hotel booking. *Journal of Business Research*, *85*, 185–196. doi:10.1016/j.jbusres.2017.12.019

Marchiori, E., & Cantoni, L. (2015). The role of prior experience in the perception of a tourism destination in user-generated content. *Journal of Destination Marketing & Management*, *4*(3), 194–201. doi:10.1016/j.jdmm.2015.06.001

Mariani, M. M., & Borghi, M. (2018). Effects of the Booking. com rating system: Bringing hotel class into the picture. *Tourism Management*, *66*, 47–52. doi:10.1016/j.tourman.2017.11.006

Marine-Roig, E., & Clavé, S. A. (2015). Tourism analytics with massive user-generated content: A case study of Barcelona. *Journal of Destination Marketing & Management*, *4*(3), 162–172. doi:10.1016/j.jdmm.2015.06.004

Marsteller, B. (2012). The Global Social Media Check-up 2011. *Lastet ned, 15*.

Moens, M. F., Li, J., & Chua, T. S. (2014). *Mining user generated content*. Chapman and Hall/CRC.

Morosan, C., & Jeong, M. (2008). Users' perceptions of two types of hotel reservation Web sites. *International Journal of Hospitality Management*, *27*(2), 284–292. doi:10.1016/j.ijhm.2007.07.023

Mostafa, M. M. (2013). More than words: Social networks' text mining for consumer brand sentiments. *Expert Systems with Applications*, *40*(10), 4241–4251. doi:10.1016/j.eswa.2013.01.019

Neuhofer, B., Buhalis, D., & Ladkin, A. (2014). A typology of technology-enhanced tourism experiences. *International Journal of Tourism Research*, *16*(4), 340–350. doi:10.1002/jtr.1958

Rheem, C. (2012). *Empowering inspiration: The future of travel search*. AMADEUS IT Group SA/PhoCus Wrights. February.

Rindova, V. P., Petkova, A. P., & Kotha, S. (2007). Standing out: How new firms in emerging markets build reputation. *Strategic Organization*, *5*(1), 31–70. doi:10.1177/1476127006074389

Schuckert, M., Liu, X., & Law, R. (2015). Hospitality and tourism online reviews: Recent trends and future directions. *Journal of Travel & Tourism Marketing*, *32*(5), 608–621. doi:10.1080/10548408.2014.933154

Shamma, H. M. (2012). Toward a comprehensive understanding of corporate reputation: Concept, measurement and implications. *International Journal of Business and Management*, *7*(16), 151. doi:10.5539/ijbm.v7n16p151

Sillence, E., Briggs, P., Harris, P. R., & Fishwick, L. (2007). How do patients evaluate and make use of online health information? *Social Science & Medicine*, *64*(9), 1853–1862. doi:10.1016/j.socscimed.2007.01.012 PMID:17328998

Sparks, B. A., & Browning, V. (2011). The impact of online reviews on hotel booking intentions and perception of trust. *Tourism Management*, *32*(6), 1310–1323. doi:10.1016/j.tourman.2010.12.011

Stringam, B. B., & Gerdes, J. Jr. (2010). An analysis of word-of-mouse ratings and guest comments of online hotel distribution sites. *Journal of Hospitality Marketing & Management*, *19*(7), 773–796. doi:10.1080/19368623.2010.508009

Tadelis, S. (2016). Reputation and feedback systems in online platform markets. *Annual Review of Economics*, *8*(1), 321–340. doi:10.1146/annurev-economics-080315-015325

TripAdvisor. (2014). Responding to Reviews Online Drives Booking Inquiries, TripAdvisor Study Reveals. Retrieved from http://globenewswire.com/news-release/2014/09/03/663342/10096925/en/Respondingto-Reviews-Online-Drives-Booking-Inquiries-TripAdvisor-Study-Reveals.html

Tsao, W. C., Hsieh, M. T., Shih, L. W., & Lin, T. M. (2015). Compliance with eWOM: The influence of hotel reviews on booking intention from the perspective of consumer conformity. *International Journal of Hospitality Management*, *46*, 99–111. doi:10.1016/j.ijhm.2015.01.008

Ukpabi, D. C., & Karjaluoto, H. (2018). What drives travelers' adoption of user-generated content? A literature review. *Tourism Management Perspectives*, *28*, 251–273. doi:10.1016/j.tmp.2018.03.006

Van Norel, N. D., Kommers, P. A., Van Hoof, J. J., & Verhoeven, J. W. (2014). Damaged corporate reputation: Can celebrity Tweets repair it? *Computers in Human Behavior*, *36*, 308–315. doi:10.1016/j.chb.2014.03.056

Vermeulen, I. E., & Seegers, D. (2009). Tried and tested: The impact of online hotel reviews on consumer consideration. *Tourism Management*, *30*(1), 123–127. doi:10.1016/j.tourman.2008.04.008

Viglia, G., Minazzi, R., & Buhalis, D. (2016). The influence of e-word-of-mouth on hotel occupancy rate. *International Journal of Contemporary Hospitality Management*, *28*(9), 2035–2051. doi:10.1108/IJCHM-05-2015-0238

Visit Porto. (n.d.). Retrieved from www.visitporto.travel

Walsh, G., Mitchell, V. W., Jackson, P. R., & Beatty, S. E. (2009). Examining the antecedents and consequences of corporate reputation: A customer perspective. *British Journal of Management*, *20*(2), 187–203. doi:10.1111/j.1467-8551.2007.00557.x

Weinberg, B. D., & Pehlivan, E. (2011). Social spending: Managing the social media mix. *Business Horizons*, *54*(3), 275–282. doi:10.1016/j.bushor.2011.01.008

Williams, M., & Buttle, F. (2011). The eight pillars of WOM management: Lessons from a multiple case study. *Australasian Marketing Journal*, *19*(2), 85–92. doi:10.1016/j.ausmj.2011.01.001

Wong, J., & Law, R. (2005). Analysing the intention to purchase on hotel websites: A study of travellers to Hong Kong. *International Journal of Hospitality Management*, *24*(3), 311–329. doi:10.1016/j.ijhm.2004.08.002

Yang, Y., Park, S., & Hu, X. (2018). Electronic word of mouth and hotel performance: A meta-analysis. *Tourism Management*, *67*, 248–260. doi:10.1016/j.tourman.2018.01.015

Yoo, K. H., & Gretzel, U. (2008). What motivates consumers to write online travel reviews? *Information Technology & Tourism*, *10*(4), 283–295. doi:10.3727/109830508788403114

Yoo, K. H., & Gretzel, U. (2010). Antecedents and impacts of trust in travel-related consumer-generated media. *Information Technology & Tourism*, *12*(2), 139–152. doi:10.3727/109830510X12887971002701

Zervas, G., Proserpio, D., & Byers, J. (2015). A first look at online reputation on Airbnb, where every stay is above average.

ADDITIONAL READING

Alrawadieh, Z., & Dincer, M. Z. (2019). Reputation management in cyberspace: Evidence from Jordan's luxury hotel market. *Journal of Hospitality and Tourism Technology*, *10*(1), 107–120. doi:10.1108/JHTT-09-2017-0093

Blomberg-Nygard, A., & Anderson, C. K.United Nations World Tourism Organization (UNWTO). (2016). United Nations world tourism organization study on online guest reviews and hotel classification systems: An integrated approach. *Service Science*, *8*(2), 139–151. doi:10.1287erv.2016.0139

Díaz, M. R., & Rodríguez, T. F. E. (2018). Determining the reliability and validity of online reputation databases for lodging: Booking. com, TripAdvisor, and HolidayCheck. *Journal of Vacation Marketing*, *24*(3), 261–274. doi:10.1177/1356766717706103

Fernandez-Cavia, J., Marchiori, E., Haven-Tang, C., & Cantoni, L. (2017). Online communication in Spanish destination marketing organizations: The view of practitioners. *Journal of Vacation Marketing*, *23*(3), 264–273. doi:10.1177/1356766716640840

Gössling, S., Zeiss, H., Hall, C. M., Martin-Rios, C., Ram, Y., & Grøtte, I. P. (2018). A cross-country comparison of accommodation manager perspectives on online review manipulation. *Current Issues in Tourism*, 1–20.

Nicoli, N., & Papadopoulou, E. (2017). TripAdvisor and reputation: A case study of the hotel industry in Cyprus. *EuroMed Journal of Business*, *12*(3), 316–334. doi:10.1108/EMJB-11-2016-0031

Phillips, P., Zigan, K., Silva, M. M. S., & Schegg, R. (2015). The interactive effects of online reviews on the determinants of Swiss hotel performance: A neural network analysis. *Tourism Management*, *50*, 130–141. doi:10.1016/j.tourman.2015.01.028

Pollák, F., Dorčák, P., Račeta, N., & Svetozarovová, N. (2016). Sustainable e-marketing of selected tourism subjects from the Mediterranean through active online reputation management. In *Smart City 360* (pp. 692–703). Cham: Springer. doi:10.1007/978-3-319-33681-7_60

Sparks, B. A., So, K. K. F., & Bradley, G. L. (2016). Responding to negative online reviews: The effects of hotel responses on customer inferences of trust and concern. *Tourism Management*, *53*, 74–85. doi:10.1016/j.tourman.2015.09.011

Chapter 21
Understanding the Role of Social Networks in Consumer Behavior in Tourism:
A Business Approach

Luís Ferreira
School of Economics and Management, University of Minho, Portugal

Bruno Barbosa Sousa
https://orcid.org/0000-0002-8588-2422
Polytechnic Institute of Cávado and Ave, Portugal

ABSTRACT

This research seeks to understand the type of influence that the hotel sector in Portugal can achieve in its consumers, being the tourism sector an area in constant growth. To that end, a qualitative methodology was adopted, using semi-structured interviews, as a data collection tool to understand the type of use that hotels attribute to social networks, as well as the results obtained from their practices. In the investigation nine hotels were analyzed, presenting a diversified sample between the participants in terms of capacity, as well as recognition, proving that, in a general way, social networks help in the divulgation of the hotel, presenting these as a direct channel for consumers, facilitating brand exposure as well as interaction with customers.

1. INTRODUCTION

Tourism is assumed as one of the major engines with more significant impact on the global development and growth of many economies. Year after year there has been an increase in competition in this sector, although this competition is established between tourism destinations (Becken & Simmons, 2002). Studies in tourism are increasing, although with different points of view and perspectives (Echtner & Jamal,

DOI: 10.4018/978-1-7998-1947-9.ch021

1997), for example on the planning of tourism destinations, local development, environmental impact, territorial brand management or even the loyalty of tourists (Getz, 1986; Backman & Crompton, 1991; Ramos, Matos, Sousa, Correia & Cascada, 2017; Ferreira, Sousa & Gonçalves, 2018).

Also in tourism, and nowadays, companies take their presence in the online world as imperative, whether it is to reach new consumers, promote the brand, better know their customers, ensure competitiveness or maintain its relevance in the market. However, maintaining the online presence may not be so easy to do, at least in a successful way. Many companies demonstrate difficulty with the use of various social media for communication purposes, especially in encouraging consumer involvement (Tsai et al., 2017). To Jafari and Aaser (1988), tourism as a study area has gained space, reflecting its increasing recognition in the academic community, an application of interdisciplinary concepts and methods. The field of tourism marketing is no exception and has faced increasing challenges caused by the diversity and demands of markets which call for new marketing approaches. New habit needs and trends in the global tourism creates more sophisticated consumers who systematically look for different and specific tourism experiences.

Digitization was a real challenge for any company, requiring cautious and well-planned action to be successful. Presently being present in social networks presents itself as a high value advantage, allowing to expose the brand, product or idea, at a low cost, to a high audience. However, there are risks if, on the one hand, exposure can increase the company's reputation, on the other hand, if it is not well achieved, there is the possibility of creating a bad image for the brand or a social controversy. The objective of this study is to understand the phenomenon of social networks in companies, especially in the hotel sector. The choice of this sector is that it has a high need to use social networks in order to be in constant contact with its customers, as well as to create a wide exposition of the establishment. Since tourism has a high internationalization load, that is, it attracts different types of nationalities to the establishment, social networks appear as a great tool of communication, since its audience is not limited to national borders. This study aims to show how the hotel sector sees and uses social networks, the level of importance attributed to them, the types of difficulties they entail and also what kind of results are obtained. For this purpose, the following research objectives were elaborated:

1. To explore the use and impact of social networks on consumer behaviour.
2. Reveal the functions of social networks within the hotel sector.
3. Understand the importance of the use of social networks for the hotel sector.
4. Provide a deeper understanding of the potential implications of social networking in the hospitality industry.

2. DIGITAL MARKETING AND SOCIAL MEDIA

It has been nearly a quarter century since the commercial use of the Internet and the World Wide Web. During this time the business landscape has changed at a frenetic pace. Large multinational corporations such as Google, Facebook, Amazon, Alibaba, eBay and Uber (Kannan, 2017). There is general agreement that digital media have greatly impacted the way a marketer reaches today's consumer. Digital media refer to electronic media that disseminate information in digital formats (Ramos & Rodrigues, 2011). This includes any media available via computers, mobile phones, smart phones, or other digital devices such as digital outdoor signs (Alves et al., 2019). The Internet is a prominent venue of digital marketing (Smith, 2011). According to Alves et al. (2019), the online relationship quality is an integrative and multidimensional concept, such as relationship quality in an offline context (Brun et al., 2014) transpose the concept of relationship quality in the offline context for the online context, approaching it in the same way. Nowadays, information and communication technologies have evolved, which has led to a sale of products and services through websites, thus allowing companies to obtain a competitive advantage over the market in which they operate. This is because online evolution has provided ever more active interactions and communications between the consumer and the business, as well as reducing the complexity of transactions and increasing the possibility of customization and customization of products and services (Alves et al., 2019). Which is why the number of companies in the online context has increased, which has raised questions about the quality of the loyalty relationship between the online customer and the company (Walsh, Hennig-Thurau, Sassenber & Bornemann, 2009).

SMM (Social Media Marketing) aims to produce content that users share in their various Social Media applications in order to increase brand exposure and broaden customer reach. There are numerous marketing techniques to apply in Social Media in order to involve the customer, some of which have costs and others do not. When using Social Media, companies can influence their client, make them feel part of it, which leads to a more personal relationship with them (Clapperton, 2009). According to Safko (2010), there are three main rules for corporate social media marketing:

1. Creating brand conversations.
2. Conversations cannot be controlled, but influenced.
3. Influence is the basis on which all economically sustainable relationships are built.

Kaplan and Haenlein (2010), called social media. First, companies must select their social network (s) to meet their purpose and their respective customers. If the company uses multiple social networks and / or combination of traditional marketing it is important that all its servers are closed and integrated, in order to avoid ambiguity. Finally, a "framework" of how to use Social Media channels is needed. Self-returns are five social media tips: Social media:

1. **Be Active**: As well as customers.
2. **Be Interesting**: Thus, giving the motivation for clients to interact as and between companies.
3. **Be Humble**: Learn to use Social Media.
4. **Do Not be Professional**: Try to mix with other users and be moist.
5. **BE Honest**: Respect the rules of using social media.

Mangold and Faulds (2009) give tips on how companies should use Social Media for their benefit:

1. Engage customers by letting them provide feedback, since in a community where feedback is accepted, they are more likely to feel encouraged to communicate and interact with other customers or with the brand.
2. Share information because when customers feel well informed about a company, product or service they are more likely to talk about it.
3. Being scandalous, since users are more likely to communicate something that is not common.
4. Give exclusivity, since customers like to feel special, what can be done by offering products, services or special promotions only to a part of the customers.

There are several types of Social Media, all of which are covered in a general way. Blogs are a type of Social Media for the user to create a personal web site, that is, to publish daily stories, opinions, ideas or own experiences, similar to a diary, thus encouraging a certain intimate sharing with their followers (Safko, 2010). Microblogs are used for a small share of content, which can take various forms such as images, phrases, videos or links (Jansen et al., 2009). Characters are received instantly, and users need to subscribe to others to receive their publications. Content Communities are applications that allow users to post their own content for other users to see. content helps to have a perspective of users' opinions on certain topics (Jin et al., 2010) .The consumer review websites are websites that allow the evaluation of certain products or services by users, they offer a wide range of tools how to send comments and photos, price comparisons, recommendations i.e., price alert, personalized purchase or even the price history of the product (Dellarocas et al., 2007). Wikis can be defined as a freely expanding collection of interlinked web pages, a hypertext system for storing and modifying information - a database, where each page is easily edited by any user (Leuf & Cunningham, 2001). Wikis are fast because there's a chance your users can read and edit at the same time. Each page is developed by the contribution of several users, filling gaps left by their predecessors. Internet Forums are websites where your users discuss topics of your interest, making this social media a great tool to have a deeper knowledge about this topic, rich in different perspectives (Carbonaro, 2011). Finally, Social Networks, these can be defined as applications that allow users to connect, by creating profiles with personal information, inviting friends and colleagues, so that they can visit your profile and send instant messages with each other (Kaplan & Haenlein, 2010). According to Boyd and Ellison (2007), its main functionalities are to support, maintain and reinforce already existing off-line relationships, while for young people it is used to socialize with friends. In this study four social networks that are highly used worldwide, Facebook, Instagram, Youtube and Twitter will be addressed.

2.1 Relationship Marketing

One of the most important experts in relationship marketing, Grönroos a representative of the Nordic school, defines relationship marketing as, establishing, maintaining, enhancing relationships with customers and other partners, at a profit, so that the objectives of the parties involved are met (Gummesson, 1994; Silva et al., 2009). This is achieved by mutual exchange and a fulfillment of promises. Therefore, relationship marketing in tourism can be defined as creating, maintaining and improving the mutually beneficial relations or possible termination of unfavorable relationships of healthcare management and medical tourism with stakeholders. This is undertaken to satisfy the stakeholder's need and to seek consensus between guest and healthcare management and tourism interests in development of tourism experiences (Sousa & Alves, 2019). Brun et al. (2016) concluded that the factors that precede and promote online relationship quality involve communication, conflict, and the perceived benefits of the relationship. And with regard to the characteristics of the website, Brun et al. (2016) have identified convenience, website design, information quality, ease of use, and security / privacy as the most relevant factors for online relationship quality (Alves et al., 2019). The factors presented above are explained below.

- **Communication:** It supports the interaction between the consumer and the company and is a decisive factor in the performance of the businesses (Vieira et al., 2008).
- **Conflict**: It is characterized by the frequency, intensity and duration of disagreement or disagreement between actors (Anderson & Narus, 1990).

- **Perceived Benefits of Relationship:** They represent the added value generated by a relation (Chou & Chen, 2018; Mukherjee & Nath, 2007). These results can be attributed to the performance of the product (Morgan & Hunt, 1994) or by the relationship itself (Hennig-Thurau et al., 2000).

Website Features

- **Convenience:** It allows consumers to reach businesses and their products and services when and how they want (Chung & Shin, 2010).
- **Website Design:** It provides the company's first impression to the consumer and should therefore make their online experience more enjoyable (Chung & Shin, 2010).
- **Quality of Information:** It concerns the authenticity and relevance of the information available on the website (Mukherjee & Nath, 2007).
- **Ease of Use:** It allows consumers to navigate the website simply and clearly, providing a pleasant online experience (Chouk & Perrien, 2004).
- **Security / Privacy:** It concerns the protection of data provided by consumers and the clarification of how the company collects and handles consumer information (Kalyanam & McIntyre, 2002). It is an essential factor in online trust building (Mukherjee & Nath, 2007).

2.2 Social Networks

Social networks have already been perceived by companies as a great mechanism to help them reach their marketing goals and goals, especially in aspects such as customer engagement, customer relationship management and communication (Saxena & Khanna, 2013). In different contexts, companies look for ways to introduce social networks into the most diverse aspects of communicating with their customers, whether through interactivity, promotion, facilitating access to information, and even improving customer buying behavior (Zeng & Gerritsen, 2014). In this study we will focus on four extremely successful social networks, such as Facebook, Instagram, Youtube and Twitter. The choice of these stands for its huge popularity worldwide as well as its ability to expand a business.

Facebook is the most widely used social networking site in the world, allowing businesses to reach many more people and more occurrences than companies using other means such as calls, emails or through meetings / meetings (Luke, 2009). Thanks to its large number of active users, as well as the high level of activity on the platform, the Facebook is considered an attractive social network for digital marketing specialists and for online ads (Francisco, 2006). These ads use social networks to connect customers to businesses, thereby creating new opportunities for customers to get to know their brands and products. To attract the attention of customers, digital marketers have commissioned social media platforms to customize their ads in order to have a greater impact on customers, resulting in a new on-line ad phase after IBM (International Business Machines) to Facebook on May 6, 2015, in an effort to create better ads on the world's largest social network, incorporating Facebook's targeting technology into IBM's services for marketers (Finley, 2015). These "personalized experiences" are associated with a number of benefits, such as reducing consumer resistance to advertising (Baek & Morimoto, 2012), increasing ad credibility (Xu, 2006) and improving brand awareness (Johns & Perrot, 2008). Making communication costs fall dramatically with Facebook and other social networks, creating opportunities for companies to communicate directly, rapidly and systematically with millions of current and potential customers (Mize, 2009). Customers can also generate WOM (Word of Mouth) interacting with each other or even on the company's Facebook page, through comments, likes, or sharing of brand-related content (Berger & Schwartz, 2011). These data are a great tool for companies, can be compared to a "crystal ball" to understand customers and communities of F7acebook contains the most relevant data for marketers (Casteleyn et al., 2009). Through these communities marketers are able to identify consumer tastes, which is essential to help create market segments, target markets and positioning strategies (Treadaway & Smith, 2012). They can get this information through the profiles of the community users, in news of the publications made in their profile pages (Casteleyn et al., 2009). However, you cannot always expect correct, accurate, and real information from users, who can create their own profiles and write what they want on Facebook (Treadaway & Smith, 2012).

Instagram is an application that allows users to take photos, apply a set of image manipulation tools and then instantly share them with friends on different social networks (Hochman & Schwartz, 2012). Compared to Facebook, Instagram is much easier to use and in terms of follower expectations users are satisfied with the sharing of some photos daily with a minimal description; already Facebook followers require more intensive participation (Miles, 2013). This network reaches a younger generation and appeals to a diversity of societies like no other social network (Abbot et al., 2013), this happens because younger generations have a greater tendency to take photos through the mobile phone and share it, on social networks. It stands out from most social networks through the application of its visual strategy (Hird, 2013), thus becoming an effective means of promoting a product, following the saying "a photo says more than a thousand words" (Silva et al., 2013). Helping also reduce brand designing costs, all images taken from a product can be edited using the Instagram functions (Herman, 2014). According to Barnes (2014), Instagram created the functionality known as "Hashtag" (#), for the convenience of indicating the relevant photos and videos of a company's products. By clicking on the hashtag, it allows users to view all relevant photos and videos (Dennis, 2014). At Instagram, leaving and receiving comments on a photo post is a good way to get feedback from customers on products. Users are sometimes interested in seeing interactions between the company and its followers; being attentive in the comments should improve the relationship with your customers (Dennis, 2014).

Within the various social networks, YouTube is among the most popular, being a database rich in information and knowledge related to markets and consumption (Dehghani, 2016). Youtubers are free to carry any kind of content created by them, including short comedies, memorable moments in life, or personal opinions on any subject imaginable. Many of them have achieved fame by increasing subscribers to their channels. Companies saw an opportunity for audience and potential market exploration with the youtubers (Wu, 2016). According to Dehghani (2016), Youtube sponsorships can take different forms, such as explicit sponsorship, where the company pays youtuber a specific amount or value for views in a video specially created to expose the brand or product; the affiliate links where purchases made by the link or coupon code provided by youtuber pay a commission to it; the supply of samples, where companies send products to youtubers, hoping that they create a general exposure to the product; and YouTube ads, where an ad is shown before (pre-roll), during (midroll) or after (post-roll) of the video. Several people visit YouTube as a form of entertainment, while others are looking for tutorials on how to do something, and a user often searches for a review of a product before buying it. This can be a strong point, because in a traditional advertisement the spokesperson presents a certain distance from the audience, whereas a video on Youtube is more like the genuine opinion of a regular person (Wu, 2016). The comment section is also very important to the consumer, this is where users interact with each other, reporting their experience of buying the product or expressing questions about the product.

Twitter is a platform that allows users to post small claims or even "microblogging", so they can transmit or share information about their day-to-day activities (Bae & Lee, 2011). Being an open, real-time, simple, free and flexible social network (Zhang, Jansen & Chowdhury, 2011), Twitter allows users to voluntarily follow another user, be it an individual, a group of people or even a brand, in order to receive future tweets on your page. The interactions are important in this platform, since a user can retweet another user's messages in order to transmit this message to his / her own followers, making this platform an excellent information dissemination tool (Bakshy et al., 2011). You can also mention or respond to another user by using the "@" symbol followed by the username and is great for companies to interact with their followers or other companies. Twitter has already been described as a tool for creating "electronic Word of Mouth" (Jansen et al., 2009), as a viral marketing mechanism (Asur & Huberman, 2010) and as an online form of "Word of Mouth branding "(Jansen et al., 2009). Although, due to its main attributes, user attention and audience size, companies value Twitter particularly as a communication tool for marketing, however, in terms of market research, the tweets database that contains comments, observations and opinions of the followers and the search engine that looks for patterns in these tweets, are equally important (Laudon & Traver, 2016). According to Thoring (2011), the main uses of Twitter in marketing go through market research and creation of feedback; advertising, branding and reputation management; be a business network; and consumer service and consumer relationship management.

3. METHODOLOGY

The choice of a research approach, whether by qualitative, quantitative or mixed methods, is not only a question of ontological and epistemological orientation, it is also a question of specific research objectives as well as research questions. This research does not seek to test a specific theory, but to find causal explanations, thus not following a deductive path, which usually requires the adoption of a quantitative approach. This study seeks to explore a phenomenon more broadly, so as to be able to provide a broader understanding by describing participants' experiences (Creswell & Creswell, 2017). In this sense, the nature of social networks demands the adoption of a constructivist perspective, that is, a perspective that involves affirmations of knowledge based on individual and collective experiences. The focus of the research thus needs to shift the researcher's point of view to the participants' point of view, from the numbers to the words, from a form of structured data collection to a less structured one, from specific data to more generalized data (Bryman et al., 2008). Qualitative research presents four main methods of data collection, according to Marshall and Rossman (2014): (a) individual interviews; (b) focus groups; (c) observation; and (d) review of documents. For this investigation the selected method was that of the interviews, more specifically individual semi-structured interviews. Semi-structured interviews consist of several key issues that help define the areas to be explored, but also allow the researcher and participant to derail to come up with an idea or answer in more detail (Silverman, 2013). The flexibility of this approach allows the discovery or elaboration of information that is important to the participants but may not have been considered relevant by the researcher. Regarding the number of participants, it is difficult to determine the ideal sample size for a qualitative study based on interviews. The opinions of several authors diverge, since the interviews are a difficult instrument to quantify and to equate, unlike, for example, the questionnaires. Guest et al. (2006) state that six to twelve interviews may be sufficient for the validity of a study. However, other authors, such as Bertaux (1981), affirm that fifteen is the minimum number of interviews to prove the validity of the study. Therefore, it can be verified that there is not yet a general consensus about the size of the qualitative sample, although it can be argued that this is usually lower than the quantitative sample (Patton, 2002). This leads to another aspect to be taken into account in relation to the qualitative sample, which is the saturation of the sample. According to van Rijnsoever (2015), the saturation of the sample occurs when new elements of the sample, in this case interviews, do not add new information to the study, that is, the ideal sample size exceeds, but this value varies according to the study objective. In this investigation will participate nine hotels with different characteristics, in order to take a more general approach and not focus too much on one type of hotel.

4. DISCUSSION AND RESULTS

While recognizing Portugal's unexplored tourist potential and considering the emergent need in enabling the universal accessibility in tourist services, it is eminent to consider the fundamental importance of a feature that distinguishes us as a nation - the know how in welcoming others. The concept of hospitality assumes the ability and the competences related to customer service, relationship management and the interaction between the hosting community and the tourists who try to enjoy the experiences offered by the tourist destination (Mendonça, 2017). For instance, and according Tur4all (2019), since 2006, the brand, Accessible Portugal, has grown to become the institution for the promotion of Accessible Tourism for All in Portugal. It its current legal format, considered to be the most appropriate for its aims, this private, non-profit Association is an entity that qualifies tourism supply and demand for all segments of the market according to their distinctive characteristics. Accessible Portugal promotes the development of areas that are gearing up to welcome a diversity of clients / tourists from a growing market, improving the quality of the lives of residents who also have a diversity of needs that must be satisfied. The promotion of Accessible Tourism from this perspective has been major success. The aging world population, particularly that of major tourist source markets, together with their growing and consolidated desire to travel, naturally require tourist destinations to gear themselves up in an endeavour to respond to the respective challenges (Alves et al., 2019). Fundación Vodafone Portugal is a self-funded, non-profit entity whose aim is to contribute to the development of the Information Society, fight the digital divide and disseminate mobile communication technologies. All projects are important because they provide technical innovation for the development of Society, but some stand out for their scope and significance and for the way in which they help to build a more sustainable and fair future. Fundación Vodafone Portugal signed a Protocol agreement with Accessible Portugal to finance the customisation of the TUR4all project in Portugal, through which both a Web Platform and Mobile App (available for iOS and Android) were developed to provide information about accessible tourism resources in Portugal (Tur4all, 2019).

Figure 1. TUR4all Portugal

In specific, and about our discussion, the present chapter aimed to understand the evolution of the use of social networks, as well as difficulties experienced in the introduction of this tool within the company and the respective results of its application. Since it is essential to use social networks today, it is of great importance to understand their adaptation to the company, as well as its use and proposed purposes. implementation of social networks in the corporate world is always something very demanding and meticulous. It is not a process that can be "rushed" because it is something that must be executed with due care, and can have an immediate influence on the way people perceive the brand / company. There are numerous social networks, although they share general functionalities, they all have their own specifics (Boyd & Ellison, 2007), leading in an initial phase of implementation to the question, which social networks to use? Kaplan and Haenlein (2010) suggest that the best way to select a social network is to identify those that best meet the purpose of the company and where most of its customers are present. According to the participants, there is a certain consensus about the choice of social networks that offer greater visibility or those that fit better in the reality of the hotel.

However, the decisions do not end after the selection of the social networks in which it is intended to operate, there are difficulties in the use of these in its premature phase. They can be of various types, from selecting the type of content to publishing, raising an audience, highlighting the competition, selecting the content to be promoted, and others (Kaplan & Haenlein, 2010). According to the participants, the greatest difficulty initially felt was to have a stable follower base, that is, to build an audience so that they could publish their content and interact with customers. This type of difficulty is frequent in an early phase of implementation of social networks, being overcome with the passage of time, where the company page is gaining some maturity and the audience is eventually built.

There are several tools that companies have digital marketing to operate in the market (Zeng & Gerritsen, 2014). Most of these tools are included in a budget, in which the most relevant are entitled to a larger share of the budget. Since social networks are only a subset within Social Media, their importance in this type of budget is not yet of great relevance for companies, seen merely as branding tools, although they are the type of Social Media that more propagates to the level of popularity (Belanche et al., 2010). It was possible to determine from the interviews that most of the hotels under study have their own marketing department (although they are small in size), although they also have a marketing team at headquarters that administers and supervises the marketing standards of the group hotels.

Thus, a small customization of the marketing of these hotels through their local departments. As for the budget available for social networks, it has been realized that it does not yet represent a significant value of the marketing budget. According to Sharma et al. (2018), there are two possible types of publications in social networks, being organic and non-organic publications.

Table 1. Illustrative citations participants 1, 5, 6, 7, 9

Participant 1	"The reason for the choice is that we are a transversal Hotel to all ages. "
Participant 5	"We chose these because they are the ones that have the greatest users and the best known. "
Participant 6	"They have the most visibility."
Participant 7	"Before we opened the accounts, we analyzed what impact it would have to the hotel, what a return it could have on sales and promotion of hotel image."
Participant 9	"We have chosen the most used ones in a global way."

Organic publications are a type of publication that does not impose on the company any kind of cost for communication, while non-organic publications, in return for the guarantee of greater exposure, entail costs for the realization of the communication. Most hotels do not invest money in non-organic publications (paid communication) and those who invest only concentrate a small percentage of the marketing value in those publications, most of which are organic publications (free communication). This causes a decrease in the possible effects of social networks, since non-organic publications have a far greater reach than organic publications (Yang et al., 2008).

There are several ways companies can work on social networks to make their brand known to the market. Tuten and Solomon (2017) point to the display of advertisements, the publication of content related to the brand, the development of online activities for customers to participate and the creation of a brand persona as the best ways to expose the brand in networks social rights. Already Zeng and Gerritsen (2014), affirms that companies seek to introduce social networks in the most diverse aspects of communication with their clients, either through interactivity, promotion, facilitating access to information and even improving consumer buying behavior. client. According to participants' responses, the type of content most used by hotels is the publication of content related to them, ie photographs of the establishment, information of events, photographs of the gastronomy that has, news related to the hotel, between others.

Another aspect to take into consideration in publishing content is the periocity. One of the tips provided by Kaplan and Haenlein (2010) to be social in Social Media applications is to be active. A daily activity is always something that reminds the users of the brand, thus creating presence. However, being too active can also lead to problems, such as overexposure to the brand, which can have negative effects on the brand image. Regarding the information provided by our participants, we can see the importance attributed to the periodicity of publications, since most hotels publish daily. However, there are some that aim to reach the daily publications, as well as others that present a fixed agenda of publications.

Zeng and Gerritsen (2014) describe the entry of companies into social networks as a way to increase their communication with customers, either through interactivity, promotion, facilitating access to information or even to improve purchasing behavior of customers. Although that interaction with customers should be cautious, as both can positively or negatively influence the brand image. As for published

Table 2. Illustrative citations participants 2, 4, 5

Participant 2	"I would say that 80% of our investment is online, and for social networks we should have about 10% of these 80%. "
Participant 4	"The value is quite variable taking into account the niche market, the of the year and the events developed in this area of business. "
Participant 5	"Investment in this area is the smallest marketing budget. "

Table 3. Illustrative citations participants 5, 6, 7

Participant 5	"Publication of our photos and news about us."
Participant 6	"Contents and activities we develop in our resorts."
Participant 7	"The content includes services that the hotel provides, such as, meeting facilities, spa and information such as the hotel's pet friendly, the fact of having a program that helps people to sleep better, among others. "

content, it plays a key role in interacting with customers. Publishing monotonous content will certainly undermine customer engagement with the brand, Mangold and Faulds (2009) have emphasized the importance of posting scandalous, "out-of-the-box" and even provocative content, as customers have a greater tendency to interact and share something that is not common. According to our participants there is great agreement that the type of content that reaches greater interaction with customers passes through photos of the establishment or the landscape in which it is inserted. According to Mangold and Faulds (2009), making the product known is also a determining factor for success in social networks, since when a consumer feels well informed of a product is more likely to recommend it to others.

However, there are several types of interaction that the client may have with these publications (Acar & Polonsky, 2007). All of them present their particularities, being some more valued than others. The main and most common types of interaction with publications are taste, comment and sharing. In this case, sharing is the most valued because the user will expose the publication to their entire network of contacts, which will create a greater reach of the publication, since people who do not follow the company page can receive the publication (Casteleyn et al., 2009). Commenting is a great tool for collecting feedback, many times customers comment on their personal opinion about the publication, which if analyzed in its entirety allows the extraction of some kind of possible improvement (Berger & Schwartz, 2011). Taste is the most common type of interaction and helps businesses get a sense whether their publication was well or poorly received (Treadaway & Smith, 2012). It was possible to verify, according to the interviews that the type of interaction most usual that the hotels receive, almost of a unanimous form, are the likes and the comments.

Customer feedback is of paramount importance, not only helps to correct errors made in the service, but also provides an opportunity for this error not to happen again in the future (Haydon et al., 2012). Often, it is necessary the perspective of someone who is enjoying the product or service to perceive some type of anomaly present. One of the most powerful tools that social networks provide to businesses is the ability for customers to give feedback. As previously noted, Mangold and Faulds (2009) emphasize the need to nurture a feedback culture in the company's pages, for which it is necessary to accept the same. Since mistaking is seen as something negative, we must also realize that it is always

Table 4. Illustrative citations participants 2, 3, 4, 5, 7

Participant 2	"We communicate daily on social networks, ..."
Participant 3	"Daily on Facebook, instagram and Twitter. Every day a new publication."
Participant 4	"The objective is to maintain a daily frequency, in exceptional cases stay active weekly."
Participant 5	"We always try to make three weekly posts with varied content. Exceptionally we do one or another extra publication, if we have some unscheduled event."
Participant 7	"We have five weekly posts, Monday through Friday."

Table 5. Illustrative citations participants 1, 2, 4

Participant 1	"Normally, likes or comments, ..."
Participant 2	"We received it anyway, but we ended up having more likes and comments."
Participant 4	"Comments, especially with emojis or tags identifying friends and tastes in publications."

an opportunity to improve. Regarding the interviews, it was possible to notice that all hotels accept and reflect on the feedback received. However, this can be positive or negative. In this regard, it has been possible to determine that, even though it is negative, hotels generally try to understand what has gone wrong and correct those lapses so that they do not repeat themselves in the future. Being the business branch mainly focused on customer satisfaction and profit, there are clearly strategies that companies adopt to achieve certain objectives. Kaplan and Haenlein (2010) give recommendations on how best to act in Social Media. The first step is to select the social network that best fits with the business profile, followed by the alignment of the information, that is, if it is present in several social networks or make a combination of digital marketing with the traditional one, be sure that all the information is aligned in the different channels and finally create a framework of how to act in social networks, with several steps to follow.

According to the Interviewees, most hotels present an overall strategy of promoting the hotel, that is, the various spaces of the establishment, its conditions, its cuisine, its events, its location and its services. This strategy, in general, refers to the goal of creating brand awareness, that is, exposing the brand and creating value for it. Since some of these hotels do not have a clearly defined fixed strategy, they seek some flexibility according to consumer reactions.

The influence of social networks worldwide is undeniable. The "secret" of creating impact and making the consumer act is something that does not exist and the development of this "special formula" that results in all cases will not arise. Human nature is something complex, cannot be simplified to a certain solution, is constantly adapting, making accommodation impossible for certain strategies, which at the moment may be very successful, but in the medium-long term they have become useless (Moon, 2004). It is important to emphasize the importance of flexibility, that it is crucial for companies to be constantly

Table 6. Illustrative citations participants 1, 2, 5, 6, 7, 9

Participant 1	"Being negative critics, we respond also and we try to perceive what happened in order to improve with the mistakes made."
Participant 2	"We always try to improve with the constructive feedback that we received."
Participant 5	"In the case of negative reviews, we always try to understand what has happened bad for the customer and improve."
Participant 6	"We respond to feedback and of course always try to improve."
Participant 7	"As for the criticisms we try to respond in a way that the relationship with the client is not compromised, trying to correct the mentioned errors, if any."
Participant 9	"We take into account all criticisms, whether positive or negative, negative."

Table 7. Illustrative citations participants 1, 2, 4, 7, 8

Participant 1	"Since Facebook and Instagram only serve to to be present in the digital world."
Participant 2	"The strategy of social networks as we ended up having a daily communication we provide lots of information, all news however small they may be, ..."
Participant 4	"This is a planning that is being adjusted according to the customer reactions. "
Participant 7	"The strategy is to promote the hotel and everything it has to offer to anyone who wants to visit the city. "
Participant 8	"Strategy for the dissemination of our products and events."

Table 8. Illustrative citations participants 1, 4, 6

Participant 1	"it is another means of exposure that the Hotel has and that way customers."
Participant 4	"Social networks offer fast visibility and notoriety. It resulted in customer acquisition and Word of Mouth."
Participant 6	"many customers and visitors came because they viewed the our contents."

learning, to perceive certain patterns of behavior and to develop a form of communication that meets these standards, taking into account that in the online world it is impossible to please everyone. According to the participants, in general, the realization of their communications by social networks has shown positive results. While attributing the importance of this success to social networks varies a bit, while some consider that this tool had a great influence on the success of influence, for others it is just another addition to the hotel communication. But in a unanimous way all respondents believe that social networking is a fast, direct and effective means of communication. The social networks, in a general way, contributed greatly in increasing the visibility and notoriety of the hotel. The creation of an online audience and the promotional publications have aroused the curiosity of followers, often leading them to experience the services of the hotel.

In the digital world, it is possible to track any type of activity developed, both in websites and in social networks. These screenings are of high importance because it is they who provide data to companies to see if a certain action taken online was successful or unsuccessful. Social networks are already incorporated with statistical tools capable of analysing the evolution of the page, and it is possible to trace the impact of certain publications, which brings a great benefit to the company. However, there is still some resistance in companies to adapt to these tools. From what was possible through the interviews, some hotels still have not incorporated into their marketing strategy the use of these analysis tools. However, most hotels that participated in the study are already starting to employ them in the management of their online campaigns or plan to do so in the future.

CONCLUSION

The present chapter highlights the importance of social networks in the influence of consumers, in a viewpoint of the hotel sector of the North Region. Thus, it aimed to explore the phenomenon of social networks, in order to create a contribution to the scientific area in question. In this sense, it was possible to investigate the use of social networks by the hotel sector. In general, results on consumer influence have proved to be promising, as social networks help to create greater exposure to the establishment globally, which is important in tourism, making it easy and quick to communicate the brand to consumers in order to create a greater desire to visit. However, it was possible to verify that there is still much progress to be made, in order to fully enjoy the potential of social networks. In conclusion, it is possible to verify that social networks have an influence on consumers. As the number of users of social networks increases drastically (Belanche et al., 2010), it is possible to assume that the level of exposure they present will tend to increase over time. However, it was also possible to see that the mechanisms offered by social networks are not yet being used in their entirety, as was the case of non-organic publications, as well as statistics offered for analysis of evolution. Thus, space for this sector will improve its level of social influence and efficiency in the use of social networks.

Regarding the limitations, at an early stage, during the review of the literature it was possible to perceive that the theme of social networks was still emerging, this being a relatively recent topic of study since social networks have not been present for a long time in marketing, as well as other scientific areas, for example consumer behaviour, also addressed in this study. Although the level of articles was not completely scarce, it was noticed that there was a great concentration around Facebook, eventually overshadowing studies related to other social networks.

Since Social Networks is an area of study still underdeveloped, it is possible to delineate some options regarding the next scientific studies, since this is still an area that has several strands to explore. The application of this study in a multisectoral perspective, where one could verify the degree of integration of social networks in companies, different techniques used in different sectors, as well as their results and effectiveness. Also, the realization of this study in the perspective of the client, that is, to use a quantitative study, through questionnaires (mainly), trying to perceive what customers think of the use of social networks by the companies, in order to analyse the types strategies that result and those that are less successful. It is also possible to carry out a focus group, in order to analyse specific strategies selected by the researcher, in order to obtain more concrete data. The literature widely suggests that the digital media have greatly impacted the way a marketer reaches today's consumer. The adaptive process enabled by the digital technologies creates value in new ways in new digital environments. Institutions enabled by digital technologies build foundational capabilities to create such value jointly for their customers and for themselves. Therefore, social tourism has been defined as the effects and phenomena that result from the participation in tourism and more specifically from the participation of low-income groups. This participation is made possible or is facilitated by initiatives of a well-defined social nature. Also, tourism for all refers to the goal to include as many people as possible into tourism and leisure. Whereas the communication function concerns the interaction between an online company and its customers. This chapter discussed also the increasing phenomena of online relationship and digital marketing in specific social tourism contexts. Based on the literature review, we developed research propositions that connect the online relationship quality dimension developed by the relational marketing (i.e. online contexts) with the consequent e-satisfaction and e-word-of-mouth (e-wom). The authors also considered some implications for management, as well as give suggestions for future lines of research (e.g. empirical study).

FUTURE RESEARCH DIRECTIONS

This study is a preliminary and exploratory research based on a qualitative approach. The conclusions are based on the researchers' perception derived from the Portuguese case study. With the insights from this study it is expected that future work can contribute to the development of empirical studies to address the developed propositions. In specific it is expected that future research explores the relationship between online relationship quality, e-trust, e-commitment, e-satisfaction and e-wom. For example, studies may look at a deeper understanding of the phenomenon in several social tourism contexts. It is important to reveal the individual decision that cause satisfaction and/or dissatisfaction, and shed light on the most determinant and critical attributes in explaining the overall social tourist experience, in the specific the case of digital marketing contexts. Studies could further develop a questionnaire to be applied to internet consumers in Portugal to capture a cross-sectional view of the relationships among the constructs gauging their impact.

REFERENCES

Abbott, W., Donaghey, J., Hare, J., & Hopkins, P. (2013). An Instagram is Worth a Thousand Words: An Industry Panel and Audience Q&A. *Library Hi Tech News*, *30*(7), 1–6. doi:10.1108/LHTN-08-2013-0047

Alves, G. M., Sousa, B. M., & Machado, A. (2019). The Role of Digital Marketing and Online Relationship Quality in Social Tourism: A Tourism for All Case Study. In J. Santos & Ó. Silva (Eds.), *Digital Marketing Strategies for Tourism, Hospitality, and Airline Industries* (pp. 49–70). Hershey, PA: IGI Global. doi:10.4018/978-1-5225-9783-4.ch003

Anderson, J., & Narus, J. (1984). A Model of the Distributor's Perspective of Distributor-Manufacturer Working Relationships. *Journal of Marketing*, 62-74.

Anderson, J. C. N., & Narus, J. A. (1990). A model of distributor firm and manufacturer firm working partnerships. *Journal of Marketing*, *54*(1), 42–58. doi:10.1177/002224299005400103

Backman, S., & Crompton, J. (1991). The usefulness of selected variables for predicting activity loyalty. *Leisure Sciences*, *13*(3), 205–220. doi:10.1080/01490409109513138

Baek, T. H., & Morimoto, M. (2012). Stay away from me. *Journal of Advertising*, *41*(1), 59–76. doi:10.2753/JOA0091-3367410105

Becken, S., & Simmons, D. (2002). Understanding energy consumption patterns of tourist attractions and activities in New Zealand. *Tourism Management*, *23*(4), 343–354. doi:10.1016/S0261-5177(01)00091-7

Berger, J., & Schwartz, E. M. (2011). What drives immediate and ongoing word of mouth? *JMR, Journal of Marketing Research*, *48*(5), 869–880. doi:10.1509/jmkr.48.5.869

Boyd, D., & Ellison, N. (2007). Social network sites: Definition, history, and scholarship. *Journal of Computer-Mediated Communication*, *13*(1), 210–230. doi:10.1111/j.1083-6101.2007.00393.x

Brun, I., Rajaobelina, L., & Ricard, L. (2014). Online relationship quality: Scale development and initial testing. *International Journal of Bank Marketing*, *32*(1), 5–27.

Brun, I., Rajaobelina, L., & Ricard, L. (2016). Online Relationship Quality: Testing an Integrative and Comprehensive Model in the Banking Industry. *Journal of Relationship Marketing*, *15*(4), 219–246. doi:10.1080/15332667.2016.1242400

Carbonaro, A. (2011). *Forum Summarization to Support Tutor and Teacher in Group Interaction Management. Semantic Web Personalization and Context Awareness: Management of Personal Identities and Social Networking* (pp. 22–31). Hershey, PA: IGI Global. doi:10.4018/978-1-61520-921-7.ch003

Casteleyn, J., Mottart, A., & Rutten, K. (2009). How to use Facebook in your market research. *International Journal of Market Research*, *51*(4), 439–447. doi:10.2501/S1470785309200669

Chou, S., & Chen, C.-W. (2018). The Influences of Relational Benefits on Repurchase Intention in Service Contexts: The Roles of Gratitude, Trust and Commitment. *Journal of Business and Industrial Marketing*, *33*(5), 680–692. doi:10.1108/JBIM-08-2017-0187

Clapperton, G. (2009). *This is social media: how to tweet, post, link and blog your way to business success*. Chichester: A Wiley Company.

Dellarocas, C., Zhang, X. M., & Awad, N. F. (2007). Exploring the value of online product reviews in forecasting sales: The case of motion pictures. *Journal of Interactive Marketing*, *21*(4), 23–45. doi:10.1002/dir.20087

Echtner, C., & Jamal, T. (1997). The disciplinary dilemma of tourism studies. *Annals of Tourism Research*, *24*(4), 868–883. doi:10.1016/S0160-7383(97)00060-1

Ferreira, J., Sousa, B. M., & Gonçalves, F. (2018). Encouraging the subsistence artisan entrepreneurship in handicraft and creative contexts. *Journal of Enterprising Communities: People and Places in the Global Economy*. doi:10.1108/JEC-09-2018-0068

Francisco, B., & Update, M. L. (2006). Cracking the social network code.

Getz, D. (1986). Models in Tourism Planning: Towards Integration of Theory and Practice. *Tourism Management*, *7*(1), 21–32. doi:10.1016/0261-5177(86)90054-3

Gummesson, E. (1994). Making relationship marketing operational. *International Journal of Service Industry Management*, *5*(5), 5–20. doi:10.1108/09564239410074349

Gummesson, E. (2017). From relationship marketing to total relationship marketing and beyond. *Journal of Services Marketing*, *31*(1), 16–19. doi:10.1108/JSM-11-2016-0398

Jafari, J., & Aaser, D. (1988). Tourism as the subject of doctoral dissertations. *Annals of Tourism Research*, *15*(3), 407–429. doi:10.1016/0160-7383(88)90030-8

Jansen, B. J., Zhang, M. M., Sobel, K., & Chowdury, A. (2009). Twitter power: Tweets as electronic word of mouth. *Journal of the American Society for Information Science and Technology*, *60*(11), 2169–2188. doi:10.1002/asi.21149

Jin, X., Gallagher, A., Cao, L., Luo, J., & Han, J. (2010). The wisdom of social multimedia: using Flickr for prediction and forecast. *Proceedings of the International Conference on Multimedia* (pp. 1235-1244). Academic Press. 10.1145/1873951.1874196

Johns, R., & Perrott, B. (2008). The impact of internet banking on business-customer relationships (are you being self-served?). *International Journal of Bank Marketing*, *26*(7), 465–482. doi:10.1108/02652320810913846

Kannan, P. K., & Li, H. A. (2017). Digital marketing: A framework, review and research agenda. *International Journal of Research in Marketing*, *34*(1), 22–45. doi:10.1016/j.ijresmar.2016.11.006

Kaplan, A. M., & Haenlein, M. (2010). Users of the world, unite! The challenges and opportunities of Social Media. *Business Horizons*, *53*(1), 59–68. doi:10.1016/j.bushor.2009.09.003

Leuf, B., & Cunningham, W. (2001). *The Wiki way: quick collaboration on the Web*. Boston: Addison-Wesley.

Luke, K. (2009). Marketing the new-fashioned way: Connect with your target market through social networking sites. *Journal of Financial Planning*, (November/December): 18–19.

Mangold, W. G., & Faulds, D. J. (2009). Social media: The new hybrid element of the promotion mix. *Business Horizons*, *52*(4), 357–365. doi:10.1016/j.bushor.2009.03.002

Mukherjee, A., & Nath, P. (2007). Role of electronic trust in online retailing: A re-examination of the commitment-trust theory. *European Journal of Marketing*, *41*(9/10), 1173–1202. doi:10.1108/03090560710773390

Narus, J. A., & Narayandas, D. (2004). *Business market management: Understanding, creating and delivering value*. Pearson Prentice Hall.

Ramos, C. M., Matos, N., Sousa, C. M., Correia, M. B., & Cascada, P. (2017, July). Marketing Intelligence and Automation–An Approach Associated with Tourism in Order to Obtain Economic Benefits for a Region. *Proceedings of the International Conference on Universal Access in Human-Computer Interaction* (pp. 402-411). Cham: Springer. 10.1007/978-3-319-58706-6_32

Ramos, C. M. Q., & Rodrigues, P. M. M. (2011). Os efeitos da internet na actividade turística. *Revista GEINTEC-Gestão. Inovação e Tecnologias*, *1*(2), 64–83. doi:10.7198/S2237-0722201100020006

Safko, L. (2010). *The social media bible: tactics, tools, and strategies for business success*. John Wiley & Sons.

Saxena, A., & Khanna, U. (2013). Advertising on social network sites: A structural equation modelling approach. *Vision*, *17*(1), 17–25. doi:10.1177/0972262912469560

Sharma, D., Gupta, A., Mateen, A., & Pratap, S. (2018). 'Making sense of the changing face of Google's search engine results page: An advertiser's perspective', Journal of Information. *Communication & Ethics in Society*, *16*(1), 90–107. doi:10.1108/JICES-06-2017-0035

Smith, K. T. (2011). Digital marketing strategies that Millennials find appealing, motivating, or just annoying. *Journal of Strategic Marketing*, *19*(6), 489–499. doi:10.1080/0965254X.2011.581383

Sousa, B., & Alves, G. (2019). The role of relationship marketing in behavioural intentions of medical tourism services and guest experiences. *Journal of Hospitality and Tourism Insights*, *2*(3), 224–240. doi:10.1108/JHTI-05-2018-0032

Tsai, W. H. S., & Men, L. R. (2017). Consumer engagement with brands on social network sites: A cross-cultural comparison of China and the USA. *Journal of Marketing Communications*, *23*(1), 2–21. doi:10.1080/13527266.2014.942678

Vieira, A. L., Winklhofer, H., & Ennew, C. T. (2008). Relationship Quality: A literature review and research agenda. *Journal of Customer Behaviour*, *7*(4), 269–291. doi:10.1362/147539208X386833

Walsh, G., Hennig-Thurau, T., Sassenberg, K., & Bornemann, D. (2010). Does relationship quality matter in e-services? A comparison of online and offline retailing. *Journal of Retailing and Consumer Services*, *17*(2), 130–142.

Xu, D. J. (2006). The influence of personalization in affecting consumer attitude toward mobile advertising in China. *Journal of Computer Information Systems*, *47*(2), 9–19.

Yang, T. A., Kim, D. J., & Dhalwani, V. (2008). Social networking as a new trend in e-marketing. In *Research and practical issues of enterprise information systems II* (pp. 847–856). Boston, MA: Springer. doi:10.1007/978-0-387-76312-5_7

Zeng, B., & Gerritsen, R. (2014). What do we know about social media in tourism? A review. *Tourism Management Perspectives, 10*, 27–36. doi:10.1016/j.tmp.2014.01.001

Zhang, M., Jansen, B. J., & Chowdhury, A. (2011). Business Engagement on Twitter: A Path Analysis. *Electronic Markets, 21*(3), 161–175. doi:10.100712525-011-0065-z

KEY TERMS AND DEFINITIONS

Consumer Behavior Online: The study of individuals, groups, or organizations and all the activities associated with the purchase, use and disposal of goods and services, including the consumer's emotional, mental and behavioral responses that precede or follow these activities in the online environment.

Digital Marketing: Is the marketing of products or services using digital technologies, mainly on the Internet, but also including mobile phones, display advertising, and any other digital medium.

Online Relationship: Is an integrative and multidimensional concept, such as relationship quality in an offline context.

e-Satisfaction: The contentment of a consumer with respect to his or her prior purchasing experiences with a given retail-oriented website.

e-WOM: Is any positive or negative statement made by potential, actual or a former customer which is available to a multitude of people via the internet.

Social Networks: Is a social structure composed of persons or organizations, connected by one or several types of relationships, which share common values and goals.

Social Media: Social media are interactive computer-mediated technologies that facilitate the creation and sharing of information, ideas, career interests and other forms of expression via virtual communities and networks.

Chapter 22
The Management of Social Media and the Relationship With the Client From the Perspective of the Managers of Small Hotels in a City in Brazil

Francieli Boaria
https://orcid.org/0000-0002-0678-7698
Universidade Federal do Rio Grande, Brazil

Ademilso Alves da Cunha
Universidade Federal do Rio Grande, Brazil

ABSTRACT

With the growth of the tourism sector and increased competition, the hotel industry is directly related to the evolution of the internet as an important communication tool capable of promoting market differentials. The objectives of this study are to evaluate the management of social media of the hotels in the city of Santa Vitória do Palmar, Brazil, as well as to analyze the interaction between the enterprises and their clients. This research had a sample of five hotels, which have demonstrated that they use social media constantly, but this practice is in the initial phase of implementation. It was noticed that the use of social media by the managers of the research sample is carried out with the same knowledge of common users, without deepening it. In order to optimize this practice and consequently for the hotel to be more profitable, the implementation of aspecialized and up-to-date management is necessary.

DOI: 10.4018/978-1-7998-1947-9.ch022

INTRODUCTION

Nowadays, Tourism shows an evident growth. According to the survey of the World Tourism Organization (WTO), world tourism registered the second best result of the last 10 years in 2018, reaching the mark of 1.4 billion international arrivals worldwide, an increase of 6% over 2017 (Tourism Department, 2019). This, in turn, has a direct impact on hotel businesses, considered as one of the fundamental services of the tourism product, as Petrocchi (2002) explains. Therefore, hotels need to assess their management strategies, especially those related to Marketing and Social Media, the focus of this study.

Such growth in tourism, and consequently in hotels, is directly related to the evolution of the Internet as an important communication tool that is capable of promoting market differentials, since, according to an EBIT study (2018), Online Tourism was the second sector with the highest nominal growth, about 17.8% in relation to 2016, a budget of around 10 billion dollars in 2017.

All segments of society are affected by the behavioral changes that the Internet provides in the forms of communication, which is shown as the potential to be explored by marketing in the relation with customers. Therefore, the hotel sector can also take strategic actions through the Internet that may boost their sales, starting from plans for consolidation of the brand before a market that is extremely competitive.

Social media is being widely explored by individuals and companies in general, such as hotel ventures. Today, Facebook, Twitter, Instagram and TripAdvisor have significant advantages in promoting business-customer interaction, especially by sharing messages and sharing photos and videos. Advantages include the organization of events, the creation of specific groups and communities, etc.

In the face of behavioral changes and market needs, the objectives of this study are to evaluate the management of social media by the hotels in Santa Vitória do Palmar, RS, Brazil, as well as to analyze how the interaction between the enterprises and their customers. An exploratory, descriptive and qualitative approach was carried out. The bibliographic research sought to explore the tools that the hotel market currently uses through searches in academic and market magazines that have in their central focus the tourism and hotel sector. This survey was conducted in December 2017 and featured a sample of five small and independent hotels management.

Social media bring a range of possibilities to be explored, such as the dissemination of content, and the interaction and communication with customers, especially the relevance that digital marketing has assumed for hotels. In addition, the intention to study this area is to identify the potential of such internet tools linked to the practice of digital marketing.

According to Boaria (2013: 138) the implementation of digital marketing "has an impact on the amount of online transactions that the hotel chain can make through the use of its tools, such as social networks, applications for mobile devices], CRM, among many other ways to approach the customer for the use of e-commerce." It is assumed that the optimized use of these tools can instrumentalize hotel managers towards better relations with their public, with competitive advantages characterized by relationships more aligned with the wishes of frequent clients, as well as potential ones.

It becomes important to sudy this issue not only because social media is efficient in considering today's consumer communication habits, but some of them can also be used for free. In this way, besides communication efficiency, one can consider the gains related to the cost savings as well as the consequent increase of the profits when satisfying the requested demands. The literature review is divided in two main parts. The first part explains about the digital marketing and and exemplifies through some studies, whereas the second part demonstrates the relationship of social media with customer relationship, deepening the theme of CRM. It is worth to mention that this study demonstrates how the interaction between companies and their customers occurs.

BACKGROUND

The purpose of digital marketing is to produce content on digital channels such as websites, blogs and social media like TripAdvisor, Facebook, Twitter, Linkedin, Slideshare, Instagram, which have the power to distribute messages in a decentralized way to disseminate content while allowing some relationship with other people (Boaria, 2013). It is important for enterprises in all market segments. Regarding the tourism context, Anjos, Boaria and Raye (2013: 1) mention that: [...] the importance of digital marketing or e-marketing for the growth and success of tourism activities is related to the fact that they have unique characteristics such as heterogeneity, intangibility and perishability, which can put information at the center of their activity.

There are a number of digital marketing tools available for businesses today to strengthen customer relationships, with a focus on costumer relationship management, as well as mobile and social media applications. It is important to note that some are available at no cost.

According to Anjos, Boaria and Raye (2013, p. 8) "Within digital marketing, we have the electronic word of mouth—e-WOM, which is classified as the most important source of information when a consumer is taking a purchase decision." With the possibility of using social media today, users have in these tools the opportunity to exchange information about products and services; these discussions and evaluations related to purchasing experiences of product and service are known as electronic word of mouth (eWOM).

Ortiz and Boaria (2016) found that consumers are increasingly using the internet to plan, choose and buy tourism services, and are often influenced by comments and opinions of their social circle. This research was carried out with tourists in an important world heritage site classified by Unesco (Unesco, 2019), the Iguassu Falls, and sought to identify how the Internet influences the choice of tourists for a certain place.

According to Torres (2010: 29) "every business benefits greatly from word of mouth from a current customer to a new customer. But on the Internet it gained scope and speed. This has turned the viral effect into a viable marketing tool." The hotel sector can enjoy the positive effect of online word of mouth to improve its image in the market where it is inserted; this dissemination of information occurs, for example, in social media such as Facebook, Instagram, Twitter and TripAdvisor.

For Gretzel and Yoo (2008) and Fotis, Buhalis and Rossides (2012), User-Generated Content is considered more relevant and more reliable than content derived from commercial information. In this way, it is essential for the managers of the lodging facilities to pay attention to the comments and/or reviews left by the clients regarding the image of their companies in the virtual environment.

On the other hand, negative evaluations can compromise the image of the hosting environment, and these evaluations, although in a smaller number, have a greater impact on the users and become determinants in the decision-making process (Rong, Vu, Law, & Li, 2012).

Some companies already understand the significant importance of user-generated content. Arruda (2016) has identified that Sofitel Copacabana Hotel uses TripAdvisor as a tool to analyze the level of quality of its services. The same media was also researched by Corrêa and Hansen (2014), where the authors sought to present the perception of the quality of services provided by restaurants in the city of São Paulo, winners of the Travelers' Choice Awards of 2013, promoted by TripAdvisor, and the result showed that the perception of the quality of services provided by the restaurants selected by the users was positive.

In addition to the studies related to TripAdvisor, the online travel agency 'Booking.com' was researched by Petry, Pickler, and Tomelin (2016) about the perception of business guests regarding the quality of services rendered in six hotels in the city of Florianópolis, Brazil. The results showed that the tangible aspects had higher priority in the users' evaluations; the authors point out that such data may assist hotels in formulating strategies to further improve these aspects.

Boaria and Santos (2018) analyzed the impact of online user-generated content by in the management of the lodging facilities in different categories of hotels in Brazil and concluded that social media and online travel agencies (OTA's) impact, on different levels, the reputation, reliability, and the increase in occupancy rate and hotel sales. From the managerial point of view, the results indicate possible actions to be taken prior to OTA membership and also how to manage users' comments.

The studies on user-generated content are of paramount importance so that consumers can obtain information and evaluations of services and products of their interest, as for the corporations that, through these indicatives, can design strategies of how to act upon the market, in addition to planning and making improvements in the services that are indicated as deficient by the users.

SOCIAL MEDIA AND CUSTOMER RELATIONSHIP MANAGEMENT (CRM)

For Gretzel, Yuan and Fensenmaier (2000), social media are the activities, practices and behaviors among communities of users who share information, exchange knowledge and opinions on the network. Kaplan and Haenlein (2010) define social media as a group of web-based applications that are based on ideological and technological foundations of Web 2.0 and that allow the creation and exchange of user-generated content. As for Cabral (2016: 02), the concept of social media is defined as: "online technologies and practices used by people or companies to disseminate content, provoking the sharing of opinions, ideas, experiences and perspectives."

Belo (2011) explains that social networks among their varied definitions can be horizontal or vertical, as well as blogging, virtual worlds, wikis and multimedia, as presented in Table 1.

The differentiation between horizontal and vertical social network definitions is associated with the content that is approached and shared among users. In horizontal social networks, users share and debate various subjects of their interests, and, in verticals, they discuss specific issues proposed by that social network.

Social media present a major role in our daily lives, which promote social interaction through the exchange of messages, sharing of photos, videos, organization of events, creation of groups and specific communities, sharing of reports of experiences, among other functions. They contribute to the interaction between client and organization and provide for the interest of potential clients and contact with frequent clients.

Table 1. Classification of Social Media

Classification	Social Media
Horizontal	Facebook, Bebo, LinkedIn, Hi5, MySpace, Orkut
Vertical	TripAdvisor, Virtual Tourist, Travelpod, LonelyPlan, Localyte
Blogging	Twitter, Blogger
Virtual Worlds	SecondLife
Wikis	Wikitravel, Wikipedia
Multimedia	Youtube, Flickr, Slideshare

Source: Adapted from Belo (2011, p. 12).

Companies can, for example, monitor and analyze the performance of actions and publications in their profile on their Facebook page. There is specialized software to manage data mining techniques, as explained by Thomaz, Biz and Bettoni (2015), who identified in their research that the Destination Management Organizations (DMO) of the city of Curitiba, Paraná (BR), do not manage the media via software, and only two DMOs out of the four surveyed use the analytical tool offered by this social media.

Besides the software tool for the managening and monitoring of consumer behavior, another tool that assists in the monitoring of customer relationships is Customer Relationship Management (CRM). It can be used by companies of all market segments, thus, its use either manually or through software provides the management of information about customers, employees and suppliers, which is of extreme importance also for the hotel sector.

Due to the characteristic of the product/service it offers, in the hotel sector it becomes relevant a closer relationship between company and customer, thus enabling the manager to better know customers and understand their wishes and needs. According to Bronze Júnior et al. (2012, p. 493) "With increasing competition and the improvement of technology, companies have begun worrying not only about winning new customers, but also about keeping those already conquered, which gives way to relationship marketing."

According to Silva and Silva (2008, p. 16) "The use of CRM strategy is of utmost importance for the hotel industry, in order to capture and use information as a means of boosting new business with customers."

Silva (2013, p. 24) states that: "In the hotel industry, CRM is crucial as it is embedded in an increasingly competitive market that forces hotels to differentiate themselves in the eyes of their customers. They need to have in-depth knowledge about their customers' needs, their behavior and their preferences, in order to improve the way their services will be delivered."

In that sense, Atlantica Hotels' senior vice-president of sales and marketing, Annie Morrissey, reports in an interview for the magazine Hosteltur Brasil (2017) that the network is working on CRM. He describes how the system allows 360-degree view of guests' consumption behavior, which enables network management to understand their tastes, preferences, schedules, and trends. Through the application of this intelligence, it is possible to add value to the lodging, offering services directed to the profile and the needs of the guest, and to generate a more pleasurable experience (Hosteltur, 2017).

Among the characteristics and possibilities of CRM, we present the Social CRM, which deals with CRM management in social media. Social CRM is a tool that combines information from social networks to traditional CRM, so a hotel can act on its customers with a database that provides vital information on the trends and preferences of consumers (Tecnohotel, 2014).

The virtues of the CRM strategy in social media are maintaining direct contact with the customer before, during and after the purchase; speaking directly to customers; possibility of offering customers service in the most varied social media channels; interaction and loyalty through the online brand; management of comments, opinions and trends; integration of Blog, web and social networks as a communication channel between the staff and the client (Tecnohotel, 2014).

Thus, CRM can be useful and important for hotel ventures to manage their customers' information, which allows the company to better relate with clients, however, it is emphasized that only the implementation of the system does not guarantee an optimization in the management of data. It is necessary for companies to have their professionals able to operate the system, as it is also necessary for companies to have a focus and to set strategies for the use of data collected through CRM. These group of practices may provide the success of the company.

Studies point to the importance of using CRM in companies, including the effectiveness of CRM in managing the relationship between companies and their employees through training and motivational activities. Oliveira (2016) studied the influence of Internal Marketing on the degree of employee motivation, satisfaction and loyalty. The results demonstrated the strong correlation between these two factors (communication and motivation) and showed that effective communication by managers leads to a high degree of motivation, potentiates the quality and perspective a better image of the organization perceived by the staff. The sample of this study is constituted of the employees of 4- and 5-star hotels in Porto and Northern Portugal. It was carried out through the application of a questionnaire.

Kanbara (2010) sought to identify which factors determine customer loyalty to a hotel business, the quality of service and the reflexes in the provision of service and what satisfies the client. For the collection of data, we used surveys with multiple choice questions, on a scale of 1 to 10 that were answered by the clients according to the services provided. The case study carried out at the Mercure Rio de Janeiro Arpoador hotel. It was confirmed that relationship marketing is a powerful tool for customer loyalty, especially in hotel developments where guests look for comfort and service experience within a closer relationship with hotel staff. On consumer behavior within a midscale hotel, it can be stated that the cost-benefit ratio is one of the main factors to mark the buying decision and customers are increasingly demanding and looking for standardization.

Silva and Silva (2008) compared, by multiple case study, the similarities and differences between two hotel chains, NH Hotels and Versare Hotel, regarding the use of CRM strategy. It was verified that the Versare Hotel Chain does not use the full potential of the CRM strategy for the development of the company and that some actions could be applied in order to enable a better result. They concluded in their investigation that in spite of the large number of employees prepared to provide a quality service to the client, those who are not prepared can cause situations that may lead to the loss of the very same customers (Silva & Silva, 2008).

Mota, Wada and Fraga (2013), through an empirical and exploratory research, made with a bibliographical survey on the main themes of hospitality, services, social media, hotels and stakeholders, aimed at clarifying the use of social media in hotels in the city of São Paulo, with specific objectives: to identify the social media inserted in the marketing strategy related to CRM processes; to determine if the practices of using platforms in social networks respond to the objectives of the hotel sector; and, to clarify trends for the use of social media in marketing strategies. The research showed that, due to the similarity in relation to the contact with the client, hotel developments should be objects of study together with digital marketing companies and online agencies, in order to establish successful practices from different industries.

In order to validate a customer satisfaction model for restaurant services, Angnes, Moyano and Lengler (2015) conducted a qualitative exploratory study to understand the relevant attributes of restaurant services and a descriptive evaluation with a quantitative approach. The main results of the exploratory phase resulted in a list of 27 judgment attributes for restaurant services and the analysis with structural equation modeling used in the validation of the model suggests that the relationship, quality and value realized by the custumer may influence their satisfaction and the loyalty with the evaluated restaurants.

METHOD

The present study is characterized by a qualitative, exploratory, descriptive approach, which seeks to evaluate the social media management of some lodging facilities, as well as to analyze how the interaction between the ventures and their clients occurs. The research was carried out in the city of Santa Vitória do Palmar, located in the state of Rio Grande do Sul, Brazil. According to IBGE (Brazilian Institute of Geography and Statistics), the latest estimate is that in 2017 there were 31,274 inhabitants in this city. In addition, the it has a territorial area of 5,243,578 km^2 (IBGE 2017). According to Schwab and Martins (2013) the city is the way for tourists that travel between Brazil and Uruguay.

In this research, it was mapped the lodging facilities located downtown due to its permanent operation (i.e., there is no seasonality in the rendering of its services, remaining open yearly). Thus, as the research population, five hotels were selected, of which all accepted to participate, thus comprising the sample. It was decided not to identify them to ensure that their reputations are not affected by the results presented, thus they are denominated as Hotels A, B, C, D and E.

As a tool to enable the achievement of the objectives proposed in this study, we opted for the adaptation of a research instrument already scientifically tested. The instrument of this study (Appendix A) is structured and has four sections. In the first section titled "The presence of the lodging facilities in social media" are nine closed questions of multiple choice. In the second section called "Point of view of the manager on social media" is a table with 20 statements, where the respondent should mark his degree of agreement or disagreement with the affirmatives through the Likert scale. The third and fourth sections of the instrument named "Manager Profile" and "Characterization of the Lodging business" comprise the identification of the data of the respondent and of the lodging business, which will be presented in the next chapter.

Characterization of the Search Sample

The Table 2 presents the characterization of the five lodging businesses participating in this research.

The largest hotel has 25 rooms and the smallest hotel has 13 rooms. Three of the sample hotels have been operating for more than 50 years. Regarding the number of employees, it is verified that proportionally Hotel A has a greater number of professionals and Hotel B has the fewest professionals in its staff.

Table 2. Characterization of the lodging facilities

Lodging Businesses	Characterization of the Lodging Businesses
Hotel A	The lodging business contains 18 housing units, with 51 beds available; it has nine employees and it was founded in 1992.
Hotel B	The lodging business contains 20 housing units, with 40 beds available; it has three employees and it was founded in 1964.
Hotel C	The lodging business contains 13 housing units, with 25 beds available; it has eight employees and it was founded in 1964.
Hotel D	The lodging business contains 26 housing units, with 60 beds available; it has six employees and it was founded in 1980.
Hotel E	The lodging business contains 25 housing units, with 55 beds available; it has six employees and it was founded in 1967.

Source: Prepared by the authors (2017).

Following, at the Table three, is the characterization of the managers participating in this research, composed of five managers.

The sample is composed of three male managers and two female managers, of whom two managers have completed higher education and one manager reported not having finished higher education.

Presence of the Lodging Business in Social Media

All five managers who comprise the entire sample of this survey responded that they use social media in their lodging facilities, as Table 4 presents.

Table 3. Characterization of managers

Lodging Facilities	Characterization of Managers
Hotel A	The manager of the accommodation is female; she is inserted in the age group from 30 to 39; she holds a graduate degree; her monthly income is up to five minimum wages; she claims to occupy the position of manager (owner) of the lodging business and has been working for six months in the respective establishment.
Hotel B	The manager of the accommodation is male; he is inserted in the age group from 30 to 39; he holds an undergraduate degree; his monthly income is up to two minimum wages; he claims to occupy the position of manager (owner) of the lodging business and has been working for 40 years in the respective establishment.
Hotel C	The manager of the accommodation is male; he is inserted in the age group more than 50; he holds a high school degree; his monthly income is more than five minimum wages; he claims to occupy the position of manager (owner) of the lodging business and has been working for 47 years in the respective establishment.
Hotel D	The manager of the accommodation is male; he is inserted in the age group more than 50; he holds a high school degree; his monthly income is more than five minimum wages; he claims to occupy the position of manager of the lodging business and has been working for 36 years in the respective establishment.
Hotel E	The manager of the accommodation is female; she is inserted in the age group from 30 to 39; she holds an undergraduate degree; her monthly income is up to five minimum wages; she claims to occupy the position of manager/ partner of the lodging business and has been working for one year in the respective establishment.

Source: Prepared by the author (2017).

Table 4. Characterization of managers

	HOTEL A	HOTEL B	HOTEL C	HOTEL D	HOTEL E
Implementation time	1 to 2 years	3 to 4 years	2 to 3 years	- than 6 months	6 months to 1 year
Social media used	Facebook WhasApp TripAdvisor	Facebook WhasApp Blog	Facebook	Facebook WhasApp	Facebook WhasApp
Most frequently used social media	WhatsApp	WhatsApp Facebook	Facebook	WhatsApp	WhatsApp Facebook
Utilization frequency of social media	Daily	Daily	Daily	Daily	From 2 to 3 times a week
Answers to commentaries at social media	Up to 48h	Up to 24h	Up to 24h	Up to 24h	Up to 24h
Target public	Tourists, institutions and touristic organizations	Tourists, potential tourists, companies in general, institutions and touristic organizations	Tourists, companies in general	Tourists, companies in general, institutions and touristic organizations	Tourists, potential tourists, companies in general, institutions and touristic organizations

Source: Prepared by the author (2017).

Manager's Point of View Social Media

We present the second session of the instrument called "The manager's point of view on social media", in which the managers express their degree of agreement or disagreement using the Likert scale of seven points. Figure 1 presents statements related to the interaction of managers with their clients.

It can be seen that, when they are approached in relation to the "customer interaction" theme that predominates in the statements, the managers demonstrated a high degree of agreement in all the affirmatives that compose the graph, a fact that is corroborated by Bronze Júnior et al. (2012) by citing that with the rise of competition and the improvement of technology, corporations are beginning to worry not only about winning new customers, but also maintaining those already loyal, thus initiating relationship marketing. Teixeira and Tomelin (2005, p. 6) add that "relationship marketing seeks a deep knowledge of customers, competitors and suppliers for strategy formation in order to maintain a lasting and profitable relationship".

Statements dealing with social media allow managers to obtain information from their clients were answered with a slight degree of disagreement of the manager of hotel B with grade three on the scale, and only in one of the statements that addresses the question of the utility of social media to obtain information about the degree of loyalty of its clients, one of the managers expressed a complete disagreement about the statement. These are facts that show a lack of understanding of what is currently being addressed by specialized magazines in hotel market such as Tecnohotel, which states in an article the relevance of Social CRM, a tool that refers to CRM management in social media. Market studies explain that the main virtues of the CRM strategy in social media presents possibilities to maintain a direct contact with customers before, during and after the purchase; to speak directly to customers; possibility of offering customers service in the most varied social media channels; interaction and loyalty through the online brand; management of comments, opinions and trends; integration of Blog, web and social media as communication channels between the team and the client (Tecnohotel, 2014).

Figure 1. Statements regarding the interaction with the client
Source: Prepared by the authors (2017).

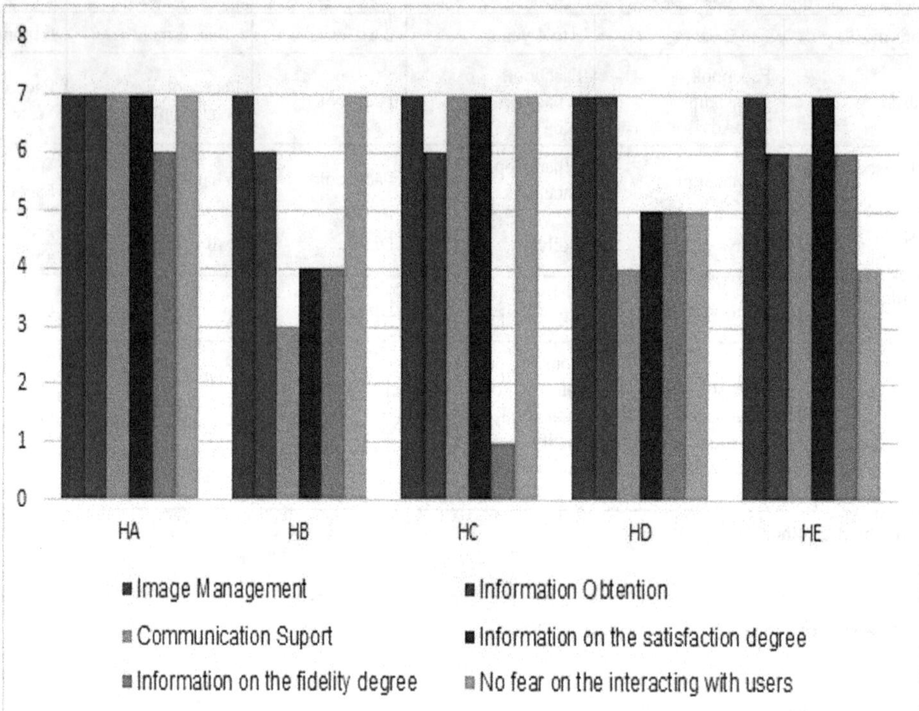

We can realise that in the sample managers understand the relevance of social media as a tool for interacting with their clients, a fact that is positive and beneficial for the lodging facilities as it can be ranked in the market and it can show efficiency in handling such tools of communication/interaction as well as be attentive to the contemporary tendencies of communication used by its clients.

Such conclusion is coherent to what Kanbara (2010) stated regarding how relationship marketing presents itself as a powerful tool for customer loyalty, especially in companies in the hotel sector, where clients seek not only comfort and servisse experiences, but also a closer relationship with hotel staff.

In order to comprehend how the managers see the usage of social media, figure 2 demonstrates the following scenario:

The theme of this section is focused on the creation and use of social media by the managers. First, in the statement "Learning how to use social media is easy", the manager of the hotel D did not indicate any degree of the scale.

The managers presented a satisfactory degree of agreement in the affirmatives of Table 2, which is highlighted by Cantinho, Correia and Ramos (2016) when citing that the creation and use of social media helps to promote customer loyalty, enables the conversation between professionals in the area, and enables dialogue between suppliers and customers. It also makes it possible to disseminate tourism through the sharing of experiences and images; presents business opportunities through discounts to its fans and followers; and allows the creation of a company's own page, where it allows customers to post their experiences and reports about the stay in a certain destination or lodging environment.

Figure 2. Statements about the creation/usage of social media
Source: Prepared by the authors (2017).

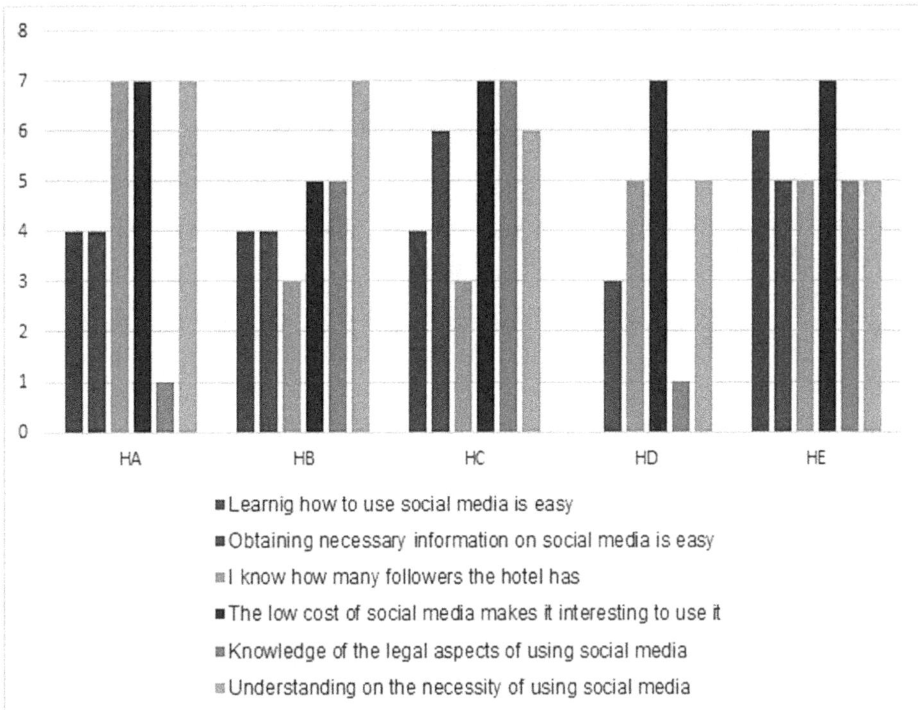

Although the managers in the broad context have agreed with the statement of the relevance of the creation and use of social media, diverge as to the legal aspects of the creation and management of social media, for some of them totally agree with it and others to disagree completely with said statement.

The managers understand the importance of creating /using social media, a fact that is positive, however, it is also perceived that they do not fully agree on the easiness in learning how to handle social media, as well as some of them demonstrate they are not clear how many followers they want to acquire with social media.

Therefore, we see a necessity in drawing a specific focus of where they want to arrive with their insertion in the social media, because, according to Boaria and Ortiz (2016), the consumer is increasingly using the internet to plan, choose and buy tourism services. Often, they are influenced by comments and opinions of their social circle. By itself, this fact enough to call the attention of the manager to this tendency and to make a satisfactory use of this contemporary practice.

With regards to the mastering of social media in the management of the lodging facilities, figure 3 presents the following panorama:

The focus of this section is related to the management of social media by the managers of the research sample, where it is possible to verify a considerable degree of agreement with the statement, but, the manager of hotel B presented slight levels of disagreement when marking grade three of the scale on the statements that relate to the efficiency of social media in terms of decisions made in its services, contribution to the loyalty of its clients and if the efforts made in the social media are being compensated.

Figure 3. Management of the lodging facilities
Source: Prepared by the author (2017).

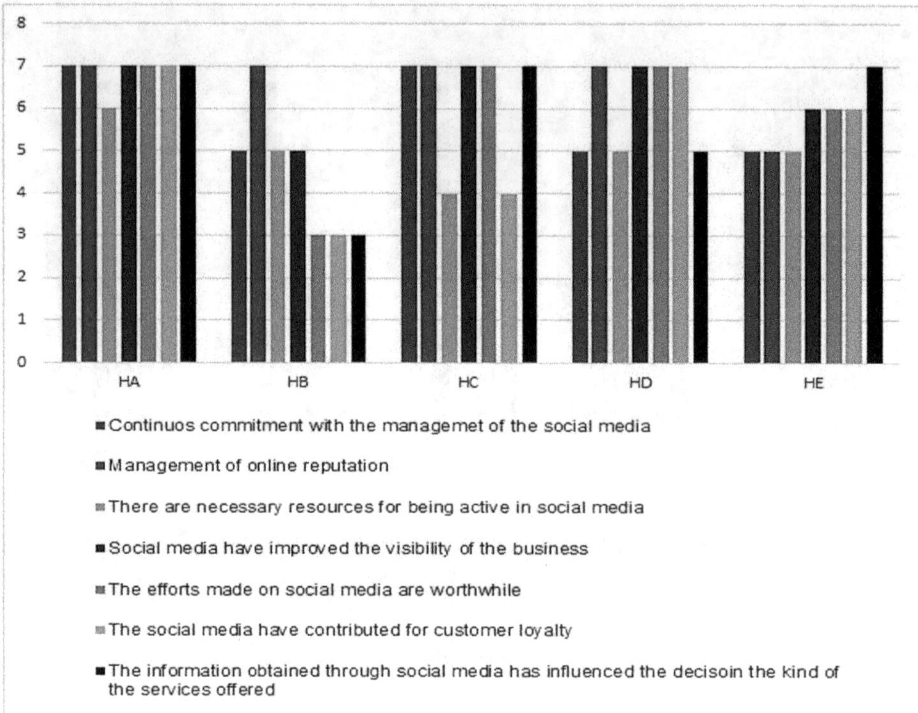

Given the potential of social media, lodging business managers need to be aware of the opportunities that these media make possible, for an efficient management by the hotel management denotes success and consolidation of the hotel in the market it is inserted in, a fact corroborated by Cantinho, Correia and Ramos (2016), who pointed to the use of social networks by hotels, more specifically for purpose of communication with customers, because it may create trust and allow managers to be updated on the trends of this segment. It is worth noting that the authors warn that the lodging facilities researched are poorly prepared to handle social media wisely, since they have not carried out an in-depth analysis due to the lack of knowledge to use the metrics programs. Such finding is in agreement with the results of Thomaz, Biz and Bettoni (2015), which points out the lack of management and monitoring of social media using software and/or data management techniques.

In this sense, it is important to emphasize that the mere presence of the lodging facilities on social media does not guarantee a good image for consumers and potential consumers.

CONCLUSION

The objective of this chapter is to evaluate the management of social media in the hotel sector in Santa Vitória do Palmar, RS, and to analyze how the interaction/communication between customers and the lodging facilities occurs on social media. The sample comprise 5 downtown hotels. It was possible to verify that social media are still in the initial phase of implementation by the hotel sector of the city, where specifically Facebook and WhatsApp appear as the most used ones by the managers. However, some of them are still confused about the legal aspects regarding the use of the media and also about the focus on the number of followers they want to obtain.

As for the relationship with the customers through social media, 4 of the 5 the managers said that the interaction/communication occurs very frequently and that this interaction facilitates the contact with the clients and contributes consequently to the loyalty.

It can be seen from a general overview that the use of social media by managers is still an incipient practice, where although the entire sample uses social media, managers focus their efforts on two specific media, Facebook and WhatsApp, and although these media are currently two of the most used, they are not the only ones with the greatest appeal among users. No sample hotel uses Twitter and Instagram as a social media tool, for example, which are two of the most used ones nowadays. Thus, it is verified that the establishments surveyed are not following the profile of consumers.

One fact that stands out is the proactivity of managers in social media. While four managers say they use social media on a daily basis, hotel manager E claims to use social media in his establishment about two to three times a week, a fact that is worrisome due to the demand for interaction that social media requires of its users/managers in relation to their customers.

Still on the interaction, another relevant fact refers to the manager of hotel A, who presents himself as the only one that does not interact online with his clients in social media, but the manager agrees that social media make communication feasible. We noticed that the interaction does not occur due to lack of knowledge, but perhaps other random factors that were not researched (lack of time or interest, for example). The time of implementation of social media can also be a factor conditioning the non-interaction of the manager of hotel A, who also stated to have implemented such practice in a period of time of one to two years, which can still be assumed to be a short time for learning.

The use of social media by the managers of the research sample is still carried out by the managers with the same knowledge of common users. In order to optimize this practice and consequently to obtain profits by the hotel management, it becomes necessary the implementation to seek specialized knowledge regarding the management of social media.

This study proves to be scientifically relevant for addressing a topic that has not yet been researched in the context of the hospitality area, especially in the academic community where it is inserted. Also, the results of this research are relevant to set a data base in relation to the theme for the hotel sector of the city and region to use it and, therefore, to optimize their actions and strategies in social media.

Finally, new studies are recommended in this area, in order to broaden the knowledge about the subject not only in the local context that was the focus of this research, but in regional and national contexts, which still lack studies that can map the scenario of the national hotel sector regarding the use of social media.

REFERENCES

Angnes, D. L., Moyano, C. A. M. & Lengler, J. F. B. (2015). Avaliação da Satisfação do Cliente em Serviços de Restaurantes com Aplicação do ACSI. *Revista RBTUR, 9*(1).

Anjos, S. G., Boaria, F., & Raye, R. L. (2013). Marketing Digital na Hotelaria: Estratégias e Atualidades. Proceedings of the X Seminário ANPTUR. Academic Press.

Belo, A. L. L. G. (2011). *O potencial dos sites de redes sociais no desempenho das pequenas e médias empresas da região do Algarve [Dissertação].* Universidade do Algarve, Faro, Portugal.

Boaria, F. (2013). *O e-commerce nos meios de hospedagem: Principais ferramentas utilizadas pelas redes hoteleiras no Brasil [Dissertação].* Universidade do Vale do Itajaí, Balneário Camboriú.

Boaria, F., & Ortiz, T. B. (2016). *O Turismo e a Web – Uma Análise sobre a Influência da Internet nas Escolhas do Turista de Foz do Iguaçu/Pr.* Fórum Internacional do Iguassu.

Boaria, F., Raye, R., & Felini, A. (2013). Mídias Sociais: uma reflexão sobre as novas ferramentas de marketing na hotelaria. In *VII Fórum Internacional de Turismo do Iguassu.* Academic Press.

Bronze Júnior, J. C. F., Sousa, C. V., Lara, J. E., Fagundes, A. F. A., & Sampaio, D. O. (2012). O setor hoteleiro e o marketing de relacionamento: Um estudo de caso na cidade de Belo Horizonte - Minas Gerais, Brasil. *Turismo & Sociedade. Curitiba, 5*(2), 489–506.

Cantinho, G. M. R., Correia, M. B., & Ramos, C. M. Q. (2016). As redes sociais online nos hotéis de 4 e 5 estrelas do Algarve. *Tourism and Hospitality International Journal, 7*(1), 27–48.

Corrêa, C. & Hansen, D. R. (2014). Qualidade de serviços em restaurantes de São Paulo premiados pelo TripAdvisor: análise do conteúdo gerado pelo usuário. *Revista Hospitalidade, 5*(11).

EBIT. (2018). Relatório Webshoppers. Retrieved from http://www.ebit.com.br/webshoppers

Fotis, J., Buhalis, D., & Rossides, N. (2012). *Social Media Use and Impact during the Holiday Travel Planning Process.* MASMI Research Group. doi:10.1007/978-3-7091-1142-0_2

Gretzel, U., & Yoo, K. H. (2008). Use and impact of online travel reviews. *Information and Communication Technologies in Tourism, 2*(1), 35–46.

Gretzel, U., Yuan, Y., & Fensenmaier, D. (2000). Preparing for the New Economy: Advertising Strategies and Change in Destination Marketing organizations. *Journal of Travel Research, 39*(2), 149–156. doi:10.1177/004728750003900204

Hoffmann, C. A. K. (2017). O comportamento do consumidor, o boca a boca eletrônico (eBAB) e as redes sociais on-line: Conhecimento atual e direções futuras. *Revista FAE, Curitiba, 20*(1), 27–41.

Hosteltur. (2017). Redes sociais: sete tendências a considerar. Retrieved from https://www.hosteltur.com/124397_redes-sociales-siete-tendencias-tener-cuenta.html?fromNewsletter=com&fromNewsletterDate=2017-1003&internalFormatClick=diario-20171003

Instituto Brasileiro de Geografia e Estatística (IBGE). (2017). Retrieved from https://cidades.ibge.gov.br/xtras/perfil.php?lang=&codmun=431730&search=||infogr%E1ficos:-informa%E7%F5es-completas

Kanbara, F. K. (2010). *O Comportamento do Consumidor na Hotelaria Corporativa e Lazer – Rede Mercure.* Rio de Janeiro: Monografia. Universidade Cândido Mendes.

Mota, A. M. G., Wada, E. K., & Fraga, L. E. (2013). Mídias Sociais e a Hospitalidade Desafios e Tendências no Século XXI. *TURyDES - Revista de Investigación em Turismo y desarrollo local, 6*(15).

Oliveira, A. M. (2016). *Marketing Interno nos Estabelecimentos Hoteleiros: A aplicabilidade do Marketing Interno como forma de obtenção de qualidade de Serviço ao Cliente* [Dissertação]. Instituto Politécnico do Porto.

Organização Mundial do Turismo (OMT). (2019). Crescimento do turismo mundial pode chegar a 4% em 2019. Retrieved from http://www.turismo.gov.br/%C3%BAltimas-not%C3%ADcias/12306-crescimento-do-turismo-mundial-pode-chegar-a-4-em-2019.html

Petrocchi, M. (2002). *Hotelaria: planejamento e gestão*. São Paulo: Futura.

Petry, T. R. E., Pickler, C. M., & Tomelin, C. A. (2016). A Percepção dos Hóspedes de Negócios quanto ao Desempenho da Qualidade dos Serviços Prestados nos Hotéis de Florianópolis: Uma Análise a Partir do Conteúdo Gerado no Website Booking. Revista Turismo - Visão e Ação, 18.

Schwab, C. B., & Martins, G. S. (2013). Turismo de Contemplação em Santa Vitória do Palmar/RS: uma pesquisa de demanda. Proceedings of the IV Seminário de Pesquisa em Turismo do Mercosul – SEMINTUR JR. Academic Press.

Silva, D. M. (2013). *Facebook - Canal de reclamações dos clientes hoteleiros [Dissertação]*. Universidade do Porto.

Silva, P. M. & Silva, L. M. (2008). Estratégia de CRM para hotelaria. *Revista Acadêmica Observatório de Inovação do Turismo, Rio de Janeiro, 3*(2).

Tecnohotel.news. (2014). CRM + Redes Sociales = Social CRM: el nuevo consumidor social. Retrieved from https://www.tecnohotelnews.com/2014/06/crm-redes-sociales-social-crm-el-nuevo-consumidor-social/#

Teixeira, A. H., & Tomelin, C. A. (2005). O Marketing de Relacionamento como Força Competitiva para as Agências de Viagens de Santa Catarina. *Proceedings of the III Seminário de Pesquisa em Turismo do Mercosul – SEMINTUR*. Academic Press.

Thomaz, G. M., Biz, A. A., & Bettoni, E. M. (2015). Análise da gestão e monitoramento em mídias sociais pelas Destination Management Organizations (DMO) de Curitiba-PR. *Proceedings of the XII Seminário ANPTUR*. Academic Press. Retrieved from http://whc.unesco.org/en/list/355

Torres, C. (2010). *Guia Prático de Marketing na Internet para Pequenas Empresas*.

KEY TERMS AND DEFINITIONS

Costumer Relationship Management: Customer relationship management (CRM) is the combination of practices, strategies and technologies that companies use to manage and analyze customer interactions and data with the goal of improving customer service relationships and assisting in customer retention and driving sales growth.

Destination Management Organizations (DMO): It's responsible for promoting a community as an attractive travel destination and enhancing its public image as a dynamic place to live and work. Facilitate the understanding of the full social and economic dimension of tourism in terms of demand (international and domestic arrivals, characteristics of their visits, and their consumption of goods and services) and supply (characteristics of local industries involved in tourism and details of their production and use of labour and capital), thereby guiding tourism policy and planning.

Digital Marketing: It's the use of the internet, mobile devices, social media, search engines to promoting and selling products and services by leveraging online marketing tactics. It's a way to connect with and influence the company potential customers.

Electronic Word of Mouth: Word of mouth sent via internet, which is made available to an unspecified number of people by using a multitude of different online mechanisms (Twitter, Facebook, FourSquare, etc.) and online review platforms. It consists of interpersonal communication which sets it apart from mass communication such as advertising. The communicators are non-commercially motivated, so communication is perceived as unbiased, genuine and honest.

Internal Marketing: It is the promotion of a company's objectives, products and services to employees within the organization. The purpose is to increase employee engagement with the company's goals and fostering brand advocacy.

Social Media: Social media refers to websites and applications that are designed to allow people to share content quickly, efficiently, and in real-time.

User Generated Content: is any form of content, such as images, videos, text and audio, that have been produced and shared by end users of an online service or website. This includes any content that is shared or produced by users that are members or subscribers of the service, but it is not produced by the website or service itself.

APPENDIX

Presence of the Lodging Facilities in Social Media

1. Does the lodging business use social media?
   a. Yes
   b. No
2. What social media are used by the lodging business?
   a. Facebook
   b. Twitter
   c. Instagram
   d. WhatsApp
   e. TripAdvisor
   f. Blog
   g. Other - Which: ___________________________________
3. What types of content is published on the social media?
   a. Tourist Attractions
   b. Cultural agenda
   c. Leisure
   d. Business and commerce
   e. Lodging
   f. The hotel brand
   g. Other - Which ones? _______________
4. Which audiences do you want to achieve through the posted content?
   a. Tourists
   b. Potential tourists
   c. Residents
   d. General companies
   e. Institutions and tourist organizations
   f. Other - Which? _____________
5. Of the aforementioned social media, which are most frequently used by the lodging business?
   a. Facebook
   b. Twitter
   c. Instagram
   d. WhatsApp
   e. TripAdvisor
   f. Blog
   g. Other - Which: ___________________________________
6. How often are social media used by the lodging business?
   a. Once a week

b. Two to three times a week
c. Four to five times a week
d. Daily
e. Greater than a week

7. Does the Manager respond to customer comments on social media?
   a. Yes, within 24 hours
   b. Yes, within 48 hours
   c. Yes, within 72 hours
   d. Yes, within a week
   e. Does not respond to comments
8. Does the Manager talk or interact online with customers in social media?
   a. Yes, every 24 hours
   b. Yes, every 48 hours
   c. Yes, every 72 hours
   d. Yes, every week
   e. We do not talk online with clients in social media
9. How long has the lodging business implemented the use of social media?
   a. Less than 06 months
   b. From 06 months to 01 year
   c. From 01 year to 02 years
   d. From 02 to 03 years
   e. From 03 to 04 years
   f. 05 years or more

Manager's Look at Social Media

Indicate your level of agreement with the following statements using the next scale: I totally agree 7 6 5 4 3 2 1 Strongly disagree

Table 5.

2.1 Social media a very useful tools for managing the image of our lodging business.	7 6 5 4 3 2 1
2.2 Social media allow the communication of my lodging business with our customers.	7 6 5 4 3 2 1
2.3 Social media allow us to obtain information of our customers.	7 6 5 4 3 2 1
2.4 Social Media are very useful for obtaining information of our customers' preferences and needs.	7 6 5 4 3 2 1
2.5 Social Media are very useful for obtaining information of our customers' satisfaction degree.	7 6 5 4 3 2 1
2.6 Social Media are very useful for obtaining information of our customers' loyalty degree.	7 6 5 4 3 2 1
2.7 Learning to use social media was easy.	7 6 5 4 3 2 1
2.8 Obtaining the information we need from social media is easy.	7 6 5 4 3 2 1
2.9 It is clear for us how many followers/fans we wanto to obtain in the different social media.	7 6 5 4 3 2 1
2.10 The use of social media is interesting because of the low cost of its implementation.	7 6 5 4 3 2 1
2.11 We know the legal aspects of the creation and management of social media.	7 6 5 4 3 2 1
2.12 At my lodging business there is the understanding of the necessity to use social media.	7 6 5 4 3 2 1
2.13 At my lodging business there is an ongoing commitment with the management of social media.	7 6 5 4 3 2 1
2.14 My lodging business is not afraid of interactions that may occur with users.	7 6 5 4 3 2 1
2.15 At my lodging business, we work to manage the online reputation adequately.	7 6 5 4 3 2 1
2.16 My lodging business gives the necessary means for an active presence on social media.	7 6 5 4 3 2 1
2.17 Social media have improve the visibility of my lodging business.	7 6 5 4 3 2 1
2.18 The efforts made on social media are worthwhile.	7 6 5 4 3 2 1
2.19 Social media have been contributing positively for making customers come back.	7 6 5 4 3 2 1
2.20 The information obtain on social media has some influence on deciding about the services offered at my lodging business.	7 6 5 4 3 2 1

Compilation of References

Abbott, W., Donaghey, J., Hare, J., & Hopkins, P. (2013). An Instagram is Worth a Thousand Words: An Industry Panel and Audience Q&A. *Library Hi Tech News*, *30*(7), 1–6. doi:10.1108/LHTN-08-2013-0047

Abbott, W., Donaghey, J., Hare, J., & Hopkins, P. (2013). An Instagram of worth a thousand words: An industry panel and audience Q&A. *Library Hi Tech News*, *30*(7), 1–6.

Abernathy, W. J., & Clark, K. B. (1985). Innovation: Mapping the winds of creative destruction. *Research Policy*, *14*(1), 3–22.

ABIA. (2017). Números do setor de alimentos e bebidas, Brasil, 2017. Retrieved from https://www.abia.org.br/vsn/anexos/faturamento2017.pdf

Abrahamson, E. (1991). Managerial fads and fashions: The diffusion and rejection of innovations. *Academy of Management Review*, *16*(3), 586–612. doi:10.5465/amr.1991.4279484

Abreu, N., & Costa, E. (2000). Estudo da viabilidade da utilização de marketing na internet no setor hoteleiro de Maceió. Retrieved from https://bit.ly/2B6HmzG

Ady, M., & Quadri-Felitti, D. (2015). Consumer research identifies how to present travel review content for more bookings. Retrieved from https://static1.squarespace.com/static/56dd934137013bc69c80ae95/t/56e6c50462cd94987f7ad826/1457964296558/Consumer+Research+Identifies+How+To+Present+Travel+Review+Content+for+More+Bookings.pdf

Afonso, C. & Borges, L. (2013). *Social Target: Da estratégia à implementação – Como tirar partido das redes sociais e potenciar o negócio*. Editora Top Books.

Afonso, C., & Borges, L. (2013). *Social target: Da estratégia à implementação – Como tirar partido das redes sociais e potenciar o negócio*. Top Books.

Akay, R. (2014). Etkinlik yönetimi uygulamalarinda yaratici rekabet ve sosyal medyanin entegrasyonu. *The Turkish Online Journal of Design, Art and Communication*, *4*(4), 55–70. doi:10.7456/10404100/005

Akehurst, G. (2009). User-generated content: The use of blogs for tourism organisations and tourism consumers. *Service Business*, *3*(1), 51–61. doi:10.100711628-008-0054-2

Alaei, A. R., Becken, S., & Stantic, B. (2019). Sentiment Analysis in Tourism: Capitalizing on Big Data. *Journal of Travel Research*, *58*(2), 175–191. doi:10.1177/0047287517747753

Alawan, A. A., Rana, N. P., Dwivedi, Y. K., & Alghatabat, R. (2017). (2017). Social Media in Marketing: A review and Analysis of the existing literature. *Telematics and Informatics*. doi:10.1016/j.tele.2017.05.008

Alexis, P. (2017). R-Tourism: Introducing the Potential Impact of Robotics and Service Automation in Tourism. *Tourism Management*, *29*(3), 458–468. doi:10.1016/j.tourman.2007.07.014

Aljandali, A. (2017). *Multivariate Methods and Forecasting with IBM SPSS Statistics*. Cham: Springer International Publishing.

Allard, T. T., Sitchon, M. L., Sawatzky, R., & Hoppa, R. D. (2005). Use of Hand-held Laser Scanning and 3D Printing for Creation of a Museum Exhibit. *Proceedings of the 6th International Symposium on Virtual Reality, Archaeology and Cultural Heritage*. Academic Press.

Alonso-Almeida, M. M., & Almeida, C. R. (2018). Use of Social Media within tourism travel behaviour. *Proceedings of the Final Conference Social Networks and Travel Behaviour*, Milan. Academic Press.

Alonso-Almeida, M.M., & Almeida, C. R. (2016). Use of social networks on travel decision making. Social Networks and Travel Behaviour. *Proceedings of the 5MC & WGs meetings*, Bucharest. Academic Press.

Altinay, Z., Saner, T., Bahçelerli, N. M., & Altinay, F. (2016). The role of social media tools: Accessible tourism for disabled citizens. *Journal of Educational Technology & Society*, *19*(1), 89–99.

Alumran, A., Hou, X. Y., & Hurst, C. (2012). Validity and reliability of instruments designed to measure factors influencing the overuse of antibiotics. *Journal of Infection and Public Health*, *5*(3), 221–232. doi:10.1016/j.jiph.2012.03.003 PMID:22632596

Alves, G. M., Sousa, B. M., & Machado, A. (2019). The Role of Digital Marketing and Online Relationship Quality in Social Tourism: A Tourism for All Case Study. In J. Santos & Ó. Silva (Eds.), *Digital Marketing Strategies for Tourism, Hospitality, and Airline Industries* (pp. 49–70). Hershey, PA: IGI Global. doi:10.4018/978-1-5225-9783-4.ch003

Amaro, S., Duarte, P., & Henriques, C. (2016). Travellers use of social media: A clustering approach. *Annals of Tourism Research*, *59*, 1–15. doi:10.1016/j.annals.2016.03.007

Amersdorffer, D., Bauhuber, F., & Oellrich, J. (2012). The Economic and Cultural Aspects of The Social Web: Implications for The Tourism Industry. *Journal of Vacation Marketing*, *18*(3), 175–184. doi:10.1177/1356766712449364

Anderson, J., & Narus, J. (1984). A Model of the Distributor's Perspective of Distributor-Manufacturer Working Relationships. *Journal of Marketing*, 62-74.

Anderson, M., & Anderson, M. (2014). 88% of consumers trust online reviews as much as personal recommendations. *Searchengineland.* Retrieved from [REMOVED HYPERLINK FIELD]http://searchengineland.com/88-consumers-trust-online-reviews-muchpersonal-recommendations-195803

Anderson, C. K. (2012). The impact of social media on lodging performance. *Cornell Hospitality Report*, *12*(15), 4–11.

Anderson, J. C. N., & Narus, J. A. (1990). A model of distributor firm and manufacturer firm working partnerships. *Journal of Marketing*, *54*(1), 42–58. doi:10.1177/002224299005400103

Andersson, T. (2007). The tourist in the experience economy. *Scandinavian Journal of Hospitality and Tourism*, *7*(1), 46–58. doi:10.1080/15022250701224035

Andrade, E. C., Delgado, R., Hauptner, B., Kerschbaum, F., List, R., Pobežalová, P., . . . Wögerbauer, S. (2015). *Inclusive technologies in museums: for a better access to culture for blind and visually impaired people*. Vienna, Austria: Project AMBAVis. Retrieved from http://www.ambavis.eu/wp-content/uploads/2016/11/AMBAVis-Broschuere_BF_V01_EN_bf.pdf

Andriole, S. J. (2010). Business impact of Web 2.0 technologies. *Communications of the ACM*, *53*(12), 67–79. doi:10.1145/1859204.1859225

Andsager, J. L., & Drzewiecka, J. A. (2002). Desirability of Differences in Destinations. *Annals of Tourism Research, 29*(2), 401–421. doi:10.1016/S0160-7383(01)00064-0

Angelova, B., & Zekiri, J. (2011). Measuring Customer Satisfaction with Service Quality Using American Customer Satisfaction Model (ACSI Model). *International Journal of Academic Research in Business and Social Sciences, 1*(3), 232–258. doi:10.6007/ijarbss.v1i2.35

Angnes, D. L., Moyano, C. A. M. & Lengler, J. F. B. (2015). Avaliação da Satisfação do Cliente em Serviços de Restaurantes com Aplicação do ACSI. *Revista RBTUR, 9*(1).

Anjos, S. G., Boaria, F., & Raye, R. L. (2013). Marketing Digital na Hotelaria: Estratégias e Atualidades. Proceedings of the X Seminário ANPTUR. Academic Press.

Antunes, N. M., Ramos, C. M., & Sousa, C. M. (2018). Who are the tourists sharing content on social media? Behaviour and characteristics. *Journal of Spatial and Organizational Dynamics, 6*(3), 237–257.

Araújo, A. M., Almeida, L. S., Costa, A. R., Alfonso, S., Conde, Á., & Deaño, M. (2015). Variáveis pessoais e socioculturais de diferenciação das expectativas académicas: Estudo com alunos do Ensino Superior do Norte de Portugal e Galiza. *Revista Portuguesa de Educação, 28*(1), 201–220. doi:10.21814/rpe.7057

Arctur, & Tourism 4.0 Consortium. (2018). What is Tourism 4.0? Retrieved from https://www.tourism4-0.org/

Arenas, A., Goh, J., & Uruena, A. (2019). How does IT affect design centricity approaches: Evidence from Spain's smart tourism ecosystem. *International Journal of Information Management, 45*, 149–162. doi:10.1016/j.ijinfomgt.2018.10.015

Argan, T. M., & Yüncü, D. (2015). *Etkinlik Pazarlama Yönetimi*. Istanbul: Detay Yayıncılık.

Argyropoulos, V. S., & Kanari, C. (2015). Re-imagining the museum through "touch": Reflections of individuals with visual disability on their experience of museum-visiting in Greece. *Alter - European Journal of Disability Research, 9*(2), 130–143. doi:10.1016/j.alter.2014.12.005

Arita, S., Edmonds, C., Croix, S. L., & Mak, J. (2011). Impact of Approved Destination Status on Chinese travel abroad: An econometric analysis. *Tourism Economics, 17*(5), 983–996. doi:10.5367/te.2011.0076

Arndt, J. (1967). Role of product-related conversations in the diffusion of a new product. *JMR, Journal of Marketing Research, 4*(3), 291–295. doi:10.1177/002224376700400308

Arnould, E. J., & Price, J. J. (1993). River magic: Extraordinary experience and the extended service encounter. *The Journal of Consumer Research, 20*(1), 24–45. doi:10.1086/209331

Arsal, I., Woosnam, K. M., Baldwin, E. D., & Backman, S. J. (2010). Residents as Travel Destination Information Providers: An Online Community Perspective. *Journal of Travel Research, 49*(4), 400–413. doi:10.1177/0047287509346856

Arsmtrong, J. S. (2001). *Planning Special Events*. Indiana: Indiana University Center on Philanthrophy.

Asakawa, S., Guerreiro, J., Ahmetovic, D., Kitani, K., & Asakawa, C. (2018). The present and future of museum accessibility for people with visual impairments. In *Proceedings of the 20th International ACM SIGACCESS Conference on Computers and Accessibility* (pp. 382–384). New York, NY: ACM; doi:10.1145/3234695.3240997

Ascaniis, S., & Cantoni, L. (2016). Information and Communication Technologies in Religious Tourism and Pilgrimage. *International Journal of Religious Tourism and Pilgrimage, 4*(3), 2.

Assistant, V. (2019) In Cambridge dictionary. Retrieved from https://dictionary.cambridge.org/pt/dicionario/ingles/virtual-assistant

Atalık, Ö., & Arslan, M. (2009). A Study to Determine the Effects of Customer Value on Customer Loyalty in Airline Companies Operating: Case of Turkish Air Travellers. *International Journal of Business and Management*, *4*(6), 154–162. doi:10.5539/ijbm.v4n6p154

Attride-Stirling, J. (2001). Thematic Networks: An Analytic Tool for Qualitative Research. *Qualitative Research*, *1*(3), 385–405. doi:10.1177/146879410100100307

Atzori, L., Iera, A., & Morabito, G. (2010). The Internet of Things: A survey. *Computer Networks*, *54*(15), 2787–2805. doi:10.1016/j.comnet.2010.05.010

Aureliano-Silva, L., Brandão, M. M., Strehlau, S., Pimenta, R. F., & Pimenta, R. F. (2015). O efeito do formato de apresentação de desconto em ambiente crowding: Um estudo experimental. *RACE - Revista de Administração Contabilidade e Economia*, *14*(2), 653. doi:10.18593/race.v14i2.6735

Aureliano-Silva, L., Strehlau, S., & Strehlau, V. (2017). Please! Wrap this gift for me: The brands effect with different promotional frames. *ICTACT Journal on Management Studies*, *3*(1), 452–456. doi:10.21917/ijms.2017.0061

Ayeh, J., Leung, D., Au, N., & Law, R. (2012). Perceptions and Strategies of Hospitality and Tourism Practitioners on Social Media: An Exploratory Study. In M. Fuchs, F. Ricci, & L. Cantoni (eds). *Information and Communication Technologies in Tourism 2012*. Springer-Verlag. 10.1007/978-3-7091-1142-0_1

Ayeh, J. K., Au, N., & Law, R. (2013). "Do we believe in TripAdvisor?" Examining credibility perceptions and online travelers' attitude toward using user-generated content. *Journal of Travel Research*, *52*(4), 437–452. doi:10.1177/0047287512475217

Ayoun, B., & Moreo, P. (2008). Does national culture affect hotel managers' approach to business strategy. *International Journal of Contemporary Hospitality Management*, *32*(1), 52–65.

Backman, S., & Crompton, J. (1991). The usefulness of selected variables for predicting activity loyalty. *Leisure Sciences*, *13*(3), 205–220. doi:10.1080/01490409109513138

Baek, T. H., & Morimoto, M. (2012). Stay away from me. *Journal of Advertising*, *41*(1), 59–76. doi:10.2753/JOA0091-3367410105

Bae, S. J., Lee, H., Suh, E., & Suh, K. (2017). Shared experience in pre-trip and experience sharing in post-trip: A survey of Airbnb users. *Information & Management*, *54*(6), 714–727. doi:10.1016/j.im.2016.12.008

Baka, V. (2016). The becoming of user-generated reviews: Looking at the past to understand the future of managing reputation in the travel sector. *Tourism Management*, *53*, 148–162. doi:10.1016/j.tourman.2015.09.004

Baker, D. (2013). Service quality and customer satisfaction in the airline industry: A comparison between legacy airlines and low-cost airlines. *American Journal of Tourism Research*, *2*(1), 67–77. doi:10.11634/216837861302317

Balandina, E., Balandin, S., Koucheryavy, Y., & Mouromtsev, D. (2015). IoT Use Cases in Healthcare and Tourism. *Proceedings of the 17th IEEE Conference on Business Informatics CBI 2015* (Vol. 2, pp. 37–44). IEEE Press. 10.1109/CBI.2015.16

Balegno, S. (2010). *2010 Social Media Marketing Benchmark Report Warren*, MarketingSherpa. Retrieved from https://bit.ly/2yspj5B

Baloglu, S., & McCleary, K. W. (1999a). A Model of Destination Image Formation. *Annals of Tourism Research*, *26*(4), 868–897. doi:10.1016/S0160-7383(99)00030-4

Baloglu, S., & McCleary, K. W. (1999b). US International Pleasure Travelers' Images of Four Mediterranean Destinations: A Comparison of Visitors and Nonvisitors. *Journal of Travel Research*, *38*(2), 144–152. doi:10.1177/004728759903800207

Bandi Tanner, M., & Hämmerli, S. J. (2018). Reviewing TripAdvisor and Co.: A quality analysis of hotel review sites. *Anatolia, 29*(4), 518–528. doi:10.1080/13032917.2018.1451350

Banerjee, S., & Chua, A. Y. (2016). In search of patterns among travellers' hotel ratings in TripAdvisor. *Tourism Management, 53*, 125–131. doi:10.1016/j.tourman.2015.09.020

Banhirov, F., Zhang, Y., & Hashim, N. H. (2019). Facebook fan page management for global airlines. *Tourism Review, 74*(3), 532–546. doi:10.1108/TR-03-2018-0045

Baran, M. L. (2016). It Is All in The Design: Creating the Foundations of a Mixed Methods Research Study. In Mixed methods research for improved scientific study (pp. 66-78). Hershey, PA: IGI Global.

Barber, N., & Scarcelli, J. M. (2010). Enhancing the assessment of tangible service quality through the creation of a cleanliness measurement scale. *Managing Service Quality: An International Journal, 20*(1), 70–88. doi:10.1108/09604521011011630

Barcala, M. F., Díaz, M. G., & Rodriguez, J. P. (2009). Factors influencing guests' hotel quality appraisals. *European Journal of Tourism Research, 2*(1), 25–40.

Barnes, S. (2017). *Understanding Virtual Reality in Marketing: Nature, Implications and Potential*. doi:10.2139srn.2909100

Barnett, G. A., Danowski, J. A., Feeley, T. H., & Stalker, J. (2010). Measuring quality incommunication doctoral education using network analysis of faculty-hiring patterns. *Journal of Communication, 60*(2), 388–411. doi:10.1111/j.1460-2466.2010.01487.x

Barnett, G. A., Danowski, J. A., & Richards, W. D. (1993). Communication networks and network analysis: a current assessment. In W. D. Richards & G. A. Barnett (Eds.), *Progress in communication science* (Vol. 12, pp. 1–19). Norwood, NJ: Ablex.

Barney, J. (1991). Firm resources and sustained competitive advantage. *Journal of Management, 17*(1), 99–120. doi:10.1177/014920639101700108

Barry, S., Murphy, K., & Drew, S. (2015). From deconstructive misalignment to constructive alignment: Exploring student uses of mobile technologies in university classrooms. *Computers & Education, 81*, 202–210. doi:10.1016/j.compedu.2014.10.014

Bassi, A., & Horn, G. (2008). Internet of Things in 2020: A Roadmap for the Future. *European Commission: Information Society and Media, 22*, 97–114.

Baumol, W. (2002). *The Free Market Innovation Machine: Analyzing the Growth Miracle of Capitalism*. Princeton, NJ: Princeton University Press. doi:10.1515/9781400851638

Bayindir, N., & Kavanagh, D. (2018). *Global Web Index - Social*. Retrieved from www.globalwebindex.com

Becken, S., & Simmons, D. (2002). Understanding energy consumption patterns of tourist attractions and activities in New Zealand. *Tourism Management, 23*(4), 343–354. doi:10.1016/S0261-5177(01)00091-7

Becker, K., Nobre, H., & Kanabar, V. (2013). Monitoring and protecting company and brand reputation on social networks: When sites are not enough. *Global Business and Economics Review, 15*(2/3), 293. doi:10.1504/GBER.2013.053075

Beerli, A., & Martin, J. D. (2004). Factors Influencing Destination Image. *Annals of Tourism Research, 31*(3), 657–681. doi:10.1016/j.annals.2004.01.010

Bekiaris, E., Loukea, M., Spanidis, P., Ewing, S., Denninghaus, M., Ambrose, I., . . . Castiglioni, R. C. (2018). Research for TRAN Committee - Transport and tourism for persons with disabilities and persons with reduced mobility. TRAN committee. Retrieved from http://www.europarl.europa.eu/RegData/etudes/STUD/2018/617465/IPOL_STU(2018)617465_EN.pdf

Belén, V. (2018). Tourism and Technology: How Tech is Revolutionizing Travel. Wearemarketing. Retrieved from https://www.wearemarketing.com/blog/tourism-and-technology-how-tech-is-revolutionizing-travel.html

Belo, A. L. L. G. (2011). *O potencial dos sites de redes sociais no desempenho das pequenas e médias empresas da região do Algarve [Dissertação].* Universidade do Algarve, Faro, Portugal.

Benwell, B., & Stokoe, E. (2006). *Discurso e identidad.* Editorial de la Universidad de Edimburgo.

Berezina, K., Bilgihan, A., Cobanoglu, C., & Okumus, F. (2016). Understanding Satisfied and Dissatisfied Hotel Customers: Text Mining of Online Hotel Reviews. *Journal of Hospitality Marketing & Management, 25*(1), 1–24. doi:10.1080/19368623.2015.983631

Berg, L. V. D., Borg, J., & Meer, J. V. (1995). *Urban Tourism: Performance and Strategies in Eight European Cities.* Avebury.

Berger, J., & Schwartz, E. M. (2011). What drives immediate and ongoing word of mouth? *JMR, Journal of Marketing Research, 48*(5), 869–880. doi:10.1509/jmkr.48.5.869

Berry, L. L., Zeithaml, V. A., & Parasuraman, A. (1990). Five Imperatives for Improving Service Quality. *Sloan Management Review, 31*(4), 29–38. Retrieved from https://search.proquest.com/openview/8b3a8e9fbf82128a35d092cf3a81003f/1?pq-origsite=gscholar&cbl=26142

Berta, D. (2009). Brands say employees atwitter for social sites. *Nation's Restaurant News, 43*(15), 1+14.

Berthon, P. R., Pitt, L. F., Plangger, K., & Shapiro, D. (2012). Marketing meets Web 2.0, social media, and creative consumers: Implications for international marketing strategy. *Business Horizons, 55*(3), 261–271. doi:10.1016/j.bushor.2012.01.007

Binucci, C., De Luca, F., Di Guiacomo, E., Liotta, G., & Montecchiani, F. (2017). Designing the content analyser of travel recommender system. *Expert Systems with Applications, 87*, 199–208. doi:10.1016/j.eswa.2017.06.028

Bizirgianni, I., & Dionysopoulou, P. (2013). The influence of tourist trends of Youth Tourism through Social Media (SM) & Information and Communication Technologies. *Procedia: Social and Behavioral Sciences, 73*, 652–660. doi:10.1016/j.sbspro.2013.02.102

Bjørkelund, E., Burnett, T. H., & Nørvåg, K. (2012, December). A study of opinion mining and visualization of hotel reviews. *Proceedings of the 14th International Conference on Information Integration and Web-based Applications & Services* (pp. 229-238). ACM. 10.1145/2428736.2428773

Blackshaw, P., & Nazzaro, M. (2006). Word of Mouth in The Age of The Web-Fortified Consumer. *Consumer-generated media (CGM), 101.*

Blain, C., Levy, S. E., & Ritchie, J. R. (2005). Destination branding: Insight and practices from destination anagement. *Journal of Travel Research, 43*(4), 328–338. doi:10.1177/0047287505274646

Blomberg-Nygard, A., & Anderson, C. K. (2016). United Nations world tourism organization study on online guest reviews and hotel classification systems: An integrated approach. *Service Science, 8*(2), 139–151. doi:10.1287erv.2016.0139

Boaria, F., Raye, R., & Felini, A. (2013). Mídias Sociais: uma reflexão sobre as novas ferramentas de marketing na hotelaria. In *VII Fórum Internacional de Turismo do Iguassu.* Academic Press.

Boaria, F. (2013). *O e-commerce nos meios de hospedagem: Principais ferramentas utilizadas pelas redes hoteleiras no Brasil [Dissertação].* Universidade do Vale do Itajaí, Balneário Camboriú.

Boaria, F., Anjos, S., & Raye, R. (2014). A aplicação do e-marketing nas redes hoteleiras do Brasil. *TMStudies*, *10*(2), 116–122.

Boaria, F., & Ortiz, T. B. (2016). *O Turismo e a Web – Uma Análise sobre a Influência da Internet nas Escolhas do Turista de Foz do Iguaçu/Pr*. Fórum Internacional do Iguassu.

Boatwright, B., Mazer, J. P., & Beach, S. (2019). The 2016 US Presidential Election and Transition Events: A Social Media Volume and Sentiment Analysis. *The Southern Communication Journal*, 1–14.

Bokunewicz, J. F., & Shulman, J. (2017). Influencer identification in Twitter networks of destination marketing organizations. *Journal of Hospitality and Tourism Technology*, *8*(2), 205–219. doi:10.1108/JHTT-09-2016-0057

Bolton, G., Greiner, B., & Ockenfels, A. (2013). Engineering trust: Reciprocity in the production of reputation information. *Management Science*, *59*(2), 265–285. doi:10.1287/mnsc.1120.1609

Bonn, M. A., Joseph, S. M., & Dai, M. (2005). International versus domestic visitors: An examination of destination image perceptions. *Journal of Travel Research*, *43*(3), 294–301. doi:10.1177/0047287504272033

Booking.com. (2019a). Ranking, Preferred Programme, Stars and Guest Reviews. Retrieved from https://www.booking.com/content/terms.en-gb.html?aid=304085&label=Hoteis-iQ35xiFDNh3rJo4ReZuiSgS14356879461%3Apl%3Ata%3Ap1%3Ap2230.000%3Aac%3Aap1t1%3Aneg&sid=ff86ec28cdd97bcd214ff482d9f74277&tmpl=docs%2Fterms-and-conditions&lang=en-gb&soz=1&lang_click=top;cdl=pt-pt;lang_changed=1

Booking.com. (2019b). Retrieved from https://www.booking.com/

Borgatti, S. P., & Everett, M. G. (1997). Network analysis of 2-mode data. *Social Networks*, *3*(3), 243–269. doi:10.1016/S0378-8733(96)00301-2

Borgatti, S. P., Everett, M. G., & Freeman, L. C. (2002). *UCINET for windows: software for social network analysis (Version 6)*. Harvard, MA: Analytic Technologies.

Bourne, R. R., Flaxman, S. R., Braithwaite, T., Cicinelli, M. V., Das, A., Jonas, J. B., ... Taylor, H. (2017). Magnitude, temporal trends, and projections of the global prevalence of blindness and distance and near vision impairment: A systematic review and meta-analysis. *The Lancet. Global Health*, *5*(9), 888–897. doi:10.1016/S2214-109X(17)30293-0 PMID:28779882

Boyd, D. M., & Ellison, N. B. (2008). Social network sites: Definition, history, and scholarship. *Journal of Computer-Mediated Communication*, *13*(1), 2010–2223.

Boyd, D., & Ellison, N. (2008). Social Network Sites: Definition, History, and Scholarship. *Journal of Computer-Mediated Communication*, *13*(1), 210–230. doi:10.1111/j.1083-6101.2007.00393.x

Brace, N., Kemp, R., & Snelgar, R. (2016). SPSS for Psychologists (6th ed.). Hampshire: Macmillan International Higher Education.

Brandau, M. (2010). *Chipotle, McD sign on with Facebook Deals*. Nation's Restaurant News.

Brandt, T., Bendler, J., & Neumann, D. (2017). Social media analytics and value creation in urban smart tourism ecosystems. *Information & Management*, *54*(6), 703–713. doi:10.1016/j.im.2017.01.004

Brandtzaeg, P., & Heim, J. (2009). Why people use social networking sites. In A. A. Ozok & P. Zaphiris (Eds.), *Online Communities and Social Computing (OCSC 2009)*. Berlin: Springer. doi:10.1007/978-3-642-02774-1_16

Brech, F. M., Messer, U., Vander Schee, B. A., Rauschnabel, P. A., & Ivens, B. S. (2017). Engaging fans and the community in social media: Interaction with institutions of higher education on Facebook. *Journal of Marketing for Higher Education*, *27*(1), 112–130. doi:10.1080/08841241.2016.1219803

Brochado, A., Rita, P., Oliveira, C., & Oliveira, F. (2019). Airline passengers' perceptions of service quality: Themes in online reviews. *International Journal of Contemporary Hospitality Management*, *31*(2), 855–873. doi:10.1108/IJCHM-09-2017-0572

Brodie, R. J., Hollebeek, L. D., Jurić, B., & Ilić, A. (2011). Customer engagement: Conceptual domain, fundamental propositions, and implications for research. *Journal of Service Research*, *14*(3), 252–271. doi:10.1177/1094670511411703

Bronze Júnior, J. C. F., Sousa, C. V., Lara, J. E., Fagundes, A. F. A., & Sampaio, D. O. (2012). O setor hoteleiro e o marketing de relacionamento: Um estudo de caso na cidade de Belo Horizonte - Minas Gerais, Brasil. *Turismo & Sociedade. Curitiba*, *5*(2), 489–506.

Brown, R. (2009). *Public Relations and the Social Web: How to use social media and web 2.0 in communications.* Kogan Page Publishers.

Brun, I., Rajaobelina, L., & Ricard, L. (2014). Online relationship quality: Scale development and initial testing. *International Journal of Bank Marketing*, *32*(1), 5–27.

Brun, I., Rajaobelina, L., & Ricard, L. (2016). Online Relationship Quality: Testing an Integrative and Comprehensive Model in the Banking Industry. *Journal of Relationship Marketing*, *15*(4), 219–246. doi:10.1080/15332667.2016.1242400

Brusha, P. (2013). Turismo e redes sociais: porque devem andar juntos? Retrieved from https://bit.ly/2OK4m0G

Buhalis, D. (2003). eTourism: Information technology for strategic tourism management. London: Pearson. doi:10.13140/2.1.2274.0804

Buhalis, D. (2019), Technology in tourism-from information communication technologies to eTourism and smart tourism towards ambient intelligence tourism: a perspective article. *Tourism Review*. doi:10.1108/TR-06-2019-0258

Buhalis, D., & Amaranggana, A. (2014). Smart tourism destinations. In Z. Xiang & I.Tussyadiah (Eds.), Information and communication technologies in tourism (pp. 553–564). Dublin:Springer.

Buhalis, D. (1998). Strategic Use of Information Technologies in The Tourism Industry. *Tourism Management*, *19*(5), 409–421. doi:10.1016/S0261-5177(98)00038-7

Buhalis, D. (2000). Marketing the competitive destination of the future. *Tourism Management*, *21*(1), 97–116. doi:10.1016/S0261-5177(99)00095-3

Buhalis, D., & Darcy, S. (Eds.). (2010). *Accessible tourism: Concepts and issues*. Channel View Publications.

Buhalis, D., & Law, R. (2008). Progress in information technology and tourism management: 20 years on and 10 years after the internet – The state of eTourism research. *Tourism Management*, *23*, 207–230. doi:10.1016/S0261-5177(01)00085-1

Buhalis, D., & Law, R. (2008). Progress in information technology and tourism management: 20 years on and 10 years after the internet - the state of eTourism research. *Tourism Management*, *24*(3), 609–623. doi:10.1016/j.tourman.2008.01.005

Buhalis, D., & Michopoulou, E. (2011). Information-enabled tourism destination marketing: Addressing the accessibility market. *Current Issues in Tourism*, *14*(2), 145–168. doi:10.1080/13683501003653361

Bundesministerium für Wirtschaft und Energie. (2016). Plattform Industrie 4.0 - Was ist Industrie 4.0?

Buonamici, F., Furferi, R., Governi, L., & Volpe, Y. (2015). Making blind people autonomous in the exploration of tactile models: a feasibility study. In M. Antona & C. Stephanidis (Eds.), *Universal Access in Human-Computer Interaction. Access to Interaction* (pp. 82–93). Cham, Switzerland: Springer International Publishing; doi:10.1007/978-3-319-20681-3_8

Burton, H. (2003). Visual cortex activity in early and late blind people. *The Journal of Neuroscience*, *23*(10), 4005–4011. doi:10.1523/JNEUROSCI.23-10-04005.2003 PMID:12764085

Butowski, B. (2011). Tourism-An Academic Discipline. *Tourism (Zagreb)*, *21*(1-2), 17–24.

Buzova, D., Sanz-Blas, S., & Cervera-Taulet, A. (2019). Does culture affect sentiments expressed in cruise tours' eWOM? *Service Industries Journal*, *39*(2), 154–173. doi:10.1080/02642069.2018.1476497

Cachia, A. (2013). Talking blind: Disability, access, and the discursive turn. *Disability Studies Quarterly*, *33*(3), 1–20. doi:10.18061/dsq.v33i3.3758

Cai, L. A. (2002). Cooperative Branding for Rural Destinations. *Annals of Tourism Research*, *29*(3), 720–742. doi:10.1016/S0160-7383(01)00080-9

Camilleri, M. A. (2018). *Travel marketing, tourism economics and the airline product. An introduction to theory and practice*. Cham: Springer International Publishing.

Campbell, A. (2013). Word-of-Mouth Communication and Percolation in Social Networks. *The American Economic Review*, *103*(6), 2466–2498. doi:10.1257/aer.103.6.2466

Candlin, F. (2003). Blindness, art and exclusion in museums and galleries. *International Journal of Art & Design Education*, *22*(1), 100–110. doi:10.1111/1468-5949.00343

Candlin, F. (2004). Don't touch! Hands off! Art, blindness and the conservation of expertise. *Body & Society*, *10*(1), 71–90. doi:10.1177/1357034X04041761

Cantallops, A., & Salvi, F. (2014). New consumer behavior: A review of research on eWOM and hotels. *International Journal of Hospitality Management*, *36*, 41–51. doi:10.1016/j.ijhm.2013.08.007

Cantinho, G. M. R., Correia, M. B., & Ramos, C. M. Q. (2016). As redes sociais online nos hotéis de 4 e 5 estrelas do Algarve. *Tourism and Hospitality International Journal*, *7*(1), 27–48.

Cantoni, V., Dondi, P., Lombardi, L., Nugrahaningsih, N., Porta, M., & Setti, A. (2018). A multi-sensory approach to cultural heritage: The Battle of Pavia exhibition. *IOP Conference Series. Materials Science and Engineering*, *364*, 1–8. doi:10.1088/1757-899X/364/1/012039

CAPA. (2018a). *A321neoLR: Ideal for new North Atlantic routes, potentially a global game changer*. Retrieved from https://centreforaviation.com/analysis/airline-leader/a321neolr-ideal-for-new-north-atlantic-routes-potentially-a-global-game-changer-439182

CAPA. (2018b). *Longhaul low cost airlines: World Airways to be US' first*. Retrieved from https://centreforaviation.com/analysis/reports/longhaul-low-cost-airlines-world-airways-to-be-us-first-405559

Caracelli, V. J., & Greene, J. C. (1993). Data analysis strategies for mixed-method evaluation designs. *Educational Evaluation and Policy Analysis*, *15*(2), 195–207. doi:10.3102/01623737015002195

Carbonaro, A. (2011). *Forum Summarization to Support Tutor and Teacher in Group Interaction Management. Semantic Web Personalization and Context Awareness: Management of Personal Identities and Social Networking* (pp. 22–31). Hershey, PA: IGI Global. doi:10.4018/978-1-61520-921-7.ch003

Carlson, N. (2010). At Last—The Full Story of How Facebook Was Founded. Business Insider. Retrieved from http://www.businessinsider.com/how-facebook-was-founded-2010-3

Carlson, R. (1997). *Experience Cognition*. New York: Lawrence Eribaum Association. doi:10.4324/9781410603265

Carl, W. J. (2006). What's all the buzz about? Everyday communication and the relational basis of word-of-mouth and buzz marketing practices. *Management Communication Quarterly*, *19*(4), 601–634. doi:10.1177/0893318905284763

Carmo, H., & Ferreira, M. M. (1998). *Metodologia de investigação. Guia para auto- aprendizagem*. Lisboa: Universidade Aberta.

Carmo, H., & Ferreira, M. M. (2008). *Metodologia da Investigação: Guia para Auto-Aprendizagem (2ª)*. Lisboa: Universidade Aberta.

Carneiro, M. J., & Costa, C. (2000). The influence of service quality on the positioning of five star hotels - the case of the Lisbon area. *Journal of Quality Assurance in Hospitality & Tourism*, *1*(4), 1–19. doi:10.1300/J162v01n04_01

Casaló, L. V., Flavian, C., Guinaliu, M., & Ekinci, Y. (2015). Do online hotel rating schemes influence booking behaviors? *International Journal of Hospitality Management*, *49*, 28–36. doi:10.1016/j.ijhm.2015.05.005

Casteleyn, J., Mottart, A., & Rutten, K. (2009). How to use Facebook in your market research. *International Journal of Market Research*, *51*(4), 439–447. doi:10.2501/S1470785309200669

Castronovo, C., & Huang, L. (2012). Social media in an alternative marketing communication model. *Journal of marketing development and competitiveness*, *6*(1), 117-134.

Cattaneo, Z., & Vecchi, T. (2011). *Blind Vision: The Neuroscience of Visual Impairment*. Cambridge, MA: MIT Press. doi:10.7551/mitpress/9780262015035.001.0001

Cavanagh, W. (2015). *Índice del Valor de la Información para ayudar a las empresas a medir su propio rendimento*. Retrieved from https://www.pwc.es/es/sala-prensa/notas-prensa/2015/Seizing-The-Information-Advantage.html

Cenamor, J., Paridab, V., & Wincent, J. (2019). How entrepreneurial SMEs compete through digital platforms: The roles of digital platform capability, network capability and ambidexterity. *Journal of Business Research*, *100*, 196–206. doi:10.1016/j.jbusres.2019.03.035

Cerro, A., Mogollon, J., & Alves, H. (2010). Marketing relacional e Turismo: A fidelização de clients no turismo rural, proposta de um modelo para o seu estudo. *Revista Turismo & Desenvolvimento*, *13/14*, 231–241.

Chaffey, D. (2019 12 Feb). Global digital report 2019. Smart Insights. Retrieved from https://www.smartinsights.com/social-media-marketing/social-media-strategy/new-global-social-media-research/

Chang, Y. T., Lu, H., & Hu, H. (2015). Persuasive messages, popularity cohesion, and message diffusion in social media marketing. *Journal of Business Research*, *68*(4), 777–782. doi:10.1016/j.jbusres.2014.11.027

Chante, K., Jessica, C., Lindsay, B., Tyler, Q., & Robert, P. D. (2014). Dermatology on Instagram. *Dermatology Online Journal*, *20*(7), 1–6. PMID:25046455

Chapleo, C., Durán, M. V. C., & Díaz, A. C. (2011). Do UK universities communicate their brands effectively through their websites? *Journal of Marketing for Higher Education*, *21*(1), 25–46. doi:10.1080/08841241.2011.569589

Chareyron, G., Da-Rugna, J., & Raimbault, T. (2014). Big data: A new challenge for tourism. *Proceedings of the 2014 IEEE International Conference on Big Data (Big Data)* (pp. 5-7). IEEE Press. 10.1109/BigData.2014.7004475

Chatzigeorgiou, C. (2017). Modelling the Impact of Social Media Influencers on Behavioral Intentions of Millennials: The Case of Tourism in Rural Areas in Greece. *Journal of Tourism. Heritage & Services Marketing*, *3*(2), 25–29.

Chaves, M. S., Gomes, R., & Pedron, C. (n.d.). Decision-making based on online reviews: The small and medium hotels management. Retrieved from https://repositorio-cientifico.essatla.pt/bitstream/20.500.12253/639/1/ecis-2012-mchaves-etal-cr.pdf

Chaves, M., Gomes, R., & Pedron, C. (2012). Analyzing reviews in the web 2.0: Small and medium hotels in Portugal. *Tourism Management*, *33*(5), 1286–1287. doi:10.1016/j.tourman.2011.11.007

Chen, C. F., & Tsai, D. (2007). How Destination Image and Evaluative Factors Affect Behavioral Intentions? *Tourism Management*, *28*(4), 1115–1122. doi:10.1016/j.tourman.2006.07.007

Chen, C. M., & Liu, H. M. (2017). Exploring the Impact of Airlines Service Quality on Customer Loyalty: Evidence from Taiwan. *International Journal of Business and Management*, *12*(5), 36–50. doi:10.5539/ijbm.v12n5p36

Chen, R., Cheung, C., & Law, R. (2012). A review of the literature on culture in hotel management research: What is the future? *International Journal of Hospitality Management*, *31*(1), 52–65. doi:10.1016/j.ijhm.2011.06.010

Chen, S. C., & Quester, P. G. (2006). Modeling store loyalty: Perceived value in market orientation practice. *Journal of Services Marketing*, *20*(3), 188–198. doi:10.1108/08876040610665643

Chen, Y., Fay, S., & Wang, Q. (2011). The role of marketing in social media: How online consumer reviews evolve. *Journal of Interactive Marketing*, *25*(2), 85–94. doi:10.1016/j.intmar.2011.01.003

Chen, Y.-R. R. (2017). Perceived values of branded mobile media, consumer engagement, business-consumer relationship quality and purchase intention: A study of WeChat in China. *Public Relations Review*, *43*(5), 945–954. doi:10.1016/j.pubrev.2017.07.005

Cherifi, B., Smith, A., Maitland, R., & Stevenson, N. (2014). Destination Images of Non-Visitors. *Annals of Tourism Research*, *49*, 190–202. doi:10.1016/j.annals.2014.09.008

Cheung, C. W., Tsang, T. I., & Wong, K. H. (2017). Robot Avatar: A Virtual Tourism Robot for People with Disabilities. *International Journal of Computer Theory and Engineering*, *9*(3), 229–234. doi:10.7763/ijcte.2017.v9.1143

Cheung, C. M. K., Chiu, P.-Y., & Lee, M. K. O. (2011). Online social networks: Why do students use Facebook? *Computers in Human Behavior*, *27*(4), 1337–1343. doi:10.1016/j.chb.2010.07.028

Cheung, M. Y., Luo, C., Sia, C. L., & Chen, H. (2009). Credibility of electronic word-of-mouth: Informational and normative determinants of on-line consumer recommendations. *International Journal of Electronic Commerce*, *13*(4), 9–38. doi:10.2753/JEC1086-4415130402

Chiang, H.-S., & Huang, T.-C. (2015). User-adapted travel planning system for personalised schedule recommendation. *Information Fusion*, *21*, 3–17. doi:10.1016/j.inffus.2013.05.011

Chi, C. G. Q., & Qu, H. (2008). Examining the Structural Relationships of Destination Image, Tourist Satisfaction and Destination Loyalty: An Integrated Approach. *Tourism Management*, *29*(4), 624–636. doi:10.1016/j.tourman.2007.06.007

Chick, A. (2017). Co-creating an Accessible, Multisensory Exhibition with the National Centre for Craft & Design and Blind and Partially Sighted Participants. In *REDO: Cumulus International Conference*. Academic Press. Retrieved from http://eprints.linclon.ac.uk/27590

Chikwendu, D., Ejem, E., & Ezenwa, A. (2012). Evaluation of service quality of Nigerian airline using servqual model. *Journal of Hospitality Management and Tourism*, *3*(6), 117–125. doi:10.5897/JHMT12.019

Chilembwe, J. M. (2014). Examination of Social Cultural Impacts of Tourism in Chembe Village in Mangochi District, Malawi. *International Journal of Business Quantitative Economics and Applied Management Research*, *1*(1), 61–92.

Chilembwe, J. M., & Mweiwa, V. R. (2014). Tour guides: Are they developers and promoters - Case study of Malawi. *Impact Journals: International Journal of Research and Business Management*, *2*(9), 29–46.

Chilembwe, J. M., & Mweiwa, V. R. (2019). Responsible Travel and Tourism Adventure: Evidence from Malawi as a Tourist Destination. In A. Sharma (Ed.), *Sustainable Tourism Development: Futuristic Approaches* (pp. 31–53). New York: Apple Academic Press.

Chilembwe, J. M., Mweiwa, V. R., & Mankhomwa, E. (2018). Role of Tour Operators in Destination Tourism Marketing in Malawi. In M. A. Camilleri (Ed.), *Strategic Perspectives in Destination Marketing* (pp. 295–321). Hershey, PA: IGI Global Publishers. doi:10.4018/978-1-5225-5835-4.ch012

Chiu, C. M., Wang, E. T., Fang, Y. H., & Huang, H. Y. (2014). Understanding customers' repeat purchase intentions in B2C e-commerce: The roles of utilitarian value, hedonic value and perceived risk. *Information Systems Journal*, *24*(1), 85–114. doi:10.1111/j.1365-2575.2012.00407.x

Chiu, C., Ip, C., & Silverman, A. (2012). Understanding social media in China. *The McKinsey Quarterly*, *2*, 78–81.

Choi, S., Lehto, X. Y., & Morrison, A. M. (2007). Destination image representation on the web: Content analysis of Macau travel related websites. *Tourism Management*, *28*(1), 118–129. doi:10.1016/j.tourman.2006.03.002

Choi, S., Lehto, X. Y., Morrison, A. M., & Jang, S. (2012). Structure of travel planning processes and information use patterns. *Journal of Travel Research*, *51*(1), 26–40. doi:10.1177/0047287510394191

Choi, Y., Hickerson, B., & Kerstetter, D. (2018). Understanding the sources of online travel information. *Journal of Travel Research*, *57*(1), 116–128. doi:10.1177/0047287516683833

Chon, K. S. (1990). The Role of Destination Image in Tourism: A Review and Discussion. *Tourism Review*, *45*(2), 2–9.

Chou, C. C., Liu, L. J., Huang, S. F., Yih, J. M., & Han, T. C. (2011). An evaluation of airline service quality using the fuzzy weighted SERVQUAL method. *Applied Soft Computing*, *11*(2), 2117–2128. doi:10.1016/j.asoc.2010.07.010

Choudhury, N. (2014). World wide web and its journey from web 1.0 to web 4.0. *International Journal of Computer Science and Information Technologies*, *5*(6), 8096–8100.

Chou, S., & Chen, C.-W. (2018). The Influences of Relational Benefits on Repurchase Intention in Service Contexts: The Roles of Gratitude, Trust and Commitment. *Journal of Business and Industrial Marketing*, *33*(5), 680–692. doi:10.1108/JBIM-08-2017-0187

Chugh, R., & Ruhi, U. (2018). Social media in higher education: A literature review of Facebook. *Education and Information Technologies*, *23*(2), 605–616. doi:10.100710639-017-9621-2

Chumpitaz, R., & Swaen, V. (2002). Service Quality and Brand Loyalty Relationships. *Proceedings of the 31st Conference of European Marketing Academy*, (pp. 1-7). Academic Press..

Chung, J. Y., & Buhalis, D. (2008). Information needs in online social networks. *Information Technology & Tourism*, *10*(4), 267–281. doi:10.3727/109830508788403123

Chung, N., & Koo, C. (2015). The use of social media in travel information search. *Telematics and Informatics*, *32*(2), 215–229. doi:10.1016/j.tele.2014.08.005

Chu, W., Choi, B., & Song, M. R. (2005). The role of on-line retailer brand and infomediary reputation in increasing consumer purchase intention. *International Journal of Electronic Commerce*, *9*(3), 115–127. doi:10.1080/10864415.2005.11044336

Cintra, F. C. (2010). Marketing Digital: A era da tecnologia on-line. *Investigação*, *10*(1), 6–12.

Clapperton, G. (2009). *This is social media: how to tweet, post, link and blog your way to business success*. Chichester: A Wiley Company.

Clarke, R. H., Twede, D., Tazelaar, J. R., & Boyer, K. K. (2006). Radio frequency identification (RFID) performance: The effect of tag orientation and package contents. *Packaging Technology & Science*, *19*(1), 45–54. doi:10.1002/pts.714

Claver-Cortes, E., Molina-Azorin, J. F., & Periera-Moliner, J. (2007). Competitiveness in mass tourism. *Annals of Tourism Research*, *34*(3), 727–745. doi:10.1016/j.annals.2007.03.010

Cleary, M. N., & Hobbs, G. D. (1983). The fifty year cycle: a look at the empirical evidence. *Long waves in the world economy*, 164-182.

Cloke, P. (2006). Conceptualizing Rurality. In P. Clocke, T. Marsden, & P. Mooney (Eds.), *The Handbook of Rural Studies*., doi:10.4135/9781848608016.n2

Cock, M., Bretton, M., Fineman, A., France, R., Madge, C., & Sharpe, M. (2018). *State of Museum Access: Does your museum website welcome and inform disabled visitors?* London, UK: Vocal Eyes. Retrieved from https://vocaleyes.co.uk/state-of-museum-access-2018

Cohen-Almagor, R. (2011). Internet history. *International Journal of Technoethics*, *2*(2), 45–64. doi:10.4018/jte.2011040104

Cooper, C. (2005). Knowledge management in tourism. *Annals of Tourism Research*, *33*(1), 47–64. doi:10.1016/j.annals.2005.04.005

Cooper, C., Fletcher, J., Gilbert, D., & Wanhill, S. (1998). *Tourism principles and practice*. London: Addison Wesley Longman.

Cormode, G., & Krishnamurthy, B. (2008). Key differences between Web 1.0 and Web 2.0. *First Monday*, *13*(6). doi:10.5210/fm.v13i6.2125

Corrêa, C. & Hansen, D. R. (2014). Qualidade de serviços em restaurantes de São Paulo premiados pelo TripAdvisor: análise do conteúdo gerado pelo usuário. *Revista Hospitalidade, 5*(11).

Cox, C., Burgess, S., Sellitto, C., & Buultjens, J. (2009). The role of user-generated content in tourists' travel planning behaviour. *Journal of Hospitality Marketing & Management*, *18*(8), 743–764. doi:10.1080/19368620903235753

Crandall, R. (2018). Industry 1.0 to 4.0: the Evolution of Smart Factories. APICS Magazine.

Creamer, M. (2011). Your Followers Are No Measure of Your Influence. *Advertising Age*, *82*(1), 1–22.

Crompton, J. L. (1979). An Assessment of The Image of Mexico As A Vacation Destination and The Influence of Geographical Location Upon That Image. *Journal of Travel Research*, *17*(4), 18–23. doi:10.1177/004728757901700404

Cronin, J. J. Jr, Brady, M. K., & Hult, G. T. M. (2000). Assessing the effects of quality, value and customer satisfaction on consumer behavioral intentions in service environments. *Journal of Retailing*, *76*(2), 193–218. doi:10.1016/S0022-4359(00)00028-2

Crotts, J. C., Mason, P. R., & Davis, B. (2009). Measuring guest satisfaction and competitive position in the hospitality and tourism industry: An application of stance-shift analysis to travel blog narratives. *Journal of Travel Research, 48*(2), 139–151. doi:10.1177/0047287508328795

Crozier, N. (1982). L'innovation dans les services.

Cunha, S., & Cunha, J. (2005). Competitividade e sustentabilidade de um cluster de turismo: Uma proposta de modelo sistémico de medida do impacto do turismo no desenvolvimento local. *Revista de Administração Contemporânea, 9*(2), 63–79. doi:10.1590/S1415-65552005000600006

Cupeto, C., Figueiredo, M., Silva, V., Santos, S., Carvalho, C., & Preguiça, C. (2014). *Empreendedorismo no turismo de experiência e emoção. In 4ª Conferência Iberica de Emprendimiento, CIEM 2014, Actas de la Conferencia: Jornada Científica*. Universidad de Vigo.

Cutler, S., & Carmichael, B. (2010). The dimensions of the tourist experience. In M. Morgan, P. Lugosi, & J. R. Brent Ritchie (Eds.), *The tourism and leisure experience: Consumer and managerial perspectives* (pp. 3–27). Bristol: Channel View Publications.

D'Agnano, F., Balletti, C., Guerra, F., & Vernier, P. (2015). Tooteko: a case study of augmented reality for an accessible cultural heritage. Digitization, 3D printing and sensors for an audio-tactile experience. In *ISPRS - International Archives of the Photogrammetry, Remote Sensing and Spatial Information Sciences* (pp. 207–213). Academic Press. doi:10.5194/isprsarchives-XL-5-W4-207-2015

Dahnil, M. I., Marzuki, K. M., Langgat, J., & Fabeil, N. F. (2014). Factors Influencing SMEs Adoption of Social Media Marketing. *Procedia: Social and Behavioral Sciences, 148*, 119–126. doi:10.1016/j.sbspro.2014.07.025

Dai, B., Jiang, Y., Yang, L., & Ma, Y. (2017). China's outbound tourism–Stages, policies and choices. *Tourism Management, 58*, 253–258. doi:10.1016/j.tourman.2016.03.009

Dandona, L., & Dandona, R. (2006). Revision of visual impairment definitions in the International Statistical Classification of Diseases. *BMC Medicine, 4*(1), 7. doi:10.1186/1741-7015-4-7 PMID:16539739

Danias, K., & Kavoura, A. (2013). The role of social media as a tool of a company's innovative communication activities. *Zeszyty Naukowe Małopolskiej Wyższej Szkoły Ekonomicznej w Tarnowie, 23*(2), 75–83.

Dann, G. M. (1996). Tourists' Images of a Destination-An Alternative Analysis. *Journal of Travel & Tourism Marketing, 5*(1-2), 41–55. doi:10.1300/J073v05n01_04

Dans, E. (2012). *Big data: una perspectiva histórica.* Retrieved from https://www.enriquedans.com/2012/05/big-data-una-perspectiva-historica.html

Dash, K., & Grohall, G. (2016). *Economic impact of creating and exhibiting 3D objects for blind and visually impaired people in museums*. Vienna, Austria: AMBAVis Project. Retrieved from http://www.ambavis.eu/wp-content/uploads/2016/09/Economic-Aspects_final-disclaimer.pdf

Davis, F. (1989), *A Technology acceptance model for empirically testing new end user information systems: theory and results* [Doctoral Dissertation]. MIT Sloan School of Management.

Davis, F. (1989). Perceived usefulness, perceived ease of use, and user acceptance of information technology. *Management Information Systems Quarterly, 13*(3), 319–340. doi:10.2307/249008

de Oliveira, S. L. I. (2007). *Desmistificando o marketing*. São Paulo: Novatec Editora.

de Souza, F. N., Costa, A. P., & Moreira, A. (2016). webQDA - Qualitative Data Analysis (versão 3.0). Aveiro: Micro IO e Universidade de Aveiro. Retrieved from www.webqda.net

Dedeoğlu, B. B., Taheri, B., Okumus, F., & Gannon, M. (2020). Understanding the importance that consumers attach to social media sharing (ISMS): Scale development and validation. *Tourism Management*, *76*, 103954. doi:10.1016/j.tourman.2019.103954

Del Vecchio, P., Mele, G., Ndou, V., & Secudo, G. (2018). Creating value from social big data: Implications for smart tourism destination. *Information Processing & Management*, *54*(5), 847–860. doi:10.1016/j.ipm.2017.10.006

Delamere, T. A., Wankel, L. M., & Hinch, T. D. (2001). Development of a scale to measure resident attitudes toward the social impacts of community festivals, Part I: Item generation and purification of the measure. *Event Management*, *7*(1), 11–24. doi:10.3727/152599501108751443

Dellarocas, C., & Wood, C. A. (2008). The sound of silence in online feedback: Estimating trading risks in the presence of reporting bias. *Management Science*, *54*(3), 460–476. doi:10.1287/mnsc.1070.0747

Dellarocas, C., Zhang, X. M., & Awad, N. F. (2007). Exploring the value of online product reviews in forecasting sales: The case of motion pictures. *Journal of Interactive Marketing*, *21*(4), 23–45. doi:10.1002/dir.20087

Desvallées, A., & Mairesse, F. (2010). *Key Concepts of Museology*. Paris, France: Armand Colin.

Devile, E., & Kastenholz, E. (2018). Accessbile tourism experiences: the voice of people with visual disabilities. *Journal of Policy Research in Tourism, Leisure and Events, 10(3),* 265-285. doi:10.1080/19407963.2018.1470183

Di Marino, E. (2008, May). The Strategic Dimension of Destination Image: An Analysis of The French Riviera Image from The Italian Tourists' Perceptions. *Proceedings of the 17th International Tourism and Leisure Symposium* (pp. 17-18). Academic Press.

Di Pietro, L., Di Virgilio, F., & Pantano, E. (2012). Social network for the choice of tourist destination: Attitude and behavioural intention. *Journal of Hospitality and Tourism Technology*, *3*(1), 60–76. doi:10.1108/17579881211206543

Dias, D., Lima, F., & Dias, E. (2016). Tourism management and automation: RFID applications in Brazilian maritime cruises. *International Journal of Internet of Things and Web Services*, *1*, 43–49. Retrieved from https://www.iaras.org/iaras/filedownloads/ijitws/2016/022-0006.pdf

Dichter, E. (1985). What's in An Image. *Journal of Consumer Marketing*, *2*(1), 75–81. doi:10.1108/eb038824

Dickinson, J. E., Ghali, K., Cherrett, T., Speed, C., Davies, N., & Norgate, S. (2014). Tourism and the smartphone app: Capabilities, emerging practice and scope in the travel domain. *Current Issues in Tourism*, *17*(1), 84–101. doi:10.1080/13683500.2012.718323

Dijkmans, C., Kerkhof, P., & Beukeboom, C. J. (2015). A stage to engage: Social media use and corporate reputation. *Tourism Management*, *47*, 58–67. doi:10.1016/j.tourman.2014.09.005

DiPietro, R. B., Crews, T. B., Gustafson, C., & Strick, S. (2012). The Use of Social Networking Sites in the Restaurant Industry: Best Practices. *Journal of Foodservice Business Research*, *15*(3), 265–284. doi:10.1080/15378020.2012.706193

Doolin, B. B., Burgess, L., & Cooper, J. (2002). Evaluating the use of the web for tourism marketing: A case study form New Zealand. *Tourism Management*, *23*(5), 557–561. doi:10.1016/S0261-5177(02)00014-6

Dorsch, J. J., & Yasin, M. M. (1998). A framework for benchmarking in the public sector. In J. J. Lennon, H. Smith, N. Cockerell, & J. Trew (Eds.), *Benchmarking National Tourism Organisations and Agencies – Understanding best practice*. Amsterdam: Elsevier. doi:10.1108/09513559810216410

Dosi, G. (1988). Technical Change and Economic Theory.

Doub, A., Levin, A., Heath, C., & LeVangie, L. (2015). Mobile app-etite: Consumer attitudes towards and use of mobile technology in the context of eating behavior. *Journal of direct, data and digital marketing practice, 17*(2), 114-129.

Douglas, A., Lubbe, B., & van Rooyen, A. (2018). Profiling business travellers who use mobile travel applications. *Ereview of Tourism Research, 15*, 2–3. Retrieved from http://ertr.tamu.edu

Drucker, P. F. (1993). *Post-Capitalist Society. Heinemann*. NY: Butterworth.

DSEC. (2017). *Tourism Statistics*. Macau: Documentation and Information Centre of Statistics and Census Service.

Duggan, M., Ellison, N. B., Lampe, C., Lenhart, A., & Madden, M. (2015). Social media update 2014. *Pew research center, 19*.

Dunteman, G. H. (1989). *Principal component analysis (Quantitative applications in the social sciences)*. Beverly Hills, CA: Sage Publication.

Dwivedi, M. (2009). Online Destination Image of India: A Consumer Based Perspective. *International Journal of Contemporary Hospitality Management, 21*(2), 226–232. doi:10.1108/09596110910935714

Dwivedi, M., Yadav, A., & Raghavjibhai Patel, V. (2009). The Online Destination Image of Goa. *Worldwide Hospitality and Tourism Themes, 1*(1), 25–39. doi:10.1108/17554210910949869

EBIT. (2018). Relatório Webshoppers. Retrieved from http://www.ebit.com.br/webshoppers

Echtner, C. M. (2002). The Content of Third World Tourism Marketing: A 4A Approach. *International Journal of Tourism Research, 4*(6), 413–434. doi:10.1002/jtr.401

Echtner, C. M., & Ritchie, J. B. (1991). The Meaning and Measurement of Destination Image. *Journal of Tourism Studies, 2*(2), 2–12.

Echtner, C. M., & Ritchie, J. B. (1993). The Measurement of Destination Image: An Empirical Assessment. *Journal of Travel Research, 31*(4), 3–13. doi:10.1177/004728759303100402

Echtner, C., & Jamal, T. (1997). The disciplinary dilemma of tourism studies. *Annals of Tourism Research, 24*(4), 868–883. doi:10.1016/S0160-7383(97)00060-1

Economic Times. (2019). Internet users in India to reach 627 million in 2019. Retrieved from https://economictimes.indiatimes.com:https://economictimes.indiatimes.com/tech/internet/internet-users-in-india-to-reach-627-million-in-2019-report/articleshow/68288868.cms?from=mdr

Edelman, D. C. (2010). Branding in the Digital Age. *Harvard Business Review, 88*(12), 6269.

Edosomwan, S., Prakasan, S. K., Kouame, D., Watson, J., & Seymour, T. (2011). The History of social media and its impact on business. *The Journal of Applied Management and Entrepreneurship, 16*(3), 79–91.

Eisenhardt, K. M., & Martin, J. A. (2000). Dynamic capabilities: What are they? *Strategic Management Journal, 21*(10-11), 1105–1121. doi:10.1002/1097-0266(200010/11)21:10/11<1105::AID-SMJ133>3.0.CO;2-E

Embacher, J., & Buttle, F. (1989). A Repertory Grid Analysis of Austria's Image as A Summer Vacation Destination. *Journal of Travel Research, 27*(3), 3–7. doi:10.1177/004728758902700302

Emrouzeh, M. P., Dewar, K., Fleet, G., & Bourgeois, Y. (2017). Implementing ICT for Tourists with Disabilities. *Proceedings of the 2017 International Conference on E-Education, E-Business and E-Technology* (pp. 50–53). Academic Press. 10.1145/3141151.3141152

Engel, J., Miniard, P., & Blackwell, R. (2006). Consumer behaviour (10th ed.). Thomson/South-Western.

Esser, S. K., Andreopoulos, A., Appuswamy, R., Datta, P., Barch, D., Amir, A., ... Chandra, S. (2013, August). Cognitive computing systems: Algorithms and applications for networks of neurosynaptic cores. *Proceedings of the 2013 International Joint Conference on Neural Networks (IJCNN)* (pp. 1-10). IEEE.

ET. (2018), Indians who order food online just can't get enough of this dish. Retrieved from https://economictimes.indiatimes.com/industry/cons-products/food/biryani-now-rules-indias-fast-food-landscape-and-heres-how-it-helped-emerge-a-lot-of-start-ups/articleshow/65273024.cms?from=mdr

Euromonitor International. (2018). *Top 100 City Destinations 2018.*

European Best Destinations. (2017). European Best Destinations 2017. Retrieved from https://www.europeanbestdestinations.com/best-of-europe/european-best-destinations-2017/

European Commission. (2013). 19 destinations across Europe win awards for accessible and sustainable tourism. Retrieved from http://europa.eu/rapid/press-release_MEMO-13-972_en.htm

Eye for Travel. (2017) The State of Data and Analytics in Travel Report 2017. Retrieved from http://1.eyefortravel.com/LP=16751

Fakeye, P. C., & Crompton, J. L. (1991). Image Differences Between Prospective, First-Time, And Repeat Visitors to The Lower Rio Grande Valley. *Journal of Travel Research, 30*(2), 10–16. doi:10.1177/004728759103000202

Falk, J. H. (2013). Understanding museum visitors motivations and learning. In I. Lundgaard & J. Jensen (Eds.), *Museums: Social Learning Spaces and Knowledge Producing Processes* (pp. 106–127). Copenhagen, Denmark: Kulturstyrelsen.

Fang, B., Ye, Q., Kucukusta, D., & Law, R. (2016). Analysis of the perceived value of online tourism reviews: Influence of readability and reviewer characteristics. *Tourism Management, 52*, 498–506. doi:10.1016/j.tourman.2015.07.018

Fantanti, N. M., & Suyandnya, I. W. (2015). Beyond User Gaze: How Instagram Creates Tourism Destination Brand? *Procedia: Social and Behavioral Sciences, 211*, 1089–1095. doi:10.1016/j.sbspro.2015.11.145

Farías, D. I. H., Patti, V., & Rosso, P. (2016). Irony Detection in Twitter: The Role of Affective Content. *ACM Transactions on Internet Technology, 16*(3), 19. doi:10.1145/2930663

Femenia-Serra, F., Neuhofer, B., & Ivars-Baidal, J. A. (2019). Towards a conceptualisation of smart tourists and their role within the smart destination scenario. *Service Industries Journal, 39*(2), 109–133. doi:10.1080/02642069.2018.1508458

Ferguson, R. (2010). Word of mouth and viral marketing: Taking the temperature of the hottest trends in marketing. *Journal of Consumer Marketing, 25*(3), 179–182. doi:10.1108/07363760810870671

Ferreira, J., Sousa, B. M., & Gonçalves, F. (2018). Encouraging the subsistence artisan entrepreneurship in handicraft and creative contexts. *Journal of Enterprising Communities: People and Places in the Global Economy*. doi:10.1108/JEC-09-2018-0068

Fesenmaier, D.R., & Zach, F. (2009). Innovation in Tourism: The Case of Destination Marketing Organizations.

Fesenmaier, D. R., Xiang, Z., Pan, B., & Law, R. (2011). A Framework of Search Engine Use for Travel Planning. *Journal of Travel Research, 50*(6), 587–601. doi:10.1177/0047287510385466

Fesenmaier, D. R., & Zach, F. (2009). Innovation in Tourism: The Case of Destination Marketing Organizations. *Ereview of Tourism Research, 7*(1), 27–35.

Field, A. (2009). Discovering Statistics Using SPSS (3rd ed.). London: SAGE Publications Ltd.

Figueiredo, E., Eusébio, C., & Kastenholz, E. (2012). How Diverse are Tourists with Disabilities? A Pilot Study on Accessible Leisure Tourism Experiences in Portugal. *International Journal of Tourism Research*, *14*(6), 531–550. doi:10.1002/jtr.1913

Filieri, R. (2016). What makes an online consumer review trustworthy? *Annals of Tourism Research*, *58*, 46–64. doi:10.1016/j.annals.2015.12.019

Filieri, R., Alguezaui, S., & McLeay, F. (2015). Why do travelers trust TripAdvisor? Antecedents of trust towards consumer-generated media and its influence on recommendation adoption and word of mouth. *Tourism Management*, *51*, 174–185. doi:10.1016/j.tourman.2015.05.007

Fishbein, M., & Ajzen, I. (1977). Belief, attitude, intention, and behavior: An introduction to theory and research. *Philosophy & Rhetoric*, *10*(2), 130–132.

Fleischer, A., & Felsenstein, D. (2000). Support for rural tourism: Does it make a difference? *Annals of Tourism Research*, *27*(4), 1007–1024. doi:10.1016/S0160-7383(99)00126-7

Fleischer, A., & Pizam, A. (1997). Rural tourism in Israel. *Tourism Management*, *18*(6), 367–372. doi:10.1016/S0261-5177(97)00034-4

Foddy, W. (1996). *Como Perguntar : teoria e prática da construção de perguntas em entrevistas e questionários*. Celta Editora.

Fodness, D., & Murray, B. (1997). Tourist information search. *Annals of Tourism Research*, *24*(3), 503–523. doi:10.1016/S0160-7383(97)00009-1

Fombrun, C. J. (1996). Reputation: Realizing value from the corporate image.

Fossey, E., Harvey, C., McDermott, F., & Davidson, L. (2002). Understanding and evaluating qualitative research. *The Australian and New Zealand Journal of Psychiatry*, *36*(6), 717–732. doi:10.1046/j.1440-1614.2002.01100.x PMID:12406114

Fotiadis, A. (2009). The role of tourism in rural development through a comparative analysis of a Greek and a Hungarian rural tourism area [PhD Thesis]. University of Pécs.

Fotis, J., Buhalis, D., & Rossides, N. (2011). Social Media impact on holiday travel planning: The case of the Russian and the FSU markets. *International Journal of Online Marketing*, *1*(4), 1–19. doi:10.4018/ijom.2011100101

Fotis, S., Buhalis, D., & Rossides, N. (2012). *Social media use and impact during the holiday travel planning process*. Berlin, Germany: Springer. doi:10.1007/978-3-7091-1142-0_2

Foxall, G. R. (2014). *Marketing in the Service Industries*. New York: Routledge.

Fradkin, A., Grewal, E., Holtz, D., & Pearson, M. (2014). Reporting bias and reciprocity in online reviews: Evidence from field experiments on Airbnb.

Fraenkel, J. R., Wallen, N. E., & Hyun, H. H. (2012). How to Design and Evaluate Research in Education (8th ed.). New York: Mc-Graw Hill.

Francisco, B., & Update, M. L. (2006). Cracking the social network code.

Francis, G., Dennis, N., Ison, S., & Humphreys, I. (2007). The transferability of the low-cost model to long-haul airline operations. *Tourism Management*, *28*(2), 391–398. doi:10.1016/j.tourman.2006.04.014

Franks, B. (2012). *Taming the Big Data Tidal Wave: Finding Opportunities in Huge Data Streams with Advanced Analytics*. John Wiley & Sons, Inc. doi:10.1002/9781119204275

Frazer, L., & Merrilees, B. (2012). Pioneering Asian Franchise Brands: Pho24 in Vietnam. *Journal of Marketing Channels, 19*(4), 295–309. doi:10.1080/1046669X.2012.700264

Freeman, A., Becker, S., Cummins, M., McKelroy, E., Giesinger, C., & Yuhnke, B. (2016). *NMC Horizon Report: 2016 Museum Edition*. Austin, TX: The New Media Consortium; doi:10.1111/j.2151-6952.2010.00064.x

Freitas, L., & Vieira, R. (2015). Exploring resources for sentiment analysis in Portuguese Language. *Brazilian Conference on Intelligent Systems*. Retrieved from https://www.researchgate.net/publication/300409116_Exploring_Resources_for_Sentiment_Analysis_in_Portuguese_Language

French, A. M., Luo, X. R., & Bose, R. (2017). Toward a holistic understanding of continued use of social networking tourism: A mixed-methods approach. *Information & Management, 54*(6), 802–813. doi:10.1016/j.im.2016.10.006

Friedman, L., & Friedman, H. (2008) The New Media Technologies: Overview and Research Framework. doi:10.2139srn.1116771

Fuchs, M., Höpken, W., & Lexhagen, M. (2014). Big data analytics for knowledge generation in tourism destinations - A case from Sweden. *Journal of Destination Marketing & Management, 3*(4), 198–209. doi:10.1016/j.jdmm.2014.08.002

Fuciu, M., Gorski, H., & Dumitrescu, L. (2016). Online social networking sites in personal and professional life - A marketing research. *Revista Economica, 68*(3), 57–73. Retrieved from http://economice.ulbsibiu.ro/revista.economica/archive/68305fuciu&gorski&dumitrescu.pdf

Fujita, H., & Herrera-Viedma, E. (2018). A Proposal for Sentiment Analysis on Twitter for Tourism-Based Applications. *A Proposal for Sentiment Analysis on Twitter for Tourism-Based Applications*. Academic Press.

Furtado, J. V., Pereira, A. R., Pereira, I., & Moreira, A. C. (2019). Does Theory Really Fit Real Life Situations? In Handbook of Research on Entrepreneurship, Innovation, and Internationalization (pp. 419–438). Hershey, PA: IGI Global. doi:10.4018/978-1-5225-8479-7.ch016

Furtado, N. G., Canto, R. S., de Oliveira, S. L. I., & Moretti, S. L. A. (2017). Perceptions in the use of technology for payments: A study of customer behavior in food and beverage sector. *Ágora - Revista de Divulgação Científica, 22*(2), 4–23.

Furtado, N. G., Furtado, J. V., Vieira, L. C., & Cunha, R. da S. (2019). The influence of technologic payment adoption in satisfaction: a study with restaurant costumers. *International Journal of Business Excellence*. doi:10.1504/IJBEX.2019.10020420

Furtado, N. G., & Cunha, R. da S. (2017). *A influência da adoção de technologia de meios de pagamentosna satisfação do consumidor: Um estudo com consumidores de restaurantes*. Universidade Anhembi Morumbi.

Furtado, N. G., & de Oliveira, S. L. I. (2016). *Expectativas e percepções no uso da tecnologia de pagamentos voltada ao cliente*. São Paulo: III - Encontro Dos Mestrados Profissionais Em Administração.

Gaál, Z., Szabó, L., Obermayer-Kovács, N., & Csepregi, A. (2015). Exploring the role of social media in knowledge sharing. *Electronic Journal of Knowledge Management, 13*(3), 185–197.

Gallegos, J. A. (2016). *What is User Generated Content (and Why You Should Be Using it)*. Tintup. Retrieved from https://www.tintup.com/blog/user-generated-content-definition/

Gallegos, J. A. (August 23, 2016). What is user generated content (and why you should be using it). Tintup. Retrieved from https://www.tintup.com/blog/user-generated-content-definition/

Gamboa, A. M., & Gonçalves, H. M. (2014). Customer loyalty through social networks: Lessons from Zara on Facebook. *Business Horizons, 57*(6), 709–717. doi:10.1016/j.bushor.2014.07.003

Gandhi-Arora, R., & Shaw, R. N. (2002). Visitor loyalty in sport tourism: An empirical investigation. *Current Issues in Tourism*, *5*(1), 45–53. doi:10.1080/13683500208667907

Gannon, A. (1994). Rural tourism as a factor in rural community economic development for economies in transition. *Journal of Sustainable Tourism*, *2*(1-2). 51-60

Gao, H., Tate, M., Zhang, H., Chen, S., & Liang, B. (2018). Social media ties strategy in international branding: An application of resource-based theory. *Journal of International Marketing*, *26*(3), 45–69. doi:10.1509/jim.17.0014

García-Palomares, J. C., Gutiérrez, J., Mínguez, C., Gutierrez, J., & Mínguez García, C. (2015). Identification of tourist hot spots based on social networks: A comparative analysis of European metropolises using photo-sharing services and GIS. *Applied Geography*, *63*, 408–417. doi:10.1016/j.apgeog.2015.08.002

Gardial, S. F., Clemons, D. S., Woodruff, R. B., Schumann, D. W., & Burns, M. J. (1994). Comparing consumers' recall of prepurchase and postpurchase product evaluation experiences. *The Journal of Consumer Research*, *20*(4), 548–560. doi:10.1086/209369

Garrod, B., Wornell, R., & Youell, R. (2006). Re-conceptualising rural resources as countryside capital: The case of rural tourism. *Journal of Rural Studies*, *22*(1), 117–128. doi:10.1016/j.jrurstud.2005.08.001

Gartner, W. C. (1986). Temporal Influences on Image Change. *Annals of Tourism Research*, *13*(4), 635–644. doi:10.1016/0160-7383(86)90006-X

Gartner, W. C. (1989). Tourism Image: Attribute Measurement of State Tourism Products Using Multidimensional Scaling Techniques. *Journal of Travel Research*, *28*(2), 16–20. doi:10.1177/004728758902800205

Gašpar, D., & Mabić, M. (2017). Strengths and Limitations of Social Media Analytics Tools. In Applying Predictive Analytics Within the Service Sector (pp. 198–219). Hershey, PA: IGI Global. doi:10.4018/978-1-5225-2148-8.ch012

Gedenk, K., & Neslin, S. A. (1999). The role of retail promotion in determining future brand loyalty: Its effect on purchase event feedback. *Journal of Retailing*, *75*(4), 433–459. doi:10.1016/S0022-4359(99)00018-4

Gedikoglu, Z.A. (2018). Exploring Destination Image Themes on Twitter, Before, During, and After Terror Attacks in Paris: An Application of Agenda Setting Theory.

Gelo, O., Braakmann, D., & Benetka, G. (2008). Quantitative and Qualitative Research: Beyond the Debate. *Integrative Psychological & Behavioral Science*, *42*(3), 266–290. doi:10.100712124-008-9078-3 PMID:18795385

Gerdes, J. Jr, & Stringam, B. (2008). Addressing researchers' quest for hospitality data: Mechanism for collecting data from web resources. *Tourism Analysis*, *13*(3), 309–315. doi:10.3727/108354208786094906

Getz, D. (1986). Models in Tourism Planning: Towards Integration of Theory and Practice. *Tourism Management*, *7*(1), 21–32. doi:10.1016/0261-5177(86)90054-3

Getz, D. (2007). *Event Studies: Theory, Research and Policy for Planned Events*. Oxford: Elseiver Butterworth-Heinemann.

Ghiglione, R., & Matalon, B. (2001). *O inquérito: teoria e prática* (4th ed.). Celta Editora. Retrieved from https://books.google.pt/books?id=_1v-SAAACAAJ

Ghose, A., Ipeirotis, P. G., & Li, B. (2012). Designing ranking systems for hotels on travel search engines by mining user-generated and crowdsourced content. *Marketing Science*, *31*(3), 493–520. doi:10.1287/mksc.1110.0700

Giglietto, F., Rossi, L., & Bennato, D. (2012). The Open Laboratory: Limits and Possibilities of Using Facebook, Twitter, And YouTube As A Research Data Source. *Journal of Technology in Human Services*, *30*(3-4), 145–159. doi:10.1080/15228835.2012.743797

Gikas, J., & Grant, M. M. (2013). Mobile computing devices in higher education: Student perspectives on learning with cellphones, smartphones & social media. *The Internet and Higher Education*, *19*, 18–26. doi:10.1016/j.iheduc.2013.06.002

Ginley, B. (2013). Museums: A Whole New World for Visually Impaired People. *Disability Studies Quarterly*, *33*(3), 1–15. doi:10.18061/dsq.v33i3.3761

Giorgi, A. (1997). The theory, practice, and evaluation of the phenomenological method as a qualitative research procedure. *Journal of Phenomenological Psychology*, *28*(2), 235–260. doi:10.1163/156916297X00103

Girardin, F., Blat, J., Calabrese, F., Dal Fiore, F., & Ratti, C. (2008). Digital footprinting: Uncovering tourists with user-generated content. *IEEE Pervasive Computing*, *7*(4), 36–43. doi:10.1109/MPRV.2008.71

Gligorijevic, B., & Luck, E. (2012). Engaging social customers–Influencing new marketing strategies for social media information sources. In *Contemporary research on e-business technology and strategy* (pp. 25–40). Berlin: Springer. doi:10.1007/978-3-642-34447-3_3

Godes, D., & Mayzlin, D. (2004). Using online conversations to study word-of-mouth communication. *Marketing Science*, *23*(4), 545–560. doi:10.1287/mksc.1040.0071

Goes, P. B., Lin, M., & Au Yeung, C. M. (2014). "Popularity effect" in user-generated content: Evidence from online product reviews. *Information Systems Research*, *25*(2), 222–238. doi:10.1287/isre.2013.0512

Goldblatt, J. (2010). *Special events: A new generation and the next frontier* (6th ed.). John Wiley & Sons.

Goldie, J. G. S. (2016). Connectivism: A knowledge learning theory for the digital age? *Medical Teacher*, *38*(10), 1064–1069. doi:10.3109/0142159X.2016.1173661 PMID:27128290

Goldsmith, R. E., & Horowitz, D. (2006). Measuring motivations for online opinion seeking. *Journal of Interactive Advertising*, *6*(2), 2–14. doi:10.1080/15252019.2006.10722114

González-Bailón, S. (2013). Social Science in The Era of Big Data. *Policy and Internet*, *5*(2), 147–160. doi:10.1002/1944-2866.POI328

Google. (n.d.). *Cloud Speech-to-Text: Speech-to-text conversion powered by machine learning and available for short-form or long-form audio*. Retrieved from https://cloud.google.com/speech-to-text/

Govers, R., & Go, F. M. (2003). Deconstructing Destination Image in The Information Age. *Information Technology & Tourism*, *6*(1), 13–29. doi:10.3727/109830503108751199

Govers, R., & Go, F. M. (2004). Projected Destination Image Online: Website Content Analysis of Pictures and Text. *Information Technology & Tourism*, *7*(2), 73–89. doi:10.3727/1098305054517327

Govers, R., Go, F. M., & Kumar, K. (2007). Promoting Tourism Destination Image. *Journal of Travel Research*, *46*(1), 15–23. doi:10.1177/0047287507302374

Govers, R., Go, F. M., & Kumar, K. (2007). Virtual Destination Image A New Measurement Approach. *Annals of Tourism Research*, *34*(4), 977–997. doi:10.1016/j.annals.2007.06.001

Grafström, M., & Falkman, L. L. (2017). Everyday narratives: CEO rhetoric on Twitter. *Journal of Organizational Change Management*, *30*(3), 312–322. doi:10.1108/JOCM-10-2016-0197

Graham, M. E., & Bansal, P. (2007). Consumers' willingness to pay for corporate reputation: The context of airline companies. *Corporate Reputation Review*, *10*(3), 189–200. doi:10.1057/palgrave.crr.1550052

Gravetter, F. J., & Forzano, L.-A. B. (2017). *Research Methods for the Behavioral Sciences* (6th ed.). Belmont, CA: Cengage Learning.

Grégoire, Y., Laufer, D., & Tripp, T. M. (2010). A comprehensive model of customer direct and indirect revenge: Understanding the effects of perceived greed and customer power. *Journal of the Academy of Marketing Science*, *38*(6), 738–758. doi:10.100711747-009-0186-5

Gregory, A., Wang, Y., & DiPietro, R. B. (2010). Towards a functional model of website evaluation: A case study of casual dining restaurants. *Worldwide Hospitality and Tourism Themes*, *2*(1), 68–85. doi:10.1108/17554211011012603

Gretzel, U., & Yoo, K.H. (2008) Use and impact of online travel reviews. In *Information and Communication Technologies in Tourism 2008* (pp. 35-46). Springer.

Gretzel, U. (2006). Consumer Generated Content–Trends and Implications for Branding. *Ereview of Tourism Research*, *4*(3), 9–11.

Gretzel, U. (2011). Intelligent systems in tourism: A social science perspective. *Annals of Tourism Research*, *38*(3), 757–779. doi:10.1016/j.annals.2011.04.014

Gretzel, U., Kang, M., & Lee, W. (2008). Differences in consumer-generated media adoption and use: A cross-national perspective. *Journal of Hospitality & Leisure Marketing*, *17*(1-2), 99–120. doi:10.1080/10507050801978240

Gretzel, U., Sigala, M., Xiang, Z., & Koo, C. (2015a). Smart tourism: Foundations and developments. *Electronic Markets*, *25*(3), 179–188. doi:10.100712525-015-0196-8

Gretzel, U., Werthner, H., Koo, C., & Lamsfus, C. (2015b). Conceptual Foundations for Understanding Smart Tourism Ecosystems. *Computers in Human Behavior*, *50*, 558–563. doi:10.1016/j.chb.2015.03.043

Gretzel, U., & Yoo, K. H. (2008). Use and impact of online travel reviews. *Information and Communication Technologies in Tourism*, *2*(1), 35–46.

Gretzel, U., Yuan, Y., & Fensenmaier, D. (2000). Preparing for the New Economy: Advertising Strategies and Change in Destination Marketing organizations. *Journal of Travel Research*, *39*(2), 149–156. doi:10.1177/004728750003900204

Groeger, L., & Buttle, F. (2014). Word-of-mouth marketing influence on offline and online communications: Evidence from case study research. *Journal of Marketing*.

Grönroos, C. (1994). From marketing mix to relationship marketing: Towards a paradigm shift in marketing. *Management Decision*, *32*(2), 4–20. doi:10.1108/00251749410054774

Gruzd, A., Lannigan, J., & Kevin Quigley, K. (2018). Examining government cross-platform engagement in social media: Instagram vs Twitter and the big lift project. *Government Information Quarterly*, *35*(4), 579–587. doi:10.1016/j.giq.2018.09.005

Guangrui, Z. (1989). Ten years of Chinese tourism: Profile and assessment. *Tourism Management*, *10*(March), 51–62. doi:10.1016/0261-5177(89)90034-4

Gudmundsson, S. (2019). Limits to the Low-Cost Niche? Finding Sustainable Strategies for Low-Cost Long-Haul Airlines. Retrieved from https://www.researchgate.net/profile/Sveinn_Gudmundsson3/publication/278966773_Limits_to_the_low-cost_niche_Sustainable_strategies_for_low-cost_long-haul_airlines/links/5d00a73792851c874c5fcac3/Limits-to-the-low-cost-niche-Sustainable-strategies-for-low-c

Guest, G., MacQueen, K. M., & Namey, E. E. (2011). *Applied Thematic Analysis*. Sage Publications.

Guiver, J., & Stanford, D. (2014). Why destination visitor travel planning falls between the cracks. *Journal of Destination Marketing & Management*, *3*(3), 140–151. doi:10.1016/j.jdmm.2013.11.005

Gu, J.-C., Lee, S.-C., & Suh, Y.-H. (2009). Y.-H. (2009). Determinants of behavioral intention to mobile banking. *Expert Systems with Applications*, *36*(9), 11605–11616. doi:10.1016/j.eswa.2009.03.024

Gujarati, D. (2003). *Basic Econometrics* (4th ed.). Boston: McGraw-Hill.

Gummesson, E. (1994). Making relationship marketing operational. *International Journal of Service Industry Management*, *5*(5), 5–20. doi:10.1108/09564239410074349

Gummesson, E. (2017). From relationship marketing to total relationship marketing and beyond. *Journal of Services Marketing*, *31*(1), 16–19. doi:10.1108/JSM-11-2016-0398

Gunn, C. A. (1972). *Vacationscape: Designing Tourist Regions (Vol. 1.)*. Austin: Bureau of Business Research, University of Texas at Austin.

Gupta, S., & Vajic, M. (1999). The contextual and dialectical nature of experiences. In J. Fitzsimmons & M. Fitzsimmons (Eds.), New Service Development (pp. 33-51). Thousand Oaks, CA: Sage.

Gupta, H. (2018). Evaluating service quality of airline industry using hybrid best worst method and VIKOR. *Journal of Air Transport Management*, *68*, 35–47. doi:10.1016/j.jairtraman.2017.06.001

Gwinner, K. P., & Eaton, J. (1999). Building brand image through event sponsorship: The role of image transfer. *Journal of Advertising*, *28*(4), 47–57. doi:10.1080/00913367.1999.10673595

Gyimóthy, S. (2013). Symbolic convergence and tourism social media. In *Tourism social media: Transformations in identity, community and culture* (pp. 55–71). Emerald Group Publishing Limited. doi:10.1108/S1571-5043(2013)0000018006

Ha, A. (2015). An Experiment: Instagram Marketing Techniques and Their Effectiveness. *Communication Studies*. Retrieved from http://digitalcommons.calpoly.edu/comssp/185

Ha, E. Y., & Lee, H. (2018). Projecting service quality: The effects of social media reviews on service perception. *International Journal of Hospitality Management*, *69*, 132–141. doi:10.1016/j.ijhm.2017.09.006

Hage, J. T. (1999). Organizational Innovation and organizational change. *Annual Review of Sociology*, *25*(1), 597–622. doi:10.1146/annurev.soc.25.1.597

Hair, J. F., Wolfinbarger, M. F., Ortinau, D. J., & Bush, R. P. (2010). Essentials of Marketing Research (2nd ed.). New York: McGraw-Hill/Irwin.

Hair, J. F. (2010). *Multivariate data analysis*. Prentice Hall.

Hair, J. F., Black, W. C., Babin, B. J., & Anderson, R. E. (2014). *Multivariate Data Analysis* (7th ed.)., Retrieved from www.pearsoned.co.uk

Hair, J., Bush, R., & Ortinau, D. (2003). *Marketing Research: Within a Changing Information Environment* (2nd ed.). New York: McGraw-Hill/Irwin.

Hajli, M. N., & Lin, X. (2014). Developing Tourism Education Through Social Media. *Tourism Planning & Development*, *11*(4), 405–414. doi:10.1080/21568316.2014.883426

Halbwachs, M. (1992). *On Collective Memory*. University of Chicago Press.

Hall, M. C., & Williams, A. M. (2008). *Tourism Innovation*. New York: Routledge. doi:10.4324/9780203938430

Han, H., Lee, K. S., Chua, B. L., Lee, S., & Kim, W. (2019). Role of airline food quality, price reasonableness, image, satisfaction, and attachment in building re-flying intention. *International Journal of Hospitality Management*, *80*, 91–100. doi:10.1016/j.ijhm.2019.01.013

Hanna, R., Rohm, A., & Crittenden, V. L. (2011). We're all connected: The power of the social media ecosystem. *Business Horizons*, *54*(3), 265–273. doi:10.1016/j.bushor.2011.01.007

Han, W., McCabe, S., Wang, Y., & Chong, A. Y. L. (2018). Evaluating user-generated content in social media: An effective approach to encourage greater proenvironmental behavior in tourism? *Journal of Sustainable Tourism*, *26*(4), 600–614. doi:10.1080/09669582.2017.1372442

Harris, L., & Rae, A. (2009). Social networks: The future of marketing for small business. *The Journal of Business Strategy*, *30*(5), 24–31. doi:10.1108/02756660910987581

Havell, R., & Ross, J. (1819). *A voyage of discovery… made for the purpose of exploring Baffin's Bay and inquiring into possibility of North-West passage*. London: John Murray.

Hawk, A. T., van den Eijnden, R. J. J. M., van Lissa, C. J., & ter Bogt, T. F. M. (2019). Narcissistic adolescents' attention-seeking following social rejection: Links with social media disclosure, problematic social media use, and smartphone stress. *Computers in Human Behavior*, *92*, 65–75. doi:10.1016/j.chb.2018.10.032

Hayhoe, S. (2017). *Blind Visitor Experiences at Art Museums*. London: Rowman & Littlefield.

Hays, S., Page, S. J., & Buhalis, D. (2013). Social media as a destination marketing tool: Its use by national tourism organisations. *Current Issues in Tourism*, *16*(3), 211–239. doi:10.1080/13683500.2012.662215

Helm, S. (2010). Viral marketing–Establishing customer relationships by word-of mouth. *Electronic Markets*, *10*(3), 158–161. doi:10.1080/10196780050177053

Hendry, J. (2000). Strategic decision making, discourse, and strategy as social practice. *Journal of Management Studies*, *37*(7), 955–978. doi:10.1111/1467-6486.00212

Henning-Thurau, T., Gwinner, K.P., Walsh, G., & Gremler, D.D. (2004). El boca a boca electrónico a través de las plataformas de opinión de los consumidores: ¿qué motiva a los consumidores a articularse en internet? *Revista de marketing interactivo, 18*(1), 38-52.

Henning-Thurun, T., Gwinner, K. P., Walsh, G., & Gremler, D. D. (2004). Electronic word of mouth via consumer-opinion platform: What motivates consumers to articulate themselves n the Internet? *Journal of Interactive Marketing*, *18*(1). doi:10.1002/dir.10073

He, W., Zha, S., & Li, L. (2013). Social media competitive analysis and text mining: A case study in the pizza industry. *International Journal of Information Management*, *33*(3), 464–472. doi:10.1016/j.ijinfomgt.2013.01.001

Hipp, C., & Grupp, H. (2000). Innovation in the service sector: The demand for service-specific innovation measurement concepts and typologies. *Research Policy*, *34*, 517–535.

Hirschman, E. C., & Holbrook, M. B. (1982). Hedonic consumption: Emerging concepts, methods and propositions. *Journal of Marketing*, *48*(3), 92–101. doi:10.1177/002224298204600314

Hjalager, A. (2015). 100 innovations that transformed tourism. *Journal of Travel Research*, *54*(1), 3–21. doi:10.1177/0047287513516390

Hjalager, A. M. (1996). Tourism and the Environment: The Innovation Connection. *Journal of Sustainable Tourism*, *4*(4), 201–217. doi:10.1080/09669589608667268

Hjalager, A. M. (1997). Innovation patterns in sustainable tourism: An analytical typology. *Tourism Management*, *18*(1), 35–41. doi:10.1016/S0261-5177(96)00096-9

Hjalager, A. M. (2002). Repairing innovation defectiveness in tourism. *Tourism Management*, *23*(5), 465–474.

Hjalager, A. M. (2002). Reparing innovation defectiveness in tourism. *Tourism Management*, *23*(5), 465–474. doi:10.1016/S0261-5177(02)00013-4

Ho, C. I., Lin, Y. C., Yuan, Y. L., & Chen, M. C. (2016). Pre-trip tourism information search by smartphones and use of alternative information channels: A conceptual model. *Cogent Social Science*, *2*, 1136100.

Ho, C.-I., Lin, M.-H., & Chen, H.-M. (2012). Web users' behavioural patterns of tourism information search: From online to offline. *Tourism Management*, *33*(6), 1468–1482. doi:10.1016/j.tourman.2012.01.016

Hoelzl, B., Pechlaner, H., & Laesser, C. (2005). Imitation processes of SMEs-A special form of innovation?

Hoffmann, C. A. K. (2017). O comportamento do consumidor, o boca a boca eletrônico (eBAB) e as redes sociais online: Conhecimento atual e direções futuras. *Revista FAE, Curitiba*, *20*(1), 27–41.

Holloway, C. (1995). *Marketing for tourism*. London: Addison Wesley Longman.

Hooley, G. J., Greenley, G. E., Cadogan, J. W., & Fahy, J. (2005). The performance impact of marketing resources. *Journal of Business Research*, *58*(1), 18–27. doi:10.1016/S0148-2963(03)00109-7

Horster, E., & Gottschalk, C. (2012). Computer-assisted webnography: A new approach to online reputation management in tourism. *Journal of Vacation Marketing*, *18*(3), 229–238. doi:10.1177/1356766712449369

Hosteltur. (2017). Redes sociais: sete tendências a considerar. Retrieved from https://www.hosteltur.com/124397_redes-sociales-siete-tendencias-tener-cuenta.html?fromNewsletter=com&fromNewsletterDate=2017-1003&internalFormatClick=diario-20171003

Hoy, M. B. (2018). Alexa, Siri, Cortana, and More: An Introduction to Voice Assistants. *Medical Reference Services Quarterly*, *37*(1), 81–88. doi:10.1080/02763869.2018.1404391 PMID:29327988

Hsieh, A., & Tsai, C. (2009). Does national culture really matter? Hotel service perceptions by Taiwan and American tourists. *International Journal of Culture, Tourism and Hospitality Research*, *3*(1), 54–69. doi:10.1108/17506180910940342

Hsu, C.-L., & Lin, J. C.-C. (2008). Acceptance of blog usage: The roles of technology acceptance, social influence and knowledge sharing motivation. *Information & Management*, *45*(1), 65–74. doi:10.1016/j.im.2007.11.001

Hsu, C., & Lin, J. (2015). What drives purchase intention for paid mobile apps? An expectation confirmation model with perceived value. *Electronic Commerce Research and Applications*, *14*(1), 46–57. doi:10.1016/j.elerap.2014.11.003

Huang, C. (2011). Facebook and Twitter key to Arab Spring uprisings: report. *The National* (Vol. 6). Retrieved from http://www.thenational.ae/news/uae-news/facebook-and-twitter-key-to-arab-spring-uprisings-report

Huang, C. D., Goo, J., Nam, K., & Yoo, C. W. (2017). Smart tourism technologies in travel planning: The role of exploration and exploitation. *Information & Management*, *54*(6), 757–770. doi:10.1016/j.im.2016.11.010

Huang, J., Huang, C., & Wu, S. (1996). National character and response to unsatisfactory hotel service. *International Journal of Hospitality Management*, *15*(3), 229–243. doi:10.1016/S0278-4319(96)00009-6

Huang, S. H., Liu, P., Mokasdar, A., & Hou, L. (2013). Additive manufacturing and its societal impact: A literature review. *International Journal of Advanced Manufacturing Technology*, *67*(5-8), 1191–1203. doi:10.100700170-012-4558-5

Huang, Y., Basu, C., & Hsu, M. (2010). Exploring motivations of travel knowledge sharing on social network sites: An empirical investigation of U.S. college students. *Journal of Hospitality Marketing & Management*, *19*(7), 717–734. do i:10.1080/19368623.2010.508002

Huck, S. W. (2012). Reading Statistics and Research (6th ed.). Boston: Pearson Education.

Hudson, S., Roth, M. S., Madden, T. J., & Hudson, R. (2015). The effects of social media on emotions, brand relationship quality, and word of mouth: An empirical study of music festival attendees. *Tourism Management*, *47*, 68–76. doi:10.1016/j.tourman.2014.09.001

Hu, H. H., Kandampully, J., & Juwaheer, T. D. (2009). Relationships and Impacts of Service Quality, Perceived Value, Customer Satisfaction, and Image: An Empirical Study. *Service Industries Journal*, *29*(2), 111–125. doi:10.1080/02642060802292932

Hunter, W. C. (2013). China's Chairman Mao: A Visual Analysis of Hunan Province Online Destination Image. *Tourism Management*, *34*, 101–111. doi:10.1016/j.tourman.2012.03.017

Hunter, W. C. (2016). The Social Construction of Tourism Online Destination Image: A Comparative Semiotic Analysis of The Visual Representation of Seoul. *Tourism Management*, *54*, 221–229. doi:10.1016/j.tourman.2015.11.012

Hunt, J. D. (1975). Image as A Factor in Tourism Development. *Journal of Travel Research*, *13*(3), 1–7. doi:10.1177/004728757501300301

Hunt, J., & Truong, D. (2019). Low-fare flights across the Atlantic: Impact of low-cost, long-haul trans-Atlantic flights on passenger choice of Carrier. *Journal of Air Transport Management*, *75*, 170–184. doi:10.1016/j.jairtraman.2018.12.005

Husain, T., Li, B., & Wang, D. (2018). What factors influence the sustainable tour process in social media usage? Examining a rural mountain region in Pakistan. *Sustainability*, *10*(7), 2220. doi:10.3390u10072220

Hutchinson, A. (2018, November 13) Social media today: Facebook releases new report into travel booking trends online [Infographic]. Socialmediatoday. Retrieved from https://www.socialmediatoday.com/news/facebook-releases-new-report-into-travel-booking-trends-online-infographic/542015/

Hutchinson, A. (2019, June 27) Social media today: The state of Instagram marketing 2019 part 4: Instagram analytics. Socialmediatoday. Retrieved from https://www.socialmediatoday.com/news/the-state-of-instagram-marketing-2019-part-4-instagram-analytics/557400/

Hutchinson, R. S., & Eardley, A. F. (2018). Museum audio description: The problem of textual fidelity. *Perspectives: Studies in Translation Theory and Practice*, *27*(1), 42–57. doi:10.1080/0907676X.2018.1473451

Huyssen, A. (2003). *Present pasts: Urban palimpsests and the politics of memory*. Stanford University Press.

IBM. (2015). What is Cloud Computing? doi:10.1007/978-3-642-13067-0_35

IBM. (2017). Vacation with IBM Watson and discover a whole new world. Retrieved from https://www.ibm.com/blogs/client-voices/vacation-ibm-watson-discover-new-world/

IDC. (2018). *Los retos de la gestión de datos*. Retrieved from https://custom.cvent.com/1E8AD1B771DA4B029B78FF1784749EF5/files/c595061048184ba49482223d066ac565.pdf

INE. (2012). *Censos 2011 Resultados Definitivos - Portugal*. Lisbon, Portugal: Instituto Nacional de Estatística.

INE. (2019). *Destaque: Crescimento nos principais indicadores do setor do alojamento turístico*. Instituto Nacional de Estatística. Retrieved from www.ine.pt

Instituto Brasileiro de Geografia e Estatística (IBGE). (2017). Retrieved from https://cidades.ibge.gov.br/xtras/perfil.php?lang=&codmun=431730&search=||infogr%E1ficos:-informa%E7%F5es-completas

Inversini, A., Marchiori, E., Dedekind, C., & Cantoni, L. (2010). Applying a conceptual framework to analyze online reputation of tourism destinations. *Information and communication technologies in tourism*, 321-332.

Ioannides, D., & Petersen, T. (2003). Tourism non-entrepreneurship in peripheral destinations: a case study of small and medium enterprises on Bornholm, Denmark. Tourism Geographies, 5(4), 408-435. In Hall, M.C. & Williams, A.M. (2008) (Eds.) Tourism Innovation. New York: Routledge.

Ip, C., Leung, R., & Law, R. (2011). Progress and development of information and communication technologies in hospitality. *International Journal of Contemporary Hospitality Management*, *23*(4), 533–551. doi:10.1108/09596111111130029

IPM. (2004). *Museus e Acessibilidade. Colecção Temas de Museologia*. Lisbon, Portugal: Instituto Português de Museus.

Israeli, A., Lee, S. A., & Karpinski, A. C. (2017). Investigating the Dynamics and the Content of Customers' Social Media Reporting after a Restaurant Service Failure. *Journal of Hospitality Marketing & Management*, *26*(6), 606–626. doi:10.1080/19368623.2017.1281193

Ivanov, S., Webster, C., & Berezina, K. (2017). Adoption of robots and service automation by tourism and hospitality companies. *Turismo & Desenvolvimento, 27*(28), 1501–1517. Retrieved from https://www.researchgate.net/publication/322635104

Ivanović, S., Milojica, V., & Roblek, V. (2017). A Holistic Approach to Innovations in Tourism. *Proceedings of the International Conference on Economic and Social Studies (ICESoS'16)* (pp. 367–380). Academic Press.

Jackson, J., & Murphy, P. (2006). Clusters in regional tourism – an Australian case. *Annals of Tourism Research*, *33*(4), 1018–1035. doi:10.1016/j.annals.2006.04.005

Jacobsen, J. K. S., & Munar, A. M. (2012). Tourist information search and destination choice in a digital age. *Tourism Management Perspectives*, *1*(1), 39–47. doi:10.1016/j.tmp.2011.12.005

Jacob, Tintoré, J., Aguiló, E., Bravo, A., & Mulet, J. (2003). Innovation in the tourism sector. Results from a pilot study in the Balearic Islands. *Tourism Economics*, *9*(3), 279–295. doi:10.1177/135481660300900303

Jafari, J., & Aaser, D. (1988). Tourism as the subject of doctoral dissertations. *Annals of Tourism Research*, *15*(3), 407–429. doi:10.1016/0160-7383(88)90030-8

Jain, D., Jain, A., Paul, R., Komarika, A., & Balakrishnan, M. (2013). A path-guided audio based indoor navigation system for persons with visual impairment. *Proceedings of the 15th International ACM SIGACCESS Conference on Computers and Accessibility* (pp. 33:1-33:2). New York, NY: ACM. 10.1145/2513383.2513410

Jalilvand, M. R., Esfahani, S. S., & Samiei, N. (2011). Electronic word-of-mouth: Challenges and opportunities. *Procedia Computer Science*, *3*, 42–46. doi:10.1016/j.procs.2010.12.008

Jalilvand, M. R., & Samiei, N. (2012). The impact of electronic word of mouth on a tourism destination choice: Testing the theory of planned behavior (TPB). *Internet Research: Electronic Networking Applications and Policy*, *22*(5), 591–612. doi:10.1108/10662241211271563

Jancenelle, V., Storrud-Barnes, S., & Javalgi, R. (2017). Augmented reality, virtual reality and 3D printing for the co-creation of value for the visitor experience at cultural heritage places. *Management Research Review*, *40*(3), 352–367.

Jansen, B. J., Zhang, M. M., Sobel, K., & Chowdury, A. (2009). Twitter power: Tweets as electronic word of mouth. *Journal of the American Society for Information Science and Technology*, *60*(11), 2169–2188. doi:10.1002/asi.21149

Jasperneite, J. (2013). *Was hinter Begriffen wie Industrie 4.0 steckt*. Internet Und Automation.

Jazdi, N. (2014). Cyber physical systems in the context of Industry 4.0. *Proceedings of the 2014 IEEE International Conference on Automation, Quality and Testing, Robotics, AQTR 2014*. 10.1109/AQTR.2014.6857843

Jeacle, I., & Carter, C. (2011). In TripAdvisor we trust: Rankings, calculative regimes and abstract systems. *Accounting, Organizations and Society*, *36*(4-5), 293–309. doi:10.1016/j.aos.2011.04.002

Jenkins, A. S., & Mazer, J. P. (2018). #Notokay: Stories of Sexual Assault in The Midst of the 2016 US Presidential Election. *Qualitative Research Reports in Communication*, *19*(1), 9–17. doi:10.1080/17459435.2017.1404487

Jenkins, H. (2010). Transmedia storytelling and entertainment: An annotated syllabus. Continuum. *Journal of Media & Cultural Studies*, *24*(6), 943–958. doi:10.1080/10304312.2010.510599

Jennings, G. (2010). *Tourism Research* (2nd ed.). John Wiley & Sons.

Jensen, J. M. (2009). Travellers' intentions to purchase travel products online: The role of shopping orientation. In Advances in tourism economics (pp. 203-215). Physica-Verlag HD.

Jensen, R. (2001). *The dream society*. London: McGraw-Hill.

Jeong, M., & Jeon, M. (2008). Customer Reviews of Hotel Experiences through Consumer Generated Media (CGM). *Journal of Hospitality & Leisure Marketing*, *17*(1-2), 121–138. doi:10.1080/10507050801978265

Jiang, H., & Zhang, Y. (2016). An investigation of service quality, customer satisfaction and loyalty in China's airline market. *Journal of Air Transport Management*, *57*, 80–88. doi:10.1016/j.jairtraman.2016.07.008

Jiang, J., Gretzel, U., & Law, R. (2014). Influence of Star Rating and Ownership Structure on Brand Image of Mainland China Hotels. *Journal of China Tourism Research*, *10*(1), 69–94. doi:10.1080/19388160.2013.870506

Jin, X., Gallagher, A., Cao, L., Luo, J., & Han, J. (2010). The wisdom of social multimedia: using Flickr for prediction and forecast. *Proceedings of the International Conference on Multimedia* (pp. 1235-1244). Academic Press. 10.1145/1873951.1874196

Joachimsthaler, E., & Aaker, D. A. (1997). Building brands without mass media. *Harvard Business Review*, *75*(1). PMID:10174453

Johanesson, J. A., Olsen, B., & Lumpkin, G. T. (2001). Innovation as newness: What is new, how is new, and new to whom? *European Journal of Innovation Management*, *4*(1), 20–31.

Johnson, J. (2018). Sensory: Please touch the art. *Art Education*, *71*(1), 12–15. https://doi.or/10.1080/00043125.2018.1389580. doi:10.1080/00043125.2018.1389580

Johns, R., & Perrott, B. (2008). The impact of internet banking on business-customer relationships (are you being self-served?). *International Journal of Bank Marketing*, *26*(7), 465–482. doi:10.1108/02652320810913846

Jørgensen, M. T., King, B. E., & Law, R. (2018). The impacts of China's policy-making and legislation on outbound tourism–perspectives from long-haul intermediaries. *Journal of Policy Research in Tourism, Leisure & Events*, 1–13. doi:10.1080/19407963.2018.1505104

Jovicic, D. Z. (2019). From the traditional understanding of tourism destination to the smart tourism destination. *Current Issues in Tourism*, *22*(3), 276–282. doi:10.1080/13683500.2017.1313203

Junco, R. (2012). The relationship between frequency of Facebook use, participation in Facebook activities, and student engagement. *Computers & Education*, *58*(1), 162–171. doi:10.1016/j.compedu.2011.08.004

Jung, T., tom Dieck, M. C., Lee, H., & Chung, N. (2016). Effects of Virtual Reality and Augmented Reality on Visitor Experiences in Museum. In Information and Communication Technologies in Tourism 2016. Springer. doi:10.1007/978-3-319-28231-2_45

Kanbara, F. K. (2010). *O Comportamento do Consumidor na Hotelaria Corporativa e Lazer – Rede Mercure*. Rio de Janeiro: Monografia. Universidade Cândido Mendes.

Kane, G. C. (2015). Enterprise social media: Current capabilities and future possibilities. *MIS Quarterly Executive*, *14*(1).

Kang, M., & Schuett, M. A. (2013). Determinants of sharing travel experiences in social media. *Journal of Travel & Tourism Marketing*, *30*(1-2), 1–15. doi:10.1080/10548408.2013.751237

Kankaew, K. (2013). Importance-Performance Analysis in Airlines Service Quality: A Case Study of legacy Airlines in Thailand. *Proceedings of the International Conference on Tourism, Transport and Logistics* (pp. 363-369). Paris: UP Organizer and Publication Co., Ltd.

Kannan, P. K., & Li, H. A. (2017). Digital marketing: A framework, review and research agenda. *International Journal of Research in Marketing*, *34*(1), 22–45. doi:10.1016/j.ijresmar.2016.11.006

Kanter, R. M. (1983). The Change Masters.

Kaplan, A. M., & Haenlein, M. (2009). The fairyland of Second Life: Virtual social worlds and how to use them. *Business Horizons*, *52*(6), 563–572. doi:10.1016/j.bushor.2009.07.002

Kaplan, A. M., & Haenlein, M. (2010). Users of the world, unite! The challenges and opportunities of social media. *Business Horizons*, *53*(1), 59–68. doi:10.1016/j.bushor.2009.09.003

Kapoor, A. P., & Vij, M. (2018). Technology at the dinner table: Ordering food online through mobile apps. *Journal of Retailing and Consumer Services*, *43*, 342–351. doi:10.1016/j.jretconser.2018.04.001

Kaur, K., & Kaur, R. (2016). Internet of Things to promote Tourism : An insight into Smart Tourism. *International Journal of Recent Trends in Engineering and Research*.

Kavoura, A. (2014). Advertizing activities in social media and the creation of a community belonging in the digital era. *Zeszyty Naukowe Małopolskiej Wyższej Szkoły Ekonomicznej w Tarnowie*, *25*(2), 97–106.

Kavoura, A., & Stavrianea, A. (2015). Following and Belonging to an Online Travel Community in Social Media, its Shared Characteristics and Gender Differences. *Procedia: Social and Behavioral Sciences*, *175*, 515–521. doi:10.1016/j.sbspro.2015.01.1231

Kennedy-Eden, H., & Gretzel, U. (2012). A taxonomy of mobile applications in tourism. *Ereview of Tourism Research*, *10*(2), 47–50.

Kietzmann, J. H., Hermkens, K., McCarthy, I. P., & Silvestre, B. S. (2011). Social media? Get serious! Understanding the functional building blocks of social media. *Business Horizons*, *54*(3), 241–251. doi:10.1016/j.bushor.2011.01.005

Kim, J., & Connolly, D. (2012). Hospitality in the age of empowerment: First annual customer technology study. *Hospitality Technology*. Retrieved from https://luiscodo.files.wordpress.com/2012/09/customer-engagement2012.pdf

Kim, D., Park, J., & Morrison, A. (2008). A model of traveller acceptance of mobile technology. *International Journal of Tourism Research*, *10*(5), 393–407. doi:10.1002/jtr.669

Kim, J., & Fesenmaier, D. R. (2017). Sharing Tourism Experiences: The Posttrip Experience. *Journal of Travel Research*, *56*(1), 28–40. doi:10.1177/0047287515620491

Kim, J., & Tussyadiah, I. P. (2013). Social networking and social Support in tourism experience: The moderating role of online Self-presentation strategies. *Journal of Travel & Tourism Marketing*, *30*(1-2), 78–92. doi:10.1080/10548408.2013.751220

Kim, S. S., & Petrick, J. F. (2005). Residents' perceptions on impacts of the FIFA 2002 World Cup: The case of Seoul as a host city. *Tourism Management*, *26*(1), 25–38. doi:10.1016/j.tourman.2003.09.013

Kim, S.-E., Lee, K. Y., Shin, S., & Yang, S.-B. (2017). Effects of tourism information quality in social media on destination image formation: The case of Sin Weibo. *Information & Management*, *54*(6), 687–702. doi:10.1016/j.im.2017.02.009

Kim, W. G., Li, J. J., & Brymer, R. A. (2016). The impact of social media reviews on restaurant performance: The moderating role of excellence certificate. *International Journal of Hospitality Management*, *55*, 41–51. doi:10.1016/j.ijhm.2016.03.001

Kim, W., Jun, H. M., Walker, M., & Drane, D. (2015). Evaluating the perceived social impacts of hosting large-scale sport tourism events: Scale development and validation. *Tourism Management*, *48*, 21–32. doi:10.1016/j.tourman.2014.10.015

Kim, Y. G., Suh, B. W., & Eves, A. (2010). The relationships between food-related personality traits, satisfaction, and loyalty among visitors attending food events and festivals. *International Journal of Hospitality Management*, *29*(2), 216–226. doi:10.1016/j.ijhm.2009.10.015

Kiral'ova, A., & Pavliceka, A. (2015). Development of social media strategies in tourism destination. *Procedia: Social and Behavioral Sciences*, *175*, 358–366. doi:10.1016/j.sbspro.2015.01.1211

Kisilevich, S., Krstajic, M., Keim, D., Andrienko, N., & Andrienko, G. (2010). Event-based analysis of people's activities and behavior using Flickr and Panoramio geotagged photo collections. *Proceedings of the International Conference on Information Visualisation* (pp. 289-296). Academic Press. 10.1109/IV.2010.94

Kislali, H., Kavaratzis, M., & Saren, M. (2016). Rethinking Destination Image Formation. *International Journal of Culture, Tourism and Hospitality Research*, *10*(1), 70–80. doi:10.1108/IJCTHR-05-2015-0050

Klein, S., Avery, M., Adams, G., Pollard, S., & Simske, S. (2014). From scan to print: 3D printing as a means for replication. Proceedings of the NIP & Digital Fabrication Conference. Academic Press.

Költringer, C., & Dickinger, A. (2015). Analyzing destination branding and image from online sources: A web content mining approach. *Journal of Business Research*, *68*(9), 1836–1843. doi:10.1016/j.jbusres.2015.01.011

Konkel, M., Leung, V., Ullmer, B., & Hu, C. (2004). Tagaboo: A collaborative children's game based upon wearable RFID technology. *Personal and Ubiquitous Computing*, *8*(5). doi:10.100700779-004-0302-y

Koo, C., Park, J., & Lee, J. N. (2017). Smart tourism: Traveler, business, and organizational perspectives. doi:10.1016/j.im.2017.04.005

Koo, C., Shin, S., Kim, K., Kim, C., & Chung, N. (2013). Smart tourism of the Korea: A case study. *Proceedings of the Pacific Asia conference on information system*. AIS Electronic Library. Retrieved from http://aisel.aisnet.org/pacis2013

Kotler, P., & Armstrong, G. (2007). *Princípios de Marketing (12a Edição)*. São Paulo: Pearson Prentice Hall.

Kotler, P., Kartajaya, H., & Setiawan, I. (2017). *Marketing 4.0 Moving from Traditional to Digital*. New Jersey: Wiley.

Kotler, P., Kartajaya, H., & Setiawan, I. (2017). *Marketing 4.0: Moving from Traditional to Digital*. New Jersey: John Wiley & Son, Inc.

Kounavis, C. D., Kasimati, A. E., & Zamani, E. D. (2012). Enhancing the tourism experience through mobile augmented reality: Challenges and prospects. *International Journal of Engineering Business Management*, *4*, 10. doi:10.5772/51644

Kourouthanassis, P. E., Mikalef, P., Pappas, I. O., & Kostagiolas, P. (2017). Explaining traveller online information satisfaction: A complexity theory approach on information needs, barriers, sources and personal characteristics. *Information & Management*, *54*(6), 814–824. doi:10.1016/j.im.2017.03.004

Kovar, J., Mouralova, K., Ksica, F., Kroupa, J., Andrs, O., & Hadas, Z. (2016). Virtual Reality in Context of Industry 4.0. *Proceedings of the 2016 17th International Conference on Mechatronics - Mechatronika (Me) 2016*. Academic Press.

Kozinets, R. V., De Valck, K., Wojnicki, A. C., & Wilner, S. J. S. (2010). Networked narratives: Understanding word-of-mouth marketing in online communities. *Journal of Marketing*, *74*(2), 71–89. doi:10.1509/jmkg.74.2.71

Krush, M. T., Agnihotri, R., Trainor, K. J., & Nowlin, E. L. (2013). Enhancing organizational sensemaking: An examination of the interactive effects of sales capabilities and marketing dashboards. *Industrial Marketing Management*, *42*(5), 824–835. doi:10.1016/j.indmarman.2013.02.017

Kuflik, T., Wecker, A. J., Lanir, J., & Stock, O. (2015). An integrative framework for extending the boundaries of the museum visit experience: Linking the pre, during and post visit phases. *Information Technology & Tourism*, *15*(1), 17–47. doi:10.100740558-014-0018-4

Kukanja, M., Gomezelj Omerzel, D., & Kodrič, B. (2017). Ensuring restaurant quality and guests' loyalty: An integrative model based on marketing (7P) approach. *Total Quality Management & Business Excellence*, *28*(13–14), 1509–1525. doi:10.1080/14783363.2016.1150172

Kurtulmuşoğlu, F. B., Can, G. F., & Tolon, M. (2016). A voice in the skies: Listening to airline passenger preferences. *Journal of Air Transport Management*, *57*, 130–137. doi:10.1016/j.jairtraman.2016.07.017

Kuzma, J. M., & Wright, W. (2013). Using social networks as a catalyst for change in global higher education marketing and recruiting. *International Journal of Continuing Engineering Education and Lifelong Learning*, *23*(1), 53–66. doi:10.1504/IJCEELL.2013.051766

Kwok, L., & Yu, B. (2013). Spreading social media messages on Facebook: An analysis of restaurant business-to-consumer communications. *Cornell Hospitality Quarterly*, *54*(1), 84–94. doi:10.1177/1938965512458360

Lancestremère, S. (2015). *Big Data, Internet de las Cosas y sus aplicaciones prácticas. Fundaciontelefonica*. Retrieved from https://espacio.fundaciontelefonica.com/evento/big-data-internet-de-las-cosas-y-sus-aplicaciones-practicas-2/

Lan, J., Ma, Y., Zhu, D., Mangalagiu, D., & Thornton, T. (2017). Enabling value co-creation in the sharing economy: The case of mobike. *Sustainability*, *9*(9), 1504. doi:10.3390u9091504

Larsen, S. (2007). Aspects of a psychology of the tourist experience. *Scandinavian Journal of Hospitality and Tourism*, *7*(1), 7–18. doi:10.1080/15022250701226014

Lashley, C. (2008). Marketing hospitality and tourism experiences. In H. Oh & A. Pizam (Eds.), *Handobook of hospitality marketing management*. Oxford, UK: Butterworth-Heineman. doi:10.1016/B978-0-08-045080-3.50004-4

Laurillard, D. (2002). *Rethinking University Teaching: A Conversational Framework for the Effective Use of Learning Technologies* (2nd ed.). London: Routledge Falmer. doi:10.4324/9780203160329

Law, R., Chan, I. C. C., & Wang, L. (2018). A comprehensive review of mobile technology use in hospitality and tourism industry. *Journal of Hospitality Marketing & Management*, *27*(6), 626–648. doi:10.1080/19368623.2018.1423251

Law, R., & Cheung, S. (2010). The Perceived Destination Image of Hong Kong As Revealed in The Travel Blogs of Mainland Chinese Tourists. *International Journal of Hospitality & Tourism Administration*, *11*(4), 303–327. doi:10.1080/15256480.2010.518521

Lawrence, J. E., & Tar, U. A. (2018). Factors that influence teachers' adoption and integration of ICT in teaching/learning process. *Educational Media International*, *55*(1), 79–105. doi:10.1080/09523987.2018.1439712

Lecture, J. S., Fehr, E., & Falk, A. (2002). Psychological foundations of incentives. *European Economic Review*, *46*(4-5), 687–724. doi:10.1016/S0014-2921(01)00208-2

Lee, H. S., & Cho, C. H. (2012). Sporting event personality: Scale development and sponsorship implications. *International Journal of Sports Marketing & Sponsorship*, *14*(1), 46–63. doi:10.1108/IJSMS-14-01-2012-B005

Lee, H. Y., Qu, H., & Kim, Y. S. (2007). A study of the impact of personal innovativeness on online travel shopping behavior—A case study of Korean travelers. *Tourism Management*, *28*(3), 886–897. doi:10.1016/j.tourman.2006.04.013

Lee, H., Law, R., & Murphy, J. (2011). Helpful reviewers in TripAdvisor, an online travel community. *Journal of Travel & Tourism Marketing*, *28*(7), 675–688. doi:10.1080/10548408.2011.611739

Lee, J., & Ko, S. (2016). Effect of the In-Flight Meal Service Quality on the Customer Value and Loyalty. *Indian Journal of Science and Technology*, *9*(26), 1–6. doi:10.17485/ijst/2016/v9i26/97276

Lee, J., Park, D., & Han, I. (2008). The effect of negative online consumer reviews on product attitude: An information processing view. *Electronic Commerce Research and Applications*, *7*(3), 341–352. doi:10.1016/j.elerap.2007.05.004

Lee, U. K. (2017). International Tourism Advertisements on Social Media: Impact of Argument Quality and Source. *Sustainability*, *9*(9), 1537–1555. doi:10.3390u9091537

Lee, W., Xiong, L., & Hu, C. (2012). The effect of Facebook users' arousal and valence on intention to go to the festival: Applying an extension of the technology acceptance model. *International Journal of Hospitality Management*, *31*(3), 819–827. doi:10.1016/j.ijhm.2011.09.018

Lehto, X. Y., Kim, D. Y., & Morrison, A. M. (2006). The effect of prior destination experience on online information search behaviour. *Tourism and Hospitality Research*, *6*(2), 160–178. doi:10.1057/palgrave.thr.6040053

Lei, S., & Law, R. (2015). Content Analysis of Tripadvisor Reviews on Restaurants: A Case Study of Macau. *Journal of tourism*, *16*(1), 17–28.

Lei, S. S. I., Nicolau, J. L., & Wang, D. (2019). The impact of distribution channels on budget hotel performance. *International Journal of Hospitality Management*, *81*, 141–149. doi:10.1016/j.ijhm.2019.03.005

Leite, R. & Azevedo, A. (2017). The Role of Digital Marketing: A perspective from Porto Hotels' managers. *International Journal of Marketing, Communication and New Media,* (2), 88-105.

Lennon, J. J., Smith, H., Cockerell, N., & Trew, J. (2006). *Benchmarking National Tourism Organisations and Agencies – Understanding best practice*. Amsterdam: Elsevier.

Leonardi, P. M., Huysman, M., & Steinfield, C. (2013). Enterprise social media: Definition, history, and prospects for the study of social technologies in organizations. *Journal of Computer-Mediated Communication*, *19*(1), 1–19. doi:10.1111/jcc4.12029

Leuf, B., & Cunningham, W. (2001). *The Wiki way: quick collaboration on the Web*. Boston: Addison-Wesley.

Leung, D., Law, R., van Hoof, H., & Buhalis, D. (2013). Social media in tourism and hospitality: A literature review. *Journal of Travel & Tourism Marketing*, *30*(1-2), 3–22. doi:10.1080/10548408.2013.750919

Leung, X. Y., Bai, B., & Stahura, K. A. (2015). The Marketing Effectiveness of Social Media in the Hotel Industry: A Comparison of Facebook and Twitter. *Journal of Hospitality & Tourism Research (Washington, D.C.)*, *39*(2), 147–169. doi:10.1177/1096348012471381

Levent, N., & Reich, C. (2012). How can my museum help visitors with vision loss? *Museum. American Association of Museums,* (July-August), 21–22.

Liehr, K., Peters, P., & Zerfaß, A. (2009). *Reputationsmessung: grundlagen und verfahren.* Leipzig: DPRG.

Lien, C. H., & Cao, Y. (2014). Examining WeChat users' motivations, trust, attitudes, and positive word-of-mouth: Evidence from China. *Computers in Human Behavior, 41*, 104–111. doi:10.1016/j.chb.2014.08.013

Lien, C. H., Cao, Y., & Zhou, X. (2017, March). Service quality, satisfaction, stickiness, and usage intentions: An exploratory evaluation in the context of WeChat services. *Computers in Human Behavior, 68*, 403–410. doi:10.1016/j.chb.2016.11.061

Li, H., Ye, Q., & Law, R. (2013). Determinants of customer satisfaction in the hotel industry: An application of online review analysis. *Asia Pacific Journal of Tourism Research, 18*(7), 784–802. doi:10.1080/10941665.2012.708351

Li, J., Xu, L., Tang, L., Wang, S., & Li, L. (2018). Big data in tourism research: A literature review. *Tourism Management, 68*, 301–323. doi:10.1016/j.tourman.2018.03.009

Liker, J. K., Haddad, G., & Karlin, J. (1999). Perspectives on technology and work organization. *Annual Review of Sociology, 25*(1), 575–596. doi:10.1146/annurev.soc.25.1.575

Limberger, P., Anjos, F., Meira, J., & Anjos, S. (2014a). Satisfaction in hospitality on TripAdvisor.com: An analysis of the correlation between evaluation criteria and overall satisfaction. *Tourism & Management Studies, 10*(1), 59–65.

Limberger, P., Boaria, F., & Anjos, S. (2014b). A relação entre a satisfaçãogeral e as variáveis da satisfação na hotelaria em hotéis de excelência. *Revista Brasileira de Pesquisa em Turismo, 8*(3), 435–455. doi:10.7784/rbtur.v8i3.803

Lim, S. S., & Tkaczynski, A. (2017). Origin and money matter: The airline service quality expectations of international students. *Journal of Hospitality and Tourism Management, 31*, 244–252. doi:10.1016/j.jhtm.2017.03.001

Lin, C. H., Morais, D. B., Kerstetter, D. L., & Hou, J. S. (2007). Examining the Role of Cognitive and Affective Image in Predicting Choice Across Natural, Developed, And Theme-Park Destinations. *Journal of Travel Research, 46*(2), 183–194. doi:10.1177/0047287507304049

Lin, Y.-W., & Chen, C.-H. (2012). College students' Facebook use characteristics, critical-thinking tendencies and information verification behaviours: National Taiwan University and Technology. *Journal of Liberal Arts and Social Sciences, 8*(1), 17–36.

Liou, J. J., Hsu, C. C., Yeh, W. C., & Lin, R. H. (2011). Using a modified grey relation method for improving airline service quality. *Tourism Management, 32*(6), 1381–1388. doi:10.1016/j.tourman.2011.01.013

Liou, J. J., & Tzeng, G. H. (2007). A non-additive model for evaluating airline service quality. *Journal of Air Transport Management, 13*(3), 131–138. doi:10.1016/j.jairtraman.2006.12.002

Litvin, S. W., Goldsmith, R. E., & Pan, B. (2018). A retrospective view of electronic word-of-mouth in hospitality and tourism management. *International Journal of Contemporary Hospitality Management, 30*(1), 313–325. doi:10.1108/IJCHM-08-2016-0461

Litvin, S., Goldsmith, R., & Pan, B. (2008). Electronic word-of-mouth in hospitality and tourism management. *Tourism Management, 29*(2), 458–468. doi:10.1016/j.tourman.2007.05.011

Liu, B. (2015). *Sentiment analysis: Mining opinions, sentiments, and emotions*. New York: Cambridge University Press. doi:10.1017/CBO9781139084789

Liu, H. (2007). In-Flight Entertainment System: State of the Art and Research Directions. *Proceedings of the Second International Workshop on Semantic Media Adaptation and Personalization (SMAP 2007)* (pp. 241-244). Uxbridge, UK: IEEE. 10.1109/SMAP.2007.37

Liu, H., Wu, L., & Li, X. R. (2019). Social Media Envy: How Experience Sharing on Social Networking Sites Drives Millennials' Aspirational Tourism Consumption. *Journal of Travel Research*, *58*(3), 355–369. doi:10.1177/0047287518761615

Liu, Z., & Park, S. (2015). What makes a useful online review? Implication for travel product websites. *Tourism Management*, *47*, 140–151. doi:10.1016/j.tourman.2014.09.020

Li, X., & Petrick, J. F. (2006). A review of festival and event motivation studies. *Event Management*, *9*(4), 239–245. doi:10.3727/152599506776771526

Li, Y., Hu, C., Huang, C., & Duan, L. (2017). The concept of smart tourism in the context of tourism information services. *Tourism Management*, *587*, 293–300. doi:10.1016/j.tourman.2016.03.014

Llodrà-Riera, I., Martínez-Ruiz, M. P., Jiménez-Zarco, A. I., & Izquierdo-Yusta, A. (2015). A Multidimensional Analysis of The Information Sources Construct and Its Relevance for Destination Image Formation. *Tourism Management*, *48*, 319–328. doi:10.1016/j.tourman.2014.11.012

Lo, I. S., McKercher, B., Lo, A., Cheung, C., & Law, R. (2011). Tourism and Online Photography. *Tourism Management*, *32*(4), 725–731. doi:10.1016/j.tourman.2010.06.001

Lonely Planet. (2018) Map of Malawi. Retrieved from https://www.lonelyplanet.com/maps/africa/malawi/

Lopes, R., Martins, J. A., & Roque, V. (2017). Perfil e Impacto Socioeconómico dos Estudantes nas Escolas Politécnicas do Interior de Portugal – O Caso de Estudo da ESTH-IPG. In *Egitânia Sciencia - Edição Especial ISITH - International Symposium on Innovation in Tourism and Hospitality* (pp. 123–143). Academic Press. Retrieved from http://www.egitaniasciencia.ipg.pt/revista.aspx?id=32

Lopes, R., Abrantes, J. L., & Kastenholz, E. (2013). Online Reviews Credibility: Implications on Traveller's Decision Making. In M. Kosak & N. Kosak (Eds.), *Aspects of Tourist Behavior*. Newcastle: Cambridge Scholars Publishing.

Lopes, R., Abrantes, J., & Kastenholz, E. (2014). Innovation, tourism and social networks. *Revista Turismo e Desenvolvimento*, *21/22*, 151–154.

López Fernández, M. C., & Serrano Bedia, A. M. (2004). Is the hotel classification system a good indicator of hotel quality? *Tourism Management*, *25*(6), 771–775. doi:10.1016/j.tourman.2004.06.007

Lorenzo-Romero, C., Alarcón-del-Amo, M., & Constantinides, E. (2014). Determinants of use of social media tools in retailing sector. *Journal of Theoretical and Applied Electronic Commerce Research*, *9*(1), 44–55. doi:10.4067/S0718-18762014000100005

Lovelock, C., & Gummesson, E. (2004). Whither services marketing? *Journal of Service Research*, *7*(1), 20–41. doi:10.1177/1094670504266131

Luca, M. (2017). Designing online marketplaces: Trust and reputation mechanisms. *Innovation Policy and the Economy*, *17*(1), 77–93. doi:10.1086/688845

Lu, D., Wu, R., & Sang, J. (2017). Overlapped user-based comparative study on photo-sharing websites. *Information Sciences*, *376*, 54–70. doi:10.1016/j.ins.2016.10.005

Luke, K. (2009). Marketing the new-fashioned way: Connect with your target market through social networking sites. *Journal of Financial Planning*, (November/December): 18–19.

Lund, N. F., Cohen, S. A., & Scarles, C. (2018). The power of social media storytelling in destination branding. *Journal of Destination Marketing & Management*, *8*, 271–280. doi:10.1016/j.jdmm.2017.05.003

Luo, Y., & Bu, J. (2018). Contextualizing international strategy by emerging market firms: A composition-based approach. *Journal of World Business*, *53*(3), 337–355. doi:10.1016/j.jwb.2017.01.007

Lu, W., & Stepchenkova, S. (2012). Ecotourism experiences reported online: Classification of satisfaction attributes. *Tourism Management*, *33*(3), 702–712. doi:10.1016/j.tourman.2011.08.003

Lu, W., & Stepchenkova, S. (2015). User-generated content as a research mode in tourism and hospitality applications: Topics, methods, and software. *Journal of Hospitality Marketing & Management*, *24*(2), 119–154. doi:10.1080/19368623.2014.907758

Lu, X., Wang, C., Yang, J. M., Pang, Y., & Zhang, L. (2010) Photo2trip: Generating travel routes from geo-tagged photos for trip planning. *Proceedings of the 18th ACM International Conference on Multimedia*, Firenze, Italy. ACM. 10.1145/1873951.1873972

Macário, J. (2017). Portugal em sexto entre os países europeus mais requisitados na Airbnb. Retrieved from https://bit.ly/2KoUK6a

MacKay, K. J., & Fesenmaier, D. R. (1997). Pictorial Element of Destination in Image Formation. *Annals of Tourism Research*, *24*(3), 537–565. doi:10.1016/S0160-7383(97)00011-X

Majchrzak, A., Rice, R. E., Malhotra, A., King, N., & Ba, S. (2000). Technology adaptation: The case of a computer-supported inter-organizational virtual team. *Management Information Systems Quarterly*, *24*(4), 569–600. doi:10.2307/3250948

Mak, A. H. (2017). Online Destination Image: Comparing National Tourism Organization's and Tourists' Perspectives. *Tourism Management*, *60*, 280–297. doi:10.1016/j.tourman.2016.12.012

Malawi Government. (2006). *Malawi Growth and Development Strategy*. Malawi Government Press.

Malhotra, N. K. (2010). *Marketing Research: An Applied Orientation* (6th ed.). Pearson.

Malthora, Y. (2002). Why Knowledge Management System Fail? Enablers and Constraints of Knowledge Management in Human Entreprises. In C. Holsapple (Ed.), *Handbook of Knowledge Management (1)* (pp. 577–599). Heidelberg: Springer Verlag.

Manaman, H. S., & Jamali, S. (2016). Online reputation measurement of companies based on user-generated content in online social networks. *Computers in Human Behavior*, *54*, 94–100. doi:10.1016/j.chb.2015.07.061

Manca, S., & Ranieri, M. (2016a). Facebook and the others. Potentials and obstacles of Social Media for teaching in higher education. *Computers & Education*, *95*, 216–230. doi:10.1016/j.compedu.2016.01.012

Manca, S., & Ranieri, M. (2016b). "Yes for sharing, no for teaching!": Social Media in academic practices. *The Internet and Higher Education*, *29*, 63–74. doi:10.1016/j.iheduc.2015.12.004

Manes, E., & Tchetchik, A. (2018). The role of electronic word of mouth in reducing information asymmetry: An empirical investigation of online hotel booking. *Journal of Business Research*, *85*, 185–196. doi:10.1016/j.jbusres.2017.12.019

Mangold, W., & Faulds, D. (2009). Social media: The new hybrid element of the promotion mix. *Business Horizons*, *52*(4), 357–365. doi:10.1016/j.bushor.2009.03.002

Mao, J. (2014). Social media for learning: A mixed methods study on high school students' technology affordances and perspectives. *Computers in Human Behavior*, *33*, 213–223. doi:10.1016/j.chb.2014.01.002

Marchiori, E., & Cantoni, L. (2015). The role of prior experience in the perception of a tourism destination in user-generated content. *Journal of Destination Marketing & Management*, *4*(3), 194–201. doi:10.1016/j.jdmm.2015.06.001

Marczyk, G., DeMatteo, D., & Festinger, D. (2005). *Essentials of Research Design and Methodology*. New Jersey: John Wiley & Sons, Inc.

Mariani, M. M. (2015). Coordination in inter-network co-opetitition: Evidence from the tourism sector. *Industrial Marketing Management*. doi:10.1016/j.indmarman.2015.11.015

Mariani, M. M., & Borghi, M. (2018). Effects of the Booking. com rating system: Bringing hotel class into the picture. *Tourism Management*, *66*, 47–52. doi:10.1016/j.tourman.2017.11.006

Mariani, M. M., & Giorgio, L. (2017). The 'Pink Night' festival revisited: Meta-events and the role of destination partnerships in staging event tourism. *Annals of Tourism Research*, *62*, 89–109. doi:10.1016/j.annals.2016.11.003

Mariani, M., Di Felice, M., & Mura, M. (2016). Facebook as a destination marketing tool: Evidence from Italian regional destination management organizations. *Tourism Management*, *54*, 321–343. doi:10.1016/j.tourman.2015.12.008

Marine-Roig, E., & Clavé, S. A. (2015). Tourism analytics with massive user-generated content: A case study of Barcelona. *Journal of Destination Marketing & Management*, *4*(3), 162–172. doi:10.1016/j.jdmm.2015.06.004

Market Research Media. (n.d.). Retrieved from https://www.marketresearchmedia.com/?p=421

Markham, A., & Buchanan, E. (2012). Ethical Decision-Making and Internet Research: Version 2.0. The Aoir Ethics Working Committee. Retrieved from aoir.org/reports/ethics2.pdf

Marktest. (2017). Os portugueses e as redes sociais 2017. Retrieved from http://www.marktest.com/wap/a/grp/p~96.aspx

Marktest. (2019). Os portugueses e as redes sociais 2019. Retrieved from http://www.marktest.com/wap/a/grp/p~96.aspx

Marques, N. (2015). Crise ajudou à expansão do alojamento local. Retrieved from https://bit.ly/2OBO7mD

Marsteller, B. (2012). The Global Social Media Check-up 2011. *Lastet ned, 15.*

Martensen, A., Brockenhuus-Schack, S., & Zahid, A. L. (2018). How Citizen Influencers Persuade Their Followers. *Journal of Fashion Marketing and Management: An International Journal*, *22*(3), 335 353. doi:10.1108/JFMM 09 2017 0095

Martin-Fuentes, E. (2016). Are guests of the same opinion as the hotel star-rate classification system? *Journal of Hospitality and Tourism Management*, *29*, 126–134. doi:10.1016/j.jhtm.2016.06.006

Martin-Fuentes, E., Mateu, C., & Fernandez, C. (2018a). Does verifying uses influence rankings? Analyzing Booking. com and TripAdvisor. *Tourism Analysis*, *23*(1), 1–15. doi:10.3727/108354218X15143857349459

Martin-Fuentes, E., Mateu, C., & Fernandez, C. (2018b). The more the merrier? Number of reviews versus score on TripAdvisor and Booking.com. *International Journal of Hospitality & Tourism Administration*, 1–14. doi:10.1080/15256480.2018.1429337

Martin-Fuentes, E., Mateu, C., & Fernandez, C. (2018c). Are users' ratings on TripAdvisor similar to hotel categories in Europe? *Cuadernos de Turismo*, *42*(42), 305–316. doi:10.6018/turismo.42.13

Martín-Santana, J. D., Beerli-Palacio, A., & Nazzareno, P. A. (2017). Antecedents and Consequences of Destination Image Gap. *Annals of Tourism Research*, *62*, 13–25. doi:10.1016/j.annals.2016.11.001

Maulisa, A. R., & Hijrah Hati, S. R. (2018). The Anomaly of Airline Passenger Behavioral Intention: A Comparison Between Low-Cost and Full-Service Airlines. *ASEAN Marketing Journal*, *10*(2), 75–90.

Ma, W. W. K., Andersson, R., & Streith, K. O. (2005). Examining user acceptance of computer technology: An empirical study of student teachers. *Journal of Computer Assisted Learning*, *21*(6), 387–395.

Mayring, P. (2000). Qualitative Content Analysis. *Forum Qualitative Sozialforschung / Forum: Qualitative. Social Research*, *1*(2), 20. Retrieved from http://nbn-resolving.de/urn:nbn:de:0114-fqs0002204

McAfee, A., Brynjolfsson, E., Davenport, T. H., Patil, D. J., & Barton, D. (2012). Big data: The management revolution. *Harvard Business Review*, *90*(10), 60–68. PMID:23074865

McCarthy, I. P., Lawrence, T. B., Wixted, B., & Gordon, B. R. (2010). A multidimensional conceptualization of environmental velocity. *Academy of Management Review*, *35*(4), 604–626.

McCole, P. (2002). The role of trust for electronic commerce in services. *International Journal of Contemporary Hospitality Management*, *14*(2), 81–87. doi:10.1108/09596110210419264

McKercher, B. (2008). A citation analysis of tourism scholars. *Tourism Management*, *29*(6), 1226–1232. doi:10.1016/j.tourman.2008.03.003

Mckillop, H. (2013). Sustainable Archaeological Tourism of the Underwater Maya Project by 3D. *Anthmpology News*, (May/June), 12–13. Retrieved from http://www.ga.lsu.edu/hmckill/publications/2013%20McKillop%20and%20Sills%20Anth%20News.pdf

McLean, D. J. (2006). *Philosophy and Leisure. In G. Kassing (Eds.), Introduction to Recreation and Leisure*. United States: Human Kinetics.

McLellan, H. (2000). Experience design. *Cyberpsychology & Behavior*, *3*(1), 59–69. doi:10.1089/109493100316238

McMillen, R., & Alter, F. (2017). Social media, social inclusion, and museum disability access. *Museums & Social Issues*, *12*(2), 115–125. doi:10.1080/15596893.2017.1361689

Medeiros, S. A. De, Gosling, M., Alves, L., & Vera, R. (2015). *Emoções em Experiências Negativas de Turismo: um estudo sobre a influência na insatisfação, 26*, 188–215.

Medina, K., & Hadwick, A. (2019). *The state of hospitality distribution: Direct bookings – A report series between Eye for Travel and Fornova*. London: Eye for Travel Publishing.

Meleddu, M., Paci, R., & Pulina, M. (2015). Repeated behaviour and destination loyalty. *Tourism Management*, *50*, 159–171. doi:10.1016/j.tourman.2015.01.032

Melián-González, S., Bulchand-Gidumal, J., & López-Valcárcel, B. (2013). Online customer reviews of hotels: As participation increases, better evaluation is obtained. *Cornell Hospitality Quarterly*, *54*(3), 274–283. doi:10.1177/1938965513481498

Meliones, A., & Sampson, D. (2018). Blind MuseumTourer: A system for self-Guided tours in museums and blind indoor navigation. *Technologies*, *6*(1), 4. doi:10.3390/technologies6010004

Mellinas, J. P., María-Dolores, S. M. M., & García, J. J. B. (2015). Booking. com: The unexpected scoring system. *Tourism Management*, *49*, 72–74. doi:10.1016/j.tourman.2014.08.019

Mellinas, J. P., María-Dolores, S. M. M., & García, J. J. B. (2016). Effects of the Booking. com scoring system. *Tourism Management*, *57*, 80–83. doi:10.1016/j.tourman.2016.05.015

Men, L. R., & Tsai, W.-H. S. (2013). Beyond liking or following: Understanding public engagement on social networking sites in China. *Public Relations Review*, *39*(1), 13–22. doi:10.1016/j.pubrev.2012.09.013

Mertler, C. A., & Reinhart, R. V. (2017). Advanced and multivariate statistical methods: Practical application and interpretation (6th ed.). New York: Routledge.

Mesquita, S., & Carneiro, M. J. (2016). Accessibility of European museums to visitors with visual impairments. *Disability & Society*, *31*(3), 373–388. doi:10.1080/09687599.2016.1167671

Messner, W. (2016). The impact of an aircraft' s service environment on perceptions of in-flight food quality. *Journal of Air Transport Management*, *53*, 123–130. doi:10.1016/j.jairtraman.2016.02.010

Metcalfe, J. S. (2005). Innovation, competition and enterprise foundation for economic evolution in learning economies. Manchester: University of Manchester, Centre for Innovation and Competition, Discussion Paper 71. In Hall, M.C. & Williams, A.M. (2008). *Tourism Innovation.* New York: Routledge.

Middleton, V., & Clarke, J. (2004). *Marketing in travel and tourism.* Oxford: Butterworth-Heinemann.

Miguéns, J., Baggio, R., & Costa, C. (2008). Social media and tourism destinations: TripAdvisor case study. *Advances in tourism research, 26*(28), 1–6.

Miguéns, J., Baggio, R., & Costa, C. (2008). Social media and tourism destinations: TripAdvisor case study. *Advances in Tourism Research*, *26*(28), 1–6.

Milicchio, F., & Prosperi, M. (2016). Accessible tourism for the deaf via mobile apps. *Proceedings of the 9th ACM International Conference on PErvasive Technologies Related to Assistive Environments - PETRA '16* (pp. 1–7). New York: ACM Press. 10.1145/2910674.2910694

Milović, B. (2012). Social Media and eCRM as a Prerequisite for Hotel Success. *Management Information Systems*, *7*(3), 26–31.

Mkono, M. (2012). Netnographic tourist research: The Internet as a virtual fieldwork site. *Tourism Analysis*, *17*(4), 553–555. doi:10.3727/108354212X13473157390966

Mkono, M. (2018). 'Troll alert!': Provocation and harassment in tourism and hospitality social media. *Current Issues in Tourism*, *21*(7), 791–804. doi:10.1080/13683500.2015.1106447

Moavenzadeh, J. & Maar, L. (2018). The Known Traveller: Unlocking the potential of digital identity for secure and seamless travel. *World Economic Forum*, (January). Retrieved from http://www3.weforum.org/docs/WEF_The_Known_Traveller_Digital_Identity_Concept.pdf

Mochón, F., & Martínez, O. S. (2014). A first approach to the implicit measurement of happiness in Latin America through the use of social networks. *IJIMAI*, *2*(5), 16–22. doi:10.9781/ijimai.2014.252

Moens, M. F., Li, J., & Chua, T. S. (2014). *Mining user generated content.* Chapman and Hall/CRC.

Moran, M., Seaman, J., & Tinti-Kane, H. (2011). *Teaching, Learning, and Sharing: How Today's Higher Education Faculty Use Social Media.* Boston. Retrieved from https://files.eric.ed.gov/fulltext/ED535130.pdf

Morgan, J. (2012). The multisensory museum. *Glasnik Etnografskog Instituta*, *60*(1), 65–77. doi:10.2298/GEI1201065M

Moro, S., & Rita, P. (2018). Brand strategies in social media in hospitality and tourism. *International Journal of Contemporary Hospitality Management*, *30*(1), 343–364. doi:10.1108/IJCHM-07-2016-0340

Morosan, C. (2013). The Impact of the Destination's Online Initiatives on Word of Mouth. *Tourism Analysis*, *18*(4), 415–428. doi:10.3727/108354213X13736372325993

Morosan, C., & Jeong, M. (2008). Users' perceptions of two types of hotel reservation Web sites. *International Journal of Hospitality Management*, *27*(2), 284–292. doi:10.1016/j.ijhm.2007.07.023

Morrell, P. (2008). Can long-haul low-cost airlines be successful? *Research in Transportation Economics*, *24*(1), 61–67. doi:10.1016/j.retrec.2009.01.003

Mossberg, L. (2007). A marketing approach to the tourist experience. *Scandinavian Journal of Hospitality and Tourism*, *7*(1), 59–74. doi:10.1080/15022250701231915

Mostafa, M. M. (2013). More than words: Social networks' text mining for consumer brand sentiments. *Expert Systems with Applications*, *40*(10), 4241–4251. doi:10.1016/j.eswa.2013.01.019

Mota, A. M. G., Wada, E. K., & Fraga, L. E. (2013). Mídias Sociais e a Hospitalidade Desafios e Tendências no Século XXI. *TURyDES - Revista de Investigación em Turismo y desarrollo local, 6*(15).

Moura, A., Mónico, L., & Mira, M. (2019). Measuring quality regarding destination marketing: Perceptions from local public stakeholders in Portugal. *Tourism & Management Studies*, *15*(1), 44–53. doi:10.18089/tms.2019.150104

Moutinho, L., Rita, P., & Cury, B. (1996). *Expert Systems in Tourism Marketing*. New York: Routledge.

Mukherjee, A., & Nath, P. (2007). Role of electronic trust in online retailing: A re-examination of the commitment-trust theory. *European Journal of Marketing*, *41*(9/10), 1173–1202. doi:10.1108/03090560710773390

Munar, A. M. (2010). Digital exhibitionism: The age of exposure. *Culture Unbound*, *2*(3), 401–422. doi:10.3384/cu.2000.1525.10223401

Munar, A. M. (2012). Social media strategies and destination management. *Scandinavian Journal of Hospitality and Tourism*, *12*(2), 101–120. doi:10.1080/15022250.2012.679047

Munar, A. M., & Jacobsen, J. K. S. (2014). Motivations for sharing tourism experiences through social media. *Tourism Management*, *43*, 46–54. doi:10.1016/j.tourman.2014.01.012

Muzellec, L., & O'Raghallaigh, E. (2018). Mobile technology and its impact on the consumer decision-making journey. *Journal of Advertising Research*, *58*(1), 12–15. doi:10.2501/JAR-2017-058

Nadeau, J., Heslop, L., O'Reilly, N., & Luk, P. (2008). Destination in A Country Image Context. *Annals of Tourism Research*, *35*(1), 84–106. doi:10.1016/j.annals.2007.06.012

Namin, A. (2017). Revisiting customers' perception of service quality in fast food restaurants. *Journal of Retailing and Consumer Services*, *34*, 70–81. doi:10.1016/j.jretconser.2016.09.008

Nam, Y., Barnett, G. A., & Kim, D. (2014). Corporate hyperlink network relationships in global corporate social responsibility system. *Quality & Quantity*, *48*(3), 1225–1242. doi:10.100711135-013-9831-7

Narangajavana Kaosiri, Y., Callarisa Fiol, L. J., Moliner Tena, M. A., Rodriguez Artola, R. M., & Sanchez Garcia, J. (2019). User-Generated Content Sources in Social Media: A New Approach to Explore Tourist Satisfaction. *Journal of Travel Research*, *58*(2), 253–265. doi:10.1177/0047287517746014

Narangajavana, Y., Fiol, L. J. C., Tena, M. A. M., Artola, R. M. R., & Garcia, J. S. (2017). The influence of social media in creating expectations. An empirical study for a tourist destination. *Annals of Tourism Research*, *65*, 60–70. doi:10.1016/j.annals.2017.05.002

Narus, J. A., & Narayandas, D. (2004). *Business market management: Understanding, creating and delivering value*. Pearson Prentice Hall.

Naslund, B. (1986). Financial Innovations. A comparison with R&D in Physical Products.

Needles, A., & Thompson, G. (2013). Social Media Use in the Restaurant Industry: A Work in Progress. *Center for Hospitality Research Publications, 13*(7), 6–16. Retrieved from http://scholarship.sha.cornell.edu/chrpubs/101

Neethu, M. S., & Rajasree, R. (2013). Sentiment analysis in Twitter using machine learning techniques. *Proceedings of the 2013 Fourth International Conference on Computing, Communications and Networking Technologies (ICCCNT)* (pp. 1–5). IEEE. 10.1109/ICCCNT.2013.6726818

Nejati, M., Nejati, M., & Shafaei, A. (2009). Ranking airlines' service quality factors using a fuzzy approach: Study of the Iranian society. *International Journal of Quality & Reliability Management, 36*(3), 247–260. doi:10.1108/02656710910936726

Neuhofer, B., Buhalis, D., & Ladkin, A. (2013). Experiences, co-creation and technology: a conceptual approach to enhance tourism experiences. In *Tourism and Global Change: on the edge of something big. CAUTHE 2013. Conference Proceedings* (pp. 546-555).

Neuhofer, B., Buhalis, D., & Ladkin, A. (2014). A Typology of Technology-Enhanced Tourism Experiences. *International Journal of Tourism Research, 16*(4), 340–350. doi:10.1002/jtr.1958

Neves, J. (2012). Multi-sensory approaches to (audio) describing the visual arts. *MonTI. Monografías de Traducción e Interpretación*, (4), 277–293.

Nevo, S., & Wade, M. R. (2010). The formation and value of IT-enabled resources: Antecedents and consequences of synergistic relationships. *Management Information Systems Quarterly, 34*(1), 163. doi:10.2307/20721419

Newman, D. (2018). Top 6 Digital Transformation Trends in Hospitality and Tourism. Forbes. Retrieved from https://www.forbes.com/sites/danielnewman/2018/01/02/top-6-digital-transformation-trends-in-hospitality-and-tourism/#23f6ddb267df

Nezakati, H., Amidi, A., Jusoh, Y., Moghadas, S., Aziz, Y., & Sohrabinezhadtalemi, R. (2015). Review of Social Media Potential on Knowledge Sharing and Collaboration in Tourism Industry. *Procedia: Social and Behavioral Sciences, 172*, 120–125. doi:10.1016/j.sbspro.2015.01.344

Nghiêm-Phú, B. (2014). A Review of Destination Image Studies From 2008 to 2012. *European Journal of Tourism Research, 8*, 35.

Nicholson, R., & Pearce, D. G. (2000). Who goes to events: A comparative analysis of the profile characteristics of visitors to four south island events. *Journal of Vacation Marketing, 6*(3), 236–253. doi:10.1177/135676670000600304

Niininen, O., March, R., & Buhalis, D. (2006). Consumer centric tourism marketing. In D. Buhalis & C. Costa (Eds.), *Tourism management dynamics: Trends, management and tools* (p. xxiii). Amsterdam: Buttterworth Heinemann. doi:10.1016/B978-0-7506-6378-6.50029-9

Nisar, T. M., & Whitehead, C. (2016). Brand interactions and social media: Enhancing user loyalty through social networking sites. *Computers in Human Behavior, 62*, 743–753. doi:10.1016/j.chb.2016.04.042

Nobre, H., & Faria, J. (2017). Exploring marketing strategies in architectural services: The case of the architecture firms in Portugal. *International Journal of Business Excellence, 12*(3), 273–293. doi:10.1504/IJBEX.2017.084438

Nobre, H., & Silva, D. (2014). Social Network Marketing Strategy and SME Strategy Benefits. *Journal of Transnational Management, 19*(1), 138–151. doi:10.1080/15475778.2014.904658

Nogueira Tomas, R., Pierre Meschgrahw, R., & Lúcia Chicarelli Alcantara, R. (2012). *As redes sociais e o comportamento de compra do consumidor: O reinado do & BOCA-A-BOCA & ESTÁ DE VOLTA?* doi:10.5585/remark.v11i2.2325

Nunes, A. (2014). *A importância e utilização da internet como fonte de informação para o enoturista que visita regiões portuguesas*. Unpublished MSc thesis, Polytechnic Institute of Porto, Porto.

Nunes, M. L., Pereira, A. C., & Alves, A. C. (2017). Smart products development approaches for Industry 4.0. *Procedia Manufacturing*, *13*, 1215–1222. doi:10.1016/j.promfg.2017.09.035

Nuno, M. (2018). As 6 redes sociais mais usadas do mundo. E-Konomista. Retrieved from https://www.e-konomista.pt/artigo/redes-sociais-mais-usadas/

Nyaribo, Y. M., & Munene, A. G. (2018). *Effect of social media predication in the workplace on employee productivity*. IJAME.

O'Connor, P. (1999). Electronic information distribution in tourism and hospitality. Wallingford: CAB.

O'Connor, P. (2010). Managing a hotel's image on TripAdvisor. *Journal of Hospitality Marketing & Management*, *19*(7), 754–772. doi:10.1080/19368623.2010.508007

OECD. (1994). *Tourism strategies and rural development*. Paris: OECD.

Öğüt, H., & Onur Taş, B. K. (2012). The influence of Internet customer reviews on the online sales and prices in hotel industry. *Service Industries Journal*, *32*(2), 197–214. doi:10.1080/02642069.2010.529436

Oh, H., & Nguyen, C. (2010). Influence of Retweets.

Oh, H., Fiore, A. M., & Jeoung, M. (2007). Measuring experience economy concepts: Tourism applications. *Journal of Travel Research*, *46*(2), 119–132. doi:10.1177/0047287507304039

Okazaki, S., Campo, S., Andreu, L., & Romero, J. (2015). A latent class analysis of Spanish travellers' mobile internet usage in travel planning and execution. *Cornell Hospitality Quarterly*, *56*(2), 191–201. doi:10.1177/1938965514540206

Okumus, A., & Duygun, A. (2008). Service Quality Measurement on Education Services Marketing and Relationship Between Perceived Service Quality and Student Satisfaction. *Anadolu University Journal of Social Sciences*, *8*(2), 17–38.

Oliveira, A. M. (2016). *Marketing Interno nos Estabelecimentos Hoteleiros: A aplicabilidade do Marketing Interno como forma de obtenção de qualidade de Serviço ao Cliente* [Dissertação]. Instituto Politécnico do Porto.

Oliveira, A., Correia, A., & Fernandes, C. (2017). A utilização das redes sociais na perspetiva das empresas de turismo em espaço rural. *Revista Pasos*, *15*(4), 807–822. doi:10.25145/j.pasos.2017.15.055

Oliveira, B., & Casais, B. (2019). The importance of user-generated photos in restaurant selection. *Journal of Hospitality and Tourism Technology*, *10*(1), 2–14. doi:10.1108/JHTT-11-2017-0130

Oliver, R. L. (1977). Effect of expectation and disconfirmation on postexposure product evaluations: An alternative interpretation. *The Journal of Applied Psychology*, *62*(4), 480–486. doi:10.1037/0021-9010.62.4.480

Önder, I., Gunter, U., & Gindl, S. (2019). Utilizing Facebook statistics in tourism demand modeling and destination marketing. *Journal of Travel Research*. doi:10.1177/0047287519835969

Oppenheim, A. N. (2000). *Questionnaire Design, Interviewing and Attitude Measurement*. Bloomsbury Academic. Retrieved from https://books.google.pt/books?id=6V4GnZS7TO4C

Organização Mundial do Turismo (OMT). (2019). Crescimento do turismo mundial pode chegar a 4% em 2019. Retrieved from http://www.turismo.gov.br/%C3%BAltimas-not%C3%ADcias/12306-crescimento-do-turismo-mundial-pode-chegar-a-4-em-2019.html

Osterman, M. (2017). Accessibility and technology: Developing a virtual access tour. *MW17: Museums and the Web.* Retrieved from https://mw17.mwconf.org/paper/accessibility-and-technology-developing-a-virtual-access-tour/

Öztayşi, B., Baysan, S., & Akpinar, F. (2009). Radio frequency identification (RFID) in hospitality. *Technovation, 29*(9), 618–624. doi:10.1016/j.technovation.2009.05.014

Pacheco, L. (2016). An analysis of online reviews by language groups: The case of hotels in Porto, Portugal. *European Journal of Tourism Research, 14*, 66–74.

Pacheco, L. (2017). Customer satisfaction in Portuguese hotels: Evidence for different regions and hotel segments. *Tourism Analysis, 22*(3), 337–347. doi:10.3727/108354217X14955605216087

Page, S. J., Brunt, P., Busby, G., & Connel, J. (2001). *Tourism: a modern synthesis.* London: Thomson Learning.

Pak, A., & Paroubek, P. (n.d.). Twitter as a corpus for sentiment analysis and opinion mining. *LREC*, 1320–1326. Retrieved from https://pdfs.semanticscholar.org/ad8a/7f620a57478ff70045f97abc7aec9687ccbd.pdf

Pala, Z., & Inanç, N. (2007). Smart parking applications using RFID technology. *2007 1st Annual RFID Eurasia*, (May), 2007–2010. doi:10.1109/RFIDEURASIA.2007.4368108

Palumbo, F., Dominici, G., & Basile, G. (2013, June). Designing a mobile app for museums according to the drivers of visitor satisfaction. In *Recent Advances in Business Management and Marketing*. Dubrovnik, Croatia: WSEAS Press.

Pan, B., & Fesenmaier, D. R. (2006). online information search: Vacation planning process. *Annals of Tourism Research, 3*(3), 809–832. doi:10.1016/j.annals.2006.03.006

Pan, B., MacLaurin, T., & Crotts, J. C. (2007). Travel Blogs and The Implications for Destination Marketing. *Journal of Travel Research, 46*(1), 35–45. doi:10.1177/0047287507302378

Pang, B., & Lee, L. (2008). Opinion mining and sentiment analysis. *Foundations and Trends in Information Retrieval, 2*(12), 1–135. doi:10.1561/1500000011

Panwar, D., Anand, S., Ali, F., & Singal, K. (2019). Consumer Decision Making Process Models and their Applications to Market Strategy. *International Management Review, 15*(1), 36–44.

Parasuraman, A., Zeithaml, V. A., & Berry, L. L. (1985). A Conceptual Model of Service Quality and Its Implications for Future Research. *Journal of Marketing, 49*(4), 41–50. doi:10.1177/002224298504900403

Parasuraman, A., Zeithaml, V. A., & Berry, L. L. (1988). Servqual: A multiple-item scale for measuring consumer perc. *Journal of Retailing, 64*(1), 12.

Parida, V., Sjödin, D., & Reim, W. (2019). Reviewing literature on digitalization, business model innovation, and sustainable industry: Past achievements and future promises. doi:10.3390u11020391

Park, D., & Lee, J. (2008). eWOM overload and its effect on consumer behavioral intention depending on consumer involvement. *Electronic Commerce Research and Applications, 7*(4), 386–398. doi:10.1016/j.elerap.2007.11.004

Park, J. H., Lee, C., Yoo, C., & Nam, Y. (2016). An analysis of the utilization of Facebook by local Korean governments for tourism development and the network of smart tourism ecosystem. *International Journal of Information Management, 36*(6), 1320–1327. doi:10.1016/j.ijinfomgt.2016.05.027

Park, J. W. (2007). Passenger perceptions of service quality: Korean and Australian case studies. *Journal of Air Transport Management, 13*(4), 238–242. doi:10.1016/j.jairtraman.2007.04.002

Patel, K. (2013). Incremental journey for World Wide Web: Introduced with Web 1.0 to recent Web 5.0 – A survey paper. *International Journal of Advanced Research in Computer Science and Software Engineering*, *3*(10), 410–417.

Peacock, J. L., Kerry, S. M., & Balise, R. R. (2017). Presenting Medical Statistics from Proposal to Publication (2nd ed.). Oxford: Oxford University Press.

Pearce, P. L. (1977). Mental souvenirs: A study of tourists and their city maps. *Australian Journal of Psychology*, *29*(3), 203–210. doi:10.1080/00049537708255282

Pearce, P. L. (1982). Perceived changes in holiday destinations. *Annals of Tourism Research*, *9*(2), 145–164. doi:10.1016/0160-7383(82)90044-5

Peceny, U. S., Urbančič, J., Mokorel, S., Kuralt, V., & Ilijaš, T. (2019). Tourism 4.0: Challenges in Marketing a Paradigm Shift. *Intech Open*, *2*, 64. doi:10.5772/32009

Pellin, V. (2005). O turismo no espaço rural como alternativa para o desenvolvimento local sustentável: o caso do Município de Rio dos Cedros. *Caderno Virtual de Turismo*, *5*(1). Retrieved from http://www.inclusaodejovens.org.br/Documentos/BIBLIOTECA/Desenvolvimento/turismo_c omo_alternativa.pdf

Peltier, D. (2018). Independent Chinese Travelers Close Gap With Tour Groups. Skift. Retrieved from https://skift.com/2018/10/29/independent-chinese-travelers-close-gap-with-tour-groups/

Perçin, S. (2018). Evaluating airline service quality using a combined fuzzy decision-making approach. *Journal of Air Transport Management*, *68*, 48–60. doi:10.1016/j.jairtraman.2017.07.004

Peres, P., Moreira, F., & Mesquita, A. (2017). Higher education teachers in Portugal use technology in their educational activities: Myth or Reality? *Proceedings of the 12th Iberian Conference on Information Systems and Technologies (CISTI)* (p. 16). Academic Press. 10.23919/CISTI.2017.7975681

Peres, P., Moreira, F., & Mesquita, A. (2018). Are Really Technologies at the Fingers of Teachers? Results from a Higher Education Institution in Portugal. *Journal of Information Systems Engineering & Management*, *3*(1), 1–9. doi:10.20897/jisem.201808

Perretti, F., & Negro, G. (2006). Filling empty seats: How status and organizational hierarchies affect exploration versus exploitation in team design. *Academy of Management Journal*, *49*(4), 759–777. doi:10.5465/amj.2006.22083032

Perry, L., & Williams, C. (2015). *The Viability of Long-Haul Low-Cost Carrier Service*. Leighfisher. Retrieved from http://www.leighfisher.com/sites/default/files/free_files/viabilityoflong-haullow-costcarrierservice.pdf

Persson, H., Åhman, H., Yngling, A. A., & Gulliksen, J. (2015). Universal design, inclusive design, accessible design, design for all: Different concepts—one goal? On the concept of accessibility—historical, methodological and philosophical aspects. *Universal Access in the Information Society*, *14*(4), 505–526. doi:10.100710209-014-0358-z

Peters, L. D., Pressey, A. D., & Greenberg, P. (2010). The impact of CRM 2.0 on customer insight. *Journal of Business and Industrial Marketing*.

Petrocchi, M. (2002). *Hotelaria: planejamento e gestão*. São Paulo: Futura.

Petry, T. R. E., Pickler, C. M., & Tomelin, C. A. (2016). A Percepção dos Hóspedes de Negócios quanto ao Desempenho da Qualidade dos Serviços Prestados nos Hotéis de Florianópolis: Uma Análise a Partir do Conteúdo Gerado no Website Booking. Revista Turismo - Visão e Ação, 18.

Pettenati, M. C., & Cigognini, M. E. (2007). Social networking theories and tools to support connectivist learning activities. *International Journal of Web-Based Learning and Teaching Technologies*, *2*(3), 42–60. doi:10.4018/jwltt.2007070103

Pew Research Center. (2015). Social media update 2014. Retrieved from http://www.pewinternet.org/2015/01/09/social-media-update-2014/

Pezenka, I. (2016). Paired Comparisons or Sorting? Comparing Web-Based Methods for Collecting Similarity Data for Large Stimulus Sets for Destination Image Positioning. *Journal of Travel Research.*

Phillips, P., Zigan, K., Silva, M. M. S., & Schegg, R. (2015). The interactive effects of online reviews on the determinants of Swiss hotel performance: A neural network analysis. *Tourism Management*, *50*, 130–141. doi:10.1016/j.tourman.2015.01.028

Pierce, R., Stacey, K., & Barkatsas, A. (2007). A scale for monitoring students' attitudes to learning mathematics with technology. *Computers & Education*, *48*(2), 285–300. doi:10.1016/j.compedu.2005.01.006

Pike, S. (2002). Destination Image Analysis—A Review of 142 Papers from 1973 to 2000. *Tourism Management*, *23*(5), 541–549. doi:10.1016/S0261-5177(02)00005-5

Pike, S., & Page, S. (2014). Destination marketing organizations and destination marketing: A narrative analysis of the literature. *Tourism Management*, *41*, 202–227. doi:10.1016/j.tourman.2013.09.009

Pillow, B., Banks, M., & Reeves, D. (2015). *Equal access for all: providing for impaired stakeholders in a museum setting*. Retrieved from http://www.byronpillow.com/uploads/7/6/7/9/7679 7585/disability_access.pdf

Pimenta, C., Ribeiro, R., Sá, V., & Belfo, F. P. (2018). Factors Influencing School Success in Undergraduate Programs at a Portuguese Higher Education Institution. In *18ª Conferência da Associação Portuguesa de Sistemas de Informação (CAPSI'2018)* (p. 33). Santarém: Associação Portuguesa de Sistemas de Informação. Retrieved from https://aisel.aisnet.org/capsi2018/16

Pinchot, G. (1985). Intrapreneuring.

Pine, B. J., & Gilmore, J. H. (1999). *The Experience Economy: Work in Theatre and Every Business a Stage: Goods and Services are no longer enough*. Boston: Harvard Business School Press.

Pine, B. J., & Gilmore, J. H. (1999). *The experience economy: work is a theatre and every business a stage*. Boston: Harvard Business School Press.

Pira, A. (2004). *Etkinlik Yönetimi*. Istanbul: MediaCat Yayınları.

Pitt, M., & Bennett, J. (2008). Workforce ownership of space in a space sharing environment. *Journal of facilities management, 6*(4), 290-302.

Pons, F., Mourali, M., & Nyeck, S. (2006). Consumer orientation toward sporting events scale development and validation. *Journal of Service Research*, *8*(3), 276–287. doi:10.1177/1094670505283931

Poon, A. (1993). *Tourism, Technology and Competitive Strategies*. Wallingford: C.A.B. International.

Popiel, M. (2016). Barriers in undertaking tourist activity by disabled people. Prace Naukowe Akademii im. *Jana Długosza w Częstochowie Kultura Fizyczna*, *15*(3), 103–110.

Porter, M. E. (2000a). Location, competition, and economic development: Local clusters in a global economy. *Economic Development Quarterly*, *14*(1), 15–34. doi:10.1177/089124240001400105

Postma, A., Zuidhoek, S., Noordzij, M., & Kappers, A. (2007). Differences between early-blind, late-blind, and blindfolded-sighted people in haptic spatial-configuration learning and resulting memory traces. *Perception*, *36*(8), 1253–1265. doi:10.1068/p5441 PMID:17972487

Preece, J., Rogers, Y., & Sharp, H. (2002). *Interaction design: beyond human-computer interaction*. New York, NY: John Wiley & Sons, Inc.

Pühretmair, F. (2004, July). It's time to make eTourism accessible. *Proceedings of the International Conference on Computers for Handicapped Persons* (pp. 272-279). Springer. doi:10.1007/978-3-540-27817-7_41

Pyo, S., Mihalik, B. J., & Uysal, M. (1989). Attraction attributes and motivations. A canonical correlation analysis. *Annals of Tourism Research*, *16*(2), 277–282. doi:10.1016/0160-7383(89)90077-7

Quan, S., & Wang, N. (2004). Towards a structural model of tourist experience: An illustration from food experiences in tourism. *Tourism Management*, *25*(3), 297–305. doi:10.1016/S0261-5177(03)00130-4

Quivy, R., & Campenhoudt, L. (1995). *Manual de Investigação em Ciências Sociais* (2nd ed.). Lisboa: Gradiva.

Racherla, P., & Friske, W. (2012). Perceived 'usefulness' of online consumer reviews: An exploratory investigation across three services categories. *Electronic Commerce Research and Applications*, *11*(6), 548–559. doi:10.1016/j.elerap.2012.06.003

Raidah Asma, A. R., & Daud, D. (2012). Perceptions of travel planning for musicians: A Malaysian perspective. *Business Management Dynamics*, *2*(6), 56–63.

Rajaguru, R. (2016). Role of value for money and service quality on behavioural intention: A study of full service and low cost airlines. *Journal of Air Transport Management*, *53*, 114–122. doi:10.1016/j.jairtraman.2016.02.008

Ramos, C. M. Q., & Rodrigues, P. M. M. (2011). Os efeitos da internet na actividade turística. *Revista GEINTEC-Gestão. Inovação e Tecnologias*, *1*(2), 64–83. doi:10.7198/S2237-07222011000200006

Ramos, C. M., Matos, N., Sousa, C. M., Correia, M. B., & Cascada, P. (2017, July). Marketing Intelligence and Automation–An Approach Associated with Tourism in Order to Obtain Economic Benefits for a Region. *Proceedings of the International Conference on Universal Access in Human-Computer Interaction* (pp. 402-411). Cham: Springer. 10.1007/978-3-319-58706-6_32

Rapp, A., Beitelspacher, L. S., Grewal, D., & Hughes, D. E. (2013). Understanding social media effects across seller, retailer, and consumer interactions. *Journal of the Academy of Marketing Science*, *41*(5), 547–566. doi:10.100711747-013-0326-9

Rapp, A., Trainor, K. J., & Agnihotri, R. (2010). Performance implications of customer-linking capabilities: Examining the complementary role of customer orientation and CRM technology. *Journal of Business Research*, *63*(11), 1229–1236. doi:10.1016/j.jbusres.2009.11.002

Ray, A. (2008). *Experiential art: Marketing imitating art imitating life*. Retrieved from https://www.experiencetheblog.com/2008/05/experiential-art-marketing-imitating.html

Rayna, T., & Striukova, L. (2010). Large-scale open innovation: Open source vs. patent pools. *International Journal of Technology Management*, *52*(3), 477–496. doi:10.1504/IJTM.2010.035986

Redman, T. C. (2017). Does Your Company Know What to Do with All Its Data? *Harvard Business Review*. Retrieved from https://hbr.org/2017/06/does-your-company-know-what-to-do-with-all-its-data

Reich, C., Lindgren-Streicher, A., Beyer, M., Levent, N., Pursley, J., & Mesiti, L. A. (2011). Speaking out on art and museums : a study on the needs and preferences of adults who are blind or have low vision. Boston, MA: Museum of Science and Art Beyond Sight.

Reichinger, S., S., Löw, C., Sportun, S., Reichl, P., & Purgathofer, W. (2016). Spaghetti, Sink and Sarcophagus: Design Explorations of Tactile Artworks for Visually Impaired People. *Proceedings of the 9th Nordic Conference on Human-Computer Interaction* (pp. 82:1-82:6). New York, NY: ACM. 10.1145/2971485.2996471

Reichinger, A., Carrizosa, H. G., & Travnicek, C. (2018). Designing an interactive tactile relief of the Meissen table fountain. In K. Miesenberger & G. Kouroupetroglou (Eds.), *Computers Helping People with Special Needs* (pp. 209–216). Cham, Switzerland: Springer International Publishing. doi:10.1007/978-3-319-94274-2_28

Reichinger, A., Fuhrmann, A., Maierhofer, S., & Purgathofer, W. (2016). Gesture-based interactive audio guide on tactile reliefs. *Proceedings of the 18th International ACM SIGACCESS Conference on Computers and Accessibility - ASSETS*, (pp. 91–100). New York, NY: ACM. 10.1145/2982142.2982176

Rennie, K. M. (1997). Exploratory And Confirmatory Rotation Strategies in Exploratory Factor Analysis. *Paper presented at the Annual Meeting Of The Southwest Educational Research Association*. Academic Press. Retrieved from http://files.eric.ed.gov/fulltext/ED406446.pdf

Reuters. (2019). *Etihad to offer wireless streaming, remove screens from narrow-body fleet*. Retrieved from https://in.reuters.com/article/gulf-travel-etihad-airways-idINKCN1S40PE

Rheem, C. (2012). *Empowering inspiration: The future of travel search*. AMADEUS IT Group SA/PhoCus Wrights. February.

Richards, V., Pritchard, A., & Morgan, N. (2010). (Re)Envisioning tourism and visual impairment. *Annals of Tourism Research*, *37*(4), 1097–1116. doi:10.1016/j.annals.2010.04.011

Rindova, V. P., Petkova, A. P., & Kotha, S. (2007). Standing out: How new firms in emerging markets build reputation. *Strategic Organization*, *5*(1), 31–70. doi:10.1177/1476127006074389

Roberts, N., & Grover, V. (2012). Investigating firm's customer agility and firm performance: The importance of aligning sense and respond capabilities. *Journal of Business Research*, *65*(5), 579–585. doi:10.1016/j.jbusres.2011.02.009

Roche, M. (1994). Mega-events and urban policy. *Annals of Tourism Research*, *21*(1), 1–19. doi:10.1016/0160-7383(94)90002-7

Rodrigues, A. (2004). *The contribution of marketing communication on destination image formation process: the case of the Alentejo region*. Unpublished MSc thesis, Lisbon University Institute (ISCTE-IUL).

Rodrigues, A. (2015). *Lake-destination image assessment: the case of the Alqueva Lake*. Unpublished doctoral dissertation, University of Algarve.

Rodríguez, A. M., & O'Connell, J. F. (2017). Can low-cost long-haul carriers replace Charter airlines in the long-haul market? A European perspective. *Tourism Economics*, *24*(1), 64–78. doi:10.1177/1354816617724017

Rogers, M. (1999), *Boomers And Technogy: An Extended Conversation*. AARP. Retrieved from https://assets.aarp.org/www.aarp.org_/articles/computers/2009_boomers_and_technology_final_report.pdf

Romeo, K., Chottin, M., Ancet, P., & Pissaloux, E. (2018). Access to Artworks and Its Mediation by and for Visually Impaired Persons. In K. Miesenberger & G. Kouroupetroglou (Eds.), *Computers Helping People with Special Needs* (pp. 233–236). Cham, Switzerland: Springer International Publishing. doi:10.1007/978-3-319-94274-2_32

Roque, V., Martins, J. A., Silva, I., Rodrigues, C., & Simões, P. (2010). The ICT Area in High Education Tourism Courses - The Portuguese Case. *Proceedings of the International Congress on Tourism - CIT2010* (pp. 225–232). Porto: Instituto Superior de Ciências Empresariais e do Turismo (ISCET).

Roque, V., & Raposo, R. (2016). Social media as a communication and marketing tool in tourism: An analysis of online activities from international key player DMO. *Anatolia*, *27*(1), 58–70. doi:10.1080/13032917.2015.1083209

Rosen, L. D., Whaling, K., Carrier, L. M., Cheever, N. A., & Rokkum, J. (2013). The media and technology usage and attitudes scale: An empirical investigation. *Computers in Human Behavior*, *29*(6), 2501–2511. doi:10.1016/j.chb.2013.06.006 PMID:25722534

Roth, M. S., & Romeo, J. B. (1992). Matching Product Category and Country Image Perceptions: A Framework for Managing Country-Of-Origin Effects. *Journal of International Business Studies*, *23*(3), 477–497. doi:10.1057/palgrave.jibs.8490276

Rothschild, P. C. (2011). Social media use in sports and entertainment venues. *International Journal of Event and Festival Management*, *2*(2), 139–150. doi:10.1108/17582951111136568

Russell, J. A., & Pratt, G. (1980). A Description of The Affective Quality Attributed to Environments. *Journal of Personality and Social Psychology*, *38*(2), 311–322. doi:10.1037/0022-3514.38.2.311

Rüßmann, M., Lorenz, M., Gerbert, P., Waldner, M., Justus, J., Engel, P., & Harnisch, M. (2015). *Industry 4.0: The Future of Productivity and Growth in Manufacturing Industries*. Boston Consulting; doi:10.100712599-014-0334-4

Ruths, D., & Pfeffer, J. (2014). Social media for large studies of behavior. *Science*, *346*(6213), 1063–1064. doi:10.1126cience.346.6213.1063 PMID:25430759

Ryan, C., & Cave, J. (2005). Structuring Destination Image: A Qualitative Approach. *Journal of Travel Research*, *44*(2), 143–150. doi:10.1177/0047287505278991

Safko, L. (2010). *The social media bible: tactics, tools, and strategies for business success*. John Wiley & Sons.

Sahin, D. R., Sener, I., & Polat, A. S. (2018). Power of universe: How do you have social media effective of organisational image of airline companies? In C. Cobanoglu, L. M. Tanh, & A. Corbaci (Eds.), *Advances in business Hospitality, and Tourism Research* (Vol. 1, pp. 66–77). Florida: Anahei Publishing.

Saldaña. J. (2009). The coding manual for qualitative researchers. London: Sage.

Salehan, M., & Negahban, A. (2013). Social network on smartphones: When mobile phone become addictive. *Computers in Human Behavior*, *29*(6), 2632–2639. doi:10.1016/j.chb.2013.07.003

Sales, R., Carvalho, C., Arruda, D., & Albuquerque, J. (2015). Clique aqui para decidir: As recomendações online na decisão de compra dos serviços hoteleiros. *Revista de Turismo y Património Cultural*, *13*(3), 541–550. doi:10.25145/j.pasos.2015.13.038

Salomon, D. (2013). Moving on from Facebook: Using Instagram to connect with Undergraduates and engage in teaching and learning. *ACRL TechConnect*, *74*(8), 408–412.

Sameh, A. (2013). A Twitter Analytic Tool to Measure Opinion, Influence and Trust. *Journal of Industrial and Intelligent Information*, *1*(1), 37–45. doi:10.12720/jiii.1.1.37-45

San Martín, H., & Del Bosque, I. A. R. (2008). Exploring the Cognitive–Affective Nature of Destination Image and The Role of Psychological Factors in Its Formation. *Tourism Management*, *29*(2), 263–277. doi:10.1016/j.tourman.2007.03.012

Sangeetha. (2019, May 6). Tamil Nadu's biryani binge keeps cash registers ringing, *Life style & Food, The Hindu*. Retrieved from https://www.thehindu.com/life-and-style/food/tamil-nadus-biryani-binge-keeps-cash-registers-ringing/article25987062.ece

Santini, F. O., Lübeck, R. M., & Hoffmann, C. S. (2015). Promoção de vendas: Uma análise dos fatores influenciadores da intenção de compra de um produto em desconto. *Revista Pensamento Contemporâneo Em Administração*, *9*(4). doi:10.12712/rpca.v9i4.490

Saritha. (2019), Foreign investors show a growing appetite for Indian biryani. *Live mint*. retrieved from https://www.livemint.com/companies/start-ups/foreign-investors-show-a-growing-appetite-for-indian-biryani-1564637303890.html

Saturno, M., Pertel, V. M., Deschamps, F., De, F. R., & Loures, E. R. (2018). Proposal of an Automation Solutions Architecture for Industry 4. *0. LogForum*, *14*(2), 185–195. doi:10.17270/J.LOG.266

Saxena, A., & Khanna, U. (2013). Advertising on social network sites: A structural equation modelling approach. *Vision*, *17*(1), 17–25. doi:10.1177/0972262912469560

Schindler, R. M., & Bickart, B. (2005). Published word of mouth: Referable, consumer-generated information on the Internet. *Online Consumer Psychology: Understanding and influencing consumer behavior in the virtual world*, *32*, 35-61.

Schlüter, F. F., Hetterscheid, E., & Henke, M. (2017). A Simulation-Based Evaluation Approach for Digitalization Scenarios in Smart SupplyChain Risk Management. *Journal of Industrial Engineering and Management Science*, (1), 179–206. doi:10.13052/jiems2446-1822.2017.009

Schmallegger, D., & Carson, D. (2008). Blogs in Tourism: Changing Approaches to Information Exchange. *Journal of Vacation Marketing*, *14*(2), 99–110. doi:10.1177/1356766707087519

Schuckert, M., Liu, X., & Law, R. (2015). Insights into suspicious online ratings: Direct evidence from tripadvisor. *Asia Pacific Journal of Tourism Research*, *21*(3), 259–272. doi:10.1080/10941665.2015.1029954

Schuckert, M., Liu, X., & Law, R. (2015a). Hospitality and tourism online reviews: Recent trends and future directions. *Journal of Travel & Tourism Marketing*, *32*(5), 608–621. doi:10.1080/10548408.2014.933154

Schuckert, M., Liu, X., & Law, R. (2015b). A segmentation of online reviews by language groups: How English and non-English speakers rate hotels differently. *International Journal of Hospitality Management*, *48*, 143–149. doi:10.1016/j.ijhm.2014.12.007

Schumpeter, J. (1939). Business Cycles.

Schumpeter, J. (1943). *Capitalism*.

Schumpeter, P. (1934). *The Theory of Economic Development*. Cambridge, MA: Harvard University Press.

Schwab, C. B., & Martins, G. S. (2013). Turismo de Contemplação em Santa Vitória do Palmar/RS: uma pesquisa de demanda. Proceedings of the IV Seminário de Pesquisa em Turismo do Mercosul – SEMINTUR JR. Academic Press.

Schwab, K. (2016). *The Fourth Industrial Revolution*. World Economic Forum.

Seaton, A. V. (1996). The Marketing Concept in Tourism. In A. V. Seaton & M. M. Bennet (Eds.), *Marketing tourism products* (pp. 350–376). London: International Thomson Business Press.

Šebjan, U., Tominc, P., & Širec, K. (2017). Multiple conceptual modelling of perceived quality of in-flight airline services. *Promet-Traffic & Transportation*, *29*(3), 311–319. doi:10.7307/ptt.v29i3.2195

Sedlar, U., Kos, A., Pustisek, M., Bester, J., Pogacnik, M., Mali, L., & Stojmenova Duh, E. (2018). Tackling the Challenges of ICT Innovation and Talents for Industry 4.0. *IPSI BGD Transactions on Internet Research*, *14*(1).

Serrano-Guerrero, J., Olivas, J. A., Romero, F. P., & Herrera-Viedma, E. (2015). Sentiment analysis: A review and comparative analysis of web services. *Information Sciences*, *311*, 18–38. doi:10.1016/j.ins.2015.03.040

Seyfi, M. (2017). *Dijital Etkinlik Yönetimi*. Istanbul: Der Yayınları.

Shamma, H. M. (2012). Toward a comprehensive understanding of corporate reputation: Concept, measurement and implications. *International Journal of Business and Management*, *7*(16), 151. doi:10.5539/ijbm.v7n16p151

Shang, S., Chen, L., Wei, Z., Jesen, C. S., Wen, J., & Kalnis, P. (2016). Collective travel planning in spatial networks. *IEE Transactions on Knowledge and Data Engineering*, *28*(5), 1132–1146. doi:10.1109/TKDE.2015.2509998

Shan, Y. (2015). How credible are online product reviews? The effects of self-generated and system-generated cues on source credibility evaluation. *Computers in Human Behavior*, *55*, 633–641. doi:10.1016/j.chb.2015.10.013

Shao, G. (2009). Understanding the appeal of user-generated media: A uses and gratification perspective. *Internet Research*, *19*(1), 7–25. doi:10.1108/10662240910927795

Sharma, D., Gupta, A., Mateen, A., & Pratap, S. (2018). 'Making sense of the changing face of Google's search engine results page: An advertiser's perspective', Journal of Information. *Communication & Ethics in Society*, *16*(1), 90–107. doi:10.1108/JICES-06-2017-0035

Sharpley, R., & Telfer, D. (2002). *Tourism and development – concepts and issues. Channel View Publication.* Clevedon: Library of Congress Cataloging in Publication Data.

Sherman, L. E., Payton, A. A., Hernandez, L. M., Greenfield, P. M., & Dapretto, M. (2016). The power of the like in adolescence: Effects of peer influence on neural and behavioural responses to social media. *Psychological Science*, *27*(7), 1027–1035. doi:10.1177/0956797616645673 PMID:27247125

Shone, A., & Parry, B. (2004). *Successful Event Management: A Practical Handbook*. Cengage Learning EMEA.

Short, D. B. (2016). Use of 3D Printing by Museums: Educational Exhibits, Artifact Education, and Artifact Restoration. *3D Printing and Additive Manufacturing*. doi:10.1089/3dp.2015.0030

Sigala, M. (2011). eCRM 2.0 applications and trends: The use and perceptions of Greek tourism firms of social networks and intelligence. *Computers in Human Behavior*, *27*(2), 655–661. doi:10.1016/j.chb.2010.03.007

Sigala, M. (2015). Applying Gamification and Assessing its Effectiveness in a Tourism Context: Behavioural and Psychological outcomes of the Tripadvisor's Gamification Users. *Asia Pacific Journal of Information System*, *25*(1), 179–210. doi:10.14329/apjis.2015.25.1.179

Sigala, M., Christou, E., & Gretzel, U. (2012). *Social Media in Travel, Tourism and Hospitality: Theory, Practice, and Cases*. UK: MPG Books Group.

Sigala, M., Christou, E., & Gretzel, U. (Eds.). (2012). *Social media in travel, tourism and hospitality: Theory, practice and cases*. Ashgate Publishing, Ltd.

Sillence, E., Briggs, P., Harris, P. R., & Fishwick, L. (2007). How do patients evaluate and make use of online health information? *Social Science & Medicine*, *64*(9), 1853–1862. doi:10.1016/j.socscimed.2007.01.012 PMID:17328998

Silva, J., & Perna, F. (2002). Turismo e desenvolvimento auto-sustentado. In J. Costa (Ed.), Compêndio de Economia Regional (pp. 451-474). Coimbra: APDR.

Silva, P. M. & Silva, L. M. (2008). Estratégia de CRM para hotelaria. *Revista Acadêmica Observatório de Inovação do Turismo, Rio de Janeiro, 3*(2).

Silva, T. (2012). Para Entender o Monitoramento de Mídias Sociais. Editora Bookess Online. Retrieved from https://bit.ly/2SY6aBY

Silva, D. M. (2013). *Facebook - Canal de reclamações dos clientes hoteleiros [Dissertação]*. Universidade do Porto.

Silva, J., & Silva, J. (2003). Inserção territorial das actividades turísticas em Portugal: Uma tipologia de caracterização. *Revista Portuguesa de Estudos Regionais.*, *1*, 53–73.

Simms, A. (2012) Online user-generated content for travel planning- Different for different kinds of trips? *e-Review of Tourism Research*, *10*(3), 76-85.

Sing, T., Sander, O., Beerenwinkel, N., & Lengauer, T. (2015). *ROCR: Visualising the Performance of Scoring Classifiers (R package)*. Retrieved from https://cran.r-project.org/package=ROCR

Singh, M., Sarfaraz, A., Sarfaraz, M., & Jenab, K. (2015). Analytical QFD model for strategic justification of advanced manufacturing technology. *International Journal of Business Excellence*, *8*(1), 20–37. doi:10.1504/IJBEX.2015.065979

Singh, P. K., & Pattanayak, J. K. (2016). Study of the Relationship among the Factors of Brand Equity: A Study on Fast-food Brands. *Global Business Review*, *17*(5), 1227–1239. doi:10.1177/0972150916656694

Sirilli, G., & Evangelista, R. (1998). Innovation in the service sector. Results from the Italian Statistical Survey. *Research Policy*, *27*, 881–899. doi:10.1016/S0048-7333(98)00084-5

Skift.com. (2014). State of travel: 2014. Retrieved from http://skift.com/wp-content/uploads/2014/08/Skift-2014-State-of-Travel-Report-Full.pdf

Small, J., Darcy, S., & Packer, T. (2012). The embodied tourist experiences of people with vision impairment: Management implications beyond the visual gaze. *Tourism Management*, *33*(4), 941–950. doi:10.1016/j.tourman.2011.09.015

Smallman, C., & Moore, K. (2010). Process Studies of Tourists'decision-Making. *Annals of Tourism Research*, *37*(2), 397–422. doi:10.1016/j.annals.2009.10.014

Smith, K. T. (2011). Digital marketing strategies that Millennials find appealing, motivating, or just annoying. *Journal of Strategic Marketing*, *19*(6), 489–499. doi:10.1080/0965254X.2011.581383

Smith, T. C., Coyle, J. R., Lightfoot, E., & Scott, A. (2007). Reconsidering models of influence:the relationship between consumer social networks and word-of-mouth effectiveness. *Journal of Advertising Research*, *47*(4), 387–397. doi:10.2501/S0021849907070407

Smits, M., & Mogos, S. (2013, June). The Impact Of Social Media On Business Performance. In ECIS 2013 Completed Research (p. 125). AIS Electronic Library.

Smyth, C., & Crabtree, M. (2012). The Use of RFID Technology At Pedestrian Crossings To assist disabled users: on-street trial results.

Soares, A. B., Francischetto, V., Dutra, B. M., Miranda, J. M., de, Nogueira, C. C. de C., & Leme, V. R. (2014). O impacto das expectativas na adaptação acadêmica dos estudantes no Ensino Superior. *Psico-USF*, *19*(1), 49–60. doi:10.1590/S1413-82712014000100006

Soares, A. M., Pinho, J. C., & Nobre, H. (2012). From Social to Marketing Interactions: The Role of Social Networks. *Journal of Transnational Management*, *17*(1), 45–62. doi:10.1080/15475778.2012.650085

Sobaih, A. E. E., Moustafa, M. A., Ghandforoush, P., & Khan, M. (2016). To use or not to use? Social media in higher education in developing countries. *Computers in Human Behavior*, *58*, 296–305. doi:10.1016/j.chb.2016.01.002

Song, H., & Liu, H. (2017). Predicting tourist demand using big data. In *Analytics in smart tourism design* (pp. 13–29). Cham: Springer International Publishing. doi:10.1007/978-3-319-44263-1_2

Sophia van Zyl, A. (2009). The impact of Social Networking 2.0 on organisations. *The Electronic Library*, *27*(6), 906–918. doi:10.1108/02640470911004020

Sotiriadis, M. D. (2017). Sharing tourism experiences in social media: A literature review and a set of suggested business strategies. *International Journal of Contemporary Hospitality Management*, *29*(1), 179–225. doi:10.1108/IJCHM-05-2016-0300

Sotiriadis, M. D., & Van Zyl, C. (2013). Electronic Word-Of-Mouth and Online Reviews in Tourism Services: The Use of Twitter by Tourists. *Electronic Commerce Research*, *13*(1), 103–124. doi:10.100710660-013-9108-1

Sousa, B., & Alves, G. (2019). The role of relationship marketing in behavioural intentions of medical tourism services and guest experiences. *Journal of Hospitality and Tourism Insights*, *2*(3), 224–240. doi:10.1108/JHTI-05-2018-0032

Souza, F., De Las Casas, D., Flores, V., Youn, S., Cha, M., Quercia, D., & Almeida, V. (2015). Dawn of the selfie era: The whos, wheres, and hows of selfies on Instagram. In *COSN 2015 - Proceedings of the 2015 ACM Conference on Online Social Networks* (pp. 221–231). ACM. doi:10.1145/2817946.2817948

Soyk, C., Ringbeck, J., & Spinler, S. (2017). Long-haul low cost airlines: Characteristics of the business model and sustainability of its cost advantages. *Transportation Research Part A, Policy and Practice*, *106*, 215–234. doi:10.1016/j.tra.2017.09.023

Sparks, B. A., & Browning, V. (2010). Complaining in cyberspace: The motive and forms of hotel guests' complaints online. *Journal of Hospitality Marketing & Management*, *19*(7), 797–818. doi:10.1080/19368623.2010.508010

Sparks, B., & Browning, V. (2011). The impact of online reviews on hotel booking intentions and perception of trust. *Tourism Management*, *32*(6), 1310–1323. doi:10.1016/j.tourman.2010.12.011

Spicer, J. (2018). 'That's their brand, their business': How police officers are interpreting County Lines. *Policing and Society*, 1–14.

Stamboulis, Y., & Skayannis, P. (2003). Innovation strategies and technology for experience-based tourism. *Tourism Management*, *24*(1), 35–43. doi:10.1016/S0261-5177(02)00047-X

StatCounter. (2019) Social Media Stats Worldwide. Retrieved from https://bit.ly/2zNzuTr

Statista. (2019). Number of social network users in India from 2015 to 2023. Retrieved from https://www.statista.com/statistics/278407/number-of-social-network-users-in-india

Statista. (2019a). *Number of global social media users 2010-2021*. Retrieved from https://www.statista.com/statistics/278414/number-of-worldwide-social-network-users/

Statista. (2019b). *Most famous social network sites 2019, by active users*. Retrieved from https://www.statista.com/statistics/272014/global-social-networks-ranked-by-number-of-users/

Stepchenkova, S., & Mills, J. E. (2010). Destination Image: A Meta-Analysis Of 2000–2007 Research. *Journal of Hospitality Marketing & Management*, *19*(6), 575–609. doi:10.1080/19368623.2010.493071

Stepchenkova, S., & Morrison, A. M. (2006). The Destination Image of Russia: From the Online Induced Perspective. *Tourism Management*, *27*(5), 943–956. doi:10.1016/j.tourman.2005.10.021

Stern, E., & Krakover, S. (1993). The Formation of a Composite Urban Image. *Geographical Analysis*, *25*(2), 130–146. doi:10.1111/j.1538-4632.1993.tb00285.x

Stevens, J. P. (2012). *Applied multivariate statistics for the social sciences*. London: Routledge.

Streubert, H., & Carpenter, D. (1999). *Qualitative research in nursing: advancing the humanistic perspective* (2nd ed.). Philadelphia: Lippincott Williams & Wilkins.

Stringam, B., & Gerdes, J. Jr. (2010). An Analysis of word-of-mouse ratings and guest comments of online hotel distribution sites. *Journal of Hospitality Marketing & Management*, *19*(7), 773–796. doi:10.1080/19368623.2010.508009

Stringam, B., Gerdes, J. Jr, & Vanleeuwen, D. (2010). Assessing the importance and relationships of ratings on user-generated traveler reviews. *Journal of Quality Assurance in Hospitality & Tourism*, *11*(2), 73–92. doi:10.1080/1528008X.2010.482000

Stumbo, N. J., & Pegg, S. (2005). Travelers and Tourists With Disabilities : A Matter of Priorities and Loyalties. *Tourism Review International*, *8*(3), 195–209. doi:10.3727/154427205774791537

Sundbo, J. (1996). Balancing Empowerment. *Technovation*, *16*(8), 397–409. doi:10.1016/0166-4972(96)00029-6

Sundbo, J. (1997). Management of Innovation in Services. *Service Industries Journal*, *17*(3), 432–455. doi:10.1080/02642069700000028

Sundbo, J. (1998). *The theory of innovation: enterpreneurs, technology and strategy*. Edward Elgar Publishing.

Sundbo, J. (1998). *The Theory of Innovation: Entrepreneurs, Technology and Strategy*. Cheltenham: Edward Elgar.

Sundbo, J. (2001). *The Strategic Management of Innovation*. Cheltenham: Elgar.

Sundbo, J., Orfila-Sintes, F., & Sørensen, F. (2007). The innovative behaviour of tourism firms-Comparative studies of Denmark and Spain. *Research Policy*, *36*(1), 88–106. doi:10.1016/j.respol.2006.08.004

Sun, S., Law, R., Schuckert, M., Kucukusta, D., & Guillet, B. D. (2017). From information technology to mobile information technology: Applications in hospitality and tourism. *Ereview of Tourism Research*.

Sussman, S., & Siegal, W. (2003). Informational influence in organizations: An integrated approach to knowledge adoption. *Information Systems Research*, *14*(1), 47–65. doi:10.1287/isre.14.1.47.14767

Tabachnick, B. G., & Fidell, L. S. (2013). Using Multivariate Statistics (6th ed.). New Jersey: Pearson Education, Inc.

Tadelis, S. (2016). Reputation and feedback systems in online platform markets. *Annual Review of Economics*, *8*(1), 321–340. doi:10.1146/annurev-economics-080315-015325

Tan, G. W. H., Lee, V. H., Lin, B., & Ooi, K. B. (2017). Mobile applications in tourism: The future of the tourism industry? *Industrial Management & Data Systems*, *117*(3), 560–581. doi:10.1108/IMDS-12-2015-0490

Tapachai, N., & Waryszak, R. (2000). An Examination of The Role of Beneficial Image in Tourist Destination Selection. *Journal of Travel Research*, *39*(1), 37–44. doi:10.1177/004728750003900105

Tasci, A. D., Gartner, W. C., & Tamer Cavusgil, S. (2007). Conceptualization and Operationalization of Destination Image. *Journal of Hospitality & Tourism Research (Washington, D.C.)*, *31*(2), 194–223. doi:10.1177/1096348006297290

Tassiopoulus, D. (2005). *Events-An Introduction. Event Management: A Professional and Developmental Approach* (D. Tassiopoulus, Ed.). Lansdowne: Juta Academic.

Tavakoli, R., & Wijesinghe, S. N. (2019). Tourism management perspectives. *Tourism Management*, *29*, 48–55.

Tecnohotel.news. (2014). CRM + Redes Sociales = Social CRM: el nuevo consumidor social. Retrieved from https://www.tecnohotelnews.com/2014/06/crm-redes-sociales-social-crm-el-nuevo-consumidor-social/#

Teece, D. J., Pisano, G., & Shuen, A. (1997). Dynamic capabilities and strategic management. *Strategic Management Journal, 18*(7), 509–533. doi:10.1002/(SICI)1097-0266(199708)18:7<509::AID-SMJ882>3.0.CO;2-Z

Teixeira, A. H., & Tomelin, C. A. (2005). O Marketing de Relacionamento como Força Competitiva para as Agências de Viagens de Santa Catarina. *Proceedings of the III Seminário de Pesquisa em Turismo do Mercosul – SEMINTUR.* Academic Press.

Telisman-Kosuta, N. (1989). Tourist Destination Image. In *Tourism Marketing and Management Handbook* (pp. 557-561). Academic Press.

Tencent. (2019). WeChat is a way of life. Retrieved from https://weixin.qq.com/

Thakran, K., & Verma, R. (2013). The emergence of hybrid online distribution channels in travel, tourism and hospitality. *Cornell Hospitality Quarterly, 54*(3), 240–247. doi:10.1177/1938965513492107

Tham, A., Croy, G., & Mair, J. (2013). Social media in destination choice: Distinctive electronic word-of-mouth dimensions. *Journal of Travel & Tourism Marketing, 30*(1-2), 144–155. doi:10.1080/10548408.2013.751272

The World Bank. (2018). China. Retrieved from https://data.worldbank.org/country/china

Thomaz, G. M., Biz, A. A., & Bettoni, E. M. (2015). Análise da gestão e monitoramento em mídias sociais pelas Destination Management Organizations (DMO) de Curitiba-PR. *Proceedings of the XII Seminário ANPTUR.* Academic Press. Retrieved from http://whc.unesco.org/en/list/355

Thurau, T; Malthouse, E.; Friege, C.; Gensler, S.; Lobschat, L.; Rangaswamy, A.; & Skiera, B., (2010). The impact of new media on customer relationships. *Journal of Service Research, 13*, 311-330. doi:. doi:10.1177/1094670510375460

Thüring, M., & Mahlke, S. (2007). Usability, aesthetics and emotions in human–technology interaction. *International Journal of Psychology, 42*(4), 253–264. doi:10.1080/00207590701396674

Tidd, J., Bessant, J., & Pavitt, K. (1977). *Managing Innovation.* Integrating Technological and Organizational Change.

Tobergte, D. R., & Curtis, S. (2013). *Einführung und Umsetzung von Industrie 4.0 - Grundlagen, Vorgehensmodell und Use Cases aus der Praxis. Journal of Chemical Information and Modeling, 53.* doi:10.1017/CBO9781107415324.004

Toffler, A. (1970). *The future shock.* USA: Bantman Books.

Toh, R. S., DeKay, C. F., & Raven, P. (2011). Travel Planning: Searching for and Booking Hotels on the Internet. *Cornell Hospitality Quarterly, 52*(4), 388–398. doi:10.1177/1938965511418779

Torres, C. (2010). *Guia Prático de Marketing na Internet para Pequenas Empresas.*

Torres, E. N., Adler, H., & Behnke, C. (2014). Stars, diamonds, and other shiny things: The use of expert and consumer feedback in the hotel industry. *Journal of Hospitality and Tourism Management, 21*, 34–43. doi:10.1016/j.jhtm.2014.04.001

Trainor, K. J. (2012). Relating social media technologies to performance: A capabilities-based perspective. *Journal of Personal Selling & Sales Management, 32*(3), 317–331. doi:10.2753/PSS0885-3134320303

Travel China Guide. (2019). 2018 China Tourism Facts & Figures. Retrieved from https://www.travelchinaguide.com/tourism/2018statistics/

Travel Lifestyle Network. (2018). Using WeChat to attract more Chinese travellers. Retrieved from https://travellifestylenetwork.com/news/using-wechat-to-attract-more-chinese-travellers/

Travel Market Report. (2018). Travel market report's outlook on social media based on the voice of the travel advisor 2018-2019. Retrieved from https://www.travelmarketreport.com/library/TMR/TMR_SocialMediaOutlook.pdf

Treem, J. W., & Leonardi, P. M. (2013). Social media use in organizations: Exploring the affordances of visibility, editability, persistence, and association. *Annals of the International Communication Association*, *36*(1), 143–189. doi:10.1080/23808985.2013.11679130

TripAdvisor. (2014). Responding to Reviews Online Drives Booking Inquiries, TripAdvisor Study Reveals. Retrieved from http://globenewswire.com/news-release/2014/09/03/663342/10096925/en/Respondingto-Reviews-Online-Drives-Booking-Inquiries-TripAdvisor-Study-Reveals.html

TripAdvisor. (2017). PT Media Center | About TripAdvisor. Retrieved from https://tripadvisor.mediaroom.com/pt-about-us

TrueSocialMetrics. (2018). Glossário. Retrieved from https://bit.ly/2ku1FP2

Tsai, C. Y., & Chung, S. H. (2012). A personalized route recommendation service for theme parks using RFID information and tourist behavior. *Decision Support Systems*, *52*(2), 514–527. doi:10.1016/j.dss.2011.10.013

Tsai, W. H. S., & Men, L. R. (2017). Consumer engagement with brands on social network sites: A cross-cultural comparison of China and the USA. *Journal of Marketing Communications*, *23*(1), 2–21. doi:10.1080/13527266.2014.942678

Tsao, W. C., Hsieh, M. T., Shih, L. W., & Lin, T. M. (2015). Compliance with eWOM: The influence of hotel reviews on booking intention from the perspective of consumer conformity. *International Journal of Hospitality Management*, *46*, 99–111. doi:10.1016/j.ijhm.2015.01.008

Tsaur, S. H., Chang, T. Y., & Yen, C. H. (2002). The evaluation of airline service quality by fuzzy MCDM. *Tourism Management*, *23*(2), 107–115. doi:10.1016/S0261-5177(01)00050-4

Tse, E., & Ho, S. (2009). Service quality in the hotel industry when cultural contexts matter. *Cornell Hospitality Quarterly*, *50*(4), 460–474. doi:10.1177/1938965509338453

Tu, Q., & Liu, A. (2014). Framework of Smart Tourism Research and Related Progress in China. *Proceedings of the International Conference on Management and Engineering (CME 2014)* (pp. 140-146). DES tech Publications.

Tussyadiah, I., & Fesenmaier, D. R. (2009). Mediating tourist experiences: Access to places via shared videos. *Annals of Tourism Research*, *36*(1), 24–40. doi:10.1016/j.annals.2008.10.001

Tussyadiah, I., & Zach, F. (2012). The role of geo-based technology in place experiences. *Annals of Tourism Research*, *39*(2), 780–780. doi:10.1016/j.annals.2011.10.003

Tussyadiah, S. P. II, Kausar, D. R., & Soesilo, P. K. M. (2018). The Effect of Engagement in Online Social Network on Susceptibility to Influence. *Journal of Hospitality & Tourism Research (Washington, D.C.)*, *42*(2), 201–223. doi:10.1177/1096348015584441

Twenge, J. M., Campbell, S. M., Hoffman, B. J., & Lance, C. E. (2010). Generational differences in work values: Leisure and extrinsic values increasing, social and intrinsic values decreasing. *Journal of Management*, *36*(5), 1117–1142. doi:10.1177/0149206309352246

Uber eats India. (n.d.). Retrieved from https://www.facebook.com/ubereatsindia/?brand_redir=696009167187917

Ukpabi, D. C., & Karjaluoto, H. (2018). What drives travelers' adoption of user-generated content? A literature review. *Tourism Management Perspectives*, *28*, 251–273. doi:10.1016/j.tmp.2018.03.006

United Nations. (2006). The UN convention on the rights of persons with disabilities (Vol. 49). Academic Press. doi:10.1057/palgrave.development.1100310

UNWTO. (1980). Manila Declaration on World Tourism. *Proceedings of the World Tourism Conference*. Academic Press.

UNWTO. (2009). Understanding Tourism: Basic Glossary. World Tourism Organization. Academic Press. doi:10.1177/1479164111417471

UNWTO. (2011). World Tourism Organization (UNWTO) Affiliate Members: AM-reports – Technology in Tourism (Vol. 1). UNWTO. Retrieved from https://bit.ly/2MyXNKi

Urbas, R., Pivar, M., & Elesini, U. S. (2016). Development of tactile floor plan for the blind and the visually impaired by 3D printing technique. *Journal of Graphic Engineering and Design*, *7*(1), 19–26. doi:10.24867/JGED-2016-1-019

Uriely, N. (2005). The tourist experience. *Annals of Tourism Research*, *32*(1), 199–216. doi:10.1016/j.annals.2004.07.008

Urry, J. (2001). *Globalizing the tourist gaze,* published by the Department of Sociology. Lancaster University. Retrieved from https://www.lancaster.ac.uk/fass/resources/sociology-online-papers/papers/urry-globalising-the-tourist-gaze.pdf

Urry, J., Larsen, J., & Axhausen, K. (2007). Networks and tourism. Mobile social life. *Annals of Tourism Research*, *34*(1), 244–262. doi:10.1016/j.annals.2006.08.002

Valdivia, A., Luzón, M. V., & Herrera, F. (2017). Sentiment analysis on TripAdvisor: Are there inconsistencies in user reviews? In F. Martínez de Pisón, R. Urraca, H. Quintián, & E. Corchado (Eds.), *Hybrid artificial intelligent systems*. Cham: Springer. doi:10.1007/978-3-319-59650-1_2

Van Doorn, J., Lemon, K. N., Mittal, V., Nass, S., Pick, D., Pirner, P., & Verhoef, P. C. (2010). Customer engagement behavior: Theoretical foundations and research directions. *Journal of Service Research*, *13*(3), 253–266. doi:10.1177/1094670510375599

Van Norel, N. D., Kommers, P. A., Van Hoof, J. J., & Verhoeven, J. W. (2014). Damaged corporate reputation: Can celebrity Tweets repair it? *Computers in Human Behavior*, *36*, 308–315. doi:10.1016/j.chb.2014.03.056

Vargas-Sánchez, A. (2018). The unavoidable disruption of the circular economy in tourism. *Worldwide Hospitality and Tourism Themes*, *10*(6), 652–661. doi:10.1108/WHATT-08-2018-0056

Venkatesh, V., Morris, M. G., Davis, G. B., & Davis, F. D. (2003). User acceptance of information technology: Toward a unified view. *Source: MIS Quarterly*, *27*(3), 425–478. doi:10.2307/30036540

Veríssimo, M., & Menezes, N. (2016). Social media as a tool to enhance customer experience in hospitality industry. *Revista Portuguesa de Marketing*, *38*(4), 23–30.

Vermeulen, I., & Seegers, D. (2009). Tried and tested: The impact of online hotel reviews on consumer consideration. *Tourism Management*, *30*(1), 123–127. doi:10.1016/j.tourman.2008.04.008

Vieira, A. L., Winklhofer, H., & Ennew, C. T. (2008). Relationship Quality: A literature review and research agenda. *Journal of Customer Behaviour*, *7*(4), 269–291. doi:10.1362/147539208X386833

Viglia, G., Minazzi, R., & Buhalis, D. (2016). The influence of e-word-of-mouth on hotel occupancy rate. *International Journal of Contemporary Hospitality Management*, *28*(9), 2035–2051. doi:10.1108/IJCHM-05-2015-0238

Villalobos, L. (2017). Novas regras fazem duplicar registos de alojamento local. Retrieved from https://bit.ly/2Ka1xlm

Virtual Assistant. (2019) Cambridge dictionary. Retrieved from https://dictionary.cambridge.org/pt/dicionario/ingles/virtual-assistant

Visit Porto. (n.d.). Retrieved from www.visitporto.travel

Vlachvei, A., & Notta, O. (2014). Social Media adoption and managers' perceptions. *International Journal of Strategic Innovative Marketing*, *1*(2), 61–73. doi:10.15556/IJSIM.01.02.001

Volo, S. (2004). Conceptualizing experience: A tourist-based approach. *Journal of Hospitality Marketing & Management, 18*(2), 111–126.

Volo, S. (2010). Bloggers reported tourist experiences: Their utility as a tourism data source and their effect on perspective tourists. *Journal of Vacation Marketing, 16*(4), 297–311. doi:10.1177/1356766710380884

Vorhies, D. W., & Morgan, N. A. (2005). Benchmarking marketing capabilities for sustainable competitive advantage. *Journal of Marketing, 69*(1), 80–94. doi:10.1509/jmkg.69.1.80.55505

Vučetić, S. (2017). The Importance of Using ICT in the Rural Tourism of the Zadar County. *DIEM – 4th Dubrovnik International Economic Meeting*. Academic Press.

Vu, H. Q., Li, G., Law, R., & Ye, B. H. (2015). Exploring the travel behaviour of inbound tourists to Hongkong using geotagged photos. *Tourism Management, 46*, 222–232. doi:10.1016/j.tourman.2014.07.003

Walls, A. R., Okumus, F., Wang, Y. R., & Kwun, D. J. (2007). An epistemological view of consumer experiences. *International Journal of Hospitality Management, 30*(1), 10–21. doi:10.1016/j.ijhm.2010.03.008

Walmsley, D. J., & Jenkins, J. M. (1992). Tourism Cognitive Mapping of Unfamiliar Environments. *Annals of Tourism Research, 19*(2), 268–286. doi:10.1016/0160-7383(92)90081-Y

Walmsley, D. J., & Young, M. (1998). Evaluative Images and Tourism: The Use of Personal Constructs to Describe the Structure of Destination Images. *Journal of Travel Research, 36*(3), 65–69. doi:10.1177/004728759803600307

Walsh, G., Hennig-Thurau, T., Sassenberg, K., & Bornemann, D. (2010). Does relationship quality matter in e-services? A comparison of online and offline retailing. *Journal of Retailing and Consumer Services, 17*(2), 130–142.

Walsh, G., Mitchell, V. W., Jackson, P. R., & Beatty, S. E. (2009). Examining the antecedents and consequences of corporate reputation: A customer perspective. *British Journal of Management, 20*(2), 187–203. doi:10.1111/j.1467-8551.2007.00557.x

Walters, P., & Davies, K. (2010). *3D printing for artists: research and creative practice*. Rapport.

Walters, T. (2016). Using Thematic Analysis in Tourism Research. *Tourism Analysis, 21*(1), 107–116. doi:10.3727/108354216X14537459509017

Wang, C. Y., & Hsu, M. K. (2010). The Relationships of Destination Image, Satisfaction, And Behavioral Intentions: An Integrated Model. *Journal of Travel & Tourism Marketing, 27*(8), 829–843. doi:10.1080/10548408.2010.527249

Wang, D., Park, S., & Fesenmaier, D. (2012). The role of smartphones inmediating the tourism experience. *Journal of Travel Research, 54*(4), 371–387. doi:10.1177/0047287511426341

Wang, D., Xiang, Z., & Fesenmaier, D. R. (2016). Smartphone use in everyday life and travel. *Journal of Travel Research, 55*(1), 52–63. doi:10.1177/0047287514535847

Wang, Y., & Fesenmaier, D. R. (2004). Modeling participation in an online travel community. *Journal of Travel Research, 42*(3), 261–270. doi:10.1177/0047287503258824

Wang, Y., & Fesenmaier, D. R. (2007). Collaborative Destination Marketing: A Case Study of Elkhart County, Indiana. *Tourism Management, 28*(3), 863–875. doi:10.1016/j.tourman.2006.02.007

Wang, Y., Yu, Q., & Fesenmaier, D. R. (2002). Defining the virtual tourist community: Implications for tourism marketing. *Tourism Management, 23*(4), 407–417. doi:10.1016/S0261-5177(01)00093-0

Wang, Z., & Kim, H. G. (2017). Can social media marketing improve customer relationship capabilities and firm performance? Dynamic capability perspective. *Journal of Interactive Marketing, 39*, 15–26. doi:10.1016/j.intmar.2017.02.004

Waschke, S. (2004). *Labeling im Barrierefreien Tourismus in Deutschland – Vergleichende Analyse auf Basis Europäischer Beispiele* [Thesis]. Universität Lüneburg.

Wearesocial.com. (2018). Gloval digital Report. Retrieved from https://wearesocial.com/blog/2018/01/global-digital-report-2018

Weber, S., & Rech, J. (2010). An overview and differentiation of the evolutionary steps of the web X.Y movement. Handbook of Research on Web 2.0, 3.0, and X.0, 12-39. Doi:10.4018/978-1-60566-384-5.ch002

Weinberg, B. D., & Pehlivan, E. (2011). Social spending: Managing the social media mix. *Business Horizons, 54*(3), 275–282. doi:10.1016/j.bushor.2011.01.008

Welbourne, D. G. (2016). Science communication on YouTube: factors that affect channel and video popularity. *public understanding of science, 25*(6), 706-718. doi:10.1177/0963662515572068

Wensveen, J. G., & Leick, R. (2009). The long-haul low-cost carrier: A unique business model. *Journal of Air Transport Management, 15*(3), 127–133. doi:10.1016/j.jairtraman.2008.11.012

Werthner, H., & Klein, S. (1999). *Information Technology and Tourism – A Challenging Relationship*. Springer-Verlag.

Werthner, H., & Klein, S. (1999). *Information technology and tourism: A challenging relationship*. Wien: Springer-Verlag. doi:10.1007/978-3-7091-6363-4

Wesgro. (2017). France: Tourism Market Insights 2017. Retrieved from http://www.wesgro.co.za/pdf_repository/Tourism Market Insights_France.pdf

Whyte, R., & Lohmann, G. (2015). Low-cost long-haul carriers: A hypothetical analysis of a "Kangaroo route." *Case Studies on Transport Policy, 3*(2), 159–165. doi:10.1016/j.cstp.2015.01.003

Williams, A. (2007). A lecture on "Tourism Innovations." In Lecture Series in Tourism at Waterloo University, Canada. Academic Press.

Williams, M., & Buttle, F. (2011). The eight pillars of WOM management: Lessons from a multiple case study. *Australasian Marketing Journal, 19*(2), 85–92. doi:10.1016/j.ausmj.2011.01.001

Wilson, A., Murphy, H., & Fierro, J. (2012). Hospitality and travel: The nature and implications of user-generated content. *Cornell Hospitality Quarterly, 53*(3), 220–228. doi:10.1177/1938965512449317

Wittmer, A. (2011). Acceptance of self-service check-in at Zurich airport. *Research in Transportation Business & Management, 1*(1), 136–143. doi:10.1016/j.rtbm.2011.06.001

WodeMaya. (2019, May 04) Do not come to China [YouTube video]. Retrieved from https://www.youtube.com/watch?v=Av0y_bZ_KPU

Wolny, J., & Charoensuksai, N. (2014). Mapping customer journeys in multichannel decision-making. Journal of Direct. *Data and Digital Marketing Practice, 15*(4), 317–326. doi:10.1057/dddmp.2014.24

Wolter, M. I., Mönnig, A., Hummel, M., Weber, E., Zika, G., Helmrich, R., ... Neuber-Pohl, C. (2015). Industrie 4.0 und die Folgen für Arbeitsmarkt und Wirtschaft: Szenario-Rechnungen im Rahmen der BIBB-IAB-Qualifikations- und Berufsfeldprojektionen. *IAB-Forschungsbericht, 8*, 68. doi:10.100710273-011-1262-2

Wong, J., & Law, R. (2005). Analysing the intention to purchase on hotel websites: A study of travellers to Hong Kong. *International Journal of Hospitality Management*, *24*(3), 311–329. doi:10.1016/j.ijhm.2004.08.002

Woodside, A. G., & Lysonski, S. (1989). A General Model of Traveler Destination Choice. *Journal of Travel Research*, *27*(4), 8–14. doi:10.1177/004728758902700402

Woodward, J. (1965). *Industrial Organization: Theory And Practice*. London: Oxford University Press.

World Economic Forum, & Accenture. (2017). Digital Transformation Initiative Aviation, Travel and Tourism Industry. Retrieved from http://reports.weforum.org/digital-transformation

World Health Organization. (2017). *Vision impairment and blindness*. Retrieved from http://www.who.int/mediacentre/factsheets/fs282/en/

World Health Organization. (2018). *International classification of diseases: the global standard for diagnostic health information*. Retrieved from https://icd.who.int/en/

World Tourism Organization. (2017). *Penetrating the Chinese Outbound Tourism Market – Successful Practices and Solutions*. Madrid: UNWTO.

World Travel & Tourism Council. (2015). Seven ways technology is changing the travel industry. Retrieved from https://medium.com/@WTTC/seven-ways-technology-is-changing-the-travel-industry-85cff79c1ece

World Wide Web Consortium. (2008). *Web Content Accessibility Guidelines (ACAG) 2.0*. Retrieved from https://www.w3.org/TR/WCAG20/

Wu, J. (2014). How WeChat, the most popular social network in China, cultivates wellbeing. *Master of Applied Positive Psychology (MAPP) Capstone Projects*. Retrieved from http://repository.upenn.edu/mapp_capstone/65

Wu, J. (2018). A surprising number of countries now accept WeChat Pay or Alipay. Techinasisa. Retrieved from https://www.techinasia.com/surprising-number-countries-accept-wechat-pay-alipay

Xiang, U., & Gretzel, U. (2010). Role of social media in online travel information search. *Tourism Management*, *31*(2), 179–188. doi:10.1016/j.tourman.2009.02.016

Xiang, Z. (2017). Digitalisation to the age of acceleration: On information technology and tourism. *Tourism Management Perspectives*, *25*, 147–150. doi:10.1016/j.tmp.2017.11.023

Xiang, Z., Du, Q., Ma, Y., & Fan, W. (2017). A comparative analysis of major online review platforms: Implications for social media analytics in hospitality and tourism. *Tourism Management*, *58*, 51–65. doi:10.1016/j.tourman.2016.10.001

Xiang, Z., & Fesenmaier, D. R. (2017). Big Data Analytics, Tourism Design and Smart Tourism. In Z. Xiang & D. R. Fesenmaier (Eds.), *Analytics in Smart Tourism Design* (pp. 299–307). Switzerland: Springer. doi:10.1007/978-3-319-44263-1_17

Xiang, Z., Magnini, V. P., & Fesenmaier, D. R. (2015). Information technology and consumer behaviour in travel and tourism: Insights from travel planning using the internet. *Journal of Retailing and Consumer Services*, *22*, 244–249. doi:10.1016/j.jretconser.2014.08.005

Xiang, Z., Wang, D., O'Leary, J. T., & Fesenmaier, D. R. (2014). Adopting the internet: Trends in travellers use of the web for trip planning. *Journal of Travel Research*. doi:10.177/0047287514522883

Xicola-Tugas, H. (2012). Heritage access for the visually impaired: An analysis and assessment of museum adaptations in Catalonia. *The International Journal of the Inclusive Museum*, *4*(4), 59–68. doi:10.18848/1835-2014/CGP/v04i04/44383

Xie, H., Miao, L., Kuo, P.-J., & Lee, B.-Y. (2011). Consumers' responses to ambivalent online hotel reviews: The role of perceived source credibility and pre-decisional disposition. *International Journal of Hospitality Management, 30*, 178–183. doi:10.1016/j.ijhm.2010.04.008

Xing, X., & Chalip, L. (2006). Effects of hosting a sport event on destination brand: A test of co-branding and match-up models. *Sport Management Review*, *9*(1), 49–78. doi:10.1016/S1441-3523(06)70019-5

Xu, D. J. (2006). The influence of personalization in affecting consumer attitude toward mobile advertising in China. *Journal of Computer Information Systems*, *47*(2), 9–19.

Xu, J., Kang, Q., Song, Z., & Clarke, C. P. (2015). Applications of mobile social media: WeChat among academic libraries in China. *Journal of Academic Librarianship*, *41*(1), 21–30. doi:10.1016/j.acalib.2014.10.012

Yaghmaie, F. (2003). Content validity and its estimation. *Journal of Medical Education*, *3*(1), 25–27. doi:10.22037/jme.v3i1.870

Yang, T. A., Kim, D. J., & Dhalwani, V. (2008). Social networking as a new trend in e-marketing. In *Research and practical issues of enterprise information systems II* (pp. 847–856). Boston, MA: Springer. doi:10.1007/978-0-387-76312-5_7

Yang, Y., Park, S., & Hu, X. (2018). Electronic word of mouth and hotel performance: A meta-analysis. *Tourism Management*, *67*, 248–260. doi:10.1016/j.tourman.2018.01.015

Yannopoulos, P. (2011). Impact of the Internet on Marketing Strategy Formulation. *International Journal of Business and Social Science*, *2*(18).

Yen, C. L. A., & Tang, C. H. H. (2015). Hotel attribute performance, eWOM motivations, and media choice. *International Journal of Hospitality Management*, *46*, 79–88. doi:10.1016/j.ijhm.2015.01.003

Yeoh, E., & Chan, J. K. (2011). Malaysian Low Cost Airlines: Key Influencing Factors on Customers' Repeat Purchase Intention. *World Applied Sciences Journal, 12,* 35-43.

Ye, Q., Law, R., & Gu, B. (2009). The impact of online user reviews on hotel room sales. *International Journal of Hospitality Management*, *28*(1), 180–182. doi:10.1016/j.ijhm.2008.06.011

Ye, Q., Law, R., Gu, B., & Chen, W. (2011). The influence of user-generated content on traveler behavior: An empirical investigation on the effects of eword-of-mouth to hotel online bookings. *Computers in Human Behavior*, *27*(2), 634–639. doi:10.1016/j.chb.2010.04.014

Yoo, K., & Gretzel, U. (2012). Use and creation of social media by travellers. In M. Sigala, E. Christou & U. Gretzel (Eds.), Social media in travel, tourism and hospitality (pp. 189-205). United Kingdom: Ashgate.

Yoo, K.H. & Gretzel, U. (2009). Comparison of deceptive and truthful travel reviews. *Information and Communication Technologies in Tourism 2009* (pp. 37-47). Springer.

Yoo, K. H., & Gretzel, U. (2008). The influence of perceived credibility on preferences for recommender systems as sources of advice. *Information Technology & Tourism*, *10*(2), 133–146. doi:10.3727/109830508784913059

Yoo, K. H., & Gretzel, U. (2008). What motivates consumers to write online travel reviews? *Information Technology & Tourism*, *10*(4), 283–295. doi:10.3727/109830508788403114

Yoo, K. H., Gretzel, U., Yoo, K.-H., Lee, Y., & Fesenmaier, D. R. (2010). Antecedents and impacts of trust in travel-related consumer-generated media. *Journal of Information Technology and Tourism*, *12*(2), 139–152. doi:10.3727/109830510X12887971002701

Yoo, K.-H., & Gretzel, U. (2008). Use and impact of online travel reviews. In P. O'Connor, W. Hopken, & U. Gretzel (Eds.), *Information and Communication Technologies in Tourism 2008* (pp. 35–46). Springer.

Yoo, K.-H., & Gretzel, U. (2011). Influence of personality on travel-related consumer-generated media creation. *Computers in Human Behavior*, *27*(2), 609–621. doi:10.1016/j.chb.2010.05.002

Yürük, P., Akyol, A., & Şimşek, G. G. (2017). Analyzing the effects of social impacts of events on satisfaction and loyalty. *Tourism Management*, *60*, 367–378. doi:10.1016/j.tourman.2016.12.016

Yusuf, N., Al-Banawi, N., & Al-imam, H. A. R. (2014). The Social Media As Echo Chamber: The Digital Impact. *Journal of Business & Economics Research*, *12*(1), 1–8. doi:10.19030/jber.v12i1.8369

Zeithaml, V. A. (1988). Consumer Perception of Price, Quality, and Value: A Means-End Model and Synthesis of Evidence. *Journal of Marketing*, *52*(3), 2–22. doi:10.1177/002224298805200302

Zeng, B., & Gerritsen, R. (2014). What do we know about social media in tourism: A Review. *Tourism Management Perspectives*, *10*, 27–36. doi:10.1016/j.tmp.2014.01.001

Zeni, N., Kiyavitskaya, N., Barbera, S., Oztaysi, B., & Mich, L. (2009). RFID-Based Action Tracking for Measuring the Impact of Cultural Events on Tourism. In Information and Communication Technologies in Tourism 2009. Springer. doi:10.1007/978-3-211-93971-0_19

Zervas, G., Proserpio, D., & Byers, J. (2015). A first look at online reputation on Airbnb, where every stay is above average.

Zhang, K. Z. K., Zhao, S. J., Cheung, C. M. K., & Lee, M. K. O. (2014). Examining the influence of online reviews on consumers' decision-making: A heuristic–systematic model. *Decision Support Systems*, *67*, 78–89. doi:10.1016/j.dss.2014.08.005

Zhang, L. (2012). Smart tourism: The coming age of customization and intelligent public services. *Journal of Tourism Tribune*, *27*(2), 3–5.

Zhang, M., Jansen, B. J., & Chowdhury, A. (2011). Business Engagement on Twitter: A Path Analysis. *Electronic Markets*, *21*(3), 161–175. doi:10.100712525-011-0065-z

Zhang, T., Omran, B. A., & Cobanoglu, C. (2017). Genaration Y's positive and negative eWOM: Use of social media and mobile technology. *International Journal of Contemporary Hospitality Management*, *29*(2), 732–761. doi:10.1108/IJCHM-10-2015-0611

Zheng, T., Youn, H., & Kincaid, C. (2009). An analysis of customers' e-complaints for luxury resort properties. *Journal of Hospitality Marketing & Management*, *18*(7), 718–729. doi:10.1080/19368620903170240

Zhou, L. (2014). Online rural destination images: Tourism and rurality. *Journal of Destination Marketing & Management*, *3*(4), 227–240. doi:10.1016/j.jdmm.2014.03.002

Zhou, X., Wang, M., & Li, D. (2017). From stay to play – A travel planning tool based on crowdsourcing user-generated contents. *Applied Geography (Sevenoaks, England)*, *78*, 1–11. doi:10.1016/j.apgeog.2016.10.002

Živković, R., Gajić, J., & Brdar, I. (2014). *The impact of social media on tourism. Sinteza 2014 - Impact of the Internet on Business Activities in Serbia and Worldwide*. Retrieved from; doi:10.15308inteza-2014-758-761

About the Contributors

Celia Ramos graduated in Computer Engineering from the University of Coimbra, obtained her master's in electrical and computer engineering from the Higher Technical Institute, Lisbon University, and a PhD in econometrics from the University of the Algarve (UALG), Faculty of Economics, Portugal. She is an Adjunct Professor at School for Management, Hospitality and Tourism, also in the UALG, where she lectures computer science. Areas of research and special interest include conception and development of information systems, tourism information systems, big data, etourism, econometric modeling, and panel-data models. Célia Ramos has published in the fields of information systems and tourism, namely, she has authored a book, ten book chapters, conference papers, and journal articles. At the level of applied research, she has participated in several funded projects.

Paula Fernandes PhD in Economics and Management, Professor of Management at the Polytechnic Institute of Bragança (IPB) - Portugal. She is a researcher and Scientific Coordinator of UNIAG (Applied Management Research Unit, since 2013) and was a researcher at the Research Unit in Business Sciences (NECE-UBI, 2010-2017). She participated in 14 international and national projects I&D (6 as Responsible Researcher), supervised one PhD thesis and several master students (over 80 master theses). She published more than 200 publications in peer-reviewed journals, book chapters and proceedings, and has several communications at international scientific conferences and won 7 scientific awards. She serves as a member of Program Board and Organizing Committees for some Scientific Committees of International Conferences and has collaborated as a reviewer with several journals. Her current research interest is in Tourism Research, Importance-Performance Analysis, Management, Artificial Neural Network, Entrepreneurship, Econometric Modelling, Marketing and Strategic Management, Market Research, Corporate Social Responsibility and Sustainable, Higher Education Quality and Applied Research Methods.

***

Sahap Akan is a PhD candidate in the Civil Aviation Department at Anadolu University. Akan is currently a research assistant in this field. Akan's research interests are airline marketing, neuromarketing and consumer behavior in the airline market.

Maria Del Mar Alonso-Almeida is a teacher of Business Administration in Autonomus University of Madrid in Spain. The author is interested in exploring sustainability in organizations. In addition, she is interested in the influence of social media, the development of circular economy and gender issues since and strategic management point of view.

Marta Amaral has received her Ph.D. in Tourism from the University of Évora, in 2013. She is a professor at the Polytechnic Institute of Beja – Portugal since 1998 and currently is the Tourism Bachelor Coordinator. She received is Master in Sociology (Human Resources and Sustainable Development) from the University of Évora in 2002. She is also graduated in the Sociology of Work from the ISCSP - University of Lisbon. Her research interests are related to cooperation and networks in tourism, sociology of tourism, innovation, human resources, sustainable development. She is a reviewer for several journals such as Tourism and Management Studies; Pasos Revista de Turismo y Patrimonio Cultural and Turismo Visão e Ação. She was also a member of the scientific committee of international conferences such as TMS Conference and IFM Conference.

Periasamy Amuthalakshm teaches social work, management, and psychology at the Madras School of Social work, which is an autonomous school of excellence affiliated to the University of Madras. Having acquired a Doctorate from Anna University for her research on Women Entrepreneurial Skills, she was awarded fellowship by the Entrepreneurial Development Institute (EDI) Ahmedabad, and the Indian Institute of Management (IIM), Ahmedabad. She had earlier served in Indore School of Social work and as Professor in Mohammed Sathak Arts & Science College, Chennai, in the Department of Social Work & Management. Dr. Amu is an ardent admirer for bringing up the economically and socially underprivileged to heights, which are above all classes, segments, divisions and statures.

Noelia Araújo Vila is PhD in Tourism Management and Planning (2013) and Bachelor in Business Administration and Management (2006). Since 2007 she collaborates in the coordination tasks of the University Master in Tourism Management and Planning and teaches as professor in the Departments of Business Organization and Marketing and Financial Economics and Accounting (University of Vigo). She participates in various research projects linked to the tourism and business sector. She is a member of the research group of the University of Vigo Emitur (Marketing and Tourism). She holds a Master in Management and Management of SMEs with a prize for the best business project.

Ozlem Atalik is an Assoc. Professor in the Aeronautical and Astronautical Faculty at the Eskisehir Technical University. Atalik completed her Ph.D. at Anadolu University and her undergraduate studies at Anadolu University. Her research interests lie in the area of airline marketing, customer relationship management and service marketing.

Kenneth Backman has been a Professor in the Department of Parks, Recreation and Tourism Management for more than twenty years and his research has focused on use of social media in tourism development and tourism development in Sub-Saharan Africa.

Sheila Backman is a Professor in the Department of Parks Recreation and Tourism Management at Clemson University. Her research interest focus on digital marketing and social media. Dr. Backman earned her PhD from Texas A & M University.

Mahmut Bakır is a PhD candidate in the Civil Aviation Department at Anadolu University. Bakir currently is a research assistant in this field. His research interests are airline marketing, service quality research, and quantitative research.

Sandeep Basnyat holds a PhD in Tourism from the University of Otago, New Zealand. He is an Assistant Professor at the Macau Institute for Tourism Studies, Macau, China. His research interests include social media and its applications in the tourism and hospitality industry, and human resources, work and working conditions in the hospitality industry.

Zelia Breda holds a PhD in Tourism, a MA in Chinese Studies (Business and International Relations) and a BSc in Tourism Management and Planning from the University of Aveiro, where she is an Assistant Professor, at the Department of Economics, Management, Industrial Engineering, and Tourism, holding the position of Director of the Master in Tourism Management and Planning. She is a member of the Research Unit 'Governance, Competitiveness and Public Policies' of the University of Aveiro; a founding member and vice-president of the Observatory of China (www.observatoriodachina.org), and a senior researcher of the spin-off company Idtour (www.idtour.pt). She is also a member of the editorial and scientific boards of a few academic national and international journals, as well a member of the organizing and scientific committees of international tourism conferences. She has authored and co-authored several national and international papers and communications on tourism. She has also been taking part in several research projects in the tourism field, both as member of the team and as a consultant.

Beatriz Casais holds a PhD in Business and Management Studies - specialization in Marketing and Strategy by the Faculty of Economics, University of Porto. Her research interests focus on social marketing, digital marketing, e-wom and social media marketing applied to tourism. In this field, she has published in Journal of Hospitality and Tourism Technology. She is a Professor at University of Minho and IPAM Porto.

Lauro César Vieira is a PhD student in Education by the University of Aveiro, Portugal. Has a master's in production engineering by the Federal University of Santa Catarina and a bachelor's degree in computer science by the Catholic University of Pernambuco. Acted as Academic Director, having participated in the accreditation of more than 10 colleges and 60 higher education courses. He coordinated several undergraduate and graduate courses in the computing area. He is currently an Institutional Research and Evaluation Coordinator at Cesar School in Recife. He has experience in the management of institutions of higher education, educational legislation and institutional evaluation. He is also a managing partner of startup in information technology and communication.

James Malitoni Chilembwe is a Lecturer at the Faculty of Tourism, Hospitality, and Management of Mzuzu University, Malawi. He is Doctor of Philosophy candidate in Tourism Management at Glasgow Caledonian University (GCU), UK. He holds an MSc. in International Tourism Enterprise obtained from GCU in 2012. He possesses a BSc. in International Travel and Tourism Management (Transportation), and a certificate in Global Distribution System. He has authored/co-authored and published peer-reviewed research journal articles and book chapters. He is a Reviewer of International Journal of Tourism and Hospitality Reviews (IJTHR). He is also a member of the Travel and Tourism Research Association (TTRA). His primary research interests include; aviation, sustainable transport, tourism development, corporate social responsibility, and information technology.

Rui Costa was born in Oliveira de Azeméis, district of Aveiro. He graduated in Management and Tourism Planning at the University of Aveiro in 1988, completed his master's degree in Innovation and Policy Development at the University of Aveiro and a PhD in Tourism at the same University in 2012. He is Assistant Professor in the Tourism Area in the Department of Economics, Management, Industrial Engineering and Tourism at University of Aveiro. He is a member of the executive board of DEGEIT and he's also an Integrated Member of the Research Unit on Governance, Competitiveness and Public Policy in the Research Group on Tourism and Development. Develops his research in the planning and project in Tourism, networks, governance and public policy, territorial dynamics of investment and financing of small and micro enterprises in the tourism sector. He participates in several projects of applied research nationally and internationally, and he is author and co-author of several articles in national and international journals. He is also Associate editor of the Journal of Tourism & Development and member of the Organizing Committee of the International Conference INVTUR.

Ademilso Cunha graduated in hospitality and is a student of tourism, both at the Federal University of Rio Grande.

Pablo de Carlos is PhD in Economics. He is a Professor in the Department of Business Organization and Marketing of the University of Vigo. He participates in various research projects linked to the tourism and business sector. He is a member of the research group of the University of Vigo Emitur (Marketing and Tourism). He teaches several management subjects, which include international strategic management and tourism management information systems. His research is focused on the economic and business analysis of tourism and sport.

Valentina Della Corte is a Full Professor of Business Management at Federico II University of Naples. She teaches Tourism Business Management, Strategic Management and Marketing Policies and Management of Cultural Heritage. She is the academic coordinator of the hospitality management degree course and member of the industry 4.0 task force in Federico II University of Naples. Author of numerous publications in national and international academic journals such as Tourism Management, International Journal of Tourism Research, British Food Journal, European Journal of Innovation Management, Corporate Ownership & Control, International Journal of Business and Management, Mercati e competitività, International Journal of Quality and Service Sciences, International Journal of Leisure and Tourism Marketing, International Journal of Marketing Studies, Journal of Management and Sustainability. Author of books published by Egea, CEDAM and Giappichelli as well as of chapters in books edited by Elgar and Mc Graw Hill. She has participated, also as chair and reviewer, to national and international conferences such as Strategic Management Society Annual Meeting, Academy of Management Annual Meeting, Italian Academy of Management Annual Meeting, International Forum of Knowledge Asset Dynamic, International Conference on Tourism (ICOT), International Marketing Trends Conference, Forum on Marketing and Markets, AIDEA Annual Meeting and Sinergie Annual Meeting. Her research is focused on strategic management and marketing and in particular resource-based theory, business networks and coopetition, dynamic capabilities, value creation and appropriation both intra-firm and inter-firm, innovative marketing in tourism and heritages.

Maria Gorete Dinis is a deputy director and professor of Tourism at the Polytechnic Institute of Portalegre (PIP) / School of Education and Social Sciences. She has a degree in Management and Planning in Tourism, a Master on Innovation, Planning and Development Policies and a PhD in Tourism, all from the University of Aveiro. She is a researcher at the Governance, Competitiveness and Public Policies Research Unit and of Centre for Interdisciplinary Research and Innovation of PIP. During recent years she has published several scientific articles, presented communications and participated in various projects in areas such as planning and management of tourist destinations, tourism indicators, big data in tourism and information and communication technologies applied to the tourism sector.

Jeet Dogra is working as an Assistant Professor in Indian Institute of Tourism and Travel Management which is an Autonomous Body under Ministry of Tourism, Govt. of India. He is a Ph.D. in Tourism Management and also Junior Research Fellow, UGC India. His research area includes destination marketing and branding.

Celeste Eusébio is an Assistant Professor of Tourism and a researcher at the GOVCOPP Research Unit at the University of Aveiro (Portugal). She holds a Degree in Tourism Management and Planning from the University of Aveiro (1995), a Master in Economics from the University of Coimbra (1998) and a Doctoral Degree in Tourism, also from the University of Aveiro (2006). Her research interests include tourism economics, tourism impacts, tourism forecasts and consumer behaviour in tourism. She is at present the coordinator of the Degree in Management and Planning in Tourism at University of Aveiro.

Julia Vasconcelos Furtado holds a bachelor's degree in social communication and Marketing by the ESPM – Escola Superior de Propaganda e Marketing, (Brazil) and an MBA in Strategic Management by FGV – Fundação Getúlio Vargas (Brazil). After more than 15 years of work as Marketing professional in Brazil, she is currently finishing a master's degree in management (Marketing and International Business) at the University of Aveiro (Portugal), and a Master Degree in Political Studies at the University of Saskatchewan (Canada).

Zeynep Gedikoglu received her Ph.D. in Parks, Recreation and Tourism Management from Clemson University in 2018. In August 2018 she was hired as a Postdoctoral Fellow by Clemson University College of Behavioral, Social and Health Sciences Department of Parks, Recreation and Tourism Management (PRTM). She writes and presents widely on issues of destination image, tourism communication, and effects of terrorism and is the author of Exploring Destination Image Themes on Twitter, Before, During, and After Terror Attacks in Paris: An Application of Agenda Setting Theory. (2018, Gedikoglu).

Frank Gondwe is an Instructor at Mzuzu University in the Tourism Department, Faculty of Tourism, Hospitality and Management. He holds BSc in Tourism Management from Mzuzu University. He also has the following professional qualifications from IATA Training & Development Institute, Montreal-Canada: IATA/UFTAA Stepping into Management Diploma in International Travel and Tourism, IATA/UFTAA Consultant Diploma in International Travel and Tourism, IATA/UFTAA Foundation Diploma in International Travel and Tourism, IATA/UFTAA Certificate in Global Distribution Systems Fares and Ticketing-Galileo and Global Distribution Systems-Galileo Certificate from Galileo Travelport-Malawi. Before joining Mzuzu University, he worked as a Travel Consultant at Skylinks Travel Bureau in Blantyre for two years. He has been teaching IATA/UFTAA Travel and Tourism Courses on a private basis for many years

Yuan Jiahui has a Master's degree in International Business in the Hospitality industry from Switzerland and is currently continuing her research in Germany.

Sachin Kale is a Lecturer in the Department of Travel and Tourism at Government Polytechnic Thane, India, and also pursuing Ph.D. from the Indian Institute of Tourism and Travel Management (IITTM), Gwalior, India. His research areas are Smart tourism, Destination marketing, and big data analysis.

Amandine Martins holds a Master's degree in Tourism Management and Planning from the University of Aveiro.

Ana Martins was born in Guimarães and graduated in Tourism from the Polytechnic Institute of Viana do Castelo, later obtaining a master's degree in Tourism Management from the Polytechnic Institute of Cávado and Ave. Her career began in the field of tourism promotion and later in the hotel sector; also having an interest in marketing. The subject of her master thesis is the result of her growing interest, during her studies, in tourism and its connection with technology.

Alexandra Matos holds a Ph.D. on Tourism, Leisure and Culture from the University of Coimbra and a Diploma on Advanced Studies in the same scientific area. Her Master in Translation Studies is from the University of Porto as well as her Licentiate Degree in French and English Modern Languages and Literatures. She is Assistant Professor at Universidade Lusófona do Porto and ISLA – Instituto Politécnico de Gestão e Tecnologia and Invited Assistant Professor at University of Coimbra where she lectures Tourism Studies, focusing mainly on the social and cultural aspects of tourism. She served as academic director and executive manager on a private polytechnic institute, in Porto, where she lectured Tourism Studies and Applied English to undergraduate and graduate students. She also tutored internships in tourism, translation and interpretation, and took part in dissertations' defense juries. She has been a full member of organizing and scientific committees of international conferences on Tourism and on Translation, where she has participated as keynote speaker, moderator and reviewer. She is Vice-President of the Scientific Committee of CEPESE - Centre for the Study of Population, Economics and Society, University of Porto, being also Research Associate in the Heritage, Culture and Tourism Research Group. Her research interests are mainly focused on Touring tourism (Ph.D. thesis), Roots tourism, Tour Guiding, Tourism Destination Image, having already published several papers on those areas. She is an editorial board member of several international journals of tourism and hospitality.

Joseph Mazer is a Professor and Chair of the Department of Communication at Clemson University.

Marica Mazurek has a research interest in competitiveness of tourism destinations, branding, marketing, place marketing and place branding, innovations in tourism. She participated in several academic exchanges and Visitor Professorship activities in Canada (Brock University, University of Waterloo), where she participated in research activities dealing with tourism marketing and branding, innovations and tourism destinations competitiveness. She is the author of several academic publications, for example in the International Journal of Emerging Markets, Journal of Internet Commerce, the book chapter in Tourism Marketing: On both Sides of the Counter edited by Cambridge Scholars Publishing (edited by Prof. Metin Kozak et al.), Kybernetes. Recently, she participated at the international conferences in Turkey, Fethiye (2nd Interdisciplinary conference in Tourism research), 4th ATMC international conference in

Maribor, Slovenia, the 5th ATMC international conference in Algarve, Portugal and the Multidisciplinary International Conference in Porvoo, Finland (Haaga-Helia University), the 5th International Conference of Marketing in Algarve, Portugal, the International Conference of Tourism, Hospitality and Recreation in Poznań, Poland, and the 3rd Interdisciplinary Tourism Research Conference in Istanbul, Turkey, conferences in Taipei, Taiwan and Joensu, Helsinki and the 2nd EATSA (The European and Asian Tourism Association) conference in Lisboa, Peniche and Coimbra. She is the author of several book chapters and Scopus articles as well as involved in several international projects (Europe, Asia). The author teaches Marketing, Digital Marketing, Economics and Management at the University of Zilina, Slovakia. She is a member of the Editorial board of the Scopus journal Communications.

Fernando Moreira is Head of the Science and Technology Department. He is graduated in Computer Science (1992), M.Sc. in Electronic Engineering (1997) and Ph.D. in Electronic Engineering (2003), both at Faculty of Engineering of the University of Porto, and Habilitation (2018). He is member of the Science and Technology Department at Portucalense University since 1992, currently as Associate Professor with Habilitation and a visiting professor at the University of Porto Business School. He teaches subjects related to undergraduate and pos-graduated studies. He supervises several Ph.D. and M.Sc. students. He is a (co-)author more than 150 scientific publications with peer-review on national and international journals and conferences. He serves as a member of Editorial Advisory Board for several journals and books. He organized several special issues from JCR journals. He has regularly served as a member of Programme and Scientific committees of national and international conferences. He was the MSc in Computation coordinator during last 10 years. He holds editorial experience, and he is co-editor of several books. He is associated with NSTICC, ACM and IEEE. His principal research areas are mobile computing, ICT in Higher Education, Mobile learning, Social business and Digital transformation.

Victor Mweiwa is a lecturer in Travel, Tourism and Hospitality Management at the Malawi Institute of Tourism (MIT). Before joining MIT, he worked as a Senior Sales and Tour Consultant for Express Air Travel and Tours in Malawi for three years. He holds a BBA in Strategic Management. He has also Tourism Management and Teacher Training qualifications obtained at Austria's Institute of Tourism and Hospitality Management (ITH). He also possesses travel agency professional qualifications and is a multi-tour guiding certificate holder. His research interest includes sustainable tourism, tourism development, tourism marketing, travel agency, and tour operations.

Giuliana Nevola is a Ph.D. student in Management at University of Naples Federico II. She holds a Master of Science in Tourism Management and a Master's in development economics and International Cooperation at University of Tor Vergata in Rome. During her academic years, her studies were focused on ethical finance and welfare policies applied on the tourism sector. Her research topic is focused on strategic tourism management and sustainability.

Ana Isabel Rodrigues is a Professor of Tourism at the Polytechnic Institute of Beja (Portugal) since 1997. Her academic background focus on two main areas: Tourism and Communication: BSc in Tourism, Estoril Higher Institute for Tourism; Degree in Social & Cultural Communication, Catholic University of Lisbon; MSc in Communication & Tourism, ISCTE/Lisbon University Institute; PhD in Tourism, Algarve University, Portugal. Her main research interests focus on communication and image, qualitative methods, visual-based methods, tourism education methodologies, experiential-based learning and CAQDAS (Computer-Assisted on Qualitative Data Analysis Software). She has papers and book chapters published in tourism and leisure. She is also a member of the Scientific Committee of CIAIQ (Ibero-American Congress on Qualitative Research), WCQR (World Congress on Qualitative Research), ATMC (Advances on Tourism Marketing Conference) and World Conference for Graduate Research in Tourism, Hospitality & Leisure.

Raquel Seabra is a Junior Consultant in Marketing and Tourism at IDTOUR. She holds a master's degree in Tourism Management and Planning from the University of Aveiro and has a basic academic background in Marketing from the same University. She has participated in several marketing and communication projects (territorial and business) related to the tourism sector, and was also involved in the organization of workshops and focus groups in the area of Local Accommodation for IDTOUR and its clients and partners.

Fabiana Sepe is a Ph.D. in Management at University of Naples Federico II. She is the author of some book chapters and articles on national and international journals in the fields of marketing and management focusing on the application of the resource-based view and innovation in tourism industry. She has attended several national and international conferences with co-authored papers.

Murat Seyfi is an assistant professor at the department of Public Relations and Advertising. He studied for master's degree in Lifelong Education Department and has works on "self-directed learning". His doctorate topic was "storytelling as a communication tool" He is conducting interdisciplinary studies on both communication and education areas. He serves as Coordinator of Awareness, Informing and Training Unit in the office of technology transfer in The University of Gaziosmanpaşa as well as his academic duty in the same university. His current research is on storytelling, social media, creativity, entrepreneurship, communication technology, digital literacy, interpersonal communication and appreciative inquiry. His publications have appeared in many international journals.

Leonor Teixeira graduated in Industrial Engineering and Management, received a M.Sc. degree in Information Management, and a PhD in Industrial Management (Information Systems area), in 2008, from the University of Aveiro, Portugal. She is currently an Assistant Professor of the Department of Economics, Management, Industrial Engineering and Tourism at the University of Aveiro and teaching subjects related to Information Systems, Information Technologies to undergraduate and post-graduate studies. She is also a researcher at the Institute of Electronics and Telematics Engineering (IEETA) of University of Aveiro. Her current research interests include Information Systems applied to Industry, focusing on Requirement Engineering, Software Engineering and Usability Engineering. She has several publications in peer-reviewed journals, book chapters and proceedings, and oral communications at international scientific conferences. She serves as a member of Program Board and Organizing Committees for several Scientific Committees of International Conferences. She is associated with IIIS, IEEE and APSI/PTAIS.

Pedro Teixeira has a master's degree in Industrial Engineering in Industrial Engineering (2019). Currently is a PHD student in Industrial Engineering, at the University of Aveiro. The main research preferences are in the area of information systems, Industry 4.0 and investigation related to technological solutions to real-world problems.

Roberto Vaz graduated in Electronic and Telecommunications Engineering at the University of Aveiro – Portugal, and holds an MSc Degree in Interactive Multimedia Communication, by the same university. He participated in one international project I&D in Brazil and developed projects and research in the areas of human-computer interaction, accessibility, technology in museums and digital media. Currently, he is a Ph.D. Research Fellow in the UT Austin | Portugal Digital Media Doctoral Program.

Index

D

E

F

G

H

P

Q

R

S

T

U

V

W

www.ingramcontent.com/pod-product-compliance
Lightning Source LLC
LaVergne TN
LVHW060352300926
849162LV00006B/1250